DARK HERO

OF THE

INFORMATION AGE

ALSO BY FLO CONWAY & JIM SIEGELMAN

SNAPPING

HOLY TERROR

DARK HERO

OF THE

INFORMATION AGE

IN SEARCH OF

NORBERT WIENER

THE FATHER OF CYBERNETICS

FLO CONWAY & JIM SIEGELMAN

BASIC
BOOKS

A Member of the Perseus Books Group
New York

Published by Basic Books,
A Member of the Perseus Books Group

Books published by Basic Books are available at special discounts for bulk purchases in the United States by corporations, institutions, and other organizations. For more information, please contact the Special Markets Department at the Perseus Books Group, 11 Cambridge Center, Cambridge MA 02142, or call (617) 252-5298 or (800) 255-1514, or e-mail special.markets@perseusbooks.com.

Designed by Jeff Williams

Library of Congress Cataloging-in-Publication Data
Conway, Flo
    Dark hero of the information age : in search of Norbert Wiener, the father of cybernetics / Flo Conway and Jim Siegelman.
        p. cm.
    Includes bibliographical references and index.
    ISBN 0-7382-0368-8 (hardcover : alk. paper)
    1. Wiener, Norbert, 1894-1964. 2. Mathematicians—United States—Biography.
3. Cybernetics. I. Siegelman, Jim. II. Title.

QA29.W497 C66 2004
510'.92—dc22

2004022116

05 06 07 08 / 10 9 8 7 6 5 4 3 2 1

CONTENTS

FOR OUR FATHERS

ROBERT PATRICK CONWAY, SR.

AND

LEONARD P. SIEGELMAN

Time present and time past
Are both perhaps present in time future
And time future contained in time past. . . .
Footfalls echo in the memory
Down the passage which we did not take
Towards the door we never opened
Into the rose-garden. My words echo
Thus, in your mind.

But to what purpose

— T. S. Eliot
Burnt Norton

PROLOGUE
TIME PAST, TIME PRESENT

He is the father of the information age. His work has shaped the lives of billions of people. His discoveries have transformed the world's economies and cultures.

He was one of the most brilliant minds of the twentieth century, a child prodigy who became a world-class genius and visionary thinker, an absentminded professor whose eccentricities assumed mythical proportions, a best-selling author whose name was a household word during America's first heyday of high technology.

His footprints are everywhere today, etched in silicon, wandering in cyberspace, and in every corner of daily life. Yet his words are muted echoes in the memory.

This is the story of a dark hero who has fallen through the cracks in the information age and his fight for human beings that is the stuff of legend.

Born on the doorstep of the twentieth century, Norbert Wiener was a descendant of Eastern European rabbis, scholars, and, purportedly, of the medieval Jewish philosopher Moses Maimonides. He entered college at eleven, received his Ph.D. from Harvard at eighteen, apprenticed with renowned European mathematicians, and, in 1919, joined the faculty at the Massachusetts Institute of Technology.

His early mathematical work solved practical problems in electronics theory that engineers had been wrestling with for decades. In the 1920s, he worked on the design of the first modern computer, and during World War II, he helped create the first intelligent automated machines. Wiener's wartime vision grew into a new interdisciplinary science of communication, computation, and automatic control, spanning the forefronts of engineering, biology, and the social sciences. His ideas attracted an eclectic group of scientists and scholars: computer pioneer John von Neumann, information theorist Claude Shannon, and anthropologists Margaret Mead and Gregory Bateson. Wiener named his new science "cybernetics"—from the Greek word for steersman.

His 1948 book *Cybernetics: Or Control and Communication in the Animal and the Machine* set off a scientific and technological revolution. In less than a decade, cybernetics transformed the day-to-day labors of workers in every industry and unleashed a flood of dazzling devices on postwar society.

Wiener gave the word "feedback" its modern meaning and introduced it into popular parlance. He was the first to perceive the essence of the new stuff called "information." He worked with eminent biologists and neurophysiologists to crack the communication codes of the human nervous system, and with the engineers who incorporated those codes into the circuits of the first programmable "electronic brains." He led the medical team that created the first bionic arm controlled by the user's own thoughts.

In his mind's eye, he saw the technical promises of the new world that was dawning and modern marvels few could imagine at the time. But, alone among his peers, Wiener also saw the darker side of the new cybernetic era. He foresaw the worldwide social, political, and economic upheavals that would begin to surface with the first large-scale applications of computers and automation. He saw a relentless momentum that would pit human beings against the seductive speed and efficiency of intelligent machines. He worried that the new time- and labor-saving technology would prompt people to surrender to machines their own purpose, their powers of mind, and their most precious power of all—their capacity to choose.

And he feared for humanity's future.

Wiener spent his later years tirelessly warning the leaders of governments, corporations, labor unions, and the public about those far-reaching changes that were coming to work and daily living. He was the first person to sound alarms about intelligent machines that could learn from experience, reproduce without limitation, and act in ways unforeseen by their human creators, and he called for greater moral and social responsibility by scientists and technicians in an age of mushrooming productive and destructive power. Wiener spoke and wrote passionately about rising threats to human values, freedoms, and spirituality that were still decades in the offing. His efforts won him the National Book Award and the National Medal of Science, the nation's highest scientific award.

Yet, even as his new ideas were taking hold in America and worldwide, Wiener's visionary science was foundering. By the late 1950s, cybernetics was being superseded by the specialized technical fields and subdisciplines it had spawned, and Wiener himself wound up on the sidelines of his own revolution. His moral stands were rejected by his peers and a gadget-happy consumer public, and his grim predictions were dismissed by many as the doomsaying of an aging, eccentric egghead. He died suddenly, at age 69, on a trip to Europe in 1964, even as so many of the things he had predicted were coming to pass.

———

Wiener's revolutionary contributions have been largely forgotten for reasons that have remained obscure until now. This book travels back to that abandoned stretch of information age history, a place removed in time but intimately connected to the technologies and social realities that affect all our lives in the twenty-first century. It chronicles Wiener's life and work from his precocious childhood through the opening shots of the cybernetics revolution and the first waves of the information age explosion that followed. But Wiener's legacy is not only a technical one. As his own writings make clear, cybernetics was not merely a narrow engineering discipline. It was a new way of thinking about the world—about life as well as technology—that was utterly different from anything that had come before.

Wiener's science provided powerful new tools for understanding all manner of modern complexities, from the workings of the human genome, to the flow of human communication, to the dynamics of today's global economy and the teeming networks of the World Wide Web. The innovations Wiener's work made possible, and the public stands he took to keep human beings in control of their new creations, made him a hero to many in his day and to a loyal few in the years since his death. But his story goes far beyond the known facts about the boy prodigy who became a world-famous scientist.

Two academic biographies from 1980 and 1990, a smattering of professional memoirs in scientific journals, and Wiener's two-volume autobiography published in the 1950s gave insights into his childhood years and later life as a mathematician. But Wiener the man has remained elusive. The official word from MIT, where he held court for forty-five years, is properly praiseful, but there are gaping holes in the record and disquieting undertones rumbling below. With his science and social warnings long gone from public view, and yet so many of his concerns rising again, the time is right to pick up the search for Norbert Wiener where history left off, to reassess the legacy of his work and the long-range accuracy of his warning shots, and to unravel the mysteries that enshrouded his life and remained unsolved four decades after his death.

Among those mysteries are unanswered questions about the effects of Wiener's high-pressure childhood on the events of his adult life; rumors about Wiener's turbulent relations with his colleagues that marred the early years of the cybernetics revolution; political questions about Wiener's activism in defense of workers and in defiance of society's "powers that be," as he called them; and deeper philosophical questions about his later spiritual excursions and his last enigmatic messages on relations between people and machines.

Equally important are questions about the fate of cybernetics. What lasting contributions can Wiener's science claim? Why, after stirring so much ferment, did cybernetics all but vanish from the American scene a decade after Wiener's death? What lost pieces of his revolution need to be reclaimed by younger generations grappling with the technical challenges and human complexities of a global information society?

Now, after extended conversations with Wiener's surviving colleagues and family members, and a thorough combing of the archives of the information age, many of those questions can be answered.

Wiener's saga is replete with testimony to his genius and fabled eccentricities. Tales abound of his gregariousness and the hungering inquisitiveness that gave impetus to his *Wienerwegs*—his meandering walks across the MIT campus, suburban Boston, and the New England countryside in search of fresh insights and an audience for his latest cogitations. And there are numerous, whimsical accounts of his deafening snoring through his colleagues' lectures (often with a lit cigar dangling precariously from his mouth).

But Wiener was no cartoon genius. Beyond the legends and comic antics lay the darker realm of Wiener the self-described "bent twig" whose fast-track upbringing turned back on him in adulthood and wreaked havoc on his relationships. His small circle of close acquaintances were privy to something few others knew: that amid his brilliance and deep concern for the devilish forces inherent in the new technology he helped to sire, Wiener spent his life combatting his own inner demons. His furies sprang from the deep psychic wounds of his youth, and from his decades-long struggle with the manic depression that followed.

During his high times, Wiener was ebullient, impulsive, and often petulant. At his low points, he fell prey to paralyzing depressions that drove him to threaten suicide frequently in the confines of his home and family, and at times among his MIT colleagues. But, in many ways, Wiener's extremes were matched by those of his wife, a fastidious *frau-professor* cast in the Old World mold. In her dutiful efforts to preserve and protect her high-strung husband, Margaret Wiener took steps to neutralize Wiener's peers, women in any proximity to him, and anyone she perceived as a threat to his prominence. One stratagem in particular backfired catastrophically on him, personally and professionally.

For a decade, Wiener worked productively with the pioneering neuroscientist Warren McCulloch and Walter Pitts, the furtive young genius of cybernetics' next generation. The sudden end of his partnership with McCulloch, Pitts, and other talented young scientists who came to MIT to take cybernetics forward was a crisis for Wiener and everyone involved. This split dealt a crippling blow to the cybernetics revolution at a crucial moment and changed the course of the new technological era in ways that redound to the present day.

Wiener's activism made him a target in politically perilous times. As newly released government records reveal, his outspoken stand against military research during the early years of the Cold War prompted the FBI to investigate his alleged "subversive activities" and "communist sympathies." Cold War fever also struck at Wiener's science. In the mid-1950s, when scientists and government officials in the Soviet Union embraced cybernetics, the CIA took steps to assess the threat and counter it. But years of secret inquiries did little to enlighten the American intelligence community about the power of cybernetics, and some government officials became openly hostile to it. At the peak of the Cold War, funding for cybernetics research in the United States dried up. The forward motion of cybernetic theory and applications slowed to a crawl and never fully recovered. That political reaction is only one among many reasons for Wiener's personal slide into obscurity, but it may be the major factor in the decline of cybernetics in America and its conspicuous absence from the knowledge base of the twenty-first century.

Time has confirmed that Wiener's work was revolutionary in the scientific sense. He identified a new set of fundamental entities of which the universe is composed: messages, information, and basic communication and control processes observable in every domain of life. He brought within the bounds of understanding phenomena of both mind and matter that had eluded philosophers and scientists for centuries. His was the first interdisciplinary scientific revolution, the first grounded, not in inanimate nature alone, but equally in the world of living things and in the everyday actions of human beings. It was also the first *American* scientific revolution, the first to originate and play out primarily in the United States.

Cybernetics has sired, inspired, or contributed to dozens of new technical and scientific fields, from artificial intelligence and cognitive science to environmental science and modern economic theory. Yet many of Wiener's contributions have been denied, dismissed, or credited to others, and some of the most profound aspects of his work remain almost wholly unexplored. More than most scientific revolutionaries, Wiener took the trouble to tell us explicitly why he was so worried about the fate of his discoveries, and to leave behind some basic instructions to help us save ourselves. He made clear that our greatest tasks ultimately would be to determine those purposes and values we want to embrace as human beings, and how we choose to share our existence with the machines we have created in our image.

Wiener's most dire predictions have not come to pass, but his legacy is still unfolding in the global society of the twenty-first century. It can be seen in the fragile bubbles that have roiled the market for new technologies—Wiener watched such bubbles form and burst for decades and he cautioned eager investors to "watch your hat and coat"—and in the global shift toward offshoring of jobs in manufacturing and the new technology industries themselves.

And another important part of his legacy is only beginning to appear. His work paved the way for the digital revolution, but Wiener's driving passions were analog. His imagination was inspired, not by strings of ones and zeros, but by automatic machines that mimicked the movements of human muscles and limbs, and by intelligent devices that emulated the feats performed by human brains and minds. The advance of digital technology put many of those analog processes out to pasture, yet today they are emerging as the dark horses of twenty-first-century science. The latest breakthroughs in biotechnology and genetic engineering, robotics and sensor technology, and the tantalizing new domain of atomic-scale nanotechnology promise to change daily living, and life itself, more profoundly than all the digital technologies to date. They are unleashing formidable new powers that can benefit humankind or, in some scenarios, extinguish it. This new analog universe is bringing Wiener's science and social concerns back to the fore, along with his early warning that cybernetic technology is "a two-edged sword, and sooner or later it will cut you deep."

———

Throughout his life and long after his death, Wiener remained a mystery even to those who were closest to him, and nowhere more so than in his own spiritual excursions. Incongruous reports that, in his later years, Wiener, the prodigal son of Maimonides and self-proclaimed agnostic, held private, weekly meetings with a Hindu swami have turned out to be accurate in substance and spirit. His lifelong interest in the cultures of the East also drew him to India in the 1950s, where, at the request of the Indian government, he laid out a long-range program for that nation's emergence as a technological power, which has put its scientists and technicians in the front ranks of today's global information economy.

This giant of the new technological age, who loved to quote dark fables from age-old cultures to dramatize his warnings to the modern world, was himself the embodiment of another genre of wisdom-rich parables. Like the Elephant's Child in Kipling's *Just So Stories,* a character he loved and in form resembled, Wiener was filled with a "'satiable curtiosity" that led him on to great things. His multi-faceted persona also evoked a second famous pachyderm parable: the Hindu tale of the elephant and the blind men who struggled in vain to describe it, each from his own isolated grasp. Indeed, the many witnesses to his life describe many different and sometimes mutually exclusive Norbert Wieners: one brilliant, one deficient, one robust, one infirm, one playful, one wrathful, one competitive, one magnanimous, one insecure, one egotistical, one self-promoting, one supremely humble. Like many historical figures, Wiener was a man of paradoxes, yet his mix was extreme even among celebrated men of genius.

Like dark heroes of old and antiheroes of contemporary culture, he flouted convention and society's superficial codes to pursue a deeper purpose and higher truth.

Like dark matter whose presence can only be inferred from its effects on the universe around it, his science and ideas continue to influence every dimension of our world.

DARK HERO

OF THE

INFORMATION AGE

PART ONE

The Elephant's Child

1

The Most Remarkable Boy in the World

There was one Elephant—a new elephant—an Elephant's Child— who was full of 'satiable curtiosity, and that means he asked ever so many questions. . . . He asked questions about everything that he saw, or heard, or felt, or smelt, or touched, and all his uncles and his aunts spanked him. And still he was full of 'satiable curtiosity!

—**Rudyard Kipling, *Just So Stories***

ON A CRISP NEW ENGLAND morning in the autumn of 1906, the first whiz kid of the twentieth century came down from his room to meet *The World*.

The reporter for Joseph Pulitzer's flagship newspaper had traveled north to Boston from New York to check out the "Youngest College Man in the History of the United States." Daily his paper's pages blared with news of newfound geniuses, brilliant discoveries, and dazzling inventions that were transforming society in the new Machine Age. A year earlier, a twenty-six-year-old Swiss patent clerk named Albert Einstein had published three abstruse papers in an obscure journal that were being hailed as harbingers of a revolution in the physical sciences. Newshounds worldwide were sniffing the air for the next big thing. This one found his big story in a small package.

"Hey, mother!" cried the child's voice from the top of the stairs, "isn't it time to go to college?"

"Yes, dear," the young matron who had greeted the reporter replied, and the patter of his feet became a clatter on the staircase.

The journalist was smitten at the first sight of the eleven-year-old "infant prodigy of Boston."

"There burst into the drawing-room a regular boy in knickerbocker stockings with the usual holes in the knee. . . shirtwaist and gold-rimmed spectacles," he would write a few days later. "Under his arm he carried a book . . . Hibben's *The Problems of Philosophy.*"

"How do you do, sir?" the boy asked, in a clipped cadence that struck the reporter as rather quaint for a child his age. He sat politely to entertain the reporter's questions.

"Why, yes, I find it a pleasure to read," he said, stealing a glance out the drawing room window at his dog beckoning him in the yard. "But I don't see why anyone is interested in me just because I am young. Other boys are young, too. I don't see anything wonderful in being fond of studying. I wouldn't study if I didn't want to study."

Tales were already spreading beyond Cambridge about the precocious child of Professor Leo Wiener, instructor in Slavic languages and literatures at Harvard. The boy, Norbert, had learned his letters at eighteen months. Under his father's tutelage, he began reading at three, reciting in Greek and Latin at five, and in German soon after. At seven he took up chemistry, by nine algebra, geometry, trigonometry, physics, botany, and zoology, and that fall, at eleven, he had entered Tufts College in the neighboring town of Medford after only three and a half years of formal schooling.

The reporter could not comprehend why an eleven-year-old would prefer Huxley and Darwin to Hansel and Gretel.

"Philosophy is more interesting than fairy tales—that's all," said young Norbert self-assuredly. "In fact, philosophy is fairyland to me." The stunned scribe scribbled away as the boy illustrated his point with a brief discourse on the popular nineteenth-century natural philosopher Ernst Heinrich Haeckel, who coined the term "ecology" and the tongue-twisting "ontogeny recapitulates phylogeny." Norbert had tackled Haeckel's natural insights in German and much preferred them to the paeans of Homer and other classical poets he had subdued in their original Greek. "Haeckel," he informed the reporter, "tried to solve the riddle of the universe. Homer only spun stories."

In due time, this young lad would do both, but for now he had enough work balancing the demands of his college studies with the fleeing pleasures of childhood. "Do I play? Of course, I do!" he said, fending off the reporter's transparent challenge to his boyish bona fides. "Swimming is my forte. But I like to study too. When I have participated in the boys' games I turn to my Huxley or my Spencer. I get suggestions from them which lead my mind to think of greater things. But I like mathematics best of all."

Petite, prim Bertha Wiener sent her son back to his room so she could speak privately with the reporter. "Of course we are proud of Norbert. What mother and father wouldn't be?" she said softly. But, she stressed, "we have tried to bring

him up as other boys are, and we have never let him think that he is any different. We want him to be just a normal boy—

"Norbert, dear," she called out, "please close your door."

"Yes, mother," said the small voice upstairs.

"I wouldn't for the world have him think that we consider him anything out
of the ordinary," she whispered. "But, of course, we do."

The reporter's rapturous story, "The Most Remarkable Boy in the World,"
took up the entire front page and then some of *The World Magazine* on Sunday,
October 7, 1906. It dwarfed that day's coverage of the launching of the
steamship *Mauretania,* which promised passage from New York to London in
only five days, and an unassuming ad offering apartments on Manhattan's Upper
East Side for $11.60 per month.

A large photoengraving, half as big as the boy himself, spanned the broadsheet
from top to bottom, portraying the preadolescent in a jaunty sailor suit, standing akimbo, hands in his knicker pockets, legs turned out in his dark stockings
and high-button shoes. The beatific shot was superimposed over an illustration
that depicted Norbert poised atop opulently bound editions of Darwin's *Origin
of Species* and Plato's *Dialogues.*

The text mirrored the image of the little Hercules, describing him as a child of
the gods, "a healthy boy . . . heavily built, almost fat. His legs and arms are thick. His
chest is broad. His skin is smooth and his muscles hard. His head is the average size."

"But his eyes tell the story," the reporter proclaimed. "They are big and black
and blazing. There is something almost uncanny in their gaze. To quote the boy's
own words, they seem already to have solved the riddle of the universe."

Next the reporter interviewed Professor Leo Wiener at his office in Harvard
Yard. "I hate to talk about the boy," Leo Wiener lied, "not because I am not
proud of him but because it might get to his ears and spoil him." In an act of
generosity he would not make publicly again, the professor praised his son's
"keen analytical mind" and "tremendous memory." "He doesn't learn by rote, as
a parrot might, but by reasoning," he said. The professor lauded his son's mastery
of Caesar, Cicero, Ovid, Virgil, and comparative philology.

"But his inclination is always toward philosophy," he declared, gainsaying his
son's expressed preference for mathematics and insisting that his son was "lazy
and doesn't study as much as the average boy his age."

The faint praise left the reporter a bit confused and uncertain of his subject's
fate. He ended his piece as he began, with bombast, and a caveat:

"Whatever he may be in the future, Norbert Wiener is the youngest college
man—beg pardon, boy—in the history of the United States, if not in the history of the world."

In that simple time when the art of overstatement was young, when newspapers
were the nation's primary source of information, radio was in its experimental

stages, and electricity itself was still a miracle to many, the child who would father the information age stepped onto the world stage and became one of the first media darlings of the American Century.

———

The infant prodigy of Boston was, in many respects, a child of the old order of knowledge. His philosophical forebears harked back to Greece and Rome. His scientific ideas derived from the patriarchs of classical physics and calculus, Newton and Leibniz, and from the trailblazers of biology in Britain during Victoria's long reign and their counterparts on the European continent.

But the Wiener boy's genome followed a different line of descent that would shape his life and his thinking in tangible ways.

"I am myself overwhelmingly of Jewish origin," he said at the outset of his autobiography, *Ex-Prodigy: My Childhood and Youth*, published in 1953. Neither he nor his father nor his father before him were religious Jews, yet he embraced his Jewish roots and the values his forefathers had carried down irrespective of ritual. The adult Wiener credited much of his success to the Jewish "attitude toward life" derived from surmounting centuries of ethnic and religious prejudice. He lauded Judaism's love of learning and traced with pride the distinguished Talmudic scholars in the Wiener line—including one celebrated Jewish sage who had lived seven hundred years before.

According to a family legend, the Wiener clan traced back to the twelfth-century Jewish philosopher and physician Moses Maimonides. The revered Moses ben Maimon, known to religious Jews as the "Second Moses" or by his Hebrew initials simply as "the Rambam," was born in Córdoba, Spain, in 1135. Like Norbert Wiener, he was a child prodigy. In 1159, his family fled Spain to escape the persecution of Córdoba's Jews by a fanatical Muslim sect and eventually settled in Egypt, where Jews were welcome. There young Maimonides earned a reputation as an adept translator and healer and was appointed personal physician to the Court of Sultan Saladin the Great in Cairo. The Rambam was a repository of the knowledge of his time and the foremost Jewish scholar of the Middle Ages. His best-known work, titled *The Guide of the Perplexed*, was read widely in the Middle East by Jews and Muslims, and in Europe, where it was a major influence on the Dominican monk Thomas Aquinas.

Norbert Wiener always acknowledged that his connection to Maimonides was tenuous. As the family history chronicled, his paternal grandfather lost the only written genealogy when a fire swept through his home in Poland in the late 1800s, and all remaining links to the past were severed when the Nazis sacked Jewish towns in the region during the Second World War. "After so much passage of time our supposed ancestry is a very shaky legend," Wiener admitted.

Yet, for all his own fierce secularism, Wiener rejoiced in the prospect of that august link through history.

Years later, a cousin of Wiener's traced their family line back before the French Revolution to the Pripet Marshes in Poland, where Leo Wiener's ancestors encamped on the flood plain of the Biala River. There other genealogists found links to several hundred modern descendants of Maimonides and gave circumstantial support to Wiener's hoary handshake with the Rambam. As it turned out, Maimonides's progeny in Eastern Poland and an adjacent region of Lithuania formed a tight triangle around the very place where Norbert Wiener's father was born—the fabled Polish village of Bialystok.

———

Europe's last bison still roamed the banks of the Biala in the late 1800s, when Bialystok and the region around it became a semi-autonomous *gubernia* of the Russian Empire. Salomon Rabinowicz, Leo Wiener's maternal grandfather, was born there in 1810. A timber merchant, he built a grand hotel at the end of Bialystok's main street, a wide tree-lined boulevard that led in from the train station to the town marketplace. Set in a prime location at the entry to the Jewish ghetto, the hotel was favored by the highest and humblest travelers, from Polish noblemen to poor Talmudic scholars, who were treated to sabbath dinners in the dining hall hosted by Salomon's wife, Rosa.

Rosa Rabinowicz, née Zabludowska, came from a family of wealthy tanners in the adjoining town of Zabludow whose members rose through the ranks of postal officials and government contractors to become "hereditary honorable citizens" of Bialystok, the highest title a Jew could aspire to in Imperial Russia. Rosa was not a typical Jewish mother. She fasted each week but also gave her children colored eggs at Easter and kept a Christmas tree behind closed curtains. The couple raised six children, and their middle daughter, Freida, would become Leo Wiener's mother.

Leo's father, Salomon Wiener, was the keeper of the family genius. Short, stocky, with strong shoulders and a wrestler's neck, he was born in 1838 in Krotoschin, a large town several hundred miles west of Bialystok in a *gubernia* that maintained close ties with its German neighbors. He was educated in Königsberg at a classical gymnasium, Europe's equivalent of high school, where he steeped himself in German language and literature, but a hearing problem kept him from a university career and he took a job in the Königsberg post office. Old Salomon Rabinowicz met young Salomon Wiener on one of his seasonal trips to ferry his timber rafts down the Vistula River. He urged the frustrated postal clerk to move to Bialystok, where local businesses were eagerly seeking German teachers and correspondents to German business houses. At that time,

Bialystok was a hotpot of languages. Most Jews in the town spoke Yiddish, a mix of German, Hebrew, and Slavic; most gentiles spoke Polish or Russian; and the children of wealthy families were schooled in German and French by private tutors and governesses.

In 1859, at the age of twenty-one, Salomon Wiener moved to Bialystok to provide services in German to the town's rich factory owners and merchants. Two years later he married Freida Rabinowicz. Salomon was tough-minded and adamant in his beliefs, as his son and grandson would be after him. He ruled his house with a "frenzied fury." A child of the Enlightenment and devotee of the German-Jewish reform movement set in motion by Moses Mendelssohn, he schooled the couple's six children at home and forbade them from speaking the mongrel language of Eastern Europe's Jews. Freida spoke only Yiddish, and years would pass before she learned enough German to communicate with her husband and children.

Salomon was a restive soul and a habitual walker, like the Wiener males who would follow in his footsteps, and he had the familial bent of absentmindedness. He would leave home for days at a time on unknown perambulations, and one day he disappeared completely, leaving his wife to raise their three boys and three girls, with whom she could barely converse.

Leo Wiener, the eldest of Salomon and Freida Wiener's six children, was born in 1862. In keeping with his father's Prussian roots and reformist principles, his first language was German, but he quickly imbibed the rich babel of Bialystok. He studied Hebrew with a local teacher, and Yiddish came naturally (and surreptitiously) soon after. Leo learned French at seven from his cousin's governess. When he was eight, an uncle taught him Russian. He began his own teaching career at nine by relaying his Russian lessons to a friend for an hourly fee of a quart of gooseberries.

After his early home schooling, Leo shuttled among specialized schools and tutors. Salomon and Freida had moved the family repeatedly to meet their son's needs, as Leo and his wife would do years later for Norbert, and Leo bloomed early as a prodigy. In 1873, at age ten, he sat for his high school entrance exams. The following year, he entered the classical gymnasium at Minsk in Byelorussia, where he learned more languages, including Latin, Greek, and German dialects. After a year in Minsk, he transferred to the gymnasium in Warsaw, where he learned Polish and Italian and picked up Dutch from another cousin who had studied in Brussels. After graduation, he studied engineering at the Berlin Polytechnique Institute, where he learned Croat from a fellow student and Danish from another relative. His inventory of tongues surpassed ten while he was still in his teens. In time, he would become fluent in more than forty languages.

Leo was small like his father, barely topping five feet and a hundred pounds, but he inherited the athletic build and physical vigor of the Wiener males. As a young man Leo carried himself proudly, his black hair, bushy mustache, and wire-rimmed spectacles enhancing his appearance of high intelligence. He was an ardent hiker, walker, and talker who tended to monopolize any conversation. He was precocious politically as well. At the gymnasium in Minsk, he mixed with the student revolutionaries who thronged the town in the 1870s. In Berlin, he joined the idealistic Tolstoy Society; swore off alcohol, tobacco, and meat; and never backslid. His family urged him to take a nice, respectable job with the Mendelssohn Bank, but he had no intention of settling for a life so mundane.

This firebrand "young Slav engineering student," as Leo called himself, recoiled at the baroque splendors of the Prussian capital, where the quest for empire was pulsing through the populace. He walked the length of Berlin's lustrous Unter den Linden, looking for a third way out of "the Russian cul-de-sac of ideals" and the drift into "German philistinism, with its juxtaposition of science and beer." He passed his nights in the Café Bauer, poring through newspapers from around the world in search of "a region where science leads to higher thoughts and where ideals materialize." All signs pointed to the "Newlands beyond the sea." At a meeting of fellow Tolstoyites, Leo hatched a visionary scheme to found a "vegetarian humanitarian socialist commune" in Belize, a British colony in the tropics of Central America. A year later, while his comrades were still bickering over the proposed language for the preamble to the colony's manifesto, Leo, now the "wild, long-haired Russian" filled with disgust over "rotten" Europe, donned his one piece of clothing with any panache, a long, double-breasted Prince Albert frock coat, and set out for the Newlands alone.

Leo sailed from Hamburg in February 1882 at age nineteen. During a layover in Liverpool, he read a dozen grammars and taught himself English. On the long crossing to Havana, he learned Spanish from the passengers who had boarded in La Coruña. He landed in New Orleans with twenty-five cents in his pocket, which he promptly spent at the fruit stands in the French Quarter on a sampling of plantains and other tropical delights.

His passage to Belize was cut short for want of funds, but Leo was ripe for the romance of simple labor. He bailed cotton in a factory, moved on to Mississippi to work on a railway, then to Kansas where, he had heard, a vegetarian commune was being built on the ruins of a progressive community founded a decade earlier by Russian intellectuals and American spiritualists. When he arrived the place was deserted. Leo single-handedly reclaimed the broken fields and tumbled-down buildings. He planted peanuts and melons, sang along with the meadowlarks, and stood in awe of the immense thunderstorms sweeping across the prairie. He wrote joyous letters back to his comrades in Berlin, beckoning them to join him, but no one came.

After a year, he cut his hair and headed back to civilization, arriving, penniless again, in Kansas City, Missouri. He found work as a janitor in a dry goods store, where he endured the "contemptuous manner of high-born ladies." He quit that job, became a lowly peddler and, one day, happened upon the city's public library. He plunged into English literature and the classics, planning for the day when he would show the families in the town's mansions that he was "their equal and more than their equal." Then, one fine day, he shook out his dusty Prince Albert and called on the superintendent of public schools. The next morning, the local papers told the story of the "swarthy laborer" who had been discovered selling peanuts and appointed to teach in the city's high school.

Leo taught his students as he would teach his son, "not according to rules learned by rote" but by imparting "the sentiments which actuated my own life"—his love of Homer and Cicero, Goethe and Schiller, philology and the diophantine solution for algebraic equations. He took his pupils on hikes through the city's hills and woods, taught them how to tell a papaw from a persimmon, and soon became a popular figure in the local Philosophical Society. He added Choctaw and Dakota to his quiver, then Chinese, Bantu, and Gaelic. Within weeks, he was named president of the town's Irish Society and dubbed the "Russian Irishman" by the doyens of Kansas City culture. He hobnobbed in high society, fulfilling his vow, and became especially enamoured of the "Browning cult," as he called it, the local chapter of a nineteenth-century cultural movement that elevated the poetry of Robert and Elizabeth Barrett Browning to oracular heights.

One evening, after delivering a talk to the Browning Ladies Club, Leo met comely young Bertha Kahn, the daughter of Henry Kahn, a department store owner in St. Joseph whose family had come from the Rhineland to the Missouri bottomlands a generation earlier. Bertha's mother was a half-Jewish Southern belle; her father's family had emigrated from Germany in the 1820s. The Kahns were thoroughly assimilated and made little attempt to hide the antagonism many German Jews showed to the later waves of poorer, ethnic Jews from Eastern Europe and Russia. Bertha was fair-skinned and soft-featured, small but hardy like Leo, and she was a proper belle herself, intent from the start on bringing her wild-headed beau into line with the more refined elements of society.

In 1892, after his first stint as a college teacher, Leo was made a professor of modern languages at the University of Missouri in Columbia. He and Bertha were married the next year in Kansas City. Their first child, a son, was born November 26, 1894 in a faculty boarding house in Columbia and named Norbert, after the leading man in Robert Browning's dramatic poem *In a Balcony*.

Little Norbert showed early signs of precocity. In the summer of 1896, when he was eighteen months old, he learned the alphabet in two days by watching his nursemaid draw the letters in the sand at a beach near the family's new home. That year Leo had lost his teaching post in a faculty reorganization. Lacking wider connections in Missouri, he decided to move his family to Boston, knowing nothing about the town save that it was home to many colleges and, therefore, seemed like a good place to ply his skills as a linguist and a teacher. In short order, he found work translating Serbian ballads for a prominent professor, who helped him to obtain an instructorship in Slavic languages and literatures at Harvard, the first appointment of its kind in the United States. To supplement his modest instructor's salary, Leo taught at Radcliffe, the new women's college up the road, and did etymological work for the Merriam-Webster Dictionary.

The family took an instant shine to New England and thrived in Cambridge's brainy environment. In the autumn of 1897, Leo moved them from their cramped walkup apartment to a nice house with a small garden on Hilliard Street off Harvard Square. Daily Bertha would read to Norbert in the garden. Among his favorite stories were the colorful jungle tales of the Indian-born English author Rudyard Kipling. By age three, Norbert was reading to her.

Norbert was a sensitive child. A walk past a hospital for "incurables" filled him with gloom. A visit to a blacksmith whose toe had been crushed by a horse instilled an abhorrence of suffering and physical deformity that would preoccupy him all his life. Among his most salient memories were the "horrible and hair-raising" tracts that cluttered his home depicting acts of cruelty to animals. They ensured at a tender age that Norbert, like his father, would be a strict, lifelong vegetarian. But in most other respects, he was a normal little boy, who romped with the neighborhood children and pulled his toy battleship by a string down Hilliard Street.

Norbert's education began on the floor of his father's study, where he played in the kneehole of Leo's massive wooden desk and grazed through his father's bookcases for works with pretty woodcuts and words he understood. Science books of every kind captivated him. On his third birthday, a family friend brought him a book on natural history. Soon after, he discovered the wonders of illustrated science magazines for children.

In the spring of 1898, Norbert's sister Constance was born, her name, too, borrowed from Browning's *In a Balcony*, and that fall, before he turned four, Norbert started kindergarten in Cambridge. The place was more a play school than a house of learning, but the hard work was already underway at home under Leo's direction. That year Norbert read *Alice in Wonderland* and feared for the little girl who dropped down the rabbit hole into a world of strange and illogical creatures. He read the *Arabian Nights* with equal concern. Its dark tale of the vengeful *djinni* in the bottle would stay with him and emerge repeatedly in his popular writings as a vivid metaphor for the dangers inherent in seemingly magical gadgets.

The new century commenced soon after Norbert turned five and, that spring, Leo returned to his roots in the land. With some capital accrued from his teaching and translating work, he bought an old farm in Foxboro, south of Boston, on a country road framed by catalpa trees. The family spent the summer on Catalpa Farm, where Norbert learned the wonders of New England's plants and animals and discovered the surprising resilience earthworms display when cut in two (the act of cruelty only twinged his conscience in the light of the fascination it held and the obvious fact that the worm suffered only a "moderate inconvenience" from the procedure). Leo took him on "tramps" through the fields and woods and introduced his son to his favorite sport of harvesting edible mushrooms. That summer also brought Norbert's first contact with his New York relatives. His *Grossmutter* Freida Wiener had immigrated to America with Leo's siblings. She and several of Norbert's aunts and cousins came from the city to breathe the fresh air of Catalpa Farm and to meet Leo's family. *Grossmutter* brought newspapers in a strange language that Norbert later learned to be Yiddish, but at the time no one told him that she, or anyone else in the family, was Jewish.

His formal schooling remained problematic. He attended the village school in Foxboro for a few days, then Leo transferred him to a little red schoolhouse farther out in the country. Although Norbert was still immature physically, his intellect was advancing at a pace far beyond his peers; and, in the fall of 1901, Leo sold Catalpa Farm and entered Norbert in the progressive Peabody School in Cambridge. He was placed in the third grade, then advanced to the fourth, but he didn't fit in there either. His reading skills were excellent; it was his math skills that were deficient. At seven, Norbert was still counting on his fingers, and he could not get his multiplication tables. Questioning his son, Leo learned that he was bored witless by the rote drills of traditional teaching methods. Certain that he could do a better job of teaching the boy himself, he withdrew Norbert from school and, for the next three years, embarked on a radical experiment in home schooling.

Two years earlier, Leo had begun training his son informally in his own fields of language and literature, starting with the Greek and Latin classics, followed by his favorite German poets and philosophers. Now Leo added Darwin, Huxley, and other scientists to his son's curriculum "in order to instill in him something of the scientific spirit." Norbert became fascinated with zoology and botany. He retraced Darwin's journeys and other famous naturalists' travels to faraway places. And he went farther afield. He studied Leo's books on psychiatry, the occult, and the excavations of Troy. He became an avid young fan of the seminal science fiction of H. G. Wells and Jules Verne. When he had devoured the volumes on his father's shelves, Leo fed him books and journals from the Harvard library: on physics and chemistry, the nature of light, and the strange new power of electricity.

Leo did not set out to engineer a boy genius, as the Wiener legend claimed, but that aim crystallized early on as Leo saw his self-styled training program yield remarkable results, and he began to distill its driving principles. Leo called his method one of "tactful compulsion." The goal, as he described it loquaciously to the local and national press, was to maintain a "constant watchfulness" over the boy's words and actions, to determine his intellectual aptitudes and interests, to abolish rote memorization, and to encourage his son to question and "constantly to think for himself." He also wanted his son to know "the blessedness of blundering." Leo held, eloquently, that "the child must be made, in a kindly manner, to work out problems, in order that he may acquire that sense of mastery, that joy of triumph, which is of itself an incentive to further effort."

———

That wasn't the way Norbert recalled it. In *Ex-Prodigy*, he described Leo as a mentor who indeed took pains to stimulate his developing intellect and imagination, but pains he felt palpably. Leo's manner in private was not kindly. He inculcated the classics, mathematics, and every other subject with military precision and a drill sergeant's demeanor. His bludgeoning approach in practice combined formal recitations with ferocious upbraiding and a premeditated program of "systematic belittling."

Five decades later, the ordeal was still vivid: "My father would be doing his homework for Harvard and I had to stand beside him and recite my lessons by memory, even in Greek, at six years old, and he would ignore me until I made the simplest mistake, then he would verbally reduce me to dust." When Norbert misspoke, Leo berated him mercilessly with shouts of "Brute!" "Ass!" "Fool!" and "Donkey!" in English, German, and the forty other languages he spoke fluently. From his earliest years, Norbert had looked up to his father from the lowly kneehole of his desk as a good-hearted but no less "austere and aloof figure." But a darker form emerged during Leo's home lessons, which could turn abruptly from tact to terror over a simple algebraic equation.

"He would begin the discussion in an easy, conversational tone," Wiener recalled. "This lasted exactly until I made the first mathematical mistake. Then the gentle and loving father was replaced by the avenger of the blood." Leo's words cut deep. "The very tone of my father's voice was calculated to bring me to a high pitch of emotion, and when this was combined with irony and sarcasm, it became a knout with many lashes." His lessons often ended in the same climactic family scene. "Father was raging, I was weeping, and my mother did her best to defend me, although hers was a losing battle." Leo taunted Norbert at the dinner table and before company, he said, to keep him humble. He condemned his "juvenile ineptitudes" until his son felt "morally raw all over."

Norbert strove to do better. He worked harder and studied longer, but in less than a year the daily demands of reading for his lessons produced a severe myopia. By age eight, the eyestrain was so bad the family's doctor ordered Norbert to stop reading for six months, fearing that he might lose his sight altogether. The harsh remedy gave Norbert some reprieve, but it threatened to disrupt his father's experiment in childhood education, which was taking on a life of its own in the academic community. Leo knew his Greek tragedies. He was not about to blind his firstborn son. But he was not prepared to abandon his experiment. Bertha Wiener was conscripted to read Norbert's lessons aloud. Once a week, Leo sent one of his Radcliffe students to the house to review his son's Latin declensions and German pronunciation. He hired a Harvard chemistry student to teach Norbert about chemical reactions and set up a makeshift laboratory in the house.

And, for six months, Leo made his son do all his reasoning, reckoning, and arithmetic in his head.

The result was profound. At age eight, Norbert learned to do algebra, geometry, and trigonometry, not on paper, but in the inner eye of his mind. He developed a near-photographic memory and, like his father, an astute ear for languages. That year, Wiener later told an MIT colleague, "I relearned the world. My mind completely opened up. I could see things I never saw before." And from that time on, his mind was able to do things that, when he looked back, "were still totally astonishing" even to him.

Indeed, already Norbert had begun his lifelong practices of traversing disciplinary boundaries and pursuing multiple projects concurrently. When he was allowed to read again, he studied children's science magazines and found similarities among animal skeletons and other structures in nature. He took a special interest in the new connections that were being made between the fledgling field of electricity and the maturing science of biology. He read with amazement the latest discovery that electrical impulses propagated along nerve fibers in an entirely different way than they did through metal wires, as he put it, in a process "analogous to the . . . fall of a train of blocks" rather than in a continuous flow of current, as electricity was depicted at the time.

Other subjects were inscrutable to young Norbert. As a budding scientist, he was fascinated by the sex lives of plants and animals. By age nine, he grasped the basics of mitosis, embryology, and the difference between an egg and a sperm, but the fine points of vertebrate reproduction eluded him, and his questions to his parents on such matters were discouraged. Somewhere he got the wild idea that he could turn a doll into a baby with the right incantations. Then, around 1903, he adopted a more realistic and, as it would turn out, lifelong desire to build what he later called "quasi-living automata"—mechanical devices that would mimic the behavior of animals and human beings.

During those years, one of the most important influences on Norbert was his father's friend, Walter Cannon, Harvard's celebrated biologist and pioneering neurophysiologist. Cannon blazed the way for American medicine's use of X-rays to diagnose and treat bodily ills. He named and explained the nervous system's innate "fight or flight" response to imminent threats and perceived dangers. In the 1920s, he would coin the term "homeostasis" to describe the body's organic mechanism for maintaining healthy, stable states of internal equilibrium through a balance of self-regulating actions and reactions: sweating to cool the body when it becomes overheated, releasing hormones to excite or inhibit the action of the body's organs and nerve cells. Years later, Cannon's principle would become a central tenet of Wiener's new science.

Norbert was captivated by his tours of Cannon's laboratory and its wondrous new scientific devices. In Cannon, his boyish interests in zoology and botany found their ultimate exemplar and champion, and his growing fascination with electricity found a new outlet for experiments and practical applications. Leo left no doubt that he wanted his son to pursue a career in philosophy, and Norbert started dutifully down that path. But quietly, following Cannon's example, he formulated his first goal that diverged from his father's directives: to become a naturalist in the grand tradition—with modern accoutrements.

Those early years in Cambridge were formative for Norbert and the Wiener family. A third child, a daughter, was born in the spring of 1902 and named Bertha after her mother. The family moved again, to Avon Street, north of Harvard Square, where Leo bought a cheery old house with a proper library and a large backyard. Two doors down lived Professor Maxime Bôcher, Harvard's eminent algebraist and geometer, who was widely viewed as the patriarch of American mathematics.

The Cambridge of that day was a preserve of eminent academicians and well-to-do Boston businessmen, but it still had the feel of a quiet country town, with unpaved roads, horse-drawn trucks, and empty lots for young boys to play in—and for Leo and Norbert to scrounge for edible mushrooms. Down those easy streets and unkempt fields, Norbert fought snowball wars with the children of other Harvard professors and, once, ran away from home with his erudite playmates to fight the Turks and save the oppressed Armenians (they got to Central Square, a few blocks east of Harvard Yard, before abandoning their mission).

In the spring of 1903, Leo's career took a turn that would affect Norbert and the entire family. He accepted a commission from a Boston publisher to translate into English the works of Tolstoy, all twenty-four volumes, for a fee of

$10,000, a meager rate per piece but a good lump sum for the time. Leo would complete the ascetic task, worthy of Tolstoy himself, in only two years, but to do so he needed a refuge and breathing room. With the publisher's advance, that summer, he bought another farm thirty miles northwest of Boston in the town of Harvard (which had no connection to the university) and moved his family back to the country.

Unlike rustic Catalpa Farm, Old Mill Farm was a real working farm, complete with cows and horses, a pre–Civil War farmhouse, a large barn, a small lake, and thirty acres of orchards and meadows bursting with wildflowers. Norbert relished the new life it laid before him as a young naturalist. He discerned each squirming species of toad, tadpole, and leech in the vicinity. He classified each flower and marsh fern. But he sorely missed his chums on Avon Street and their waggish exploits together. He made friends with some boys on neighboring farms and began a new round of escapades. One time, they nearly electrocuted themselves conducting amateur radio experiments. But for all his bucolic adventures, Norbert felt alone and isolated at Old Mill Farm.

Leo was busy with farm chores, the long daily commute into Cambridge, and translating Tolstoy at locomotive speed. He no longer had the time to micromanage his son's education. He searched schools in the area, and in the fall of 1904, Norbert, then age nine, was admitted to the public high school in the nearby town of Ayer. The schoolmarm started him as a sophomore. He proved his mastery of Latin, German, and English literature, and sat patiently through his classes in algebra and geometry, which thanks to Leo's thundering pedagogy he now had down pat. At home, Leo continued to compel his son to recite his lessons aloud every evening while he pounded out his Tolstoy translations on the typewriter. Somehow, he still managed to catch all his son's mistakes and pummel him verbally without looking up from his papers. At the end of Norbert's first year in high school he was advanced to the senior class.

That summer, at age ten, he wrote his first philosophical paper, a treatise on the incompleteness of all knowledge, and titled it with pluck, "The Theory of Ignorance." While far from rigorous in its argument, and in its exposition still clearly the work of a child, his "theory" foreshadowed more profound themes. Already, at that early age, young Norbert spoke with conviction about "the impossibility of man's being certain of anything." He vigorously disputed "man's presumption in declaring that his knowledge has no limits," and he analyzed the many causes of "uncertainty" through which "an idea has to pass to get lodged in the mind." He asserted that "Philosophy is worthless without due consideration" of the problem of uncertainty, which, he said, "has been far too often disregarded." In science, too, uncertainty exists "with every experiment," he insisted. He concluded that, "In fact all human knowledge is based on an approximation."

To be sure, his theory was only one well-formed expression of a cocksure lad, but his confident premise picked out a straw in the wind that would soon shatter the foundations of twentieth-century philosophy and science.

His agile mind was advancing at a record pace, but as it raced ahead it was straining his development physically, emotionally, and socially, and in one arena of childhood development things were beginning to get awkward.

In his senior year at Ayer High, when he was eleven, Norbert fell in love with a freckle-faced girl of fifteen who played the piano at the school's concerts. Decades later, he still recalled that first smiting as more than a mere childish infatuation. "Futile as it was, it was real love, and not the almost sexless affection of undeveloped children," he insisted. But his desire was quenched by the reality that, for all his intellectual prowess, he was a greenhorn in affairs of the heart and clueless about the new forces knocking at his body. His parents picked up on his budding passion and made inquiries to assure themselves "that this girl was not leading . . . my soul to perdition—though there had never been the slightest danger."

But Leo and Bertha were of little help to their son in that department. The age discrepancy had been a constant problem in Norbert's early friendships. Now, in his first hopeless fling at romance, it posed an insurmountable barrier. "Little as I wished to grow up," he recalled, he found himself "rushing toward maturity" without any interplay with girls his own age. The problem would affect him long afterward, he admitted, "and when I entered my twenties I was still not nearly out of the woods."

Early in 1906, Norbert's last sibling, a brother, Frederic, or Fritz, as he became known, was born. That spring Norbert graduated from Ayer High at the head of his class but far behind his classmates by every other measure. His parents threw a graduation party at the farm but he felt like "an outsider at the feast." As the older boys and girls danced in the parlor, Norbert huddled in the kneehole of his father's desk and watched the courting ritual he was too young to take part in from the safety of the familiar box that could now barely contain him. There, curled up on the floor while the company danced above and beyond him, he spent his last moments of innocence.

———

In his wisdom, Leo was reluctant to subject his son, who was already a minor celebrity in Ayer and around Cambridge, to the glare of being a pre-teen prodigy at Harvard. Instead, he chose Tufts, a fine college in nearby Medford strategically located just two miles north of Harvard Square. Leo bought a bright new house on the Medford Hillside, from which Norbert could walk to school and he could take the trolley into Cambridge, and, in September 1906, Norbert started college still wearing the short pants that separated the men from the boys of that day.

A month later, the *New York World* anointed Norbert the youngest "college man" in history, but there was something wrong with the rotogravure. The Most Remarkable Boy in the World was not the picture-perfect brainchild with the hard body, self-assured stance, and piercing stare profiled in Joe Pulitzer's Sunday supplement. Physical problems had plagued him for most of his young life. He was "heavily built" and "almost fat," as the *World* acknowledged, probably from hewing to a high-starch vegetarian diet at too early an age, with a paunch that would soon balloon to a great bandwith. His eyesight had survived his feast of early reading, but his myopia was extreme, and he was clumsy, with poor motor control and a lack of muscular coordination that made him inept in his movements indoors and out. Since his salad days with his chums on Avon Street, his agility had degenerated to the point where he could not catch a ball, in part because he could not see it coming.

More crippling were the scars from his father's relentless, multilingual tongue-lashings. Years of verbal abuse had left the world's youngest college man feeling that his precocious mind was "no good" and with a paralyzing insecurity that would burden him for the rest of his life.

Despite Leo's proddings toward philosophy, Norbert was determined to major in zoology, but that path was strewn with stumbling blocks. His tubbiness, myopia, and utter clumsiness made him a washout at laboratory work and every other activity that required manual skill. His handwriting was legible but "crabbed," and "drawing was a bugbear." As he saw it, putting his physical impediments in the best light possible, his mind worked faster than his body and simply overwhelmed his external "effectors." His lab work in chemistry was completed "at probably the greatest cost in apparatus per experiment ever run up by a Tufts undergraduate." In biology, this gentle vegetarian who shrank at the thought of suffering and mutilation cut a trail of botched dissections through dogfish, cats, and guinea pigs. He came out of the course a lifelong foe of vivisection.

To appease his father, in his sophomore year Norbert took several courses in philosophy. He attended lectures at Harvard given by William James, the university's *eminence grise* of philosophy and its newborn stepchild psychology. James had founded the first experimental psychology laboratory in America. He also co-founded the philosophical school of pragmatism, the new thoroughly modern, thoroughly American doctrine that rejected bloated philosophical theories in favor of practical actions evaluated solely on the basis of their utility for daily living. James had been an early influence on Norbert indirectly, through Leo, who incorporated James's education theories into his own eclectic training methods. Leo even arranged a private audience for his son at James's home, but Norbert said later he admired James's colorful style far more than his logic, which even as a pre-teen he found weak and haphazardly organized.

Norbert's zest for science kept pulling him away from philosophy. He spent his free time tinkering with electricity, the new century's emblem of power. With older classmates who performed the manual labor, he assembled a small substation of rudimentary electrical devices. And, after his own record-breaking laboratory disasters, Norbert began to feel a new attraction to mathematics, which, as he noted with satisfaction, was "a field in which one's blunders . . . can be corrected . . . with a stroke of the pencil." After conquering Tufts's most difficult math course on advanced algebraic problems, he changed his major. He mastered calculus and differential equations, with their mind-stretching mélange of abstract symbols and mathematical functions. Before long, his math professor turned the class over to him.

In the spring of 1909, after only three years, Norbert graduated from college with a bachelor's degree in mathematics, but his sprint through Tufts took a toll. For the first time in his high-performing young life, Norbert was exhausted physically and mentally. "I could not stop the wheels from going around, and I could not rest," he wrote later. He passed that summer in a low-grade fever. Emotionally, he felt "at loose ends," cast out of childhood prematurely, catapulted through college, and filled with foreboding about the next round to come. The momentary joy that attended his commencement was crushed by the specter that shadows every new college graduate and stalks prodigies with a special vengeance, but to Norbert it seemed his burden alone: "What should I do in the future, and what hopes might I have of success?"

The question pounded in his head. Already, in his early teens, it was beginning to strike at his health and steal his few scraps of pride in his historic achievement. He had graduated with honors, but he had not been elected to the prestigious Phi Beta Kappa Society, he was told, because of "doubt as to whether the future of an infant prodigy would justify the honor." The insult shook him to his bones. "This was the first time that I became fully aware of the fact that I was considered a freak of nature, and I began to suspect that some of those about me might be awaiting my failure."

Fading fast was the manly confidence he had presented to the *World* three swift-moving years earlier. Behind his last reserves of youthful swagger, he was the first to question his widely publicized potentials. Years later, he put words on the premonition he felt that summer and the bleak hypothesis the years ahead would affirm or refute, that "the child who makes an early start is intellectually drawing on his life capital of energy and is doomed to an early collapse and a permanent second-rateness, if not to the breadline and the madhouse."

The prospect haunted him, filling him with an acute sense of impending failure and his own mortality. He factored his fourteen years into morbid projections of his probable lifespan. His fear of death was compounded by fears of retribution—if not divine, on some celestial plane—for past sins that rebounded from every dark corner of his mind: sins of suffering he had inflicted on the animals in his biology classes, sins of sexual awakening that were doubly taboo amid his parents' silence on such matters. The lowering clouds of doom unleashed a wave of misery with all the modern warning signs—panic attacks, obsession with death, persistent, unexplained physical illness—but, in 1909, the Most Remarkable Boy in the World had nowhere to turn and no one to talk to. His first bout of depression did not end suddenly that summer, but "rather it petered out."

This was not the best place from which to embark, at the fragile age of fourteen, on a high-pressure graduate career at Harvard.

In September, with his father's blessing, Norbert registered at the Harvard Graduate School. To finance another family move back into Cambridge, Leo sold Old Mill Farm and the house in Medford and, with the proceeds, built a stately new home a few blocks west of Harvard Square. As Norbert trudged along the brick walks of Brattle Street toward the Yard, savoring the late summer air, he was certain of some things: his love for the outdoors and nature in its every expression, his ample knowledge of the world's dominant languages and literatures, and his growing skill at mathematics, which he perceived at the time as "a sword with which I could storm the gates of success."

But there was heavy traffic at those gates, genteel Harvard men brandishing keen blades of their own, and barbarians massing outside Harvard's ivy-covered walls.

America's oldest institution of higher learning was no pushover. Historic Harvard Yard, a twenty-two-acre tract of towering trees, grassy plots, and foot-worn paths angling among solemn stone and red brick buildings, was enclosed by a spiked iron fence and arching entranceways opening warily onto Cambridge's plebeian streets. The university within, Leo Wiener's incursion aside, remained the cloister of a high priestly class, its catalog still crammed with courses compatible with the Puritan philosophy of Massachusetts's first colonists, who had established the college's mission in service to the ministry.

Norbert was one of four child prodigies who entered Harvard in that auspicious fall of 1909. The others were William James Sidis, age eleven, son of Dr. Boris Sidis, a psychiatrist who, like Leo Wiener, was applying his own prodigy-making poultice to his children; Adolf Augustus Berle, Jr., age fourteen, son of a local Congregational clergyman who was grooming his son to become a statesman; and Cedric Wing Houghton, age fifteen, a scion of Harvard and Boston's blueblood society. A fifth prodigy, Roger Sessions, a musical genius from New York City's up-and-coming borough of Brooklyn, arrived the following year, also at age fourteen.

When the press realized there was, not one, but a pride of prodigies foraging in idyllic Harvard Yard, they descended on the young lions—and the Elephant's Child—in a feeding frenzy of their own. Years later, Norbert recalled dodging swarms of reporters from "the slicks," the tabloid tell-alls of the era, "who were eager to sell my birthright at a penny a line." Their presence only added to the palpable hostility Norbert felt from the day he set foot on Harvard's hallowed ground, but he quickly learned to spot his pursuers cutting across the Yard and to give them the slip by ducking out a side gate and vanishing into the back alleys of Harvard Square.

Leo Wiener was consistent in his efforts to keep his gun-shy son out of the media's sights, but Leo was not deaf to the slicks' siren song. Although he was building a solid reputation in academia for his high-quality translations and scholarly publications, he was frustrated at Harvard, where, he felt, his pioneering work in Slavic languages and literatures was not fully appreciated. With help from the slicks and his own talent for self-promotion, Leo saw a chance to tout his child-training techniques in the public eye. Soon after the 1909 school year began, Boston's *Sunday Herald* hailed "Harvard's Four Child Students" as "Products of Their Fathers' New Systems of Education" and teased "Parents Declare Others Might Do Likewise." The article paid the most attention, not to Norbert Wiener, or to William Sidis and his psychiatrist father, but to Leo Wiener.

Leo seized the day to expound on his education methods, which, he claimed, were producing spectacular results with all his children.

"My children are *not* anomalies. They are not geniuses. . . . They are not even exceptionally bright," Leo insisted, then he tossed a gratuitous slap at his son. "Norbert could have been ready for Harvard at 8. He was not forced. He is even lazy," he said, repeating the charge he leveled in *The World* in 1906. Leo excoriated his son in the press as he did in private. He even advertised his son's frequent errors in his math equations as testimony to the power of his training methods, and as proof of his point that "a conscious blunder is a grand thing."

That academic year could not end soon enough. The long break in the summer of 1910 was needed relief and important for Norbert in another way. It marked the Wiener family's first summer sojourn in Sandwich, New Hampshire, a hamlet in the White Mountains between sprawling Lake Winnepesaukee to the south and the imposing Sandwich Range to the north. The whole family was charmed by the stunning scenery, the cool mountain air, and the local country folk, whom Norbert lauded later for their "sober dignity, reserve," and legendary "reticence" that mark the humble pride of the New Englander. That summer, with Leo and his sister Conta, as Constance came to be called, Norbert, now fully grown at five-foot-six and weighing 146 pounds, hiked the granite-topped Mt. Whiteface, a 4,000-foot peak in the Sandwich Range. The craggy old mountain proved impervious to his clumsiness and a tonic for his spirit. The trio then made the 6,288-foot trek up Mt. Washington, the pinnacle of the 2,000-mile Appalachian Trail. By the time they reached the summit, Norbert was hooked on "tramping." Two summers later, he joined the Appalachian Mountain Club and took part in the group's outings that cut trails through the White Mountain wilderness that are still trafficked today. His love for hiking and for his little corner of New Hampshire were fused, and he would return to the region each summer of his life.

————

The prodigies did not break under Harvard's boughs or bend to the slicks' cat-calls for competition among them. They crossed paths frequently and even in-dulged in some boyish capers together, but the five never flourished as friends. As Wiener wrote later, "the sharing of a precocious school career is no sounder a basis for one's companionship than the sharing of a mutilation."

The fates divvied up the harvest of Harvard's *annus mirabilis*. They smiled on young Adolf Berle, who went on to become a member of President Franklin D. Roosevelt's Brain Trust. Roger Sessions, the musical prodigy, be-came one of the twentieth century's most admired composers and collected three Pulitzer Prizes. Sadly, Cedric Wing Houghton never got a chance to prove himself; a ruptured appendix took his life just before he was scheduled to graduate. William Sidis soared briefly then stripped his gears. He suffered a nervous breakdown and died suddenly of a cerebral hemorrhage at the age of forty-six, following the appearance of a biting profile in *The New Yorker* mag-azine that ridiculed his "life of wandering irresponsibility" and dismal failure to deliver on his potential.

Wiener fumed for years about the abuse Sidis suffered at the hands of his fa-ther and the press. In *Ex-Prodigy*, he flayed the *New Yorker* piece as "a cruel and quite uncalled-for" attack that "pilloried [Sidis] like a side-show freak for fools to gape at." In 1952, a Sunday supplement with a national circulation dredged up the Sidis story again in a feature titled, "You Can Make Your Child a Ge-nius." The crass piece gave Wiener one last opportunity to rebut the slicks' claim, and the unanswered assertions of fathers like Leo Wiener and Boris Sidis that the correct training formula can produce a prodigy every time.

> So you can make your child a genius, can you? Yes, as you can make a blank can-vas into a painting by Leonardo, or a ream of clean paper into a play by Shake-speare. [But] let those who choose to carve a human soul to their own measure be sure that they have a worthy image after which to carve it, and let them know that the power of molding an emerging intellect is a power of death as well as a power of life.

At Harvard, Wiener felt the competitive pressure intensely. He also felt a wall of "latent hostility" erected by many in the university community who seemed eager to see him and his fellow prodigies founder, and to watch their Pygmalion professor-fathers get their comeuppance. After a year of frustration in the laboratory and constant torment by the slicks, at Leo's insistence, Norbert turned in earnest to philosophy. He applied for a one-year fellowship to Cornell, in rural Ithaca, New York, where a colleague of Leo's from Missouri held a prestigious chair in the philosophy department. He bitterly resented his father's latest intrusion in his career decisions, but he was glad to get away from Harvard. That year marked the start of an enduring animosity he would harbor toward the ivy-covered university and scholars associated with it.

But his year at Cornell unleashed shocks of its own and one mighty surprise.

———

Leo Wiener and the slicks were not Norbert's only tormentors. His mother compounded the trauma of her son's tortuous passage to young adulthood. Bertha Kahn Wiener deferred to her husband on education matters, but as the determined daughter of a Southern belle and a thoroughly assimilated German Jew, she shaped her children's upbringing in other ways: by social climbing in local academic circles, and by assenting to the vicious anti-Semitism of the day. A longtime friend described Bertha as "a dear sweet bit of arsenic and old lace." Largely at her prompting, the Wieners lived as gentiles in turn-of-the-century Cambridge, perhaps to advance their social standing, perhaps to spare their children the hurt of New England's entrenched ethnic and religious biases. And, for whatever reasons, Bertha went a step further. By Wiener's account, throughout his childhood, not only did his mother condone anti-Semitism, she kept up her own running commentary of scornful invective, attributing loathsome qualities to individual Jews and the Jewish people.

Neither parent told their child-prodigy son that he was, not simply of Jewish lineage, but a descendant of eminent Eastern European rabbis, Talmudic scholars and, as the old family legend maintained, of the revered Rambam, Moses Maimonides.

At age fifteen, as Norbert was beginning his graduate work in philosophy at Cornell, he learned the truth by accident. In a conversation in Norbert's presence, Leo's friend Frank Thilly, Cornell's professor of ethics, made reference to that "much earlier philosopher" in the Wiener family, Maimonides, and to the old family records documenting the link that Leo's father had lost in a fire. The reference sent Norbert racing to the nearest encyclopedia, since he did not know Maimonides's name or the first thing about Jewish traditions, and then into solemn conversation with his father about the family history of the Wieners and the Kahns—which he soon learned was a variant of Cohen, the Hebrew name for the high priests of the Israelites.

The news turned Norbert's world upside down. He felt, in his own words, "disinherited." The shock of his Jewish roots hit him "where it really hurt" in the matter of his own "internal spiritual security." He stepped through the ugly logic forced upon him by his mother's words and deeds. "As I reasoned it out to myself, I was a Jew, and if the Jews were marked by those characteristics which my mother found so hateful, why I must have those characteristics myself, and share them with all those dear to me. . . . The meaning of this was clear to me: I could not accept myself as a person of any value." His stark syllogism was fraught with paradox and practical impossibilities. "To be at once a Jew and to have had inculcated in me . . . hostile or depreciating attitudes to the Jews was a morally impossible position. . . . I had neither the possibility nor the wish to live a lie. Any anti–Semitism on my part must be self-hatred, nothing less. . . . Yet it was equally impossible for me to come into the fold of Judaism. I had never been there, and in my entire earlier education I had seen the Jewish community only from the outside and had the very vaguest idea of its rites and customs."

The impossible dilemma left him no choice but to reject the faith of his fore-fathers, and religion generally, from a reasoned position as a young scholar and scientific thinker. As he defended his decision years later: "There is something against the grain in the . . . wholesale acceptance of any creed, whether in religion, in science, or in politics. The attitude of the scholar is to reserve the right to change his opinion at any time on the basis of evidence produced." And the evidence, for Norbert, was inescapable. "I could not believe in the old-line New Englanders as a Chosen People: but not even the vast weight of the Jewish tradition could persuade me to believe in Israel as a Chosen People. . . . What, then, could I do?"

The answer he arrived at, after years of internal wrestling, was to embrace a universal spirituality and humanitarian creed that would form the cornerstone of his personal philosophy and, later, provide a human foundation for his science and social activism. In *Ex-Prodigy*, he retraced his journey from Judaism to humanism that brought him full circle back to his misplaced ancestor Maimonides, with whom he felt a deep, even spiritual, connection:

> One thing became clear very early: that anti-Jewish prejudice was not alone in the world. . . . I had read enough of Kipling to know the English imperialist atti-tude. . . . My Chinese friends spoke with me very frankly concerning the aggres-sions of the Western nations in China, and I had only to use my eyes and ears to know something of the situation of the Negro in this country . . . I was quite ad-equately informed concerning the mutual bitterness between the old Bostonian

and the rising Irish. . . . The net result was that I could only feel at peace with myself if I hated anti-Jewish prejudice as prejudice without having the first emphasis on the fact that it was directed against the group to which I belonged. . . . In resisting prejudice against the Oriental, prejudice against the Catholic, prejudice against the immigrant, prejudice against the Negro, I felt that I had a sound basis on which to resist prejudice against the Jew as well.

But that wisdom would not be his for a while. His academic performance plummeted that year, owing both to his spiritual angst and to his poor study habits, which surfaced the moment he felt some slack in his father's tether. His fellowship was not renewed and Leo yanked him back to Boston. Norbert came home that summer dazed by the year's discoveries, shorn of the last threads of his identity and youthful confidence, and peering into an uncertain future that opened before him as "a turbid and depressing pool."

In that summer of 1911 Leo landed another blow. As Norbert was preparing for his third year of doctoral study, Leo again laid claim to his son's intellect and achievements, along with his two sisters' impressive beginnings. In an article in the popular *American Magazine,* Leo repeated to a nationwide audience his earlier assertions in the *Boston Herald.* "It is nonsense to say, as some people do, that Norbert, Constance and Bertha are unusually gifted children," said Leo. "If they know more than other children of their age, it is because they have been trained differently." The practiced quote was news to Norbert—the first iteration that broke through the media shield his parents had erected around him. Appearing at that moment, not in a cheap slick, but on the pages of a respected national magazine, it devastated him. "When this was written down in ineffaceable printer's ink . . . it declared to the public that my failures were my own but my successes were my father's."

That year, Leo was appointed to the tenured position of full professor, becoming the first Jew to attain that rank at Harvard; and, in September, Norbert resumed his studies as a doctoral candidate in philosophy. William James had died the year before. Now Norbert immersed himself in Emerson Hall, in the heart of the Yard, to contemplate James's shadow and study under Harvard's new leading lights: George Santayana, the Spanish-born philosopher and Jamesian disciple who lectured on morals and aesthetics; and Josiah Royce, the theologian, philosophical idealist, and no less substantive realist, who introduced Norbert to the rigors of the scientific method.

Both men, Santayana and Royce, inducted Norbert into the twentieth century's dawning scientific worldview that was becoming known among philosophers as the "new realism," and into the new school of "mathematical logic" being developed by the British philosopher-mathematicians Bertrand Russell and Alfred North Whitehead. Russell and Whitehead's daunting *Principia Mathematica,* a three-volume tour de force whose first part appeared the preceding year, sought to realize a dream of mathematicians since the time of Leibniz: to gather up all the scattered branches of mathematics, break them down into their fundamental components, and rebuild the entire edifice, systematically, from bedrock principles of logic.

Norbert was drawn to the new realism instinctively. His philosophy papers during those formative years touched on ideas he would bring to his scientific work decades later: radical reconceptions of reality and the physical foundations of rational thinking, as well as new scientific approaches to the study of purpose and design in nature. Like his essay on ignorance at age ten, his doctoral ideas were not works of genius nor of a fully mature philosophical thinker. But by Norbert's mid-teens, those primordial pieces of the century's coming scientific revolutions were seeded in his mind.

Norbert did well in the new philosophies he learned at Harvard. He wrote a commendable dissertation on Russell and Whitehead's *Principia* and passed his written exams without difficulty, yet he approached his oral exams in a "trance" of dread. Determined to see his son over the last hump of his grand experiment, Leo walked him endlessly around the streets of Cambridge, calming him, transfusing confidence, and talking him through every conceivable examiner's contingency. Norbert survived the ordeal and donned his doctoral hood at the unripe age of eighteen.

He did not have history on his side. Prodigies like Norbert Wiener were seen in the prophetic tradition as omens "of impending change in the world," as "fabulous monsters," as Wiener himself was loath to observe. Some modern minds viewed their rare appearances as signs of a greater "gambit" on nature's part. Through a confluence of innate talent, early influences, and fortuitous timing, the prodigy stood as an "exemplar and beacon" who "may have something to tell us that might help tip the balance toward our continued existence on this earth." Such individuals usually showed extreme talent in one specific field of endeavor, but young Wiener was an anomaly among anomalies. He was, in the taxonomy of monsters, an "omnibus prodigy" who showed "evidence of extreme precocity in languages, mathematics, natural science," and who—like Kipling's Elephant's Child—was "curious almost beyond belief." But that high status did not guarantee him a high-achieving future, a fact Wiener knew well. As centuries of evidence confirmed, "rarely do the young sprigs blossom into anything resembling true genius," mature minds who, by consensus, bring about "the transformation of a field of knowledge that is fundamental and irreversible."

In the warm Cambridge spring of 1913, as the youngest Ph.D. in Harvard's history strode across the stage of Sanders Theater to accept his shingle, the purblind, browbeaten boy wonder was riddled with confusion and soul-deep doubts about his intellect, identity, and self-worth—and he was not the only one. To the most important people in his life, his parents swelling with pride at their own achievement, and his younger siblings fidgeting in their seats, he would remain their beloved, bumbling "Nubbins," an affectionate nickname that nonetheless meant: *n. 1. a small or imperfect ear of corn. 2. Anything stunted or imperfectly developed.*

2

Young Wiener

Oh, if I only could have moods, could shift the blame on the weather, a third person . . . then only half the weight of this unbearable burden of discontent would rest on me. Miserable me! . . . Am I not the same who not long ago was intoxicated by the abundance of his emotions, and who stepped into a paradise wherever he went, with a heart ready longingly to embrace the whole world?

—J. W. von Goethe, *The Sorrows of Young Werther*

YOUNG NORBERT WIENER, THE FRESHLY minted teenage doctor of philosophy, stepped smartly down a wormhole in Cambridge, Massachusetts, bound for Cambridge, England, on a clear day in September 1913. With heavy suitcases in hand, he descended the stairs of the new subway station beneath Harvard Square and emerged, minutes later, in East Boston, where a small steamship of the British Leyland Line was receiving passengers on a dilapidated dock in Boston harbor.

He had been awarded one of Harvard's prized graduate fellowships for a year of unfettered travel and study abroad. He was looking forward to beginning life on his own, far away from his domineering parents, but his escape would not be easy or unencumbered.

His whole family was sailing with him. Leo Wiener had used the opportunity of his son's year abroad to take a sabbatical on the Continent. He was still hungering to win respect from European scholars for his eclectic philological studies, and both Leo and Bertha were eager to advance their daughters' education in the cultural centers of Europe. Norbert was weighed down doubly as he hauled his own valises and the family's baggage across the dock, while his father barked a barrage of "conflicting and self-contradictory" orders at him as if he were the family's porter, not its celebrated prodigy.

Boarding the boat was "a heavenly relief." Norbert watched the waves for hours and studied the ship's wireless officer with rapt attention, marveling at his radio skills and his deft fielding of the "flirtation of an old sea captain's daughter." When the ship docked in Liverpool, Leo arranged the train transfers to London and sent the rest of the family on to Munich, their designated base camp on the Continent.

————

Young Wiener strode through the great gate of Trinity College, Cambridge, the Mecca of modern philosophy and the new mathematical logic, with his father in lock-step beside him. They passed through the massive stone portal and gazed through the English drizzle at the expanse of Trinity's Great Court. They walked briskly along the cobblestone paths, past the cold monastic rooms where Isaac Newton invented calculus and performed his historic optics experiments, past the soot-stained tower where young Lord Byron kept his pet bear, past the stone stairway by the college dining hall that Tennyson took in a single bound.

It was one of their better tramps together.

In a spacious suite across the courtyard, Leo Wiener hand-delivered his son to Bertrand Russell, the foremost philosopher in the Anglo-American world, in a hail of accolades about the boy's genius and with no small amount of self-promotion. It was Leo's boldest stroke yet in his effort to procure for his son the most renowned teachers of his day in each field for which he showed a propensity. The lanky, lordly Lord Russell, then in his early forties, drew on his pipe and sized up his new charge, and his new charge sized him up in return.

Young Wiener had prepared for his studies with Russell. He had written his dissertation on the first volumes of the *Principia* and spent the summer of 1913 boning up on his mathematics. But that work, and all the rigors of doctoral study at Harvard, did not prepare him for the fierce scholarly scene he stepped into at Trinity—or for Russell's fiercer personality. In a letter to his father written only days after Leo left Cambridge for Munich, Norbert ruefully reported his tense interactions with Russell in their first private tutorials. "Russell's attitude seems to be one of utter indifference mingled with contempt. . . . I think that I shall be quite content with what I shall see of him at lectures."

He attended two of Russell's courses, one on the *Principia* and the other on "sense data," at the time, a burning issue in the philosophy of mind. Both were well within the young postdoc's grasp, but that wasn't the impression he got from his eminent mentor. Apparently, young Wiener did not sense data or do philosophy the way the titan of Trinity prescribed it. A month into the term, Norbert confessed to Leo:

My course-work under Mr. Russell is all right, but I am completely discouraged about the work I am doing under him privately. I guess I am a failure as a philosopher. . . . I made a botch of my argument. Russell seemed very dissatisfied . . . with my philosophical ability, and with me personally. He spoke of my views as "horrible fog," said that my exposition of them was even worse than the views themselves, and . . . accused me of too much self-confidence and cock-sureness. . . . His language, though he excused himself, it is true, was most violent.

Russell's piercing intellect and haughty style made Leo's harsh training methods seem like blunt instruments by comparison. His blustering attacks knocked the wind out of young Wiener's luffing self-confidence, but after the earlier ordeals he had suffered and survived, Russell's hazing inspired a new response: defiance. A week later, Norbert sent Leo a scathing counter-opinion of Russell that prefigured later feelings he would express about all overly logical minds and machines:

I have a great dislike for Russell. . . . I feel a detestation for the man. . . . He is an iceberg. His mind impresses one as a keen, cold, narrow logical machine, that cuts the universe into neat little packets, that measure, as it were, just three inches each way. His type of mathematical analysis he applies as a sort of Procrustean bed to the facts, and those that contain more than his system provides for, he lops short, and those that contain less, he draws out.

So not to seem as cruel as his tutor, he threw Russell a sop: "He is, nevertheless, within his limitations, a wonderfully accurate thinker."

Antipathies ran deep on both sides of the mentor-student relationship. In a letter to a friend written that same week, Russell assailed the "infant prodigy named Wiener":

The youth has been flattered, and thinks himself God Almighty—there is a perpetual contest between him and me as to which is to do the teaching.

At the outset, young Wiener, too, was downcast in Cambridge. In keeping with agreements between Harvard and Trinity, he had advanced-student standing, but he was not permitted to live in the college. Leo had left him with an ill-tempered landlady, "a slovenly, mean little woman" who agreed to provide lodging, along with his daily diet of vegetables and cheese, at a minimal price, but he was "hopelessly and utterly lonely" during those first weeks. He walked forlorn through the colleges' stone-faced courtyards. He roamed the sylvan Backs along the Cam and, eventually, he made contact with some other human beings. He went to tea with the undergraduates and attended their ritual parties known as "squashes"—so named for the crowds that piled into the old colleges' undersized rooms. Grad students and fellow postdocs befriended him in a big-brotherly fashion, but he felt their entrenched English snobbery and anti-Americanism "in such a way to comb a Yankee's hair the wrong way."

As the year progressed, he found more to like at Cambridge than the splendid scenery. He took sport in arguing his ideas in debates at the Moral Sciences Club. He made friends with older scholars and, by the end of his first term, he felt "happier and more of a man than I had ever been." He also warmed considerably to his fellow Cantabrigians. After many subsequent visits, he confessed to feeling "a very close and permanent bond between myself and England, and more especially between myself and Cambridge," and he confirmed that "the English were very different . . . once one had penetrated the protective layer which they assumed against Americans and other foreigners."

And young Wiener learned something more at Cambridge. He saw all around that stately bastion of high intellect and fading aristocracy a range of eccentricity that raised peculiarity to an art form. The three philosophy dons who made up the fabled "Mad Tea Party of Trinity"—Russell, G. E. Moore, the apostle of the new "common sense" philosophy, and J. E. McTaggart, the last holdout of Hegelian idealism in Britain—were characters straight from Lewis Carroll's *Alice in Wonderland*. Russell was a matchless Mad Hatter, a gangly, wild patrician who cavorted with the Bloomsbury crowd and romped through scandalous love affairs with fetching Englishwomen and Americans. In Moore, Wiener found "the perfect March Hare," a disheveled don whose "gown was always covered with chalk . . . and his hair was a tangle which had never known the brush within man's memory." The sluggish McTaggart, "with his pudgy hands, his . . . sleepy air, and his sidelong walk," said Wiener, "could only be the Dormouse."

Wiener loved the new freedom of mind and personality he encountered at Cambridge, where for once he did not feel like a freak set apart from the crowd. He contrasted the local scene with that of Harvard, which, as he asserted, "has always hated the eccentric and the individual, while . . . in Cambridge eccentricity is so highly valued that those who do not really possess it are forced to assume it for the sake of appearances." Years later, when tales of his own eccentricities gained world renown, Wiener played cat-and-mouse with longtime English friends, who charged that some of his outlandish ways were drawn deliberately from the living caricatures he observed at Trinity.

One Trinity man Wiener held in the highest regard was Godfrey Harold Hardy, England's foremost pure mathematician of that period. With his fair features and sporting passions, Hardy had the air of a sophomore, but he was in his mid-thirties and a fragile, shy man. Hardy's course was "a revelation" to Wiener, whose earlier forays into higher mathematics—the realm beyond beginning calculus—had been frustrating and, in the main, futile. He was no longer count-ing on his fingers, but the abstract concepts of number theory, functional analy-sis, differential equations, and vector geometry just didn't add up. Hardy walked him through the underlying logic of those higher domains and schooled him in the methods of mathematical proof, and, suddenly, his analytical engine began to spark.

Wiener thanked Hardy, not Russell, for grounding him in the new tools of modern mathematics: the functions of real and complex variables, the upper branches of calculus dealing with the measurement of smooth and irregular curves, and, particularly, the Lebesgue integral, a statistical widget for measuring odd geometric shapes and sets of points scattered across infinite spaces. A decade later, Wiener would make his initial triumphs as a mathematician using those basic tools of statistical theory, and he took his first toddling steps in that direc-tion under Hardy at Trinity. In November 1913, he wrote his first published paper. "Looking back . . . I do not think it was particularly good," he said later. "Still, it gave me my first taste of printer's ink, and this is a powerful stimulant for a rising young scholar." His next paper for Hardy, on a topic in mathemati-cal logic, marked a significant contribution to the field.

Their paths diverged soon after. Wiener was becoming increasingly practical-minded and eager to apply the new mathematical tools he was acquiring to real-world problems, while Hardy cherished his stature as a pure mathematician. The two men became close friends, but they kept up a running dispute for the next thirty years over the virtues of pure and applied mathematics. Hardy called the engineering bent of Wiener's mature work "humbug," while Wiener staunchly defended his role as a practical mathematician seeking answers to real problems with weighty consequences.

On that count Wiener and Russell were congruent. Russell had no qualms about making practical applications of mathematics, and while he showed little outward affection for his headstrong student, Russell performed two great favors for Wiener's early professional life. In their weekly tutorials, he took Wiener be-yond purely logical realms and alerted him to exciting developments that were shaking the foundations of the physical sciences in the first decades of the twen-tieth century—the advancing electron theory of matter and the new quantum physics emerging on the Continent. In particular, Russell directed Wiener to the three papers by Albert Einstein, published in 1905, that the scientific world was still scrambling to comprehend: his special theory of relativity, which posited in-timate physical relationships linking matter and energy and space and time; his photoelectric theory that explained the tendency of metals, when struck by light, to fling off electrons and generate electric current; and the historic, albeit esoteric, paper in which Einstein solved the mystery of Brownian motion. In that third paper, Einstein considered the uncanny movements of tiny particles in fluids, first observed by the Scottish botanist Robert Brown, and distilled the significance of those erratic actions for thermodynamics and all of physics. That haunting paper would become a springboard for Wiener's own mathematical work.

In exchange for Russell's favors, young Wiener rejected entirely the premise of his mentor's masterpiece, the *Principia Mathematica*. In the damp English winter of 1914, Wiener, still in his teens, cast a cold eye on the ideas Russell launched in his leviathan work. His analysis indicated that Russell's logic was lacking, and that his claim to have devised a formula for deducing the whole of mathematics from logic alone was fundamentally flawed. Indeed, in his studies at Trinity, Wiener came to believe that the claim of completeness for *any* self-contained system of logic was bound to fail, and he said as much in a short paper published the following year in the *Journal of Philosophy, Psychology and Scientific Method*:

> It appears to me unlikely that such an amplification of Mr. Russell's set of postulates . . . would be possible. . . . It is in any case highly probable that we can get no certainty that is absolute in the propositions of logic and mathematics, at any rate in those that derive their validity from the postulates of logic.

Wiener's critique echoed in a more mature form the embryonic opinions he had expressed at age ten in his first philosophical paper, "The Theory of Ignorance." And it anticipated more explicitly Austrian mathematician Kurt Gödel's Incompleteness Theorem, which would appear two decades later and render obsolete Russell and Whitehead's bold program. Wiener's glancing blow had no such impact, but coming as it did on Russell's turf at the height of his influence, it was tantamount to calling the emperor unclothed.

Many years would pass before Wiener was ready to acknowledge all he had learned from Russell during that first fast-moving term at Cambridge. Four decades later, he conceded, "It is not very easy for me even at this distance to write of my contact with Bertrand Russell and of the work I did under him." Yet he allowed, of his tussles with Russell as he did of his hard lessons under Leo, "I benefited enormously by them." Russell, too, thought better of young Wiener than he let on in his letters. In his private papers, he noted approvingly, after receiving an unsolicited letter from Leo detailing the training methods he had applied to his son, "Nevertheless he turned out well." And Wiener himself, in a note to one of his professors at Harvard, revealed that, after Russell read his dissertation, he commended it as "a very good technical piece of work," and gave him as a gift a copy of the third volume of the *Principia*.

Wiener latched onto his mentor's directives. In the spring of 1914, Russell took off to lecture in America and Wiener took off for the Continent. He stopped briefly in Munich to visit his family, then rode the train north to Göttingen, another centuries-old university town nestled in the rolling hills of Lower Saxony. There he entered an intellectual force field whose pull extended far beyond the old town's fortified walls. In the 1800s, Göttingen's giants, Carl Friedrich Gauss and Bernhard Riemann, laid the cornerstones of modern mathematics. Their heirs and disciples carried the field into the twentieth century and were now rattling the physical foundations of matter itself.

Wiener's time at Göttingen was brief, only a term, but it would prove crucial to his development as a mathematician and a scientist. There, a decade before the quantum revolution burst from the university's halls of physics, Wiener studied differential equations with David Hilbert, the crowning figure of German mathematics, whom he would laud years later as "the one really universal genius of mathematics" he had met. Following Russell's recommendations, he read Einstein's 1905 papers in their native language. He got his first close-up look at the new atomic theories, and he took his first steps to apply the new mathematical concepts he was mastering to practical phenomena in the physical world. To complete his immersion in the mainstream of German scholarship, he attended courses taught by Edmund Husserl, Germany's last great philosopher in the tradition of Kant and Hegel, but, already, at nineteen, Wiener was moving beyond philosophy for philosophy's sake. As an aspiring applied mathematician, he realized that he did "not have the type of philosophical mind that feels at home in abstractions unless a ready bridge is made from these to the concrete observations or computations of some field of science."

His work at the university's mathematical society took his mind across that bridge. The society's reading room housed the greatest collection of mathematical books and periodicals in the world, and he sampled them freely. Hilbert presided over the weekly meetings in the seminar room, where new research papers were read and dissected with precision and passion. After the meetings, the students and professors would trek to a café on a hill above the picturesque town and mix their mathematics with fraternizing. The genial mixture smacked of the ingrained German "juxtaposition of science and beer" that had driven Leo Wiener in disgust to America. But the younger Wiener, though still years shy of his fellow postdoctoral students, enjoyed the spirited discussions, held his own in them, and the interplay helped greatly to improve his social skills.

Young Wiener made another heady breakthrough at Göttingen. While puzzling over some fine points of mathematical logic, an idea came to him that, he quickly realized, could be applied to a wide class of logical systems with a large number of dimensions. The idea possessed him for a week as he worked out its details, breaking only long enough to steal sustaining bites of black bread and Tilsiter cheese. He recalled the agony of his first extended act of creation as a mathematician. "I soon became aware that I had something good, but the unresolved ideas were a positive torture to me until I had finally written them down and got them out of my system." The paper he produced, which was published the next year in a journal back at Cambridge, he considered one of his best early works.

The experience infused him with a love of original research in mathematics that he likened to the work of an artist or sculptor driven by a divine inspiration. "To see a difficult, uncompromising material take living shape and meaning . . . whether the material is stone or hard, stonelike logic . . . is to share the work of a demiurge," he said. In the collegial air of Göttingen, at nineteen, young Wiener's mind began to find its mettle. He discovered the maturing tools of his art, his capacious memory and his "free-flowing, kaleidoscope-like train of imagination," which rolled out carloads of new ideas for solving complicated mathematical problems "more or less by itself." He caught his first sight of himself, no longer as a trained bear cub, or indolent charge of his father or Bertrand Russell, but as a self-starter with a ripening potential to produce genuine works of genius.

Wiener's season at Göttingen was not all work. After classes, he took in the pastoral delights of Lower Saxony and indulged in his favorite outdoor pastimes: tramping in the woods south of the town and bathing lazily in the students' swimming hole on the river Leine. At night, he dined in the town's vegetarian restaurants and dived into the free-flowing *Kneipen,* or drinking bouts, in beer halls frequented by the students. Unlike the constrained squashes at Cambridge, the beer bashes at Göttingen were "long, moist, and harmonious." One evening, he joined in a raucous songfest at a popular *bierhaus* and sang "indiscriminately" in German and English until the carousers were evicted by the police.

As the summer unfolded, political tensions that had been building across the Continent broke out violently. On June 28, 1914, Austria's Archduke Franz Ferdinand was assassinated in Sarajevo by Serbian nationalists. A month later, Austria declared war on Russia, Serbia's Slavic ally, and within days Germany joined the action, declaring war on Russia and France in a thrust for land and power that Germany's generals and industrial barons had been readying for decades.

Leo Wiener had returned to America with his family months before. Now, as the ethnic skirmish in the Balkans consumed the Continent, young Wiener headed for home on the Hamburg-America line. Two days after his ship cleared the British coast, the captain relayed the news received over the ship's radio that Germany and England were at war. Packs of German U-boats prowled the sea lanes of the North Atlantic, and the crossing was tense until his ship sailed safely into Boston harbor.

———

Within weeks he headed back to England, braving his first solo crossing of the Atlantic as the U-boat traffic was quickening and trenches were being dug in the fields of the Western front. His benefactors at Harvard had renewed his traveling fellowship for the academic year 1914–15, and Russell had written from Cambridge assuring him that England was safe. But the idyllic college town Wiener returned to had been transformed. Though far removed from the front, Cambridge was immersed in the action. Many of Britain's best young minds died in the first wave of trench warfare and long lists of the latest casualties were posted daily in the Union. Stretches of the Backs had been turned into makeshift hospitals for the wounded. The town was blacked out by night and shrouded in gloom by day. And Lord Russell himself, an outspoken objector to the slaughter, was becoming an outlaw in the eyes of the British people and their government, already traveling a path toward expulsion from Trinity and eventual imprisonment.

Wiener stayed in Cambridge through the fall term, then fled to London for the holidays. There he met up with another traveling fellow he knew from Harvard, the young philosopher and poet T. S. Eliot, who was finding the mood at Oxford to be as oppressive as the gloom at Cambridge. The two men shared "a not too hilarious Christmas dinner" at a restaurant in Bloomsbury. Although Eliot was six years older than Wiener and far more worldly, they had some things in common. Eliot had already begun writing the modernist poetry that would make him famous, but at that point he was still concentrating in philosophy. He had met Russell at Harvard that spring and taken his course on mathematical logic. Afterwards, Eliot went to Oxford to rewrite his dissertation in the fading field of metaphysics, and he showed a genuine interest in the new fields Wiener was tilling at Trinity.

Wiener had written four more papers that year, including "The Highest Good," an essay on ethics, and "Relativism," an effort to reconcile the last remnants of idealism with the new realism and mathematical logic that were vying for primacy over the modern mind. He sent preprints to Eliot, who responded with measured enthusiasm. At their bleak Christmas dinner, served as the fighting was heating up across the Channel, the Elephant's Child and the young literary lion chewed gamely on the new philosophical trends—two young men mulling over the demise of the old and the shock of the new.

"You seem to be doing philosophy rather than math," Eliot observed before their dinner, pleased to see that Wiener had not fallen under the spell of Russell's new logic, which he could not fathom.

Indeed, in Wiener's essay on relativism, the notion that all knowledge is relative and dependent on other facts and contexts, he argued once again that "no knowledge is self-sufficient . . . there is no absolutely certain knowledge at all." His thesis reprised his childhood essay, but now with pointed references to the assertions of certainty and "completeness" being made by the new mathematical logicians. "*All* philosophies are . . . relativisms," Wiener insisted. "We are never . . . done with our labor of comparing one concept with another." He expressed the same view in his essay on ethics, in which he rejected the idea at the root of ethics that a "highest good" could be discerned and attained. "The hypothesis that a highest good exists may well be doubted," he said, and, again, his logic was sound. "If our highest ideal is capable of definitive attainment, then on its attainment moral progress ceases and . . . if [it] can not be attained, then . . . morality is a perpetual failure."

Eliot was a devout realist, but that Christmas he was frustrated with sputtering debates between the old idealism and the new realism—indeed, with philosophy itself. "In a sense of course, the philosophising is a *perversion* of reality . . . an attempt to organize the confused and contradictory world of common sense . . . which invariably meets with partial failure . . . and . . . partial success," he told Wiener. "Almost every philosophy seems to begin as a revolt of common sense against some other theory, and end . . . by itself becoming equally preposterous to everyone but its author."

Wiener had a better idea. He had already begun to reframe his philosophical beliefs in scientific terms, using principles from the new physics which he had first learned about from Russell. Eliot was exploring an alternative path to truth through poetry. "I am quite ready to admit that the lesson of relativism is: to avoid philosophy and devote oneself to either *real* art or *real* science," he said. "There is art, and there is science. And there are works of art, and perhaps of science, which would soon have occurred had not many people been under the impressions that there was philosophy."

Their actions spoke louder than their words. Not long after their grim dinner, each fled the field of philosophy in pursuit of more meaningful pastures: Eliot committing himself to poetry and literary criticism, and Wiener gravitating to more tangible problems of applied mathematics. Wiener's essay on ethics went on to win Harvard's prestigious Bowdoin Prize, and Eliot, too, gave Wiener a rave review, but at that point no one, including Wiener himself, could fully appreciate what he had accomplished. Like his graduate essays at Harvard, his first published philosophical papers sowed seeds that would blossom in his work decades later.

On his return to Cambridge, Wiener found a telegram from his father bearing war news the British papers were not reporting—that the submarine war in the North Atlantic was intensifying—and urging him to come home on the next available boat. In February 1915, Germany declared a submarine blockade of Britain. The university all but shut down, and Wiener booked a rough winter passage on an old steamer that "took it green up to the bridge" from Liverpool to New York harbor. Weeks later, a German submarine sank the British passenger liner *Lusitania* in the same sea lane, killing 1,195 passengers, including 128 Americans.

Europe was erupting and America would soon join the fray, but for the moment young Wiener was enjoying a new maturity and an inner peace he had never known. His two years abroad marked his mind's "emancipation." He returned from his time in Cambridge and Göttingen feeling himself "much more a citizen of the world," one who had been initiated into the circles of international science and discovered that "it did not seem utterly hopeless that I should accomplish something there." He had passed muster with Russell and matched wits with "the cream of the intellectual crop" of Europe. He had learned to care for himself and socialize with reasonable success. For the first time, he felt fully recovered from the "Slough of Despond" he had fallen into after college and the string of shocks to his confidence and identity he experienced during his Harvard years.

But his wanderings were only beginning, and his re-entry to America would prove difficult in ways he never anticipated.

The Great War years were an unsettled time for young Wiener. With one term to go on his postdoctoral fellowship, he fulfilled his commitment on safer soil in New York City where, acting again on Russell's advice, he studied philosophy at Columbia University with John Dewey. Dewey, a Vermont native, had picked up William James's mantle of Yankee pragmatism and applied it broadly to the theory of knowledge and to his practical interest in education. But, despite Wiener's protests to Eliot and to Russell himself, he found that Dewey's philosophical theories, cast in the old style of philosophical discourse, were not sufficiently systematic to impress a young mind sharpened on the new British and European schools of logical analysis and scientific thinking.

And life in Morningside Heights, on Manhattan's erudite Upper West Side, paled in comparison to his emancipating experiences in Europe. At Columbia, Wiener lived in a soulless graduate dormitory and felt no intellectual affinity with the older students. After his high times overseas, he was taken aback by his renewed feelings of awkwardness among his own countrymen, and he criticized them in ways he later realized were "completely lacking in tact." In an inept effort to win their respect, he barraged his fellow graduate students with arcane facts and theories from the new knowledge he had gathered abroad. He barged unbidden into their bridge games. In return, one group of irked grad students bothered him back by setting fire to his newspapers while he was reading them. For solace, he went tramping. He walked the length of Manhattan, from the Battery to Spuyten Duyvil, and hiked the wild Palisades across the Hudson River. On weekends, he would dutifully visit his aging *Grossmutter* Wiener in her apartment across town, but cautiously, knowing his mother hated the thought of her son tangling with his Jewish roots and the ethnic ways of his New York cousins.

His term at Columbia marked the "low point" of his life as a postdoc. He had come back riding a wave, eager to begin his career as a philosopher skilled in the new logic and the latest sciences, only to find that all his knowledge would be turned back on him in a scourge of academic chauvinism. When he began scouting the field for a teaching appointment, he was shocked to learn that he did not qualify for a permanent position at Harvard, or any other major university, because he had honed his talents at Cambridge and Göttingen and failed to study "men of the American scene." That scene, to his mind, was for the most part "facile" and "suffered from a certain thinness of texture" that would not thicken until America's entry into the war demanded more exacting applications. But, in the fall of 1915, Americans were still clinging to their isolationist fallacies, and the unemployed ex-prodigy, who had felt so manly scaling the academic capitals of Europe, soon found himself back in Boston living under the roof of his overprotective mother and the thumb of his domineering father.

That fall, Wiener did return to Harvard as a "docent lecturer" in philosophy, a perfunctory honor granted upon request to every Harvard Ph.D. during that era, but he did not fare well. His maiden lectures were nervous, rambling, and "pontifical," draped in Old World cadences that harked back to Victorian times. His odd, meandering mode of oratory, which he later chalked up to his "youthful loquacity" and "gift for gab," presaged a formal speaking and writing style that would persist throughout his career.

That first manifestation of the adult Wiener's legendary eccentricity—his anachronistic manner of speaking and writing—may have looked like an ill-conceived attempt to compensate for his relative youth in his first academic position, but it was no affectation. It traced back to his recitations under Leo's pedantic tutelage, and to his immersion in European academic culture in its fading days before the Great War. His nervous chattering was endemic; like his physical ineptitude in the laboratory, his branching sentences reflected, not a disorganized mind, but one that made connections faster than he could express them. To his listeners, however, he was simply grating, and his pontifical style failed to mask the old feelings of awkwardness and insecurity his return to America had reawakened.

Wiener's evaluators were not impressed by his performance, and he failed to win a permanent appointment at Harvard. He shouldered his share of the blame, confessing, "I was protected by my very inexperience from the awareness of what a show I was making of myself," but he offered other defenses. He felt a lingering resentment among many at Harvard, aimed at both him and his father, that dated back to his prodigy years, when Leo was unabashed in his self-promotions and Norbert, by his own admission, was "an aggressive youngster," "inconsiderate," "not a very amiable young man," and "certainly no model of the social graces." He cited Harvard's overt anti-Semitism of that era as another reason he was passed over, and the charge had some merit. Anti-Jewish attitudes at Harvard had intensified with the arrival of Harvard's patrician President Abbott Lawrence Lowell in 1909. Lowell later called for setting strict quotas on the numbers of Jewish students and faculty at the university, a bias that both Leo and Bertha Wiener, to her credit, opposed wholeheartedly.

G. D. Birkhoff, a talented young mathematician whom even Wiener acknowledged as "a star of the first magnitude," and other faculty members of German descent or leanings during the war, may have actively opposed giving him an appointment on ethnic grounds. Wiener claimed he won Birkhoff's "special antipathy, both as a Jew and, ultimately, as a possible rival," and he believed that similar resentments aimed at his father hurt him as well. Leo had become bitterly anti-German after his chilly sabbatical in Munich and incensed by the rising specter of German militarism. He voiced his views loudly, in English and German, at every opportunity, and no doubt alienated Harvard's many well-placed Germanophiles.

For young Wiener the irony was almost unbearable that, like every major American mathematician since Bôcher, he had traveled to Germany to get the latest ideas and training from the world's leading figures in the field. Yet now he was being ousted from Harvard, on his home turf, by a prejudiced German-American professor who had reached the pinnacle of American mathematics "without the benefit of any foreign training whatever."

And, with that, the prodigal son of Harvard was sent out into the wilderness.

———

That summer and the next Wiener went hiking with Conta in the White Mountains. In the spring of 1916, as the Great War drew closer to America, an officers' training corps was formed in Cambridge to recruit able bodies and minds in preparation for the inevitable deployment. Eager to serve his country in any capacity, Wiener signed up immediately and began training with the other raw recruits in the military arts of marching and musketry. With the help of a sympathetic drill instructor, he even attained the rank of sharpshooter, although he knew full well "my eyesight did not permit me to hit a barn out of a flock of barns." He attended an officer's training camp in upstate New York, hoping to win a commission in the army. His mountain tramps had given him endurance for the long marches and mock battles in the forests, but his shooting skills were revealed for the sham they were, and when the camp ended he was turned down for a commission.

Despite his stellar credentials, his job prospects in academia were next to nil. On Leo's orders, he signed up with a local teacher's agency and, in the fall of 1916, he took a job at the University of Maine in Orono, north of Bangor, teaching mathematics to "a strapping lot of young farmers and lumberjacks, who managed to be quite as idle . . . as the students of universities of the Ivy League, but at one-third the expense." His teaching style was as ineffectual in Maine as it was at Harvard, but his students were not nearly so polite, and his fumfering lectures rang with the sound of pennies being dropped by bored country boys trying to rattle their stodgy young professor. Through Maine's blustery winter, he trudged to class on snowshoes, dreading every minute of his first experience as a paid professional. On his breaks, he went home to Boston to visit his family and take in the new marvel of mass entertainment that would leave him forever spellbound—motion pictures.

But the real world beckoned and the war raged on. In the spring of 1917, Wiener tried again to enlist in any branch of the service that would have him, but, as he put it, "my eyes were against me everywhere." He returned to Harvard to vie for a post in the newly established Reserve Officers' Training Corps, but he flunked the physical and fell off an old horse during his riding exam. Reality, like the war, was quickly becoming unavoidable, and he turned his attention to finding some capacity in which he could serve his country as a civilian.

He interviewed at the General Electric factory in Lynn, north of Boston, where his math know-how earned him a starting position as an apprentice engineer in the company's turbine division. Although far from academia, the work was just what he had been seeking, a chance to make a contribution to the war effort and, at the same time, to learn applied mathematics, thermodynamics, and tangible engineering skills on the job. He loved the earthy work and practical new talents he acquired solving basic engineering problems—until his father intervened in his new career. Leo Wiener insisted that Norbert was too clumsy to be an engineer. He also was appalled by the prospect that his grand experiment would end with his son becoming a lowly factory worker, and he forced him to tender his "shamefaced resignation."

Leo took it upon himself to find Norbert a more suitable job. He made inquiries with his contacts and secured Norbert a staff writing post at the prestigious *Encyclopedia Americana,* headquartered in Albany, New York. After so many of his own failed attempts to find a position befitting his education, Norbert now felt too dependent on his father "to dare to contravene his orders," and he headed for Albany. It was high-grade hack work but, to his surprise, he enjoyed writing short articles in the *Encyclopedia's* breezy, authoritative style. He researched his topics in local libraries and collated his references meticulously (though he later admitted to authoring several pieces that were "utter balderdash").

Then the war hit home and turned young Wiener's destiny as only a war can. In the summer of 1918, Wiener was growing restless in his hack job. He had been trying in vain to restart his frustrated search for a good academic position, when he received an urgent telegram from the distinguished mathematician Oswald Veblen, the head of Princeton's math department. Veblen had been put in charge of the U.S. Army's Aberdeen Proving Ground, a new weapons training and testing facility in Maryland, on the western shore of Chesapeake Bay. There, under his auspices, every available American mathematician was being enlisted to compute the crucial artillery firing range tables needed to support American troops as they entered the escalating European war.

Wiener left for Aberdeen on the next train.

Suddenly mathematicians, of all people, those most abstract of scholars who were viewed by their fellow citizens as "useless fumblers with symbols," had a vital role to play in the nation's defense. This first mass mobilization of mathematical minds was required to compute the wide range of possible trajectories for each different type of artillery used on land and sea, with corrections for every conceivable variable: the gun's angle of elevation, the weight of the ammunition, the force of the winds blowing over the battlefield—even the earth's rotation during the shell's time of flight.

However, as Veblen and his math brigades soon discovered, the old methods of computation, using pencil and paper, imprecise slide rules, and the first mechanical desktop calculating machines, were too slow and limited mathematically to meet the technical needs of modern warfare. And they were all but useless against the twentieth century's first superweapons: young daredevils in their newfangled flying machines. The task of using artillery effectively as an antiaircraft weapon—to hit a speeding target moving in three dimensions on a looping, zigzagging aerial trajectory—transcended the proven techniques of ballistics by several orders of magnitude. It required a knowledge of gunnery, aerodynamics, and complex differential equations that only a team of specialized scientists and higher mathematicians could muster. This war effort would lay the foundations for that work and bigger challenges to come.

During the First World War, Aberdeen was to American science what Los Alamos in New Mexico would become a generation later, a place where the nation's most adept mathematical minds, young and old, mixed with a mission and camaraderie unknown to that time. The terrain of the proving ground, secluded in the marshlands of the Chesapeake estuary, was a bog of muddy streets and hastily constructed wooden huts, but Wiener was as happy there as he had been since his liberating years in Europe. He reveled in his vital role as a "computer"—the moniker given to the men and a handful of women hired to aid in Aberdeen's war effort. He loved the practical work, the lively collaborative climate, and the nonstop shoptalk among so many mathematicians. His first letters home from Aberdeen beamed with excitement for the work he was doing ten hours a day in squalid conditions thick with "dust, flies, heat—everything but water." He boasted of his first field assignment timing the release of test firings downrange. "I see more shell and shrapnel bursts in a day than I might see in a week at the western front."

He also one-upped some of the world's most renowned mathematicians. Soon after Wiener arrived at Aberdeen, he devised a new method of interpolation—computing values between known coordinates—that surpassed the prevailing method developed by J. E. Littlewood, a younger colleague of Hardy's at Trinity and one of England's best mathematical analysts.

But theoretical victories alone were not enough for Wiener at twenty-three. After toiling for three months as a computer, he still longed to serve his country bodily. He was determined "not to be a slacker," and, in the autumn of 1918, he made one last bid to enlist in the army. The prospect thrilled him more than mathematics and showed a new desire on his part to place his lot with the common man. "I should consider myself a pretty cheap kind of a swine if I were willing to be an officer but unwilling to be a soldier," he wrote his parents. The same letter suggested he also was seeking something even more basic that Leo and all his mentors had failed to instill, a degree of *self*-discipline, forced by the military's reflex-like drills and demands of unerring precision, that he hoped would spill over into his professional endeavors. "I have a special talent for making blunders," he confessed in another letter home that fall, echoing Leo's constant refrain. "None of mine have had any fatal effect . . . but my damnable carelessness is hard to overcome."

To his delight, he was accepted by the army and, in October 1918, he shipped out to basic training. Suddenly, the reality of his humble dream hit him. "I was appalled by the irretrievability of the step I had taken," he said. "I felt as if I had been sentenced to the penitentiary." All the old torments of his earlier bids to serve, his poor eyesight, the inhospitable living conditions, the rudeness of the other men in the company, doubled back on him. When the camp was over, he was assigned to a military unit, the 21st Recruit Company based—of all places—back at Aberdeen Proving Ground.

Two days after he returned to Aberdeen, the Armistice was signed.

The peace that followed was his punishment from the gods. He served stoically on a gunnery team and took his turns on guard duty carrying a rifle with a fixed bayonet. But as the war effort wound down he began to feel like a useless, and increasingly foolish, appendage to the rapidly demobilizing military. Indeed, on patrol with the troops at Aberdeen, Private Wiener stood out like a sore thumb in his thick glasses and uniform "strained by my corpulence." His fellow soldiers found this short, stout nubbins still surprisingly naive for a world-traveled scholar who walked his rounds discoursing on Aristotle and medieval philosophy. He became the butt of his comrades' crude talk and the fall guy for their jokes. At one point they even tricked him into shaving off his beloved mustache.

Around that time, in his letters home, Wiener first talked explicitly about suffering lasting sieges of depression. The symptoms had been coming on for years. They trailed back to the traumas of his childhood, to the physical exhaustion and depressed states he suffered after his tear through Tufts and his mad dash through Harvard. His time in Europe seemed to provide an antidote. At no time in his letters home from abroad—not during his worst savagings by Russell at Cambridge or as the war descended—did he write openly of depression. But, late in his twenty-third year, the Slough of Despond again befell him. What's more, those earlier passing states of fatigue and nonspecific despair seemed to be congealing into a deeper, *cyclical* affliction, one characterized by recurring bouts of depression that could shift swiftly into periods of happiness and exultant feelings of well-being.

Wiener's letters to his family gave ample evidence that he was already experiencing such moodswings in the fall of 1918. "Dear Dad: I am well and happy," he assured Leo in October.

Days later he told a different story to his sister. "Dear Conta: I'm O.K., although I have just got over a slight spell of depression."

A month later, he was upbeat again in a letter to his mother. "Dear Ma: I am well and happy here."

Two months after that, he told his mother of another nosedive and tacked on a reassuring update. "Dear Ma: . . . I have delayed in writing to you because I had been rather depressed before I received the good tidings of my coming liberty. I had been on guard, and was dog-tired and disgusted. This mood has passed, and I am happy."

———

Aberdeen passed too, and, in February 1919, Wiener went home to Cambridge and moved back in with his parents and younger siblings. With the war's end, Leo sold the family's new house, bought an older, smaller house two blocks away, and with the difference purchased a little apple farm in Groton, a few miles north of Ayer. That summer, Leo made all his children work the farm and the family garden. Norbert took it as evidence that his parents were again moving to control his life, and he resumed his search for teaching jobs with renewed dedication.

The postwar era was smiling on America's war-hero mathematicians, and the human computers of Aberdeen were in demand at the nation's best universities. Wiener had high hopes that Veblen would offer him a position at Princeton, as he had other Aberdeen vets, but "there were many good candidates, and I was not among his selections," he noted matter-of-factly. His technical triumphs notwithstanding, he reflected on his time at Aberdeen as one in which he had made "many blunders" mathematically and personally. Those mistakes, he believed, gave Veblen and other mathematicians around him "a disagreeable impression of my whole personality." Now, it appeared, he was paying the price.

After watching his son drift in his spirits and professional direction, Leo tapped another contact, a summer neighbor in New Hampshire who happened to be the publisher of the *Boston Herald*. On the strength of Norbert's work at the *Encyclopedia Americana,* he landed a job as a feature writer for the *Herald's* Sunday edition. In his first big assignment, he was sent north to the town of Lawrence to report on labor strife among immigrant workers in the town's textile mills, which were then the nation's largest producers of woolen goods. There he saw skilled Old World craftsmen, living in wretched housing and subjected to even more oppressive working conditions, being reduced systematically to cogs in America's new assembly-line mass-production machinery.

Before the war, 30,000 of the town's workers drew national attention, and made labor movement history, when they staged a successful strike over low wages and unsafe working conditions in the mills. But, seven years later, Wiener found that little had changed for the new wave of Greek and Italian immigrants pouring into the region after the devastating conflict in Europe. Wiener attended local "Americanization" classes and found them "completely out of touch with the educated elements in the immigrant communities." Their textbooks "exhorted the workers to love and to honor the boss, and to obey the foreman as if he were Jehovah himself. It was the sort of humiliating tripe which was bound to alienate any workman with a trace of character and independence." He wrote the story as he saw it, raised public awareness of the workers' worsening plight, and ruffled the feathers of the mills' wealthy owners and Boston's ruling classes. But he balked at his next assignment, a puff job to write articles promoting an aspiring presidential candidate favored by the *Herald's* owner, and he was summarily sacked for insubordination. The boot smarted, but he was "not unready to leave the paper."

Through the spring and summer of 1919, he pursued his search for an academic appointment. The door to Harvard remained closed to him and to virtually all Jewish applicants. He registered with another teacher's agency and explored an opening at the Case School of Science in Cleveland, Ohio, but it didn't pan out.

Finally, acting on a tip tossed his way by W. F. Osgood, a Harvard mathematics professor and another friend of his father's who "did not have a particularly high opinion of me or the job," Wiener interviewed for an open instructor's position, with the possibility that it might turn into something more permanent, depending on his performance. He got the job, not at Harvard, but at an upstart technical school down the road called the Massachusetts Institute of Technology.

3

The *Wunderkind* and the *Frau-Professor*

I count life just a stuff
To try the soul's strength on, educe the man. . . .
So I will seize and use all means to prove
And show this soul of mine

—Robert Browning, "In a Balcony"

THE COMMUTE TOOK MERE SECONDS by subway and only minutes by streetcar, but Wiener walked the two miles from Harvard Square to MIT's main portal almost as fast. He was a young man in a hurry, excited and more than a little trepidatious, as he passed through MIT's imposing façade into the maze he would roam for the next forty-five years.

The Massachusetts Institute of Technology, known simply as "Tech" to its friends, was a fresh-faced kid compared to Harvard, and its mission diverged markedly. Its purpose was not to churn out proper church deacons but to train the new engineers of the industrial age to construct America. When the school opened its doors on February 20, 1865, the nation's love affair with technology was young. The steam engine had only recently been yoked to riverboats, locomotives, and factory boilers in the New World. The first transcontinental telegraph line had been connected only four years earlier.

For its first fifty years, the new school of "industrial science" conducted its operations in a few buildings scattered around Boston's Copley Square. During that time, America's factories became heavy industries, electricity brought the first flashes of modern life, and new modes of transportation and communication—automobiles, airplanes, telephones, radio—transformed people's relations to space and time, and to one another. Fresh armies of engineers were needed, equipped with applied knowledge and technical skills that were nonexistent a generation earlier, and the new gentry of MIT rose to the challenge. In the fall of 1916, the institute relocated across the river in Cambridge, to a spacious campus built atop fifty acres of landfill from the newly excavated Charles River Basin.

MIT's new quarters were a temple to technology. Its U-shaped main building, formed from a half dozen connected structures fronting on the Charles, was a paean to the high industrial age neoclassical style, with a domed library at its center flanked by the broad pavilions of the departmental buildings faced in sand-colored crushed limestone. A lofty frieze snaked around its attics, carved with the names of a hundred renowned scientists and mathematicians from antiquity to the twentieth century. But behind its soaring entrance on Massachusetts Avenue, MIT's interior was strikingly plain. A single, labyrinthine corridor cut through it like a cubist alimentary canal. Its structure and function were intended to mimic the method of New England's gerrymandered textile mills, to insulate the building's inhabitants from the region's rainy days and frigid winters, and by way of that lone, heavily trafficked thoroughfare, to promote interaction among the institute's students and faculty as they went about their specialized technical chores. The adjoining buildings were numbered, not named. Each turn of MIT's "infinite corridor" opened onto an identical assemblage of classrooms and offices devoid of decorative touches.

Tech brooked no nonsense. It had no chapel. Nor, at that juncture, did it have much of a mathematics department. In the autumn of 1919, MIT's math department was strictly a service facility staffed by gentleman professors who sufficed to provide the institute's undergraduates with basic algebra, trigonometry, geometry, and just enough calculus to complete their assigned engineering tasks. However, along with the new engineering skills that the postwar industrial and technological boom demanded, a new order of mathematics instructors, with greater knowledge and a modern sensibility, was needed to teach the cadres more complicated numerical maneuvers.

Wiener entered onto this quickening turf schooled in the latest math and logic, the unfolding wonders of the electron, and the shifting foundations of atomic theory. He was twenty-four years old, practical minded but not yet technically oriented, neither an engineer nor even a competent handyman. But he was itching to join in the new postwar world of tangible tasks and rigorously applied mathematics. He was also one of several new hires at Tech who were determined to pursue research at the forward edge of mathematics, while "running interference for the engineering backs."

Within Tech's endless corridor and unadorned halls, barely beyond his father's reach, Wiener found a sanctuary. There he had room to grow and the time he needed to get his feet under him professionally. He learned to slow his speaking style and he overcame his stage fright in the lecture hall. He became an able teacher and, after his many harsh tutorials, a caring mentor himself. From the outset, he liked the young MIT students. As he saw it, the "Tech boys" wanted to work and he got on well with them.

More important, at MIT, Norbert Wiener found a fertile pasture for his fecund mind. In short order, he began tackling tantalizing problems in pure and applied mathematics, solving them, and publishing papers that turned heads in the inner circles of higher mathematics around the world.

———

From his first term at Tech, Wiener, the perpetual walker and irrepressible nature lover, veered off the infinite corridor and turned to the great outdoors, wandering the school's byways and beaten paths in search of fresh insights and real-world questions to ponder. One blowy day, while admiring the splendor of the Charles from the window of his office in Building 2, he became fascinated by the froth of waves whipping across the broad basin of the river. The roiling action buoyed his thoughts and gave rise to a mathematical quandary:

> How could one bring to a mathematical regularity the study of the mass of ever shifting ripples and waves, for was not the highest destiny of mathematics the discovery of order among disorder? At one time the waves ran high, flecked with patches of foam, while at another they were barely noticeable ripples. Sometimes the lengths of the waves were to be measured in inches, and again they might be many yards long. What descriptive language could I use that would portray these clearly visible facts without involving me in the inextricable complexity of a complete description of the water surface?

He began fiddling with formulas that might subdue the river's thrashings, hoping to parse the impulsive forces that crested in the dance of wind and waves with the same analytical mind he had applied to factoring artillery trajectories during the war. But as other agile minds were discovering, the trajectories of unruly waves were of another order of complexity altogether, one comprising more interacting dynamic forces, environmental influences, and extenuating circumstances than mathematicians to that time could even enumerate, let alone calculate, with precision.

Wiener's "problem of the waves," as he called it, was well known to mathematicians and scientists as the vexing phenomenon of fluid turbulence. The problem had been around for two centuries when he first encountered it in his tutorials at Cambridge in the writings of G. I. Taylor, whose watershed studies of fluid turbulence would later inspire the science of chaos theory. Drawing on Taylor's ideas, and on his own studies at Trinity with G. H. Hardy of the

Lebesgue integral, a statistical method invented a few years earlier by the Frenchman Henri Lebesgue to measure complex sets of scattered points and curves, Wiener began to pry open the conundrum.

To Wiener, those stock items in the mathematician's inventory—points and curves—resembled the spray of water droplets and skittering waves on the river. He tried plotting their complex paths using simple statistical methods of averaging and approximation, the most basic mathematical methods for analyzing random or uncertain events, and Lebesgue's newer, more sophisticated statistical method. But neither solved the problem to Wiener's satisfaction. As he gazed upon the waters, Wiener spied the first intimations of a bold new approach to wave problems and a raft of related phenomena. He also fished from the froth the unifying principle that would guide his career as a mathematician: *"I came to see that the mathematical tool for which I was seeking was one suitable to the description of nature, and I grew ever more aware that it was within nature itself that I must seek the language and the problems of my mathematical investigations."*

His quest led him to a fledgling field of physical science founded early in the twentieth century by Josiah Willard Gibbs. Gibbs was America's first home-grown, world-class scientific theorist. In 1863, Yale awarded Gibbs the first doctoral degree in engineering conferred in the United States. He spent the next three years studying thermodynamics—the new science of steam engines and energy—in Germany and France, then he returned to Yale and was promptly appointed professor of mathematical physics. There, building on physical principles of thermodynamics developed by James Clerk Maxwell in England and Ludwig Boltzmann in Austria, Gibbs founded his own new science he called *statistical mechanics*—a set of mathematical tools for analyzing the actions of water molecules in steam engines and other minute particles in random motion.

Gibbs's *Elementary Principles of Statistical Mechanics*, published in 1902, dealt a major blow to the Newtonian worldview that had dominated Western science for more than three centuries. His new statistical approach to the physical problems of minute particles under pressure unveiled a realm of nature that did not conform to Newton's immutable laws of motion. Gibbs's mechanics injected the new element of chance or *probability* into the physical universe and established it as the governing principle of events at the molecular level. In contrast to Newton's deterministic world, where the path of a planet or any object moving on a fixed trajectory could be calculated precisely, the paths of the energized particles careening through Gibbs's probabilistic world could not be pinned down with precision. His probabilistic approach determined, not what *will* happen to a particle or other object in motion at any given instant, but what *may* happen to an individual particle, and to particles collectively, under specific physical conditions. Yet those chancy mechanics nonetheless gave scientists tangible information about physical events that could not be measured by existing instruments or even observed directly.

Statistical mechanics would figure prominently in the work of atomic theorists in Europe, and in their new "quantum" mechanics of events transpiring at subatomic levels, but, at MIT, the rookie instructor Wiener saw other practical uses for Gibbs's probabilistics. He found in Gibbs's formulas new tools for computing the *probable paths in space* traveled by point-sized projectiles and other objects of all kinds and sizes. Discovering Gibbs's work was "an intellectual landmark in my life," he said later.

The new paths that preoccupied Wiener were not the simple parabolas of artillery shells, but wildly irregular motions, like the crazed paths of those water particles bobbling on the waves of the Charles or ricocheting around the cauldron of a boiling steam turbine, or the tortuous course of a bee in flight or—Wiener's favorite example—the path of a drunken man walking across a large deserted playing field. In each case, he observed, the object's future position bore little discernible relation to its past course or even its current direction. Its speed could vary capriciously from one point to another, and, more important, the observer's knowledge of the object's location at any moment was utterly uncertain.

Was it possible to plot such disorderly paths systematically? What could be said with *any* degree of scientific conviction about those erratic motions? It was Wiener's kind of problem, a fine puzzle of probability to busy a bright, young, real-world oriented mathematician, and a suitable challenge for an ex-prodigy eager to prove himself in his first full-time academic position. Now those questions of certainty and uncertainty that had intrigued him since childhood would propel his career forward, as the mental exercise that began with a passing rumination on a seemingly superfluous everyday occurrence revealed an entire dimension of nature that had not been described or analyzed before. The exercise led Wiener to undertake a broad investigation of all uncertain and irregular motions, and of the "essential irregularity of the universe," as he set out to produce a meaningful mathematical description for such irregularities wherever they might occur.

As Wiener contemplated the problem, he recalled Einstein's famous paper on Brownian motion, those highly irregular, seemingly random paths traced by particles in a fluid at rest in apparent defiance of the laws of motion, gravity, and even thermodynamics. The phenomenon—first observed in 1827 by the Scottish botanist Robert Brown while he was looking under a microscope at pollen particles suspended in still water and saw them jiggling erratically without any detectable force acting on them—remained inscrutable until 1905, when Einstein solved the mystery. Einstein explained that the visible movement of particles in Brownian motion was the result of myriad collisions occurring at submicroscopic levels with the smaller molecules that made up the fluid itself. Those molecules, Einstein confirmed, were in states of perpetual agitation caused by the kinetic—heat—energy innate in all matter, and he supplied the mathematical calculations that proved it.

Now Wiener took the next step. He saw a whole new order of complexity in Brown's hopped-up pollen particles. Using Gibbs's statistical principles, of which Einstein was unaware in 1905, Wiener set out to describe and compute the probable paths followed by a single particle in Brownian motion, and the probabilities of the entire family of such possible paths. His first paper in the field of mathematical analysis, written in 1920, followed by his major work on Brownian motion published the following year, unearthed a striking paradox in nature. Wiener proved that all the paths traced by particles in Brownian motion are continuous, with no inexplicable gaps or jumps along the way, yet always mathematically "strange," *infinitely* irregular, and, indeed, utterly *im*probable—at no time do the individual particles assume any definite, statistically predictable direction. The mathematics Wiener offered as an alternative, a new formula that combined the statistical methods of both Gibbs and Lebesgue, provided the first rigorous proof that probability laws governed Brown's phenomenon after all. His new formula describing the probable trajectories, not for individual particles, but for groups of particles in Brownian motion, marked a major advance in probability theory that came to be known as the "Wiener measure" and, later, in the new rubric used by Hilbert and his disciples at Göttingen, as the "Wiener space."

Wiener found his next intellectual challenge when the institute's electrical engineers came to him for help with their growing needs in the theory and practice of electrical signal transmission.

The technical field had been developing in practical devices for nearly a century. In the 1830s, Samuel Morse in America and two engineers in England, William Cook and Charles Wheatstone, had independently invented the telegraph, a simple device for transmitting messages, coded in patterns of electrical signals, over long distances along a wire. In 1876, Alexander Graham Bell harnessed the acoustic pressure of sound waves to convey voices along those same wires by telephone. In 1895, the Italian Guglielmo Marconi used invisible electromagnetic waves to transmit messages in Morse's code, and in 1914 the first "wireless" voice message was broadcast.

By 1920, the industrial revolution had been joined by the electrical movement of messages. The new devices were opening whole new realms of invention, commerce, and human experience, yet, remarkably, there was almost no hard science underlying the new technology. Streams of electronic signals were gushing uncapped through the world at a time when the nature of electrons themselves was growing more mystifying by the day, and telephone and radio engineers had no systematic knowledge to aid them in the intricate work of designing the new electronic devices and networks.

The main body of working rules and engineering methods at their disposal was the obscure, idiosyncratic "operational calculus" devised in the 1880s by the eccentric British scientist and mathematician Oliver Heaviside, whose untidy theories and formulas made long-distance telephone communication possible. Now MIT's engineers were appealing to Wiener to put a solid scientific foundation under Heaviside's "new and powerful communication techniques," as he had for the particles and paths of Brownian motion. In his widening search for new mathematical descriptions of predictable phenomena that, somehow, emerge from nature's underlying chaos and incorrigible irregularity, Wiener was intrigued and eager to take on the challenge.

In this new realm of electronic technology where science and the everyday world were converging, a seemingly superfluous phenomenon, once again, caught Wiener's attention.

The so-called shot effect occurred among streams of electrons moving in currents along copper wires or through the airless space of the vacuum tubes that empowered the new wireless technology. Engineers at the fore of the telephone and radio industries had found, to their surprise, that the speeding streams of charged particles that conveyed the new electronic sounds and signals were not smooth-flowing and regular but tended to bunch like buckshot as they passed through any conducting medium, be it metal or empty space. In fact, the paths of electrons moving through those infinitesimal channels were as strange and irregular as the jerks and jiggles of Brownian motion. The shot effect posed few practical problems for telephone operators and users, who could not hear such minute irregularities with the naked ear. But it began to cause real trouble in the 1920s with the introduction of vacuum-tube amplifiers in radios that magnified those irregularities thousands and, in some cases, millions of times over, to the point where they produced audible noise and the staticky fits of early radio broadcasts.

Wiener approached the shot effect, not as an engineer, but as a mathematician and philosopher pondering the new electronic dimension at its most minuscule levels. As he came to understand it, the shot effect was a function of the fundamental "discreteness of the universe"—the irreducible space between the individual atoms and subatomic particles that make up the physical world and our experience of it. However, as Wiener and Europe's new quantum scientists were learning concurrently, those discrete particles also behaved simultaneously, uncannily, like . . . *waves*. The new telephone and radio signals conveyed their lifelike messages, not in discrete dots and dashes of Morse code, but in a continuous flow of fluxing electromagnetic energy. The wave connection drew Wiener back to his earlier wave problem, borne on the frothy tides of the Charles, and to the irregular wave motions characteristic of fluid turbulence.

The shot effect propelled Wiener to take an entirely new approach to the waves of electric currents. When the 1920s let out their first roar of electric sound, Wiener set out to make the first "functional analysis" of those new man-made signal-carrying waves with a rigor that would explain exactly how a bunch of erratic electrons passing through a paltry telephone line or an impalpable radio wave could faithfully convey "everything from a groan to a squeak."

He found the starting point for his new study of irregular electronic waves in the well-trod mathematics of regular waveforms, and a fitting real-world analogy in the overlapping frequencies of audible sound that rippled down the length of vibrating strings. Mathematicians since Pythagoras had studied the sinuous waveforms of vibrating strings and their naturally occurring overtones or harmonics. Then, early in the nineteenth century, the French mathematician and physicist Jean-Baptiste-Joseph Fourier, a scientific adviser to Napoleon, observed that, like sound, heat and other forms of electromagnetic energy also radiate in waves through air and metal. He devised a series of mathematical formulas, called transforms, that could parse a diffuse heat wave, or *any* complex waveform, into a manageable sum of simple, utterly regular, sine waves.

Fourier's elegant wave equations defined a new field of "harmonic analysis," but his work drew critics who insisted that his math left "something to be desired on the score of generality and even rigour." As a result, for nearly a century, harmonic analysis remained separated from the physical phenomena that inspired it. Early telephone and radio engineers were left to grapple in the dark with the real-world problems Fourier's formulas modeled—until Wiener reformulated Fourier's work and reestablished its real-world connections. With Fourier's simple sine waves undulating in his mind, Wiener replicated the Frenchman's maneuver, applying Fourier transforms to the waves of electromagnetic radiation that were being harnessed to convey the complex sounds and signals of the new electronic era. It was a stroke of genius on Wiener's part: to see those invisible, irregular electric currents flowing down a wire, across the empty space in a vacuum tube, and from beaming broadcast antennas as so many variations on a vibrating string.

Then, as he had with Einstein's analysis of Brownian motion, Wiener took the next step and expanded Fourier's classical analysis of the wave motion of heat into many mediums and mathematical dimensions. His exacting new formulas applied to continuous wave motions of all kinds, and to the more problematical *discontinuous* waves that electrical engineers were struggling to work with in the world, like the on-off signals of telegraph keys, and the audible sounds of telephone and radio messages that traveled from one point to another and terminated, or were followed by an entirely different assortment of disjointed electronic waves. In his first practical contribution to electronic theory, Wiener showed how those fluxing electronic signals could be captured and frozen in time, measured physically, and analyzed mathematically in ways that could be really useful to electrical engineers.

Wiener's new statistical methods of "generalized harmonic analysis," which he began publishing piecemeal in the early 1920s, together with his work on Brownian motion and probability theory, laid the ground for all the new sciences of the information age. However, Wiener's most important work was, in his word, "stillborn" among mathematicians and engineers in the United States. A decade would pass before his contributions to electronic theory would be recognized, and another would go by before electrical engineers would fully grasp the practical applications of his new harmonic methods and begin to use them to shape and control electronic signals with unprecedented precision.

In the interim, he blazed new trails in other arenas of urgent interest to mathematicians and electrical engineers. His foray into potential theory, another hot topic in the 1920s with practical applications to engineering problems, put him back in proximity with Harvard, where he consulted with Professor O. D. Kellogg, a leading figure in the field. Kellogg pointed him toward a good problem in the mathematics of electrical potentials, which Wiener undertook to solve on his own. However, when he showed his solution to Kellogg, he was told that two younger charges of Kellogg's studying at Princeton were pursuing similar work for their dissertations and that Wiener should erase from his mind the work he had already done "to clear the tracks for the two doctoral candidates."

The sparks began to fly. Wiener took umbrage at Kellogg's request. Kellogg and his colleague G. D. Birkhoff, Wiener's old nemesis, "thundered" at Wiener and urged him to abandon his work in potential theory, but Wiener went forward. He published his paper in MIT's new mathematics journal and wrote several others on the subject. The clash prompted Wiener to confront his own competitiveness, and his persisting feeling of being an outsider on the American mathematics scene, and it led him to make a stern, unapologetic resolution for his professional future:

> I was more or less repelled by the high pressure of work which was bound to be competitive from the start, yet I knew very well that I was competitive beyond the run of younger mathematicians, and I knew equally that this was not a very pretty attitude. However, it was not an attitude which I was free to assume or to reject. I was quite aware that I was an out among ins and that I would get no shred of recognition that I didn't force. If I was not to be welcomed, well then, let me be too dangerous to be ignored.

During those first years at MIT, Wiener was still living at home, where he remained the target of his father's chidings and both his parents' ongoing control operations. His compressed life as a child prodigy, his wartime wanderings, and the new pressures of getting his teaching career off to a good start had left him years behind others his age in the "social graces." He had few female friends and no prospects for marriage. However, in this too, Leo and Bertha Wiener stood ready to intervene.

Soon after the Great War had ended, Bertha Wiener stepped up her longtime custom of holding teas on Sundays in the family's home for Leo's students, visiting scholars, and other favored guests. Bertha's Sunday teas also served as a social training ground for the Wiener children and, later, as a discreet setting for screening suitable partners for them. The older ones, especially, Norbert, Constance, and Bertha, formed friendships and professional acquaintances at those genial gatherings that served them throughout their careers, and all three met their future spouses there.

But Wiener's opening moves in the dating game were halting ones. He did poorly under the family's close scrutiny, and his parents exercised "a complete right of veto" over the young women he selected. Their verdicts, Wiener said, were determined "more by what my parents conceived to be the girls' reaction to the rest of the family than by any factor directly concerning me." Then, one afternoon in the autumn of 1920, Wiener's parents introduced him to the briskly formal young woman who would become his wife.

Marguerite Engemann had come to America from Silesia, in eastern Germany, a dozen years earlier at the age of fourteen, and made her way to Radcliffe, where she majored in romance languages and studied Russian literature with Leo Wiener. She came to tea with her older brother Herbert, who happened to be a student in one of Norbert's classes at MIT. The two siblings recently had moved with their mother from Utah to the Boston suburb of Waverley. Their mother Hedwig Engemann came from a landed family and married an innkeeper in the Silesian Alps who prospered and, then, died suddenly while still in his thirties. Hedwig had emigrated alone to the American West, and pursued as many careers as Leo Wiener had in his Wild West days, in order to bring over her three sons and young daughter Marguerite. A fifth child, the eldest and another daughter, remained with relatives in Germany.

Margaret, as the Engemann girl came to be known in America (although she continued to use Marguerite with Wiener's parents), had been groomed in the finest *Hoch-Deutsch* High German dialect and manners, although she was not Germanic in her appearance. Her hair was dark, almost black, her skin olive, her eyes hazel. She was two months older than Wiener, two inches shorter, slight of frame, and possessed of a different temperament entirely. By all accounts, Margaret was highly intelligent, sensible, and serious of purpose, with only trace elements of play and humor in her makeup. She was soft-spoken and refined, a proper Radcliffe girl shored up by her Teutonic timbre and the Victorian mores the Wiener family valued, which Norbert too proclaimed devoutly during those years.

In the hope-filled eyes of Leo and Bertha and Norbert's sisters, Margaret was the perfect mate for their socially inept Nubbins. But she was not Wiener's first choice or even his second. He wrote tenderly of his earlier love for a Radcliffe student he met shortly after he started teaching at MIT, a comely French major who "was beautiful in a pre-Raphaelite way." He courted her for more than a year, until she told him she was engaged to another man. "I did not take this with good grace, but it was not a graceful situation," he said. In fact, he was heartsick.

As he swerved back toward the Slough of Despond, his family put their united energies into promoting the Engemann girl. Bertha Wiener, who had always viewed herself as more German than Jewish, felt strongly that Margaret's curtly formal, Old World style would conform with the family's ways, and with the continuing needs of Norbert's care as she perceived them, and she proceeded with the social maneuvering. Early in the winter of 1921, Norbert and Margaret began seeing one another, but Wiener was wary of his parents' machinations and loath to climb aboard the family bandwagon rolling in Margaret's direction. His parents "were not silent in their approval," he recalled. "I felt greatly embarrassed by their obvious reaction in her favor, and my response was to keep away from Margaret for the time being."

————

Wiener retreated to Europe in the summers of 1922, 1924, and 1925, longing for his old friends and looking to make new contacts among mathematicians in England, France, and Germany, where the climate was warmer to his ideas. On his first trip, Wiener showed Kellogg and his Harvard cohorts that he was not to be trifled with in academic turf battles. In France, he defiantly delivered one of his disputed papers on potential theory to the renowned Henri Lebesgue himself, who published it in the journal of the French Academy of Sciences with a preface lauding Wiener's work as a significant contribution to the field.

On his next trips Wiener took up matters that would prove even more significant. The quantum revolution on the Continent had kicked into high gear. In 1923, Max Born, a physicist at Göttingen working on the forefront of the new atomic theories, realized that, if his Danish counterpart Niels Bohr's advancing theory of atomic structure was correct, "the whole system of concepts of physics must be restructured from the ground up." The following year, Born coined the term "quantum mechanics" to describe this new physics that was yet to be devised, and the young French physicist Louis de Broglie, in his doctoral dissertation at the Sorbonne, put forth the truly radical notion that matter, like energy and heat and light, could take the form of waves as well as particles.

The same year, 1924, Wiener was promoted to the post of assistant professor of mathematics at MIT, and, that summer, he journeyed to Göttingen for the first time since he had studied there as a traveling fellow a decade earlier. He was delighted to discover that his "stillborn" ideas were attracting attention among the university's mathematicians and physicists, and he returned to Göttingen the following summer. During that summer of 1925, Wiener gave a seminar on wave problems and his new mathematical methods of harmonic analysis to a se-lect group at the university. Although Wiener had just crossed into his thirties, he was received as a fellow *wunderkind* among the many bright young stars who were gravitating to Göttingen to formalize the new quantum mechanics.

Wiener's talk at Göttingen, while it did not address quantum theory explic-itly, keyed in on the quandary at the center of the new physics: the problem of observing and measuring vibrating waves of electrons moving through space and time, and the limits of those observations at submicroscopic levels. Wiener described the quandary and his new harmonic theory in terms Born, who had a passion for the piano, and other cultured Göttingers could relate to—music.

My talk in Göttingen, like quantum theory . . . concerned harmonic analysis—in other words, the breaking up of complicated motions into sums of simple oscilla-tions [like] the motion of the string of a musical instrument. . . . The . . . vibra-tions can be characterized in two independent ways . . . according to frequency, and . . . duration in time. [But] the frequency of a note and its timing interact in a very complicated manner. . . . Precision in time means a certain vagueness in pitch, just as precision in pitch involves an indifference to time. . . . The consider-ations are not only theoretically important but correspond to a real limitation of what the musician can do. You can't play a jig on the lowest register of the organ. If you take a note oscillating at a rate of sixteen times a second, and continue it only for one twentieth of a second . . . it will not sound to the ear like a note but rather like . . . no music at all.

Once again, as in his precocious "Theory of Ignorance" and his early papers pointing out the limits of Russell's sweeping system of mathematical logic, at Göttingen Wiener explained that there were also limitations to the new wave harmonics, intervals of scale below which all primary vibrations, whether of strings or organ pipes or electrons, could not be observed and measured with scientific precision. His harmonic theory resonated with Born's star pupil Werner Heisenberg, who was sitting in Wiener's seminar that day. In 1927,

Heisenberg would express the same conclusion more formally in his celebrated Uncertainty Principle—which stated that one cannot know with certainty both the precise position and momentum of a particle at the quantum level—using the same methods of mathematical analysis Wiener had "presented to the Göttingers . . . years before."

In the fall of 1925, Born came to MIT for a term to lecture on the new physics and to work directly with Wiener. Born sought Wiener's help in his effort to reconcile the faltering particle model of atomic structure and the brash new wave hypothesis. The joint paper the two men produced became a stepping stone and "notable contribution" in the advance of quantum theory, although Born confessed that he had not fully understood Wiener's computations and "hardly assimilated" the underlying concepts of his wave harmonics. Years later, Born would receive a Nobel Prize in physics for his "statistical interpretation of the wavefunction," and he would formally acknowledge Wiener as "an excellent collaborator."

Wiener made many important acquaintances at Göttingen during that triumphant summer of '25. He also fell in love again, this time with a young astronomer Cecilia Helena Payne, a porcelain beauty from England who had received her doctorate from Harvard that spring, and who was visiting at Göttingen's astrophysical society while Wiener was holding court in the mathematical society. According to one report, Wiener was especially taken with Payne because she was "a person more like himself in temperament and interests," and when they returned to America he "pursued this romantic interest with hope of matrimony." But Payne was already engaged to her task of carving a career as the first woman in her field, and she would not be diverted by Wiener or any suitor.

When that romance flamed out, Bertha and Leo pressed for a quick close on the merger with the Engemann girl, which they had been advocating, but Wiener would not be steamrollered in affairs of the head or the heart. He and Margaret had been seeing each other on and off for five years. During that time, while Wiener was making his name in mathematics at MIT and across Europe, Margaret earned her master's degree and took a job teaching modern languages at a small women's college in Pennsylvania. The two met midway at friends' homes and saw each other between terms, but their extended courtship was, as Wiener put it, "a bit too intermittent to suit us."

After his impressive performance at Göttingen, Wiener was invited jointly by Born and Hilbert, and by Hilbert's protégé, Richard Courant, the new head of the Göttingen mathematics department, to spend the next year at the university. Early in 1926, he was awarded a grant from the newly formed Guggenheim

Foundation in New York to support his appointment. He promptly asked Margaret to marry him and she accepted, but their timetables were in conflict. Wiener was due in Göttingen by April and Margaret wanted to finish her school year in Pennsylvania. They agreed to marry in the spring and rendezvous in Germany several months later.

At last, his work and his love life were coming together, and Wiener was thrilled to a fault with his good fortune. Of his coming year at Göttingen he recalled, "I was in a very exulted mood at what I conceived to be the first wholehearted recognition that had come my way, and I am afraid that I talked more of it to the newspapers than was strictly becoming. . . . I must have been an insufferable young man in my boasting and gloating." Of his coming wedding, too, he took heart that he had traversed the obstacle course of parental plotting and emotional detours that marked his way to the altar.

Norbert and Margaret married in Philadelphia in March 1926, both newlyweds more than ripe at age thirty-one, and they honeymooned briefly in Atlantic City. For those few days, the two were giddy lovebirds. "Dear Parents: Marguerite and I are having a wonderful honeymoon," Norbert wrote to Leo and Bertha on hotel stationery. "She is a darling girl, and I am a very lucky man." Wiener began calling Margaret by her German diminutive Gretel, a term of endearment that reflected the genuine affection he felt for his new bride. The newlyweds traveled together to New York, where Wiener embarked for Europe. Days afterward, on board the *S. S. Minnekahda,* he was the king of the world, flaunting his filial liberation. "I am delighted with my honeymoon," he wrote to his parents on the ship's frugal notepaper, and signed off, "Your (no longer) dutiful son Norbert."

By the time he made landfall, the primroses were flowering in the fields of Devon and he went off to Göttingen flush with ardor and optimism. Two months later, Margaret embarked in equally high spirits for her first return to her homeland, assuming that, as the German-born wife of a distinguished visiting professor and an accomplished scholar herself, she would be in her element at Göttingen, a place steeped in Old World academic and social traditions.

But something happened to both newlyweds in Göttingen that forever changed their life together. In contrast to Wiener's gratifying visits there the two summers before, as he settled in for his extended stay he ran into the germinating Nazism that was taking hold over Germany's universities. Nationalist extremists dogged his scholarly and social interactions, convinced that the young Harvard-educated Guggenheim fellow whose boastful remarks had been reprinted in the German press was the outspoken elder Professor Wiener from Harvard whose name had been emblazoned on the Brownshirts' blacklists for his bitter opposition to Germany's actions in the Great War.

And Wiener met a second obstacle this time around. Richard Courant, the new head of Göttingen's math department, was dogging him, too. Courant, a gifted Polish-born Jewish mathematician, had founded a new mathematics institute at the university several years earlier, and, when Wiener arrived, he was actively courting the favor of Wiener's nemesis from Harvard, G. D. Birkhoff. Birkhoff, too, was in Göttingen that spring on a reconnaissance mission for the Rockefeller Foundation, which was looking to fund worthwhile projects in Weimar Germany after its ruin in the Great War, and he was considering Courant's institute as a potential beneficiary.

Courant was torn between the two arch-foe mathematicians, Birkhoff and Wiener, and his own institutional interests, and he was pressed further as a Jew by the new Nazi zeal rising at Göttingen and within the ranks of the German government, his principal patron. Squeezed from all sides, Courant turned against Wiener. As Wiener retold the events that followed, Courant lambasted him for his bragging comments to the press. He withheld official recognition of Wiener's appointment as a visiting lecturer, and rescinded, or granted only grudgingly, perquisites Wiener had been promised. The combined strains of the Nazis' jackboot tactics and Courant's academic indignities burst Wiener's buoyant mood and brought him to the verge of a nervous breakdown.

The strain was obvious in his presentations. His brilliant ideas poured out in disarray, his German became jumbled, and his lecture series was pitifully attended. Those in the audience could only wonder how Wiener ever ranked as a *wunderkind*. To keep his balance, Wiener turned to his American and English friends at the university, who nursed him through his "blues," but their help was in vain.

By the time Margaret arrived Wiener was frantic and, by his own admission, in the throes of a "black depression."

Margaret was not amused by the scene she entered at Göttingen. She was shocked and overwhelmed by the wreck of a man she met up with—a man she had not seen during their long courtship and short honeymoon. To make matters worse, soon after Margaret arrived, Leo and Bertha Wiener dropped by Göttingen on their summer trip to Europe, as Wiener recalled, "partly to share in my supposed success and partly to keep a supervising eye over the newly married couple." Wiener's torments were compounded by the problem of telling his father that Leo's expressed hostility toward Germany, and the Nazis' confusing of father and son, were primary causes of the malice that had engulfed him at the university. The two couples spent a miserable week together before Leo and Bertha left for their other destinations on the Continent.

Wiener and Margaret departed soon after to tend to Wiener's wounds at an alpine retreat in Switzerland. When they arrived, they received a letter from Leo and Bertha summoning them to Austria for the next round of family togetherness at Innsbruck, and, in the pattern that had persisted since his childhood, Wiener in his depths was too weak to resist his parents' compulsions. At Innsbruck, Margaret watched the family dynamic through newly opened eyes and resolved to wrest her husband from the emotional clutches of his parents. Politely they bid adieu to Leo and Bertha. Then the bedraggled newlyweds headed south to Italy to commence their long-delayed European honeymoon.

But that honeymoon, too, was over before it began. Wiener's black depression stalked the couple through the romantic canals of Venice and the Renaissance splendors of Florence. Forty years later, he was still apologizing for that trip and the marital trials that followed. "It was no pleasant experience for Margaret to become involved with the problems of a neurotic husband at his very lowest emotional level," he offered, along with a wider indictment. "I had become even more of a problem, because my parents had made a policy of glossing over my emotional difficulties, instead of confronting Margaret with the real task she had undertaken in marrying me."

Those frank admissions gave only a hint of the family drama that began with Margaret's realization of her husband's turbulent emotional states. In letters Norbert and Margaret wrote to Leo and Bertha, both partners gave indications of the strains tearing at their months-old marriage. Wiener was cryptic about the couple's need "to work out our problems of adjustment." Margaret stated explicitly that Wiener had suffered a major episode of depression at Göttingen and said she was "really afraid that he might have a nervous breakdown." She also sent her new in-laws some tough words of assurance. "Norbert needs absolute quiet and I shall see that he gets it."

From that moment on, Margaret Wiener resigned herself to a role that veered profoundly from the blithe *frau-professorship* she had envisioned. She vowed to become the caretaker and protector of her high-strung husband as he careened through his accelerating career en route to international scientific stardom.

4

Weak Currents, Light Computers

Where are the thunderbolts of Zeus? . . . Electra, forsaken,
braves the storm alone.

—Sophocles, *Electra*

WIENER AND MARGARET LEFT GÖTTINGEN behind and went to Copenhagen, where
he spent the remainder of his Guggenheim grant working with Denmark's foremost mathematician, Harald Bohr, the younger brother of Niels. The couple's
time in Copenhagen lifted both their spirits and bolstered Wiener for the return
trip to the States.

They set sail on wintry seas and made landfall in Boston, just behind an
ornery Nor'easter, in mid-January 1927. While Boston dug out from the blizzard, Wiener and Margaret went house-hunting. They found a pleasant apartment on Pleasant Street, just over the Arlington town line, and Wiener
commenced valorously to adapt himself "to a life of handyman domesticity, for
which I had no particular qualifications." He learned to varnish furniture and
tend a furnace, but not much more. Proficiency on that plane of complexity, he
said, "was never my métier."

His métier was mathematics, and his talents were sorely needed by MIT's
electrical engineers and their colleagues across America. By 1927, there was
chaos in the nation's communication channels. In only a few decades, the telephone had grown from a conversation piece into the culture's most ubiquitous
appliance. At the booming American Telephone and Telegraph Company, the
demand for telephone operators had grown so fast that, only a few years earlier,
the company's projections had warned that every young woman in America
might have to be employed as a telephone operator just to run the system. Dial
telephones were put in service and the threat to American women was averted,

but the exploding volume of local and long-distance traffic compounded the system's complexities and created enormous engineering challenges.

The new radio technology, too, was growing exponentially, fueled by the sale of inexpensive home receivers and the launching of the first nationwide broadcast networks in the early 1920s. Across America, 500 stations were broadcasting on a small band of radio-frequency channels that often interfered with one another's signals, and systems for relaying radio programs to local stations over long-distance telephone lines were overwhelming the lines' capacity. Telephone and radio engineers alike needed more powerful components, and more effective methods of designing and building circuits, to keep their sprawling networks from overloading and breaking out in annoying "crosstalk" and screeching oscillations.

And other new communication contraptions were coming on the line that would put more strain on the circuitry: high-speed telegraphic stock tickers to feed the decade's speculative market fevers; and printing "teletypewriters" and "wirephoto" facsimile machines for transmitting text and images to the nation's newspapers. In 1927, electronic sound recording became practical, "talking movies" were born, the first experimental long-distance television transmission took place, and those advances posed still more technical challenges. At the Bell Telephone Laboratories in New York City, the prototype of the modern corporate research center formed in 1925 by the merging of AT&T's in-house engineering departments, the lab's staff scientists and mathematicians made strides in the development of telephone, radio, and sound recording technologies. Bell Labs scientists, working with the company's engineers, devised new technical tools to improve the nation's telephone system: automatic call switching, "multiplexing" (sending multiple calls simultaneously over one set of telephone wires by transmitting each call on a different frequency of electronic signals), and specialized electronic "wave filters" to sort out the different signals and route each call to its destination.

But despite those gains, the burgeoning terrain of electronic communication remained a barren outpost, scientifically speaking. Wiener's new formulas for the harmonic analysis of electronic signals had yet to be applied in everyday engineering practice. Systematic methods had yet to be devised for designing electronic circuits and accomplishing many other routine engineering tasks. Persisting technical problems yielded no practical solutions.

And a new problem was emerging that was difficult even to articulate. As early as 1924, Bell system engineers had observed that the transmission of radio programs over AT&T's telephone lines required about twice as much "message" as the transmission of speech. To make the fullest use of AT&T's equipment, to maximize the speed of signal transmissions and the company's profits, their engineers needed to determine *"the size of the message,"* the best coding methods for the *"transmission of intelligence,"* and better ways to describe and measure scientifically *"the commodity to be transported by a telephone system."*

But what was that nebulous "commodity"? At that point, Bell's scientists and engineers didn't have the foggiest idea what it was they were transmitting, let alone how to measure it. It was the most fundamental question of the new electronic age, but for all their know-how and engineering proficiency, the fledgling wizards of communication technology at AT&T and every electronics company had yet to grasp the big picture of their new technical enterprise. The field had no Einstein in house to perform soaring thought experiments, no Russell to recreate it from the ground up on physical principles and logical propositions.

There simply was no rigorous science of communication before Norbert Wiener took up the subject in earnest in the late 1920s. Surveying the chaos, Wiener resolved to take his work on harmonic analysis to the next level, applying his stillborn mathematics to the practical problems of the new communication technology.

From the start of his investigations, he observed that the new electronic signals that were transforming the sights and sounds of the Roaring Twenties were comprised of barely detectable "weak" currents whose subtle electrical actions marked a world apart from the "strong" currents of electrical power engineering. The distinction between strong and weak currents was made by engineers in Europe, but not in the United States, where power engineers, telephone engineers, and radio engineers were referred to interchangeably as "electrical engineers." The two fields would later come to be known as "electrical" and "electronics" engineering, but for Wiener the European distinction was the one that mattered.

Those meek currents flowing through the nation's telegraph and telephone lines moved the quickening traffic in modern messages with ridiculously small sums of electromotive force—only a few volts and watts compared to the kilovolts and megawatts of raw energy that coursed through power transmission lines and heavy-metal electricity grids. And the new wireless radio waves were more ephemeral still as they passed invisibly through the air and were picked up by the ethereal vacuum-tube "electron valves" that formed the floodgates of the rising mass communication culture.

As Wiener saw early on, those quirky weak currents were completely unknown quantities. They were not hard objects moving through the world on fixed trajectories, or fleeting subatomic phenomena governed by the new principles of quantum mechanics. Their actions were more harmonious with the waltz of waves on a river, the jerky polka of pollen particles in still water, the quavering vibrations of violin strings, and Wiener was uniquely positioned among technicians and mathematicians to subdue them scientifically.

Building on his earlier work, Wiener began to formulate a radically new approach to the entire enterprise of communication engineering—a *statistical* approach. He drew particular inspiration from Gibbs's all-encompassing system of statistical mechanics, which was based on the mathematical concept of probability and seemed to Wiener to hold the master key to the new domain of electronic signals. But Wiener was surprised to find, on closer examination, that the fundamental assumption of Gibbs's statistical mechanics—that even the most irregular and seemingly random systems behave over time in a manner that can be predicted by the principles of probability and modern statistical methods—was first stated mathematically "in a form not merely inadequate, but impossible."

Marching to his own beat, step by step and line by line, Wiener corrected the errors of old and began to weave the threads of something new. Through the late 1920s and early 1930s, which Wiener himself extolled as his "years of growth and progress," he built his new logical and mathematical foundation for communication engineering as a statistical science. In more than a dozen papers published in European and American journals, he joined the two grand traditions in which he was steeped—European science and Yankee ingenuity—and made basic connections between the different equations used by communication engineers on both sides of the Atlantic. That creative wave crested in his seminal paper, titled simply "Generalized Harmonic Analysis," published in 1930 in the prestigious Swedish journal *Acta Mathematica*. In it, Wiener named and formally established his new, unified, rigorously statistical approach to his decade-long "wave problem" and all wave problems across the spectrum of communication engineering.

His new statistical perspective marked the full flowering of his youthful "Theory of Ignorance," with his unequivocal conclusion that "in fact all human knowledge is based on an approximation." It also mirrored and reframed in the most general scientific and mathematical terms all the earth-shaking "uncertainty" principles, "incompleteness" theorems, and probability theories that other distinguished scientists and mathematicians were putting forth within a few years of one another.

In the field of communication engineering, his new statistical approach described the movement of message-carrying electronic waves, like the movement of pollen particles in Brownian motion, and of electrons singly and in bunches, not as a certainty but as *a probability expressed mathematically on a scale between the numbers 0 and 1*, where 0 marked the measure of total uncertainty and improbability, and 1 marked the measure of complete certainty and predictability. As he explained his new thinking years later, the power of his statistical theory flowed from the imprecision inherent in all scientific data, a fact of life most engineers and scientists generally were still reluctant to acknowledge. He knew the new statistical methods offered profound advantages, not only for measuring and analyzing irregular electronic waves but for all scientific procedures requiring accurate measurements and mathematical computations, from weather prediction to economic forecasting.

In the field of communication, Wiener's new statistical formulas for the harmonic analysis of electronic signals gave engineers a way to grip and harness all the irregular weak currents of the day, everything from the complex vibrations of the human voice to the daily fare of the new broadcast entertainment networks—the sounds of symphony orchestras, the crazy new rhythms of jazz—and to measure them mathematically in manageable sums as pure and perfectly proportioned as the pluck of a single violin string. And Wiener's new math went further. His generalized formulas formed a universal tool kit for parsing and analyzing all *possible* vibrations, whether they were comprised of audible sound, visible light, or invisible electromagnetic radiation, whether they were undulating in metal wires, thin air, or vacuum tubes, in two, three, or, hypothetically, an infinite number of dimensions.

Though he was not an engineer himself, in his papers Wiener showed exactly how his new statistical methods could be used to separate the flow of signals through electronic circuits from the audible "noise" their irregular motions created. In a breakthrough for electronic theory, he showed that when those noisy signals were fed through an appropriate wave filter, the output was indeed statistical in nature and gave "a valuable means of measuring the electronic charge." His new methods could be applied to every kind of electronic technology. They provided the first tools for the analysis of electronic signals as "series of data" moving in space and time on their assigned transmission frequencies. And, in a first stab at clarifying and quantifying the new commodity communication engineers were trafficking in, Wiener devised an ingenious new method for measuring and accurately predicting the flow of electronic data made up of "an infinite sequence of choices," using a novel "binary" system to represent the data as a series of two equally probable choices—designated by either a 0 or a 1.

In time, all those revolutionary concepts would emerge as central elements of the modern science of communication. And when it was published in 1930, Wiener's "Generalized Harmonic Analysis" delivered one more breakthrough. His broad scientific framework defined a general theory of random processes—those infinitely irregular occurrences in nature and human affairs—and provided the statistical methods needed to tame them. His work launched the study of probabilistic or "stochastic" processes, which he and others would develop at the forefronts of mathematics, physics, engineering, and disciplines farther afield in the years to come. In the process, Wiener had started what he proudly proclaimed to be a new "movement" in scientific thinking.

Wiener's mathematics alone was ahead of its time by a decade or two and, in some parts, three. Before long those two tiny numbers that contained the whole universe of probability—0 and 1—and the new binary logic and electronic processes they represented, would rule over the entire domain of the new communication technologies. But almost nobody perceived that at the time. In truth, most communication engineers in America were not even aware of the practical work Wiener was doing in their field, and those who were could barely begin to make sense of his new statistical theories and methods, let alone put them to use.

Like his earlier mathematical papers, Wiener's strong work on weak currents made scarcely a ripple beyond his home base at MIT.

———

Back on Pleasant Street in the spring of 1927, Wiener and Margaret were getting acclimated to their new home life. The torments of Göttingen had receded, and for once, to Wiener's surprise, he was happier in America than in Europe.

Margaret resolutely assumed her new duties as a *frau-professor* in residence. She made a comfortable, well-kept home on Wiener's meager assistant professor's salary, and she managed the household's finances with penny-perfect precision—as she said, "Norbert does the math and I do the arithmetic." She was also an excellent cook and, though not a vegetarian herself, she quickly became a proficient preparer of Wiener's favorite vegetarian dishes. Margaret doted on her husband in businesslike ways. She laid out his suitclothes on the bed in the morning while he was shaving. She handed Wiener his lunch money as he walked out the door. By now, Wiener was ready to relax the Victorian restraints of his youth and was "very demonstratively affectionate" toward Margaret. She, however, was not prone to displays of affection in public or private.

Wiener discovered, and even liked, the superior being he called "my new personality as a married man." He also liked his new autonomy within the wider Wiener family. Leo and Bertha Wiener were delighted to see their son starting a home life of his own, but they made plain their expectation that he and Margaret would comply with the family regimens they had set down for their children and all their spouses. Wiener was pleased to find "that in fact this was never a possibility." He and Margaret were equally determined to prove that Leo and Bertha "had taken a great deal too much for granted" if they supposed that Wiener's marriage to Margaret would ensure "an indefinite prolongation" of his "family captivity."

That summer, Wiener took his new colleague Dirk Jan Struik to his favorite haunts in New Hampshire and introduced him to mountain climbing. Struik was a talented Dutch mathematician Wiener had met at Göttingen in 1925. The two men became fast friends, and, a year later, Wiener enticed Struik to come to America to strengthen MIT's math department. Now they lit out for a week to make the rugged trip north through the White Mountain wilderness up the wind-whipped peak of Mt. Washington that Wiener had first climbed as a lad of fifteen. Both men returned hardy and heavily bearded. Struik's beard "made him look as if he had been painted by Rembrandt," Wiener said. Margaret trimmed Wiener's down to the goatee he never surrendered.

Later that summer, Wiener and Margaret began looking for a place of their own in the New Hampshire mountains. They found a good prospect in the tiny village of South Tamworth nine miles east of Sandwich. The white frame farmhouse was a pre–Civil War fixer upper, like Wiener's boyhood home at Old Mill Farm, situated on a knoll that looked out across green pastures in a valley between two mountain ridges. The house was ringed by towering hundred-year-old maples, which gave the little place a lofty air and set it apart from the dense spruce and pine that packed the woods nearby and covered the mountains far into the distance. The house was graceful and roomy, with wide floorboards and ample light throughout. The living room had a great brick fireplace and a cozy study at one end. Two small bedrooms nestled in the front, and a short stairway led up to an attic with low ceilings sloping down to the eaves. A large screened-in porch hung off the back of the house and opened onto a spectacular view of the noble Sandwich Range.

The place filled Margaret with warm memories of her childhood in the Riesengebirge, the "colossal mountains" of eastern Silesia. It gave Wiener endless room to roam and a quiet site, far from the bustle of Cambridge, to think and write during his long summer vacations. He viewed his yearly hiatus in New Hampshire, not as an academician's luxury, but as a mental and emotional necessity. His work demanded constant interchange with his MIT colleagues, but when his research was done and the moment came to write up his results, he needed sufficient time and space, "when my life is a simple alternation between concentrated intellectual effort and . . . completely non-intellectual pleasures." Since the summer sojourns of his youth, those simple delights had consisted of walking the hills and pastures, mingling with friends he had made in the region, taking long swims, and sunning on the beach of New Hampshire's gentle mountain lakes.

And a place in the country would serve another nurturing purpose. That summer Margaret had informed her husband that she was pregnant. Wiener was not a zealous Arcadian like Leo. He had no pretensions to agriculture or conscripting his own children as farmhands. But his naturalist urges and life-long love of the outdoors compelled him to ensure that his children would know more than merely a life in the city and Boston's middle-class suburbs. The old farmhouse and its resplendent environs, Wiener knew, would "give our children the experience of the country and the freedom of living which we feel to be the birthright of every child." Sylvan Bear Camp Pond just up the road, with its short beach and long run out into the warm water, provided a natural sandbox and a safe harbor for young children, and served as an egalitarian gathering spot for the locals and the summer crowd of Boston-area professionals and their families.

Late in the summer, Wiener made a winning bid on the old house and thirty-five acres of surrounding woods and scrub. When told of the impending purchase, Leo Wiener commended his son's choice and, in September 1927, he sent the couple a check to help pay for the property.

The following February, the couple's first child, a daughter, was born. The exultant new parents named their little girl Barbara. It was the only logical choice. Wiener took the name from the first mood of the first figure of the syllogisms, a medieval mnemonic he had learned as a boy that assigned word patterns to the classical rules of deductive reasoning. In the old ditty, the three "a" vowels of the first mnemonic, *Barbara*, encoded the three affirmative propositions that constituted the "pure universal," the primordial syllogism: "If all a's are b's and all b's are c's, then all a's are c's." (E.g., "If all Greeks are logicians, and all logicians are philosophers, then all Greeks are philosophers.")

To Wiener's mind, his new baby girl was not only his first progeny but the highest form of argument—a pure universal. As he watched the tiny newborn crying in her crib, he chanted to her the soothing scat of the four figures as he had recited them for his father:

> *Barbara, Celarent, Darii, Ferio*
> *Cesare, Camestres, Festino, Baroco . . .*

Wiener looked on his child as a naturalist might behold a wonder of nature, as he embarked on a studious new phase of his life "as a very clumsy pupil in the art of baby sitting and of hanging out a long signal hoist of diapers." That summer, the family moved a good portion of their possessions from the apartment on Pleasant Street to their new home in South Tamworth. In July, Wiener wrote to Leo from New Hampshire: "Everything is going first rate here—baby, Gretel, and all."

Their second child, another daughter, was born in December 1929. There was no argument, logical or otherwise, about her name. She was named Margaret, to appease her mother, who had been hoping for a boy and was now doubly disappointed; although like many a little Margaret, she soon became Peggy, a pearl. Wiener had put his affirming stamp on his firstborn, but this child was marked indelibly as Margaret's, and her mother would embrace her with a fondness she would not show to Barbara.

Wiener passed on to his pure universal and his pearl the stories his mother had read to him in the yard on Hilliard Street:

> One fine morning . . . this 'satiable Elephant's Child asked a new fine question that he had never asked before. He asked, "What does the Crocodile have for dinner?" . . . Then Kolokolo Bird said, with a mournful cry, "Go to the banks of the great grey-green, greasy Limpopo River, all set about with fever-trees, and find out. . . ."

He declaimed to his girls in a babel of tongues, as he had to his own father in command performances three decades before. He recited by heart Homer's mellifluous catalog of ships, from Book II of the *Iliad,* in the ancient Greek. He sang out Horace's dazzling, drunken ode celebrating the news of Cleopatra's death and the return of the Roman troops from the climactic battle at Actium, with its *pede libero*—its "featly tread"—that made the earth shake.

"I can still hear his Latin verse that echoed the tramp of Caesar's legions as they marched through Rome. I fell in love with the sound long before I could translate the words," Barbara remembered, years after the echoes had faded.

———

While Wiener and Margaret were making a home and a family, the proud dad was helping to bring another new creation into the world. Several years earlier, in the fall of 1925, Wiener had begun a lively consultation with an MIT colleague who had set out to build the first modern computing machine.

The idea of mechanized mathematics had been incubating for centuries. In 1642, the nineteen-year-old French mathematician Blaise Pascal assembled a mechanical adding machine that counted numbers arrayed on a bank of ratchet-driven wheels with saw-toothed spokes that carried over sums from one position to the next, and a little window above the gearbox that displayed the total. Thirty years later, Leibniz built a better calculator that could also multiply numbers by repeated addition. The same principles were still being employed in the clunky mechanical calculators Wiener and his fellow fumblers with numbers used at Aberdeen Proving Ground during the Great War.

Automated computing was another concept altogether. A dream machine had been on the drawing boards in Britain since 1819, when the prodigal Trinity man Charles Babbage drew up his initial design for a "Difference Engine," a steam engine made of brass, pewter, and gun metal that would compute celestial navigation tables. In 1833, he unveiled his new design for an improved "Analytical Engine" that could perform calculations in a prescribed sequence of up to a thousand numbers containing fifty digits apiece. His new steam machine incorporated all the basic components of a modern computer, but Babbage's vision was just a tad beyond the technology available in his day.

A hundred years later, Vannevar Bush, a rising star in MIT's engineering department, began work on a new machine that recast Babbage's contraption in twentieth-century technology. Like most electrical engineers in the 1920s, Bush was a power man, not a communication man. He wanted to build a computing machine that could solve differential equations—multipart mathematical expressions used widely by engineers to model real-world problems in which one or more physical factors change continuously over time. Bush's "differential analyzer" was not a digital device but strictly analog. Its computations were carried out not by counting, as desktop calculators did it, but by analogy, using real physical forces that changed in proportion to the quantities they represented. His machine worked out its differences mechanically, in a manner remotely akin to Babbage's, using an intricate assemblage of motor-driven discs and gears connected by rotating steel shafts.

Bush was not bashful about seeking Wiener's ideas for his machine, and Wiener, who had dreamed of making mechanical automata since his childhood, was happy to take Bush up on his invitation. Before he left for Göttingen in the spring of 1926, as Wiener remembered it, he was "closely associated" with Bush on the project, and "I tried to do what I could in designing computational apparatus on my own account."

The two men had a gentleman's agreement over their differing visions of computing. Bush's prototype electromechanical machine could parse only *ordinary* differential equations in which a single variable changed continuously, a common occurrence in power engineering problems concerned with the ebb and flow of heavy electrical currents. Wiener envisioned a machine that could parse more complicated *partial* differential equations—monsters of the mathematical deep in which two or more factors changed simultaneously, a situation that often arose in weak-current technology. Those complex "partials" were essential problem-solving tools for telephone and radio engineers wrestling with message-carrying electromagnetic waves that fluctuated irregularly in both time and space. They also happened to be the same brain-curdling equations Wiener was writing into his developing theory of generalized harmonic analysis.

By the spring of 1926, Wiener was already worried about the speed and reliability of Bush's clunky crankshafts and their fitness for cracking impenetrable partials. "The number of operations through which we must go to [solve] a partial-differential equation is simply enormous," he wrote later, and Bush's machine was too slow to be of any practical value. As an alternative, Wiener conceived a radical design for a new kind of computing machine that made use of the rapidly developing vacuum-tube technology. Since their first commercial applications, he had been captivated by the potentials of the tubes' bodiless electron beams, which traveled at the speed of light, unimpeded by friction and the inertia physical parts were prone to—and he was not just talking telephones and radios. The prospect of television, which was first demonstrated as a viable idea a few years earlier, had grabbed Wiener's imagination. Soon after he heard about the new live-picture transmission technology, it struck him that an optically based computing machine could solve partial differential equations and intricate problems of harmonic analysis more quickly and accurately than Bush's mechanical analyzer.

Wiener recalled how his new computing concept came to him in a darkened theater in downtown Boston. "I was visiting the show at the old Copley Theatre [and] an idea came into my mind which simply distracted all my attention from the performance. It was the notion of an optical computing machine for harmonic analysis. I had already learned not to disregard these stray ideas, no matter when they came to my attention, and I promptly left the theater to work out some of the details of my new plan. The next day I consulted with Bush."

Wiener's earlier work on the harmonic analysis of light waves convinced him that television's novel optical scanning technology could be used for purposes far beyond the mere transmission of pictures. Wiener advised Bush to scrap his mechanical apparatus made of century-old metal shafts and gears in favor of the cutting-edge weak-current technology of television, and to incorporate the new scanning technology into an all-electronic *optical* computer.

At the outset, Bush was receptive. "The idea was valid and we made a couple of attempts to put it into working form," Wiener recalled, but he held no illusions about his own nonexistent engineering abilities. "My contribution was wholly intellectual, for I am among the clumsiest of men and it is utterly beyond me even to put two wires together so they will make a satisfactory contact." In the months that followed, Wiener continued to consult with Bush on their two utterly different computer designs. In the fall of 1926, while traveling in Europe with Margaret after his debacle in Göttingen, he wrote buoyantly to his parents reporting that he "got a nice letter from Bush about my machine & his." When he returned, he jumped back into their joint computer initiatives, which Wiener hoped to apply to his new work in the statistical analysis of electronic signals.

During that brief interlude in the history of computing, Bush and his crew of crackerjack engineering grad students built both prototypes—Bush's electromechanical "Product Integraph" and Wiener's optical "Cinema Integraph."

However, Wiener's more sophisticated machine required precise measurements of the light beams passing through the system, which proved difficult to achieve using the technology then available.

Bush did not opt for optical computing in the 1920s, but he did take Wiener's advice about the value of vacuum tubes. His crowning achievement as a computerman, a top-secret, 100-ton analog computer he began designing for the Army in 1934 and delivered to Aberdeen Proving Ground in 1942, comprised 200 miles of wire, 150 electric motors—and 2,000 vacuum tubes. The machine could compute in fifteen minutes a firing table that previously took a human computer twenty hours to figure on a desktop calculator.

And there were other important by-products of Wiener's collaboration with Vannevar Bush. In 1929, Bush published a slim book on circuit theory for electrical engineers. In it, Bush became the first prominent engineer to recognize that Wiener's new mathematics of harmonic analysis could be "of considerable practical value." Bush sought Wiener's advice on many of the chapters and also asked him to write an appendix on the application of his new statistical methods. The highly technical work inspired both men and, though their personalities operated on different wavelengths, there was good humor between them. Wiener fondly recalled "the fun we had in working together." Bush, too, in the book's preface, professed that before they joined forces he "did not know an engineer and a mathematician could have such good times together."

In the years to come, Bush's talents would take him high into the bureaucracy of American science. Wiener would have more advice for him on the best directions for the advancing technology of computing. Bush would reject Wiener's ideas again, and again Wiener's vision would prove to be more prescient. Wiener's call for vacuum-tube computing came two decades early, but his foresight in the mid-1920s correctly anticipated the epic shift from mechanical to electronic machines. As he affirmed in the 1950s, "the high-speed computing machines of the present day follow very closely along the lines which I then suggested to Bush." Reflecting on his idea for optical computing decades before the advent of laser scanners and other everyday optical data-processing devices, Wiener said, "I was convinced that the scanning technique would prove socially more important in computing machines and their close relatives than in the television industry itself." On that count, too, Wiener said, "The future development of computing machines and control machines has, I believe, borne me out in this opinion."

————

By the late 1920s, the era of the lone scientist-inventor was coming to an end. The new American template of research and development was the Bell Laboratories model, in which a munificent corporation assembled the best minds from across the pure and applied sciences, lifted their teaching loads, turned them loose to pursue assigned projects and their own flights of fancy, then harvested the patent rights to their discoveries and inventions.

Wiener was not that kind of mind. He was not a company man, but he wasn't a loner either. Deep down, he longed to be a factor in some larger matrix of minds like the one he was part of at Aberdeen. During his first years at MIT, he was happy to carve out his humble corner of Building 2 and, in the process, help raise MIT's reputation to that of a top-tier research institution. But advancing MIT's organizational goals was not nearly enough purpose for an ex-prodigy haunted by high expectations and his own fierce competitive urges. His dilemma was not a simple one: How could he take his mathematics forward and make a greater contribution to science in an era that was becoming, increasingly, the province of large-scale institutional endeavors, and, at the same time, walk to the beat of his own drum?

In his thirties Wiener found a third way—his way. Along with his talents as a mathematician and trailblazer on the frontiers of the new electronic era, he began to evince an equal talent for finding smart collaborators who complemented his skills, and who opened new vistas where greater achievements became possible. That collaborative bent would become a hallmark of Wiener's mathematics, the communication sciences, and the scientific revolution his work would spawn.

His projects with Max Born and Vannevar Bush were two early, eminent examples. His collaboration with the Austrian mathematician Eberhard Hopf was another. In 1930, Hopf came to the United States to study celestial mechanics—the physics of the stars—at the Harvard Observatory. He met Wiener, and, soon after, they embarked on a separate effort to apply Wiener's new statistical mechanics to problems in modern astronomy that could not be solved by the old Newtonian methods. The "Wiener-Hopf" equations the two men published two years later calculated the physical signature of stars, the precise mix of atomic matter that makes up a star's mass and the radiating atomic energy that constitutes its sunlight.

Like Wiener's earlier work on Brownian motion, the Wiener-Hopf equations were a pioneering piece of statistics. They presented one of the first theories of prediction in which a statistical formula containing physical measures of past knowledge was used to predict future events, and they expanded Wiener's tool kit of statistical methods into realms of science where no man or method had gone before. A decade later, during the Second World War, those new statistical tools would be applied in ways Wiener embraced and in other ways he deplored: in the former, to advance his own war research projects and enable wartime weapons engineers to accurately predict future warplane positions for antiaircraft gun targeting; in the latter, to compute the "bursting size" of the first manmade star, the atomic bomb.

Wiener also began another important collaboration with one of his doctoral students, a nimble young engineer from China named Yuk Wing Lee. Wiener linked up with Lee in the late twenties with an ambitious mission in mind. After developing his statistical methods for analyzing message-carrying electronic signals, Wiener saw an opportunity to take the next step in the science of communication engineering: to devise a whole new approach to the haphazard business of electronic circuit design using his new statistical methods.

Wiener found a worthy accomplice in the mild-mannered and meticulous Lee. "His steadiness and judgment have furnished exactly the balance wheel I have needed," Wiener said with relief. To kick off their deliberations, Wiener gave Lee a rough sketch of his idea for a new kind of adjustable electronic circuit that would filter high-frequency electronic signals using applied principles of Wiener's generalized harmonic analysis, and pass only desired frequencies of a specified bandwidth through the system. Lee saw quickly that Wiener's idea as he had sketched it was novel and viable, but "at the cost of a great wastefulness of parts." He reworked Wiener's conception, using more versatile electronic components that could perform multiple functions simultaneously, and "reduced a great, sprawling, piece of apparatus into a well-designed, economical network."

The flexible network could be put to use in a wide range of practical applications, from channeling traffic through telephone systems, to refining and amplifying the weak currents of radio signals, to improving the quality of electronic sound recordings. Its design surpassed the makeshift methods used by earlier circuit designers, and Wiener's statistical theory ensured that it would produce the best circuitry of its kind that was physically attainable. The two innovators were so confident of their achievement that they hired a patent attorney, and a patent application was filed on September 2, 1931, with Wiener's name listed first and Lee's second. While they waded through the long patent process, Lee formalized Wiener's new design methods in his doctoral dissertation.

Amar Bose, another MIT engineer who studied with both men years later, relayed Lee's recollections of his work under Wiener that marked the beginning of modern network theory. "Now Lee wasn't a mathematician, and every time he tried to apply Wiener's theory it didn't work," said Bose, "so Lee went to Wiener and said, 'You know, something must be wrong with the theory.' It took him a year to understand what Wiener was saying, but he just kept at it and finally it came. When it did, the idea was so revolutionary that no one believed it."

Bose retold the events of Lee's electrifying dissertation defense. "When Lee made his doctoral presentation there were twenty faculty members in electrical engineering and all of them were present. They pounced all over him, they pounded Lee into the ground, all the big wheels at MIT, and Lee, this very cultured Chinese, just began to close up. Finally, Wiener, who was himself a young man at the time, got up and said, 'Gentlemen, I suggest you take this document home and study it and you will find it is correct.' Those were his words and that ended the examination."

Bose was appalled by the grudging acknowledgment MIT's professors gave to Lee, and indirectly to Wiener, when they rendered the verdict in Lee's trial. "Two weeks later, Lee got a little note in the mail. 'You passed.' They didn't have the dignity to say, 'Holy smokes, what a contribution!' That's all he got."

Lee's dissertation became a landmark in the field of electrical engineering and established Wiener's new communication theories and mathematics as the heirs to Heaviside's haphazard circuit-design methods. The two men would press ahead with their invention and their collaboration and, between them, put a broad foundation of theory and practical methods under the fledgling science of communication engineering.

———

The stock market crash of October 1929 and the Great Depression that descended in its wake crushed millions, but it had little impact on the already frugal Wieners. That year Wiener was promoted from assistant to associate professor. The small pay raise he received put him over the $5,000 mark in annual income, enough for his little family's modest lifestyle but still well below the pay scale of professors at Harvard and other Ivy League universities.

His own burst of creativity had propelled him to the fore of the international mathematics scene. His reputation was growing as a prolific mind with whom leading mathematicians, physicists, and engineers were eager to collaborate. But still Wiener felt that his American peers did not appreciate, or even comprehend, his work with the enthusiasm of his European and Asian colleagues. His closest friend in America was a Dutchman. His business partner was Chinese. And his biggest fan was the Russian mathematician J. D. Tamarkin, whom Wiener had met at Göttingen. Tamarkin fled Russia and settled at Brown University in Providence, Rhode Island. In 1929, he invited Wiener to lecture at Brown as an exchange professor, and he began to promote Wiener's work enthusiastically.

Wiener's other catalyst in America was his mentor from Trinity, G. H. Hardy, who toured the United States and touted his former student's work at every opportunity. Through the efforts of Tamarkin and Hardy, Wiener believed, "I began to be heard of in this country," but that recognition was still muted. Even politically isolated mathematicians in the strife-torn Soviet Union acknowledged Wiener's groundbreaking mathematics before his American colleagues did. Wiener agonized over those academic and geographic disparities. He felt confident that the more mature and congenial personality Margaret had helped him to discover had also "made it possible to allay some part of the hostility with which I had been received in mathematical circles" in America, and he began exploring his options. He applied for open chairs in mathematics at Kings College in London and the University of Melbourne in Australia, but did not get them.

Those rejections did not help his confidence, but the mere act of applying sent a shudder down Tech's corridor. "Another university is apparently quite desirous of your services," MIT's new President Karl Compton wrote mistakenly to Wiener in December 1930. "We are even more desirous of having you remain here." Compton's personal attention and his enlightened new regime, with its vocal commitment to elevating science and pure research at Tech to top-drawer status, were encouraging to Wiener but not appeasing. Two months after he received Compton's letter, Wiener applied for a job at Princeton's illustrious new Institute for Advanced Study, but again he was passed over by his old wartime commander Oswald Veblen, the IAS's founding director.

The men who mattered in American science, those who knew Wiener in his fumbling days and who were still put off by his abrasive ways, could not perceive him as a rising star of the first magnitude. Longing for the wider and more welcoming world of international science, Wiener packed up his family in the fall of 1931 and sailed for England, where he had arranged to spend the year as a visiting scholar at his old sanctum, Trinity College.

Wiener never earned a degree during his two terms there as a traveling scholar, and he felt obliged to "wear the arms of Trinity College with the difference of a bar sinister"—on his left side—as a sign that he was "a Cambridge man only on the left hand." But the university embraced him with both arms, and he resumed its centuries-old customs with newfound ease. He dined weekly at Trinity's high table and often joined the official dons in a game of bowls in the fellows' garden by the Backs. He saw his champion Hardy frequently, and they continued their amicable sparring over the virtues of pure and applied mathematics. Hardy held his former student in high regard, and he arranged for Wiener to publish his first book-length work, a treatise on Fourier analysis and its practical applications, under the regal imprint of the Cambridge University Press.

By 1931, the famed Tea Party of Trinity was over. Russell was in exile from Cambridge and would not return until 1944. For stimulation, Wiener repaired to the old library of the Cambridge Philosophical Society, where his first published papers were gathering dust with the ancients on the shelves. There he had a chance meeting that would flower into a cherished friendship with another fine mind and one of the freest scientific spirits of the day.

Wiener was already an avid reader of science fiction and detective stories when he read a sci-fi story by the English biologist J. B. S. Haldane in a popular magazine and saw, on the cover, a photograph of the "tall, powerfully built, beetle-browed man" he had observed frequently in the Philosophical Society library. The next time he spied Haldane in the stacks, Wiener introduced himself.

He praised Haldane's story but picked one nit in the piece, pointing out that Haldane "had used a Danish name for a character supposed to be an Icelander." From that day on, they became the best of friends.

"I have never met a man with better conversation or more varied knowledge than J. B. S. Haldane," said Wiener. Haldane, like Wiener, possessed an encyclopedic knowledge of science, history, and literature. He was an eclectic, activist scientist who experimented on himself and took controversial stands in the British press on topics ranging from the proper use of chemical weapons to the prospects for test-tube babies. At the time, he was also well on his way to becoming one of England's most ardent and articulate Communists.

For Wiener, whose father was a "wild, long-haired Russian" utopian in his youth, Haldane's politics were unremarkable. He and Haldane shared important traits. Wiener was a mathematician with a lifelong love of biology. Haldane was a biologist with a passion for mathematics, and Haldane like Wiener was hamfisted in the laboratory, a plight that turned him from manipulating the limbs of fruit flies to probing the probabilities of genetics using modern statistical methods. And Haldane was as much an outsider at Cambridge as Wiener was at MIT, known for "discoursing loudly at table on all sorts of unmentionable subjects." Wiener's work in probability theory was of immediate concern to Haldane, and Haldane's science revived Wiener's interest in biology, which had been inactive since his father forced him into philosophy. The two maverick Trinity men never collaborated on any project, but they exchanged ideas copiously during Wiener's year at Cambridge and for three decades afterwards, in frequent letters and get-togethers in England and America.

During that academic year 1931–32, Wiener took many vigorous walks over England's green hills. On one outing he nearly walked himself to death. After a long tramp through the Lake District in the damp of April, he came down with a virulent case of scarlet fever, which ravaged him in his weakened condition and put him in the Cambridge Hospital for Contagious Diseases. He stayed there several weeks and spent several months convalescing. By the time he recovered, the term was over and summer was in full swing.

He did not venture out on any long tramps for a while, but he and Haldane went swimming together in a tranquil stretch of the Cam that flowed past Haldane's lawn. Among his many signature moves, Haldane smoked his pipe while swimming. "Following his example, I smoked a cigar and, as has always been my habit, wore my glasses," Wiener recalled, confessing to another eccentricity he acquired at Cambridge. "We must have appeared to boaters on the river like a couple of great water animals, a long and a short walrus, let us say, bobbing up and down in the stream." Wiener, the protean Elephant Child, would propel himself around many ponds and pools to come, puffing on his stubby smokestack of a cigar, his spectacles shimmering on the swells.

———

Wiener and his family passed their holidays on the Continent, where Wiener had received numerous invitations from European colleagues to lecture on his work. He made an ambitious tour over the winter break and visited prominent mathematicians in Germany who would soon join in the great migration of European scientists fleeing the Nazis' persecutions. Those nationalist forces that had dogged Wiener at Göttingen six years before were building toward a triumphal seizure of power and an avenging sweep across Europe. For Wiener, the wise Old World he loved, with its rich scholarly and scientific traditions, seemed to be crumbling beneath his feet. But Margaret's time in Germany was a homecoming. From Berlin, she took the girls to Breslau to meet her older sister and other relatives she had not seen in years, while Wiener continued on to Austria before reuniting with his family in Prague.

Late in the summer of '32, after a last stopover on the Continent to attend the International Mathematical Congress in Zurich, Wiener and his family steamed home to Boston. Upon their return, Margaret set about securing the trappings for "a reasonably routine life." She found a new home for the family in Belmont, a stylish suburb just west of Cambridge with good streetcar connections to MIT. Belmont was a big step up for working professionals with growing families, and many Harvard and MIT professors were making their way there, although on Wiener's salary he could only afford a rented house on a quiet street in a more modest part of town.

Suddenly, the world around them was no longer looking so upwardly mobile. The family had returned to a country mired in depression, and Wiener's spirits were sinking along with the national mood. With Europe in danger, the persisting snubs on his home turf hurt and frustrated him more deeply than ever. They fired old feelings from his prodigy days and dark memories of his early adult traumas. So much pain and confusion, now inextricably intertwined with his professional achievements and pride in his marriage and fatherhood, set Wiener's head spinning and his emotions swinging widely.

Dirk Struik was mindful of Wiener's oscillations. Early on, Struik saw Wiener as a man of "extreme sensitivity," but he ascribed that trait as much to Wiener's strengths as to his weaknesses. "A man with heart and mind so close to nature and the technique of his time must have had very fine antennae," Struik maintained, but those sensitive receptors also made Wiener "a man of moods," as he put it ever so gently. "He could be terribly depressed by a real or a supposed want of recognition. At other times, he could be exhilarated to a high pitch by a new idea, an act of recognition, a success—not only as achieved by himself, but by a pupil, a friend, a colleague."

By the early 1930s, Wiener's patterns of alternating ebullience and depression were beginning to cycle with greater frequency. In the fall of 1932, a young mathematics star from Trinity, Raymond E. A. C. Paley, came to MIT to resume work they had begun at Cambridge, but Paley's sudden death in a skiing accident in April 1933 threw Wiener into a prolonged depression. "It took me some time to come back to a mental equilibrium sufficient to permit my further work and my proper attention to the environment about me," he wrote later, in the euphemistic way he acknowledged his worst depressions.

When he regained his equilibrium, Wiener completed their joint work by himself and published it the following year. He generously credited Paley as the senior author of his second book, *Fourier Transforms in the Complex Domain*, and of his new equations for the "Paley-Wiener criteria for a realizable filter," which, like the Wiener-Hopf equations before them, would become standard tools for communication engineers over the next two decades.

During those first years of the Great Depression, Wiener finally began to win some recognition from his American colleagues. In 1932, he was made a full professor at MIT. In 1933, he was elected to the elite National Academy of Sciences. Also that year, he was awarded the American Mathematical Society's Bôcher prize—named after his old neighbor on Avon Street—for the best work in mathematical analysis in the United States.

He had a secure job, a devoted wife, two growing children—even a place in the country. It seemed he had beaten the odds stacked against ex-prodigies. Or had he? Singly and in collaboration, during his years of growth and progress, Wiener had combined the theoretical strengths of European science with the American knack for applied knowledge and practical invention. He had conceived a rigorous new theory and mathematics for communication engineering, contributed to the creation of the first modern computers, and designed sophisticated electronic networks and filters for refining the signals of America's burgeoning mass communication culture. But in that deep place where his insecurities dwelled, he knew that his achievements were piecemeal, and that even his most impressive accomplishments were unacknowledged and still confined largely to paper. Given the persisting slights and resistance to his work by many of his American peers, it remained to be seen if he would, or could, complete his omnibus journey and transform a field of knowledge fundamentally and irreversibly.

5

Wienerwalks

I must get into the open air; I roam about the fields. To climb a
steep mountain is then my joy, working my way through pathless for-
est. . . . Then I feel some relief. Some!

—J. W. von Goethe

The Sorrows of Young Werther

H E WAS A MAN IN near perpetual motion. Inquisitive, gregarious, garrulous,
Wiener made a habit of briskly walking MIT's maze inside and out. By the mid-
1930s, the entire campus had adapted to the daily spectacle of the bespectacled
Wiener waddling along the university's byways and beaten paths, waving his
ever-present cigar, expounding in his booming voice on the most near and farfetched topics, his head thrown back like a trumpeting elephant, one chronicler
reported, "as he tossed peanuts into the air and caught them in his mouth."

In time, those meanderings became known among Tech's denizens as Norbert's *Wienerwegs* (from German, *weg*, walk, wandering). Wiener's friend Dirk
Struik called them "safaris." "At MIT he used to walk all the corridors and but-
tonhole everybody he saw, colleagues, students sometimes, and tell them about
his latest ideas. I got buttonholed often. Sometimes he would spout the most
complete nonsense. At other times he would be almost prophetic."

And peripatetic. Often on his wanderings, Wiener was so immersed in his internal musings he seemed oblivious to the world around him. One of the most
famous anecdotes in the Wiener fabulary happened one afternoon in the 1930s.
While crossing the courtyard between MIT's long corridor and Walker Memorial, the festooned sarcophagus that served as the main dining hall and social
complex for the campus, Wiener accosted Ivan A. Getting, a first-year physics
student he had met previously in an electronics laboratory. According to Get-

ting, "He stopped me halfway, we happened to be going in opposite directions, and he raised some question he wanted to discuss. When we finished talking he started to walk away and then he turned around suddenly, came back and asked, 'By the way, which way was I headed when we met?' I said, 'You were going toward Building 8.' And he said, 'Thanks, that means I've already had my lunch.'"

Sometimes he seemed wholly unaware of the person to whom he was talking. Sometimes he asked earnestly to be reintroduced to people he had already met. Invariably, he barged into his colleagues' offices without bothering to knock. During a conversation in the corridor with another mathematician, "Wiener needed to write, so he walked right into the nearest office and proceeded to use the blackboard, while the occupant, a physics professor, looked on incredulously." Wiener's tireless *wegs* were the vehicle of his mind and a vent for his agitations. The incessant talk was his lubricant and liniment, the boil-off of his ideas on mathematics, electronics, and ancillary topics.

At MIT, Wiener also went about his other business of educating the next generation of American engineers, and, this, too, he did in his inimitable manner. He was an inconsistent lecturer and sometimes a "famously bad" one. Steve J. Heims, a physicist who wrote two scholarly books about Wiener and his contemporaries, offered this description of Wiener's teaching style:

> In class, while presumably deriving a theorem on the blackboard, Wiener in his intuitive way . . . skips over so many steps that by the time he arrives at the result and writes it down on the board, it is impossible for the students to follow the proof. One frustrated student . . . tactfully asks Wiener if he might show the class still another proof. . . . Wiener cheerfully indicates, "Yes, of course," and proceeds to work out another proof, but again in his head. After a few minutes of silence he merely places a check after the answer on the blackboard, leaving the class no wiser.

"He could range among the worst and among the best," Dirk Struik observed. "It all depended on whether he chose to put his whole soul into the task. . . . He could lull his audience to sleep, but I have also seen him holding groups . . . at breathless attention, while he manipulated his ideas in flares of vision."

In performance, Wiener wielded the mystic symbology of mathematics with blazing speed and surprising dexterity. MIT legends depicted him scribbling away furiously and ambidextrously on math department blackboards, solving simultaneous sets of complex equations—one with each hand. Former students attested to his mathematical prowess and fecundity as he reeled out five-foot-long equations without notes. His memory for mathematics, like his cache of Greek classics, was immense, the legacy of his eyestrained childhood years when he was forbidden to read for months and his powers of storage and recall exploded.

Yet his steel trap could become a sieve on matters of professorial decorum. Often Wiener's students had to hunt him down in the corridor, where he had struck up a conversation with someone on his way to class and lost all track of time. Accounts of his absentmindedness rivaled those of his genius. While lecturing, he might be seen picking his nose "energetically," with no apparent concern for his lack of social grace. On at least one occasion, he strode into the wrong classroom and delivered a rousing lecture to an audience of baffled students. Once he walked into a packed lecture hall (the right one this time), wrote a large "4" on the blackboard and walked out. Only later did his students figure out that he was leaving town for four weeks.

In recompense for those transgressions, and all that had come before, Wiener ensured that his classes were cakewalks. After his own travails as a battered pupil of his father and Bertrand Russell, he doled out grades like war reparations for those who toughed out his classes. One observer of the MIT scene confirmed, "Every student in his class got an A. There was no such thing as a bell curve."

His kindnesses as a mentor did not stop there. In the fall of 1933, only months after the death of Wiener's young collaborator Paley, Norman Levinson, an engineering undergraduate, enrolled in one of Wiener's math classes. Levinson found Wiener to be "a most stimulating teacher," and Wiener found Levinson to be a most promising student. "As soon as I displayed a slight comprehension of what he was doing," Levinson recalled, "he handed me the manuscript of Paley-Wiener for revision. I found a gap in a proof and . . . set it right. Wiener thereupon sat down at his typewriter, typed my [proof], affixed my name and sent it off to a journal. A prominent professor does not often act as secretary for a young student."

By the spring, Wiener had convinced Levinson to change his major from electrical engineering to mathematics. He also arranged for Levinson to spend the next academic year in England studying higher mathematics, as he had, with G. H. Hardy at Cambridge. Levinson's widow, Fagi, remembered Wiener's benevolence to Levinson and, equally, to Levinson's parents.

"Norman came from a very poor family of Russian-Jewish immigrants who had come to the United States in steerage, and when they heard that their son was going back to Europe, they were just in a panic. So Norbert Wiener went to the little working class slum where Norman's parents lived to reassure them. The whole time Norman was in England, Norbert came to their house, usually on Saturdays, and he would talk to them, not about his theorems, but about nice, practical things, about England, landladies, tea time, high table."

Levinson returned from England, earned his Ph.D., and accepted a teaching position at MIT. In the years that followed, he would become a star in his own right in Tech's math department, and an expert explicator of Wiener's mathematical ideas and their engineering applications.

Wiener's *wegs* took him beyond MIT into wider academic circles. In the thirties, he even teamed up with his old rival, G. D. Birkhoff, to co-chair the joint Harvard-MIT Mathematics Colloquium. Yet, for all his intellectual energy, still he tired quickly. Members of the group recalled that, the moment Wiener sat down in their seminars, "He would lean back and start to snore." One regular attendee witnessed Wiener's uncanny ability to perceive lucidly while he dozed. "He always sat in the front row on the extreme right, and he always brought a magazine, usually *Life*," he recalled. "He alternated between seeming to fall asleep and, then, every once in a while he'd wake up and turn the page. But this was real Norbert, because he frequently made some comment which showed that he had not been asleep at all but had been listening intently to what was going on. Actually, he could fall asleep *and* talk!"

Another encounter at Harvard led Wiener to the man who would become his most important partner in science. In 1933, he met Dr. Arturo Rosenblueth, a neurophysiologist from Mexico and professor at Harvard Medical School, who hosted a monthly dinner seminar on scientific matters for Harvard med students and a select group of outside participants. Rosenblueth was the protégé and right-hand man of Harvard's trailblazing biologist Walter Cannon, who had inspired Wiener's childhood goal to become a naturalist equipped with modern investigative tools. In Rosenblueth Wiener found a *compadre* and a scientific soulmate. The two shared a deep interest in scientific methodology and a conviction that the traditional distinctions between the sciences were merely administrative conveniences that scientists should be free to ignore when their investigations required expertise from other fields. Both men viewed science as a collaborative venture.

Wiener became a regular attendee at Rosenblueth's scientific dinners, excited by the prospect of staking a new claim on his forsaken turf of biology, "a field into which few real mathematicians had strayed before him." His formidable presence quickly turned Rosenblueth's seminars into a multidisciplinary enterprise with Wiener's distinctive imprint. "Ultimately the theme of the many discussions which Rosenblueth and I had privately and in our seminar came to be the application of mathematics, and in particular of communication theory, to physiological method," Wiener chronicled, with a wishful prediction. "We laid out a policy of joint effort in these fields for a future when we might work closer together."

During those dinner seminars, Wiener and Rosenblueth embarked on an auspicious journey into the "no-man's land between the various established fields." They were among the first American adherents of a new interdisciplinary movement that was taking root worldwide in the early 1930s, and, before long, the two men would begin to apply their findings from the life sciences to tangible problems of modern technology.

———

Wiener stood by his other students and collaborators while he cultivated his new friendships and working relations. He tried to help his young Chinese colleague Yuk Wing Lee find a good job in the electronics industry, but in those early days of American electronics, the industry, like the rest of American society, had not yet opened its doors to Asians. The "sales resistance," as Wiener termed it, "was more than we could overcome," and Lee returned to China to seek a position in industry or academics.

As Lee took leave across the Pacific, the first tremors of the coming cataclysm in Europe were rippling across the Atlantic. In 1933, the Nazis took power in Germany and Jewish scholars were dismissed *en masse* from German universities. As the turmoil in Europe intensified, the mood in America grew adamantly isolationist, but American scientists and universities bucked the trend and opened their arms to their threatened colleagues. By the mid-1930s, a steady stream of émigrés from Germany and across Europe, both Jews and gentiles, was infusing American science and mathematics. Einstein left Germany late in 1932 and took up residence at Princeton's prestigious new Institute for Advanced Study. The young math prodigy John von Neumann from Göttingen left Germany in 1933 and joined Einstein in residence at the institute. Kurt Gödel came from Vienna in '34 to lecture at Princeton and he, too, would move to the institute a few years later.

Wiener was actively involved in the massive relocation effort. He became a high-profile member of the Emergency Committee in the Aid of Displaced German Scholars, which fought to loosen restrictive immigration policies. He knew the work of American scientists "was cut out for us and that we should have to get together and make a systematic effort to find jobs and a possibility of life for many a displaced scholar . . . some of whom went through my hands." He helped many prominent and lesser-known European mathematicians find new homes in American universities, including the Hungarian analysts George Pólya and Gabor Szegö, whom he situated at Stanford; the German number theorist Hans Rademacher, who went first to Swarthmore and, later, to the University of Pennsylvania; and Gödel's mentor Karl Menger, the Viennese master of probabilistic geometry, whom Wiener helped to place at Notre Dame.

By 1935, Wiener was drawing offers for his own services from another part of the world. He accepted two invitations for visiting lectureships in the Far East: the first, a joint invitation from Japan's venerable universities in Tokyo and Osaka, and the second from China's National Tsing Hua University in Beijing. The appointments in Japan were arranged by Shikao Ikehara, another of Wiener's gifted

Asian doctoral students who, like Lee, ran into "desperate straits" in the American job market and returned to his homeland. Wiener's appointment in Beijing for the 1935–36 academic year was tendered by Chinese authorities at the suggestion of Lee himself, who was by then a professor at the national university.

With so many European trips behind him, the prospect of a great tramp to Asia did wonders for Wiener's wanderlust and fed his curiosity to see another hemisphere of cultures. The trip also appealed to the universal humanism Wiener had embraced since childhood and which, he maintained, his wife shared with him harmoniously:

> I have never felt the advantage of European culture over any of the great cultures of the Orient as anything more than a temporary episode in history, and I was eager to see these extra-European countries with my own eyes and to observe their modes of life and thought by direct inspection. In this I was thoroughly seconded by my wife, to whom national and racial prejudice have always been as foreign as they have been to me.

Wiener began the summer of '35 in New Hampshire, prepping for the long sea voyage in the placid waters of Bear Camp Pond, while Margaret put the family's affairs in order. In July, Wiener and Margaret and their girls, now ages seven and five, traveled by train to San Francisco, where they boarded a ship for the Pacific crossing. At Yokohama, Wiener and his entourage were met by Ikehara, who escorted them to Tokyo and then on to Osaka. Wiener's interactions with Japanese mathematicians were cordial, although he found the Tokyo professors a bit snobbish and "affected by the rigidity that so often taints a university which is sure of its position as the first in the land." He preferred the friendlier, more progressive minds he met at the university in Osaka, where Ikehara was camped, and which would bring forth Shizuo Kakutani, Kosaku Yosida, and other world-class mathematicians.

However, on his first odyssey in the Far East, Wiener found the atmosphere oppressive—as bad in this ancient world as it was becoming in Europe. Japan's leaders, in the quest to enlarge their empire, were stoking a new militarism and culture-wide xenophobia. All foreigners were put under surveillance. Every waiter and shopkeeper was a potential spy for the imperial authorities, poised "to report our every word and attitude to the management and eventually to the police." Under watchful eyes and sweltering summer heat, Wiener and his family made the best of their time in Japan, and even managed to have a good time touring some of the country's cultural landmarks, but they were eager to reach the mainland.

They sailed for several days through winding westward passages and, early in August 1935, they docked at Beijing's port city of Tangku. Lee was waiting on the dock to greet them. He guided them across China's sprawling capital to the South Compound, a development of modern bungalows near the university. There they met Lee's wife, Betty, who, as Wiener was surprised to learn, was not Chinese but a slender, attractive Canadian whom Lee had met in America and brought to China when he found a teaching position. The Lees found "a tall, dignified, elderly gentlemen in a long gown" who knew enough English to tutor the family in Chinese. He appeared daily, like an apparition, with his thin white beard and long blue coat. After three weeks, Wiener began to get the hang of the language. He even learned to steer his rickshaw boy through Beijing's tangled streets, in the local custom, by calling out the directions by the points of the compass.

Wiener was enchanted by the local architecture and the luxurious grounds of Beijing's imperial palaces, but he was more impressed with the Chinese people themselves. He made friends easily among Chinese in all the city's social strata, from his household servants to his colleagues at the university, and he felt the current of his own universal humanism running through the populace in the deep values they had derived from their millennia-long devotion to Buddhism. "There was common to almost all that love of the whole world rather than of any specific humanity, which is so characteristic of Buddhism," he observed.

Wiener quickly reached a point where he could confidently lecture in Chinese, but he took no chances with his mathematics and delivered most of those lectures in English. In his off hours as a visiting professor, he sipped tea with his students and tried without success to master the intricate game of Go, a 3,000-year-old Chinese board game which was then all the rage among the in crowd of American mathematicians.

Wiener's focus was not on board games but on electronic circuits, and he and Lee picked up their joint venture to build a new kind of "electrical network system" where they had left off at MIT four years earlier. In September 1935, the two men received word in Beijing that AT&T had offered to purchase the patent rights to their invention for the modest sum of $5,000. They accepted the terms, knowing that, if AT&T were to use their design in the company's devices, the royalties would be substantial. When their first patent application was approved, they filed two more patents that spelled out in detail "the great advantage of this type of filter over all known types" and proudly declared their invention "available for purposes of mass production."

But Wiener and Lee's dream of commercial success was not to become a reality. As the two inventors learned later, AT&T's Bell Laboratories division had already patented a similar device, and the company had only bought the rights to the Wiener-Lee patents to prevent others from capitalizing on Wiener and Lee's innovations. "All this effort we had made went into a paper patent . . . which the Bell people never intended to use but simply to hold *in terrorem* against competitors," said Wiener, crying foul. His three patents with Lee never earned another penny, but they earned AT&T Wiener's lifelong antipathy.

Undaunted, Wiener and Lee were already at work on another ambitious circuit. While their wives got acquainted in the Lees' salon, the two inventors went to work in the den to design an all-electronic analog computer. Their goal was to build an entirely new kind of machine that would go beyond the hulking electromechanical contraption on which Wiener had consulted with Vannevar Bush, and beyond the abortive optical computing machine Wiener himself had conceived at MIT a decade earlier. Now Wiener and Lee conceived a machine to perform complex computations, including the dreaded partial differential equations of Wiener's generalized harmonic analysis, that was realizable, in principle at least, using existing electronic technology. But early into the project Wiener and Lee hit a snag. They could not master the design of a specialized circuit that was needed to carry out the repeating calculations of Wiener's harmonic formulas, in which the result of one computation was fed back into the circuit as the starting point for the next.

In 1935, Wiener's irksome "feedback" problem exceeded Lee's know-how technically and Wiener's own grasp conceptually. Years later, he reflected on the problem he and Lee failed to solve in China:

> What was lacking in our work was a thorough understanding of the problems of designing an apparatus in which part of the output motion is fed back again to the beginning of the process as a new input. . . . What I should have done was to attack the problem from the beginning and develop on my own initiative a fairly comprehensive theory of feedback mechanisms. I did not do this at the time, and failure was the consequence.

Those technical setbacks did not diminish Wiener's zest for adventure in China. The family traveled by rickshaw into central Beijing, where Margaret and the girls shopped for souvenirs in the bazaars and Wiener went on his *wegs* through the city's "mixture of glamor and squalor." As he recalled, "It was intriguing to walk down ill-paved alleys which seemed to lead from one slum to another, but where the vermilion moon-gates often opened into a little gem of a courtyard and garden surrounded by pavilions of taste and beauty." In the spring, the Wieners and the Lees toured the Chinese countryside and made a pilgrimage to the Great Wall.

Early in the summer of '36, the Japanese army began its incursions on the Chinese mainland. The country was plummeting into political turmoil and, for safety's sake alone, it was time to move on. Wiener said goodbye to Lee, not knowing when, or if, he would see him again, and the family departed hastily. They sailed a roundabout route south and westward through the South China

Sea and the Indian Ocean, north through the Suez Canal (with a quick stopover for a tour of Egypt's Great Pyramids), and on into the Mediterranean. At Marseille, Wiener jumped ship and hopped a night train to Paris to meet with the Polish-born mathematician Szolem Mandelbrojt, with whom he had arranged to make a joint presentation at the upcoming International Mathematical Congress in Oslo. Margaret went on to England with the girls, where she found a children's boarding camp that would keep them for a month, then she rejoined Wiener for the trip to Norway.

In Oslo, the couple dined luxuriously at the congress's tables and Wiener discoursed with his European peers and American colleagues he had not seen in a year. He and Mandelbrojt delivered their paper, and Wiener capped off his talks with old friends on walks in the midnight sun. When the conclave ended, he and Margaret went their separate ways and into two vastly different experiences. Margaret stole away for a few weeks to visit her relatives in Germany, where the new Nazi order was imposing its pageantry on the 1936 Olympic games in Berlin. Wiener, well aware of the worsening plight of Jews in the new Germany, returned to England for a respite with the Haldanes at their country home in Wiltshire.

Late in August, the couple reunited, retrieved their daughters, and the family sailed home to Boston. Ironically, Wiener's first walk around the world, through the poisonous political climates brewing in Asia and Europe, was filled with many happy memories and poignant moments. Years later, he reflected on the closing of that great circuit as a turning point in his family life and his career:

> Margaret and I now had a large stock of common experiences to enjoy together. My children were more than babies and had begun to be companions. . . . I had begun to see the fruition of my work . . . as a body of learning which could no longer be ignored. If I were to take any specific boundary point in my career as a journeyman in science and as in some degree an independent master of the craft, I should pick out 1935, the year of my China trip, as that point.

———

On their return from China, the family again was in need of lodgings, and Margaret found a roomy double-decker apartment on the top floors of a two-family house on Oakley Road in Belmont, in the nicer section of town. The family occupied the second floor. Margaret's mother, Hedwig Engemann, moved in with them and took the attic to herself.

And Wiener resumed his meanderings around MIT and its environs. As he approached middle age, his worsening eyesight made just finding his way a problem, and some of his eccentric doings came down to the physical fact that

Wiener often could not see clearly where he was going. To be sure, no one had trouble seeing Wiener. His bulbous form and ducklike waddle made him easily recognizable from one end of MIT's corridor to the other and far across its grassy paths. He was a sizeable presence at faculty picnics, where he was a menace at horseshoes. He was a good sport, but his bulk and blindness made him a frustrating opponent.

Ivan Getting recalled a second encounter with Wiener during his undergraduate days that turned into a withering ordeal. "He stopped me in a hall and asked me if I played tennis, and I said yes, so he asked if he could play tennis with me. I said, 'Sure.' We met on the court next to Walker Memorial. He had a tennis racket, I had a tennis racket. He was on one side of the net, I was on the other. I would lob a ball over to him gently, and about three or four seconds after it passed him he would swing his tennis racket. We soon ran out of balls. They were all behind him. Some students who had stopped to watch this miracle came into the court and threw the balls back to me, and I would hit them again, three balls, six balls." After swatting the air in vain as a hundred serves passed him by, Wiener approached the net and asked Getting, "Why don't we exchange rackets?"

Wiener often could be found holding court in the MIT pool, expounding on whatever topic was preoccupying him at the moment, tirelessly swimming his adipose breast-stroke in the buff (no bathing suits were required during those men-only sessions), a thin line of smoke trailing from the lit cigar in his mouth. He was undaunted by bad weather. During a brutal New England blizzard in the 1930s, the clumsy, absentminded professor, who presumably didn't know which way he was walking, was the only commuter who showed up on the snowbound MIT campus—after trudging seven miles on snowshoes from his home in Belmont.

His *Wienerwegs* fanned out across greater Cambridge, through suburban Boston, to farm towns near the family's country retreat in New Hampshire. In every venue, Wiener's store of knowledge was open for business. On his walks to the local markets in Belmont, he addressed the town's ethnic Greek and Italian grocers in the historical tongues of their homelands. On his wanderings around town with his young daughters, Wiener's passion for the classics became an issue among the local clergy. "He liked to recite Greek and Latin at the top of his voice," Barbara recalled. "We would go past St. Joseph's, and one day one of the priests came out and told me that every time my father came by St. Joseph's shouting in Latin the congregation stopped following him and began following my father. He asked me if I could stop him, but I couldn't do anything to make him stop."

Wiener's wanderlust and need to communicate crossed all borders and time zones. On quiet Sunday mornings in Belmont, he often walked unannounced into his neighbors' homes and waited for them to wake up and indulge him. "He'd come around . . . when nobody was dressed and we hadn't finished breakfast," one MIT colleague who lived nearby remembered. "He was patient. He'd sit in the living room and muse to himself. . . . Then he would talk about everything, world events, things that were bothering him about the government and . . . about the trends in science. His insights were always fascinating."

When Wiener was in his low states he could be downright exhausting. Dirk Struik lived only two streets away from him in Belmont. In the thirties, Wiener would walk into his neighbor's home and talk for hours, "unburdening himself" to Struik and his wife, Ruth, "until sometimes it became too much and they sent him home." The two men were best friends, but, as Struik attested, "He was a friend who could get on your nerves."

Wiener was happiest wandering the hills and valleys of New Hampshire, and he often led the children of the Sandwich valley on educational tramps over the mountain trails. On most summer days, he walked from his home down the dirt road for a swim and a cigar in Bear Camp Pond, where he would glide the quarter mile to the pond's tiny island and back, belly up to the breeze.

The summer after the family returned from China, Wiener made a few improvements on the property in South Tamworth. He hired a local contractor to tear down an old barn beside the house and replace it with a new two-car garage, which was joined to the house by a connecting wing with a proper kitchen and an oil stove. Wiener made no attempt to help with the construction, and with good reason; neighbors showed around a photograph someone took of him standing forlorn on the porch, brush in hand—painted into a corner.

Wiener's wanderings were fodder for his legend and countless anecdotes. After delivering a lecture at Brown University, Wiener caught the train from Providence and arrived on time at Boston's South Street Station. He called Margaret to come pick him up.

"But you *drove* down!" his nonplussed *frau-professor* said.

Another incident gave rise to the most famous tale of Wiener's *wegs:* On his way home from MIT, Wiener, lost in thought, stopped to ask a young child for directions, not realizing it was his own daughter. The tale had been retold and embellished endlessly as the last word on Wiener's absent mind, but it was only half true. The family had recently moved again, from the Oakley Road duplex to a new house on Cedar Road a few blocks away. Wiener went to the old address by force of habit. Barbara corroborated that part of the story. "When we moved to the Cedar Road house he came home from MIT to the Oakley Road house. He didn't remember that we had moved." But Peggy debunked the punchline. She said, "It is entirely possible he was going to the old house without thinking. It is also entirely possible that Mother would have sent one of us out to find him. But it is totally *im*possible that he wouldn't have recognized either of us."

By the late-1930s, however, more serious matters were pressing on Wiener and his family. First, there was the question of the girls' education. Years earlier, when Wiener was a "bar-sinister" don at Cambridge, a teacher in the English nursery school Barbara attended discovered that she was already reading on her own, like her father, at age three. When she informed Barbara's parents of her discovery, "the news sent shock waves through my family," Barbara recalled. "There was much talk about, 'Is this child precocious or not?' and, 'If she is precocious, what kind of education do we give her?' My parents concluded that I was not precocious. I don't think either of them wanted me to be, because that raised more problems for them than it solved." Peggy, too, showed signs of early ability. When she started school the year the family was in China, she was already a proficient reader with absorbent powers of learning. "My report card from the Yenching School said, 'Peggy soaks up knowledge the way a sponge does water,'" she remembered.

His daughters' early aptitudes reignited Wiener's memories of his own childhood. Unlike his own father, Wiener made no effort to nurture his little girls' precocity. He even publicly denied that his daughters merited a special education. He stated explicitly the terms on which he would impose on any youth the intensive training he had received. "Before I should even think of subjecting any child, boy or girl, to such a training I should have to be convinced not only of the intelligence of the child, but of its physical, mental, and moral stamina," he wrote, adding bluntly, "With my own children, indications of the need for such a highly specialized procedure have not occurred."

Barbara was hurt and angered by such statements. "I find my father's often-repeated claim that he was sparing Peggy and me by not educating us as he was educated a specious, self-justifying fabrication," she said. "He could not bring himself to even approach the subject of our education, in part, because he didn't have the energy left over after his work, and because he lived with the fear that he was his father's creation. His thoughts and feelings on his own education were in constant turmoil, and he resolved the conflict by concluding that we were not really of that caliber." Peggy bore no resentment of Wiener's child-rearing practices and pronouncements, but she agreed that Wiener was "desperate not to have happen to us what happened to him."

———

Other tensions began to surface. Side by side with his professional triumphs as a mathematician and his growing reputation on the stage of international science, Wiener's deepening cycles of ebullience and depression were becoming a cause of concern and whispering around MIT. However, the main outlet for his waves of sadness and resurgent insecurity was not in public but in private. Long after the fact, Barbara recalled Wiener's turbulent emotional states and offered an intimate look at life inside the Wiener family home.

"My father was never one person. He was many people serially and they were very contradictory," Wiener's pure universal remembered. "He was a man of dazzling intellectual creativity, but there were times when he was a dark and remote figure, tossed about in intense emotions, subject to precipitous depressions, and wide swings from elation to exhaustion and back again." That dark side of Wiener was "like the elephant in the living room," she said. "Everyone knew it was there but nobody talked about it."

From a tender age, Barbara witnessed her father's combustible states at close range. She recalled his "sudden explosions" around the house, and more severe responses which she described as "emotional storms." Wiener's emotional life during those years revolved around his professional frustrations, which he felt deeply despite his successes. "The world was against him, his colleagues were betraying him, he was going to resign his professorship or resign from this or that position," said Barbara.

But often Wiener's storms were too capricious to forecast, erupting into chaos from harmless flutterings in the breeze. "Sometimes it got very dangerous to talk about poetry around my father. He would start to recite some poem and then he would start ranting and then he would start crying and he would be in a total state. He was unable to control his emotions. They just carried him off." Each poet, language, and culture would trigger its own distinctive emotional response. "With English verse and the classics he was usually okay, but when he would recite in German, he would get caught up in the emotional swing of the language and go entirely out of control."

Like his father before him, Wiener loved the lyric poetry of Goethe and Heinrich Heine, the baptized German Jew. He recited Heine's verses loudly and with gut-wrenching emotion, no doubt reliving his own spiritually confused childhood. Barbara recalled Wiener descending into Heine's poem, "Princess Sabbath," in which the poet struggled to reconcile the Jews' traditional piety with their historical role as pariahs. Heine portrayed the Jew as a nobleman turned by witchcraft into a dog, a groveling creature throughout the week, until each Friday evening, when the cur came forth as a prince to usher in the "sabbath bride." The poem tore Wiener apart.

The storms that followed were an inversion on the howling family scenes of his own childhood—only now, it fell to his young daughter, not to his mother or even to his wife, to keep the grown-up Wiener from going over the brink. "He would stand in the front hall shouting and crying while I tried to pull him back onto solid ground before he went over the cliff," Barbara recalled. "My mother hid in the kitchen, but I tried to stick it out with him."

During those stormy times, Wiener more than once threatened suicide. "He would be tearing his hair. Then he would pack his suitcase and say he was going to go to a hotel and he had packed a gun and he was going to shoot himself because he was 'no good to anybody.'"

Barbara never saw her father brandish a gun, but as a child she was in no position to quibble. "I do not remember my father possessing any sort of firearm, but I do remember him saying one was packed in his suitcase. I saw him bring downstairs a bag that was already packed and carry it with him to the front porch. On a couple of occasions he actually left the house, but I don't think he stayed away for more than a few hours."

Wiener's cycles "ran about nine months from top to bottom," Barbara recalled, "but you never knew when a storm would come. When he was up he was really creative. When he finished a big project he was vulnerable and more prone to sudden outbursts. When the storms came close together, I knew he was heading for a crash and another prolonged period of depression."

As the younger child, Peggy's recollections of Wiener during that time were sparse, but she recalled his "temper tantrums"—as she called Wiener's tempests—that uncorked "a couple or three times a year. They made me physically sick. I'd be woken up by the shouting, and I would go sit in the staircase. I was terrified."

While Margaret hid in the kitchen and Peggy shuddered on the stairs, Barbara stood by her father. In time, she found the way to steer Wiener through his "tailspins."

"My father always required vast amounts of praise and reassurance, but he required even more when he was in one of his tailspins. Eventually I learned that he got the most reassurance when I pointed out to him that he had had similar periods of despair before and that he had always been able to pull out of them. I think he wanted to be told he was not going to just keep on spiraling until he hit the ground."

Eventually Wiener's tempest would play itself out. "After the storm crested, he usually would return to speaking English and retire to his room with a cheese sandwich and a glass of milk."

When Wiener's emotional storms made landfall in the family's foyer, time and again, Margaret sent Barbara to pull him back from the brink, but she could not grasp her empathetic child's ability to calm and reassure Wiener in his time of turmoil. Instead, Margaret perversely misread her daughter's saving grace.

"She would come out of the kitchen and tell me that the only reason I was able to cope with him when he was in such a state was because I was encouraging his 'unnatural' feelings for me." It was a strange comment for a mother to make to her little girl, but clearly Margaret was insinuating that Barbara's fortifying effect on her father had a sexual component. "I was supposed to calm him down, but if I succeeded it could only be because, in her eyes, I was using some sexual technique."

Barbara offered an explanation for Margaret's reproof. She believed her mother suffered from a disorder of her own that Barbara called "emotional deafness." "She could not perceive other people's feelings, and she didn't have a clue about my father's feelings. If I was any comfort to my father during his emotional storms, it had to be magic of some sort and that was the first magic that came to her mind."

Peggy, too, rejected her mother's insinuations that Barbara was playing to her father's "unnatural" responses. "My father's responses were *perfectly* natural," she said. "Mother was the one with the dirty mind." She seconded her sister's view that their mother was at a loss to understand Wiener's storms, let alone to assuage them. "In truth, Mother hadn't a clue how to cope with them. She would try to calm him by being logical, which was totally ineffective."

Wiener acknowledged that he was in a tumultuous state in the late 1930s. In his autobiography, he described the forces that gave rise to an admitted "internal storm" during those years. "This was a period in which I was subject to a great number of separate emotional strains," he wrote. "I was in a state of confusion. . . . The burden of many years of hard life had started to tell on me. . . . Under the circumstances it is scarcely astonishing that I needed psychoanalytic help."

At Margaret's urging, he sought counseling from, not one, but two Freudian psychoanalysts, but the effort foundered when he failed to find a Freudian who could appreciate his atypical childhood. He went through the prescribed psychoanalytic rituals with one practitioner, which he supplemented with insights into his deep urges toward creative work in mathematics, and his love of literature and poetry; and he described the emotions that overwhelmed him when he recited his favorite passages from Heine. But he was appalled at the man's response to his heartfelt testimony, which, the psychoanalyst said, was not a genuine expression of Wiener's subconscious but, instead, clear evidence of Wiener's "resistance" to the Freudian method. Wiener conceded that, in the strict terms of psychoanalysis, which focused solely on early childhood experiences, his testimony about his scientific work and poetic passions was considered irrelevant and uncooperative, but he insisted that those facts held important clues to his experience, and "to much that was at the bottom of my spiritual make-up." He quit the effort after half a year, convinced that the psychoanalyst would never have "very much of a notion of what made me tick."

Slowly an understanding grew among those who knew him well that Wiener was manic depressive, but that clinical distinction was still tentative in the 1930s and remained elusive for years afterward. The psychiatric diagnosis of manic depression was established only in 1913, the year Wiener received his Ph.D. He had experienced bouts of depression since his early teens, and he made no bones about his near breakdown at Göttingen in 1926, but the onset of his manic tendencies was harder to pin down. He complained of chronic exhaustion, yet he tapped deep energy reserves for his mountain trampings and his creative bursts of mathematical research and writing. His bubbling joy over his marriage and the prestigious Guggenheim grant he won in the spring of '26, when he experienced "a very exulted mood," were the first evidence in his own words that his elation could go to extremes with professional repercussions. By the early 1930s, his manic symptoms were on display. Wiener's students and colleagues recalled him besieging them with his erudition, continually striving to outpace, impress, and overwhelm them with his energy, his ideas, and his enthusiasms.

Like depression, bursts of energy and high productivity were common among creative minds, but the preponderance of testimony, and subsequent clinical research, suggested that, by age forty if not sooner, Wiener's cycling manic and depressive states were being steered by some underlying neurochemical mechanism. The modern medical view of "affective" disorders like manic depression classified the fluxing highs and lows Wiener experienced as a presentation of *bipolarity*, a psychiatric condition defined by major texts in the field as "a disorder of the brain marked by extreme changes in mood . . . persistent states of extreme elation . . . alternating with persistent states of extreme sadness." Other textbook symptoms of mania and depression include "people-seeking," "pressure to keep talking," "thoughts of suicide," "feelings of worthlessness," and intense "affective storms." The illness was presumed to have a strong genetic component, but environmental factors determined whether it occurred in a particular individual, when the genetic tendency was pushed over a threshold neurochemically by traumatic childhood experiences.

The serial stresses of Wiener's prodigy years fit the clinical profile of a neurochemical disorder brought on by early adverse life experiences. His hefty weight contributed to an intractable case of apnea—the cessation of breathing during sleep, which may awaken a sleeper hundreds of times a night. Severe apnea had been implicated as a trigger of manic behavior, and may have been another physical cause of his shift from mere depression to manic depression.

One more factor weighed heavy on Wiener. In the mid-1920s, his younger brother, Fritz, was institutionalized with a diagnosis of schizophrenia. The fall of Fritz, who had been positioned to surpass Norbert as Leo Wiener's greatest achievement in prodigy making, stirred Wiener's deep-rooted fears that something in his family's genes had marked him for failure, mediocrity, or, worse still, madness. In a consoling letter to Fritz written soon after his own brush with breakdown at Göttingen in 1926, Wiener voiced his fear that, "We are really very much alike. . . . We are neither of us very socially adaptable. . . . We are both of us . . . introspective and moody. . . . We both work ourselves up into a state of panic at our own actions." Even after he had secured his career and home life, Wiener agonized over Fritz's slide into mental illness and worried constantly that he would suffer a similar fate.

"My father had a real terror of turning out like Fritz," Barbara said. Peggy concurred. "I think he did believe that, and I rather think Mother nurtured it." Indeed, in her "emotional deafness," Margaret Wiener sometimes triggered her husband's depressed states and often exacerbated them. "He would come to her, obviously looking for approval or support, and then she would think of something he had done wrong and throw it at him in a gentle, seemingly understanding tone," Barbara remembered. "Sometimes, even when he was feeling good, she would stick the pin in the balloon. She would wait until he was just about to go out, and then tell him, 'Your hair's uncombed,' or 'You shouldn't be wearing that suit,' or 'Why don't you get a clean shirt?'"

Barbara did not see those takedowns as intentional. "I don't think she knew what she was doing. My mother was defective in her perception of other people's feelings, and to her *any* show of feeling was a threat." Peggy saw both sides of the family dynamic. She saw her father's emotional distress and also acknowledged her mother's frustrations and repressed anger at Wiener and at his parents, who from the beginning downplayed Wiener's emotional problems. However, Peggy felt that Margaret, too, had stopped short of full disclosure. "Dad was a very warm and generous person who needed expressions of love he never got from Mother," she said, "but their marriage did not give Mother what she wanted either. She had come from a middle-class German family and, as soon as she entered this country, she was socially at the bottom of the heap. I think one of her driving motives was to regain her social position. She believed that becoming a *frau-professor* would secure that position, yet her position with my father was well below what it would have been in Germany, and I think she was very disappointed by what she ended up with."

Wiener's history fit the pattern of manic depression, yet through all his ebullient moods, emotional storms, and repeated threats of suicide, he remained surprisingly productive. In clinical terms, his symptoms more closely conformed to a diagnosis of "cyclothymia," a strain of manic depression in which the cycles are less severe and less debilitating day to day, and never cross the line into delusions, hallucinations, and outright psychosis.

Peggy's perceptions as a child were not unlike those of Wiener's MIT colleagues who saw only glimpses of Wiener's inner turmoil in their dealings with him day to day. "Looking back, I see a badly depressed man who was striking out, not at any person, but at the world in general; however, to a small child, I saw this only as a personal flaw. It was not until very much later that I realized what was going on with him and how miserable he must have been." Reflecting on that time of Wiener's secret storms, professional triumphs, and all that would come after, she offered a self-evident assessment. "I think it's amazing he survived and was at all functional. He must have been incredibly strong."

———

Through his ups and downs, Wiener did his best to balance the demands of his work and his family, and to manage the mounting load on his time and energy and emotions. He took frequent naps during the day and went to bed early, at seven or eight most nights, but he would often awake around three in the morning, trundle down to the kitchen for a bowl of cold cereal and milk, and then go back to bed. Around the house, "there were loose pages with mathematical notes scribbled all over them, and generous supplies of pencils. He wanted to be ready if an idea came to him," Barbara recalled. At night, a pencil and paper sat waiting on his bed table.

Since his youth, Wiener was mindful that his best ideas originated in a place beneath his awareness, "at a level of consciousness so low that much of it happens in my sleep." He described the process by which ideas would come to him in sudden flashes of insight and dreamlike, hypnoid states:

> Very often these moments seem to arise on waking up; but probably this really means that sometime during the night I have undergone the process of deconfusion which is necessary to establish my ideas. . . . It is probably more usual for it to take place in the so-called hypnoidal state in which one is awaiting sleep, and it is closely associated with those hypnagogic images which have some of the sensory solidity of hallucinations.

The subterranean process convinced him that "when I think, my ideas are my masters rather than my servants."

Barbara corroborated her father's observation. "He frequently did not know how he came by his answers. They would sneak up on him in the middle of the night or descend out of a cloud," she said. Yet, because Wiener's mental processes were elusive even to him, "he lived in fear that ideas would lose interest in him and wander off to present themselves to somebody else."

Still, at times, life seemed almost normal in the Wiener home. The family played games together, with mixed results. "Dad played bridge, very badly," Peggy remembered. "His mind was always wandering, he always recklessly overbid and Mother underbid. Barbara and I beat the pants off them." But it was at mealtimes that the little Wiener clan shone. Wiener, an incorrigible punster, loved to engage his girls in wordplay during dinner. He would make a pun, often a bad one, and laugh heartily at his own joke with a deep rumble of a laugh—*"Haw, haw, haw."* Wiener's daughters caught his drift and replied in kind, but Margaret seldom joined in the fun. "My mother could not get irony," said Barbara, reflecting on another consequence of Margaret's emotional deafness. Peggy agreed. "She was an intelligent woman, obviously, but she didn't have that quickness of mind we shared with Dad, and I think she felt very left out of the conversation. I think it made her feel a bit lost." Margaret, for her part, was not wholly lacking in humor, but "her sense of humor tended more towards *Schadenfreude,*" said Peggy. "Very German."

Something else in Margaret's German background was manifesting itself. Like her kinsmen in Silesia and millions of their countrymen, she was forming an abiding attraction to Germany's new leader, Adolf Hitler, and his Nazi ideology. Margaret spoke frankly with her daughters about her infatuation.

"One day she told us that the members of her family in Germany had been certified as *Judenrein*—'free of Jewish taint.' She thought we'd be pleased to know. She was very pro-Nazi in the early days," said Barbara. Other disturbing comments followed as the Nazi oppression of Jews in Germany intensified. "She said I should not feel sorry for the Jews of Germany because they were 'not very nice people,'" Barbara recalled. At Christmas time, as she decorated the family's tree with her half-Jewish daughter, "She told me that Jesus was the son of a German mercenary stationed in Jerusalem, and this had been scientifically proven," Barbara said.

By the late thirties, Margaret was openly sympathetic to the Nazi cause. "Two books decorated her dresser, one in German and one in English. They were copies of *Mein Kampf,*" Barbara remembered. "My mother often pushed them on her friends by explaining that Hitler and Germany were 'terribly misunderstood.'" Barbara read the English version of *Mein Kampf* "in bits and snatches." Appropriately alarmed, she brought her concerns directly to her mother. "I mentioned Hitler's plans for the Jews to my mother a few times. She said I didn't understand that the Jews of Germany were not like the Jews I knew but were 'dangerous and destructive to the rest of the Germans.' She had a lovely voice and she would say these ugly things in such a tone that you wondered if she was mad or you were."

Wiener, too, knew of Margaret's views and found them to be "very painful," Barbara said. "My father refused to talk to me about my mother's politics, though I could hear them fighting about it behind closed doors. When I tried to bring up the subject, he would throw his hands in the air and say, 'Why did I bring you into the world to suffer?'"

Others outside the Wiener home witnessed Margaret's extreme views and Wiener's outrage, which he could barely contain. Fagi Levinson saw the discord on display when she and her husband socialized with the Wieners. "Margaret spoke very openly about the fact that her relatives in Germany were members of the Nazi party," Fagi remembered. "She said, 'After all, how else are they going to keep a job?' Norbert would turn red but would never say anything, because she was his wife and she was faithful and she stood his tempers and all the rest."

Wiener never learned the full extent to which his wife embraced the Nazi ideology in conversation with her children, but he knew at least that Margaret was not a person "to whom national and racial prejudice have always been as foreign as they have been to me," as he once claimed. In private, he voiced strong feelings about Margaret's ethnic and political opinions. Publicly, though, Wiener made only oblique references to those painful tensions in his marriage. When he made his first unsatisfying visits to psychiatrists, he cited first among the many emotional stresses he was bearing "the fact that Nazism threatened to dominate the world," and he admitted that his knotted Jewish roots "rendered my emotional situation somewhat ambivalent." He wrestled with the "continual nightmare . . . that somewhere in the world we were being threatened with extermination" and noted "that Nazi anti-Semitism had provoked an echoed anti-Semitism in some American quarters."

In the summer of 1939, Hedwig Engemann went into decline at the family's home in New Hampshire. Unlike her daughter, she had no love for Adolf Hitler and Nazism, and she prayed that she might be spared another war in Europe. Late in July, Hedwig died quietly in her sleep in her upstairs room of the old farmhouse, as Hitler's troops were massing on the Polish border.

Wiener's father died a few months later in Boston, at age seventy-seven, after suffering a stroke several years earlier. Only after his death could Wiener fully acknowledge his father's abiding influence on his life and the many gifts he bestowed: his love of the classics and German romantic poetry, his humanitarian values of compassion and caring for the oppressed and undervalued elements of society, which he took from Tolstoy, his qualities of "*élan* . . . of glorious and effective effort, of drinking deep of life and the emotions thereof," and his belief "that scholarship is a calling and a consecration, not a job." Above all, Wiener took into his marrow his father's uncompromising intellectual honesty and "fierce hatred of all bluff and intellectual pretense." Those paternal gifts Leo hammered into his son stayed with him. "It was because of this, because my taskmaster was at the same time my hero," said Wiener, "that I was not bent down into mere sullen ineffectiveness by the arduous course of discipline through which I went."

Throughout the 1930s, Wiener built broadly on the mathematical foundations of his new communication theory. Along with his two books, he published forty papers that decade and continued his explorations in physiology and circuit theory, while he waited for time and technology to catch up to him. In New Hampshire, the spring after Hedwig Engemann and Leo Wiener died, he purchased a little wooden cabin, about ten feet on a side, with a peaked roof and three small windows. He had it placed a short walk from the family's house, at the edge of the woods, and he went there daily to think and write uninterrupted. Within its confines, not much bigger than the cubbyhole beneath his father's desk, gazing through the windows at the woods around him and the mountains in the distance, the Elephant's Child, fatherless now, would ponder and produce his next round of books and papers that would light the way for the coming revolution in communication science and technology.

Early in the summer of 1940, Wiener drove alone from New Hampshire to attend a meeting of the American Mathematical Society in Madison, Wisconsin. On the road, he learned that the war most Americans wanted no part of was coming inexorably closer. After annexing its neighbors to the south and east, Germany was advancing in a fury to the north and west. "It was an experience curiously reminiscent of that time twenty-four years before, when the First World War had come to me on another trip, as a passenger on a German boat in the middle of the North Atlantic."

After the conference, Wiener drove back east with a colleague from England. On the way, the two men stopped to pick grapes in a vineyard in upstate New York and "to take stock of our emotions and expectations."

6

Birth of a Science

Genius is present in every age, but the men carrying it within them remain benumbed unless extraordinary events occur to heat up and melt the mass so that it flows forth.

—Denis Diderot

BACKFLOATING IN BEAR CAMP POND in the summer of 1940, Norbert Wiener cruised out and back across the tranquil waters of the little lake, as he had for so many summers before, eyeglasses glistening, cigar clenched in his mouth, his great beach ball of a belly rolling in the waves. But his mind was four thousand miles away.

In Europe, nationalist forces backed by Nazi Germany's newest weapons had vanquished the republic's defenders in Spain's civil war. Fascist Italy had signed a "pact of steel" with Germany, and Hitler's *Wehrmacht* war machine was rolling across the Continent with crushing force. By the spring of 1940, Austria, Czechoslovakia, and Poland had been conquered. Denmark and Norway and the Low Countries were subdued. In June, 330,000 Allied troops were driven from the Continent at Dunkirk. France fell, and, a month later, *Luftwaffe* warplanes crossed the English Channel and began the pounding Battle of Britain.

The Axis was pulling the world apart from both poles. Soon after Wiener left China, the Japanese incursion had erupted into a full-blown war on the Asian mainland. Three hundred thousand Chinese peasants were massacred in Nanjing, and a million more died in fighting and floods on the Yellow River. To thwart Japan's aggressions, in May 1940, President Roosevelt moved the U.S. Navy's Pacific fleet to Hawaii as a deterrent.

Wiener watched each development with a sense of impending catastrophe. As the summer passed uneasily, he savored his lazing laps around Bear Camp

Pond and his cool walks through the White Mountains. He welcomed colleagues from the besieged lands who came calling. He tried to keep a cheerful mood around his girls as they went about their usual country outings. Yet, already, more than a year before the Japanese attack on Pearl Harbor, he and his colleagues knew that America's entry into a second world war was a foregone conclusion, and that the new war would pose unprecedented scientific and technological challenges.

MIT's vice president, Vannevar Bush, had relocated to Washington the year before to head up the Carnegie Institution, a private foundation, devoted to basic research, that had become the premier patron of American science. From that place high in the nation's intellectual firmament, the no-nonsense New Englander could advise government officials directly on scientific aspects of military matters. On June 12, 1940, a week after the last chaotic evacuation of Allied troops from the Dunkirk dunes, Bush met with Roosevelt at the White House. He emerged with an executive mandate to form and direct a new National Research Defense Committee, charged with initiating and coordinating scientific research for the nation's accelerating defense effort. His first initiative was to organize a network of 700 universities and research institutions whose services he would enlist for war-related scientific and technical projects. He also sent personal letters to his peers soliciting their suggestions for the best use of scientists in the coming war.

Wiener was happy to supply some.

In a memo to Bush written in September 1940, the first of several they would exchange in the coming weeks and months, Wiener drew on his experiences at Aberdeen Proving Ground during World War I, his many collaborations with specialized scientists and engineers, and his ongoing discussions with Arturo Rosenblueth on the value of interdisciplinary ventures in science. He advised Bush to adopt a bold new war research and development strategy based on "the organization of small mobile teams of scientists from different fields, which would make joint attacks on their problems . . . pass their work over to a development group and go on in a body to the next problem."

Bush never responded to Wiener's suggestion. Wiener concluded that nothing came of his proposal for an interdisciplinary team approach to wartime scientific research, but he may have rushed to judgment. Months later, after another meeting between Bush and Roosevelt, a second research organization, the Office of Scientific Research and Development, was chartered by Congress as an independent federal agency with Bush as its director. Together, Bush's NDRC and OSRD coordinated the activities of 6,000 American scientists engaged in war-related research, including much of the theoretical and technical work for the supersecret atomic undertaking that became known as the Manhattan Project. Many of those scientists worked together, precisely as Wiener had proposed, in small, mobile, multidisciplinary teams that solved one knotty problem after another until the war was won.

Wiener had other matters on his mind that would directly affect the war effort. On September 11, 1940, ten days before he sent his memo to Bush, he traveled to Dartmouth College in New Hampshire to attend the late-summer meeting of the American Mathematical Society. The usual shoptalk was displaced by somber discussion of the conflict raging in Europe, but one technical display captured Wiener's attention. In a hallway outside a meeting room, a small teletype machine was connected by a long-distance telephone line to the new "complex number calculator" at American Telephone and Telegraph Company headquarters in New York City. The machine's inventor, thirty-six-year-old Bell Laboratories mathematician George R. Stibitz, described the device to those who began gathering around him. Then he gave his audience of skeptical academicians the first practical demonstration of remote computing.

Like Bush's analog differential analyzer, the machine Stibitz built for Bell Labs was not yet a true computer. It had no memory or internal system for processing a program of logical operating instructions. But it was the next step in that direction: the world's first *digital* electronic calculator—one that operated on discrete, not continuous, quantities. It was capable of performing basic mathematical operations with complex numbers on an array of 450, two-position telephone relay switches—known as "flip-flops"—in less than a minute. Two onlookers grasped the meaning of the machine almost as fast. One was John Mauchley of the Moore School of Engineering at the University of Pennsylvania, who would soon begin work with his colleague J. Presper Eckert on the first programmable digital computer, the ENIAC.

The other was Wiener. That first vivid demonstration of the Bell Labs Model 1 relay calculator reignited his thinking on the whole subject of automatic computation, which had been dormant since his failed effort with Lee in China to build an electronic analog computer. When Wiener returned to MIT, he gathered his thoughts on the need for better computing machines to support scientists and engineers in every sector of the war effort. In a long memo dated September 20, 1940, he framed five succinct directives on computer design that expanded on ideas he had offered to Bush more than a decade earlier. His new directives set down one of the first systematic descriptions—and perhaps the first set of technical specifications—for a fully functioning computer in the modern sense.

Wiener stated his preference for digital calculating methods over analog ones, advising Bush that the "apparatus of the computing machine should be numerical . . . rather than on a basis of measurement," as in Bush's differential analyzer. He pressed his case for vacuum tubes, as he had since the 1920s, emphasizing that "in order to secure quicker action" computers "should depend on electronic tubes rather than on gears or mechanical relays," like those in the new Bell Labs device he had seen only a few days before. He heartily endorsed Stibitz's binary system of computation, based on "the scale of two . . . rather than the scale of ten," a method Wiener himself had first applied ten years before to manage long series of mathematical data. Next, Wiener proposed a new system of electronic memory, in the form of microscopic magnetic marks made on a spool of wire or metallic tape, that would endow the new computing machines with "an apparatus for the storage of data which should record them quickly, hold them firmly until erasure, read them quickly, erase them quickly, and then be immediately available for the storage of new material."

And, finally, in so many words, Wiener called for the invention of computer software—a program of logical instructions to direct the computer's operations—although at that point in his conception the program was more hard than soft. He advised Bush:

> That the entire sequence of operations be laid out on the machine itself so that there should be no human intervention from the time the data were entered until the final results should be taken off, and that all logical decisions necessary for this should be built into the machine itself.

Those recommendations, as Wiener made clear to Bush, were not just airy notions in his head but feasible ideas based on practical computing methods being explored, in principle or in prototype, by mathematicians and engineers around the world. The English mathematician Alan Turing first postulated a "universal" machine for computing numbers in a paper he wrote in 1936, and the American engineer John V. Atanasoff, of Iowa State College, had built a prototype with some of those features in 1939—although Wiener was the first person to pool all those scattered concepts into a concrete proposal for an all-electronic digital computer with an internal logical program.

Wiener got an acknowledgment from Bush on that memo, but, after several weeks, he sent Wiener his interim opinion that he was not yet convinced of the design's practicability. Late in December 1940, Bush turned down Wiener's proposal. He rebuffed Wiener's offer to build an all-electronic digital computer and refused to assign anyone else to such a project, insisting that "it is undoubtedly of the long-range type and it appears essential that at the present time those individuals who are particularly qualified along these general lines be employed as far as possible on matters of more immediate promise."

At the time, neither man knew that before the war's end every idea in Wiener's prescient memo would be operational, or actively in development, in one all-inclusive machine.

In 1940, computers were Bush's babies and the pet projects of a tiny handful of mathematicians and electrical engineers. Wiener knew computer building would not be his domain of wartime science, but he was eager to contribute to the war effort. He wrote to Bush: "I hope you may find some corner of the activity in which I may be of use during the emergency." Then he set out to find a niche where his special talents, new mathematical methods, and pioneering communication theories could best be put to work solving practical problems of modern warfare.

The autumn brought another ominous turn of events in Europe. The Battle of Britain, which began when Hitler launched his *blitzkrieg* "lightning war" of night bombing attacks on cities across the English countryside, was now converging on London in a rain of fire that came to be known simply as the Blitz. From September to November, German bombers dropped 13,000 tons of bombs and incendiary devices on the city in the climactic phase of the German game plan to terrorize, demoralize, and ultimately break the British people. But the Brits would not be broken.

Soon after the Blitz began, a high-level delegation of British technicians and military officials traveled to Washington carrying their top-secret defensive weapon: a newly developed "cavity magnetron" microwave radar apparatus. The high-energy, high-resolution radar device offered improvements American scientists had only begun to contemplate as the next step beyond the era's existing long-wave radar technology. The British offered their electronic treasure to Bush and his NDRC team in exchange for American help to meet their three most urgent tactical and technological needs: airborne microwave radar systems to help British night fighters intercept incoming German bombers, long-range ground radar systems to guide British bombers to and from their targets on the Continent, and improved "gunlaying" radar and fire control systems for antiaircraft artillery to thwart the Blitz and give Allied forces the edge in aerial warfare. That last item proved irresistible to Bush, who believed defense against air attack was not only England's need but America's foremost problem of wartime preparedness.

Within weeks, as word of the NDRC's new projects went out discreetly through the ranks of scientists Bush had arrayed, Wiener found his wartime challenge: to discover a better way to direct and control the firing of anti-aircraft artillery. His work began humbly and unheralded. Like every NDRC project, it was classified secret and subject to tight wartime security restrictions. Then Wiener drew a bead on his target. He knew that, since the start of aerial warfare in World War I, the increased speed and maneuverability of modern warplanes had made antiaircraft firing far more difficult. He likened the gunner's challenge to the sport of shooting "ducks on the wing," where the successful shooter learned to "lead" the target, to aim ahead of the bird's observed position to the estimated place it would be when the shot arrived, hopefully, at the moment the two flight paths of the targeted bird and targeting shell converged.

This intricate war work involved much more than computing mundane artillery range tables. Wiener's main task now was one of prediction: to predict the future position of a fast-moving warplane based on the best information available about its past and ever-changing present positions, then to calculate the crucial range and targeting factors, to take aim and, finally, to fire the antiaircraft gun with split-second timing and precision. It was the second time he had sought to design an electronic system that would emulate and, as he said, "usurp a specifically human function." The first function he had been striving to usurp since the 1920s was the complex mental work of computation. Now he was trying to design an electronic system capable of performing the supreme feat of human intelligence—envisioning and predicting events in the future—and then acting on that foresight.

That problem was formidable enough, but there was another obstacle. Unlike the olden days of World War I, when artillery could be aimed in simple trajectories at sitting ducks and tortoise-like targets, the nimble new warplanes of the Second World War were steered by ace pilots trained in evasive action. Their zigzagging courses were not only nonlinear but highly irregular, providing no certain clues to the plane's future position—like the flight of a bumble bee or the path of a drunken man crossing a football field. With each maneuver, the enemy pilot veered off on an entirely new trajectory, and each new trajectory pointed to not one but many equally possible future positions.

The technical challenge was custom made for Wiener's proven talents and expertise. The task of plotting and predicting enemy flight paths called on his earlier mathematical work plotting besotted particles in Brownian motion, on the new statistical tools of filtering and prediction he had pioneered in the Wiener-Hopf and Paley-Wiener equations, on his early work with Bush on analog computation and his later work with Lee on electronic circuits and the design of complex corrective networks.

And Wiener had another powerful new tool at his disposal. The advent of radar in the mid-1930s had opened new vistas for the practical use of radio waves. Radar (for RAdio Detection And Ranging) harnessed weak currents of electromagnetic radiation to do more than transmit electronic signals carrying audible sounds. With the discovery a decade earlier that radio waves bounced off any metallic object in their path, radio engineers learned that those reflected beams could be recaptured and analyzed electronically to yield detailed information about the distance and direction of moving objects they encountered.

The new British microwave radar advanced the state of the art by leaps and bounds. Its short-wave, high-frequency impulses could be generated by compact mobile units and detect small objects flying at a wide range of altitudes. Its powerful signals passed effortlessly through cloud cover and were equally effective by day or night.

For Wiener and other scientists contemplating the new technology's potentials, the prospect was tantalizing that reflected radar waves might be tapped and harnessed to predict a warplane's future position with a high probability, then coupled electronically to an antiaircraft gun's targeting mechanism to aim the gun and fire it dead on the mark—automatically.

Early in November 1940, Wiener proposed just such a notion to Professor Samuel H. Caldwell, the MIT engineer charged with applying the institute's Bush's Differential Analyzer to wartime problems. Caldwell immediately put a cloak around the idea, which to Wiener's dismay forbade him from talking about it with outsiders. Wiener devised a trial formulation of the aircraft prediction problem, and he and Caldwell and another MIT engineer spent the next three weeks conducting test runs of Wiener's new mathematical prediction theory on the Bush analyzer. The results were promising but purely hypothetical. Wiener and Caldwell wrote up a description of an experimental device that would translate Wiener's math into mechanical action, and, on November 22, 1940, Caldwell submitted a formal proposal to Section D-2, the fire control division of the National Defense Research Committee.

Shortly before Christmas, the proposal was approved and Wiener's wartime project got under way.

———

Julian Bigelow, a polite, painstakingly precise, young MIT man, was the chief engineer on the project. In 1936, Bigelow earned his masters in engineering at MIT and went to work for the International Business Machines Company—IBM. Four years later, when President Roosevelt announced the first peacetime draft, Bigelow's local draft board scheduled him for early call-up. He made a quick trip back to MIT's engineering department to gather his credentials and, he hoped, to obtain a deferment.

Just before Christmas 1940, at age twenty-seven, he was called to the office of the department's chairman, Karl Wildes. Wildes not only agreed to Bigelow's request for a deferment, he demanded it.

"He said, 'You *can't* go into the service. We need you to stay on here and work with Wiener and find out what he's trying to do, or what he actually *is* doing. Nobody has any idea what he's talking about,'" Bigelow recalled. Wildes sent Bigelow to meet Warren Weaver, director of the Rockefeller Institute in New York and a respected mathematician in his own right, who had been put in charge of fire control research for Section D-2 of Bush's NDRC. After a brief chat, Weaver, too, was convinced that Bigelow was needed at MIT. Bigelow met Wiener by happenstance during one of Wiener's safaris through Tech's engineering department. They struck up a conversation and "in the course of our discussions I came to understand his idea for a predictor that was more flexible than the existing ones."

He was the right man for the job. Bigelow understood Wiener's technical talk and, equally important, the unflappable New Englander was the perfect foil for Wiener's volatile personality. Early in 1941, the two men took over an empty classroom on the second floor of the mathematics department, Building 2, Room 244, and set to work on the blackboard. Wiener outlined the fire control problem for his new collaborator. He instructed Bigelow in the mathematics of shooting ducks on the wing, scribbling diagrams and differential equations on the board as he talked. He schooled Bigelow in his theoretical work and the formulas he had tested on the Bush analyzer. Then Wiener and Bigelow set out to study fire control problems as gunners actually encountered them in the field.

The two men traveled south to army bases on the Virginia and North Carolina coasts, where they observed warplanes in flight, the latest ordnance, and new weapons being developed and tested at the army's antiaircraft command at Ft. Monroe, Virginia. On those trips, they learned the facts of real-world fire control situations that were far more complicated than duck hunting, and they saw the widening gap between Axis air power and American armaments that had barely advanced since the Great War.

The best German bombers blackening the skies above Europe flew over their targets at speeds exceeding 300 miles per hour, and at altitudes as high as 30,000 feet. The flight time of an artillery shell to that height could be as long as twenty seconds, and firing that shot accurately—at a point in space nearly two miles downrange from where the plane had been targeted—was no simple task. A crew of up to fourteen men was required to spot the plane through binoculars and keep it constantly in view, then to relay its changing positions to human computers, who performed crude manual computations of the plane's projected location and passed those coordinates to the gunners, who, in turn, rotated their heavy turrets into position using hand cranks and, finally, fired a volley of shells toward their designated rendezvous point.

Wiener found the whole process fascinating, absurdly cumbersome, and ripe for improvement.

––––––––

When they completed their field research, Wiener and Bigelow returned to MIT to rethink the fire control problem from the ground up. Back in Room 2-244, Wiener took to the blackboard.

"To some extent this is a purely geometrical problem," he said, thinking out loud. "The future position of the plane must be estimated from the observed past positions, what we call a problem of *extrapolation*. The simplest method is to extrapolate the present course of the plane along a straight line." He put two dots on the board, representing the plane's past and present observed positions, and drew a straight line through them pointing to its location in the future. "On purely mathematical grounds," he went on, "this has much to recommend it."

But that method had one obvious problem. As soon as they encounter the first burst of antiaircraft fire, Wiener observed, the pilots "will probably zigzag, stunt, or in some other way take evasive action." He drew a segmented line veering off in a series of sharply angled zigs and zags. Once that human factor was introduced into the equation, Wiener said, "only a prophet with the knowledge of the mind of the aviator could predict the future position of an airplane with absolute certainty." Fortunately, Bigelow, who was an amateur aviator himself, pointed out to Wiener that the warplane's flight path also was governed by some obvious physical constraints on the pilot's range of evasive action.

So, Wiener ruminated, "the pilot does *not* have complete freedom to maneuver at his will. For one thing, he is in a plane going at an exceedingly high speed, and any too sudden deviation from his course will render him unconscious and may disintegrate the plane. Then, too, he can control the plane only by moving his control surfaces, and the new regimen of flow takes some time to develop."

Wiener had no difficulty folding Bigelow's high-flying aerodynamics into his analysis. He erased the zigzag line and drew in a more smoothly undulating curve. Bigelow was impressed.

"Wiener was a philosopher and he had a good feel for process," Bigelow recalled. "He grasped the fire control problem very well conceptually, that if you're flying in an airplane and you make a decision to change direction, the dynamics of the plane itself limit how fast you can do that. You make a decision that's instantaneous, a step function, but the airplane responds with a lag and a smoothing of that process of change owing to the dynamics of the air flow around the plane and limits on the controls of the plane."

The resulting flight path was highly irregular but not purely capricious. Those waving lines on the blackboard were old friends to Wiener. "There are," he said, in a rare display of understatement, "in fact, means which will allow one to accomplish the minor task of a quite correct prediction."

Wiener was right. He had spent two decades solving partial differential equations—those formidable functions in which two or more factors change continuously. That solid grounding helped him to model the changing coordinates of a warplane moving swiftly in space and time. Then, drawing on his special expertise in the tricky business of Brownian motion, he plotted the pilot's evasive moves as a random function, like the movements of particles traveling on fast-changing

Brownian trajectories, which he used to compute the plane's most probable future positions. From there, for this ex-prodigy, finding the projected convergence point of the looping plane and an arcing artillery shell was relative child's play.

But solving the prediction problem was only half the challenge. The team's wartime assignment was to design and *construct* an automatic antiaircraft prediction and targeting device, and that tangible work fell to Bigelow. As a first step toward building a testable model, Bigelow needed to devise an experimental setup for generating the complex paths in space traversed by Axis warplanes and Allied antiaircraft fire. He sketched a rough embodiment of Wiener's vision and assembled the initial contraption from a tangle of electrical resistances, wires, condensers, magnetic coils, and two light sources coupled to small electric motors. One motor beamed a spot of white light on the ceiling of Room 2-244, representing a warplane moving on a more or less circular course. A second light, representing the antiaircraft gunner's tracking beam, was guided by a crank attached to an assembly of weights and springs designed to mimic the mechanical controls of a heavy antiaircraft gun turret.

"The handle of the controller beam was red and I connected it to a small red spotlight," Bigelow recalled. "The idea was for a human operator to try to make the red spot follow the white spot around the room."

The idea was good, but it had one serious flaw: the circular path that the motor-driven white light painted on the ceiling was smooth and unvarying. To truly reflect Wiener's vision, Bigelow had to generate the kinds of irregular movement Wiener had factored into his complex prediction equation, paths that emulated the evasive maneuvers pilots under fire could be expected to perform. Bigelow came up with an ingenious solution. He lowered the white light beam's impact point from the ceiling to the walls of Room 2-244.

He described the major improvement that minor adjustment achieved.

"The white light went around the walls of the room, and because the room was square it jumped as it went around each corner. Then, using some delay circuits, I built a lag factor into the controller that resembled the actual lag of an antiaircraft gun's tracking response. The idea was for the operator to move the handle and make the red light beam follow the white beam, which was circling the room and jumping as it went around the corners in a non-sinusoidal, non-smooth way."

Wiener was delighted. He and Bigelow had found a way to simulate the irregular motion of an enemy warplane in flight under fire from ground-based artillery. The model's "lag operator" was so true to the actual response of a gun turret, Bigelow recalled, that "the controller had to anticipate the plane's behavior and move the handle well ahead of the actual motion of the red light in order to catch up with the white light, so our model had a degree of similarity to the problem in the field."

In some ways, the similarity was too good. As work on their prototype progressed, Wiener made an "interesting and exciting, and in fact not unexpected" discovery. The pieces of their apparatus that proved best at following the smooth-curve portions of their test flight path turned out to be oversensitive and driven into "violent oscillation" at the corners of Room 2-244. Wiener pondered the problem and arrived at a familiar paradox with profound implications for his project. He wrote:

> Perhaps this difficulty is in the order of things, and there is no way in which I can overcome it. Perhaps it belongs to the nature of prediction that an accurate apparatus for smooth curves is an excessively sensitive apparatus for rough curves. Perhaps we have here the example of the same sort of malice of nature which appears in Heisenberg's principle, which forbids us to say precisely and simultaneously both where a particle is and how fast it is going.

The more Wiener studied the paradox, the more he became convinced that the problem he and Bigelow were facing was fundamental. His solution was equally profound. He wrote:

> If then we could not . . . develop a perfect universal predictor . . . we should have to cut our clothes to fit our cloth and develop the best predictor that mathematics allowed us to. The only question was: . . . If errors of inaccuracy and errors of hypersensitivity always seemed to be in opposite directions, on what basis could we make a compromise between those two errors?
> *The answer was that we could make such a compromise only on a statistical basis.*

Once again, Wiener's focus on a practical, real-world problem had led him into that paradoxical realm of nature where there was no certainty, only probabilities, compromises—and statistical conclusions. In lieu of strictly linear prediction methods that were dumb to the minds of enemy pilots and the diverse evasions made by warplanes under fire, Wiener pulled a powerful weapon from his quiver, a classic statistical tool known as the "mean square error" method. The result gave a more realistic, and far more accurate, set of coordinates for the rendezvous point antiaircraft gunners should shoot for.

———

They had made good progress and potentially important discoveries, but Wiener and Bigelow were not working in a vacuum. During the first months of their project, the blitzing of London stepped up another notch. In May 1941, the British launched a series of aggressive counterattacks, but they were still greatly overpowered, in dire need of improved radar and fire control devices, and urgently awaiting the results of their technical requests to the NDRC.

After five months of research and theoretical work, Wiener's overseer in Section D-2, Warren Weaver, arranged a meeting between Wiener and Bigelow and a team of engineers at Bell Laboratories, who were also working on the fire control problem. The Bell Labs team was hurrying to produce an electronic version of Bell's M-6 mechanical antiaircraft predictor, the principal apparatus used by the American military at that time, but Bell's engineers were still thinking in old-time engineering terms—and still running the same linear prediction formulas first used by manual plane spotters and plotters in World War I. "They had no random variables in them at all and took no account of evasive action or even the natural curvature of the plane's flight path," Bigelow confirmed.

Weaver had a hunch that the two men from Tech had found a better way.

Wiener and Bigelow went to the Bell Labs facility in Whippany, New Jersey, on June 4, 1941. At that meeting, Wiener offered the Bell team his new statistical prediction method that more accurately reflected the actual problem of targeting a maneuvering warplane in flight. His prediction method went far beyond existing fire control devices or anything else then in development, but in a cool, technocratic scenario that was itself becoming all too predictable to Wiener, the Bell team rejected his highly unconventional approach.

Bell's engineers were skeptical and did not fully grasp Wiener's statistical theory. "They were pretty smart people," Bigelow conceded. "They understood what he was driving at, but the thing they did not believe was that an ensemble of possible flight-path curves *existed* from which to choose the most probable one." Of course, Wiener had established the existence of that ensemble of curves, and put his name on it, two decades before, but Bell's mathematicians and engineers were not yet attuned to Wiener's ensemble and his new mode of mathematical thinking.

After Wiener and Bigelow completed their presentation, the Bell Labs team briefed them on their project to develop a simpler, strictly linear, automatic fire director "which did not involve statistical concepts in any form." They explained that "the scheme they proposed was based upon the fundamental urgency of the situation and . . . had intentionally been restricted to functions that could be accomplished by means of apparatus and tools already available." Then the group adjourned to the model room, where the Bell prototype was revealed.

"We didn't think it was so hot," said Bigelow of Bell's machine—or what he and Wiener saw of it. In hindsight, he felt, "They told us very little. They wanted us just to tell them what Wiener's ideas were and how to realize them." The Bell group expressed no further interest in working with the two men from MIT.

Nor did they reveal to Wiener and Bigelow that they already had an outside partner in their project—from MIT. The other team from Tech was part of a top-secret radar research lab that had been founded at MIT in October 1940 and given the deliberately misleading name of the "Radiation Laboratory." The Rad Lab, as it became known, was organized by Vannevar Bush and his inner circle of NDRC colleagues with one urgent mission: to research and develop radar technology for wartime applications. One of its first assignments was to develop an automatic radar-guided antiaircraft tracking device using the new British microwave radar.

The secret Rad Lab team had started work in a wooden shack on the roof of MIT's Building 6 in January 1941, the same month Wiener and Bigelow commenced their collaboration in the adjacent Building 2. By the spring, spurred on by the escalating Blitz of London, the Rad Lab team had assembled a prototype microwave antenna, and, late in May, the device locked onto and automatically tracked its first test plane flying above the MIT campus. That month, the Bell Labs and Rad Lab teams linked up and two Rad Lab staff members came to work full time at Bell Labs in New Jersey. By fall, the two projects had merged for all practical purposes, and in December 1941—just days after the Japanese attack on Pearl Harbor—the Rad Lab's experimental XT-1 tracking radar and Bell Labs' M-9 predictor and antiaircraft fire director were formally joined and developed in tandem.

Bigelow and Wiener were not aware of any related fire control devices in development at MIT or elsewhere when they began their work together. Nor were they told of the Rad Lab's working radar-tracking device at their meeting with Bell Labs engineers five months later, although they eventually got wind of the work in progress on the roof of Building 6.

———

After getting a cold shoulder at Bell Labs, Wiener and Bigelow returned to Tech to continue work on their statistical predictor. They worked through the summer refining their calculating methods and laying plans to build their prototype predictor "in the metal."

Their "little laboratory," as Wiener called Room 2-244, had been started up on a shoestring. The initial funding the NDRC allotted to his project totaled $2,325. (In contrast, MIT's top-secret Rad Lab, with its diverse radar projects, was launched with a staff of thirty to forty physicists and technicians, and a first-year budget of $815,000.) Wiener's modest stipend provided resources sufficient for his work with Bigelow and a support staff of two: a skilled machinist-electrician, who, as Wiener marveled, "put our ideas into the metal almost as fast as we could conceive them," and a human computer who had been an accountant before the wartime mobilization began.

In December 1941, Japan's sneak attack on Pearl Harbor came "as much more of a shame and a humiliation than a surprise" to Wiener. For months, like many Americans, he had been poised with the feeling that "something was about to blow off" in the Pacific. When it did, Wiener and his colleagues could not have stepped up their efforts any further; they were already working flat out. In the weeks before the attack, while he was waiting for Bigelow and his tiny crew to construct their prototype predictor, Wiener had begun drafting a written report on his new fire control theory. Bigelow had requested a thorough explication of Wiener's mathematics to aid in his engineering work, but as usual Wiener had his eye on the bigger picture.

On February 1, 1942, he dispatched his first formal report to the NDRC—the only technical paper he wrote for his wartime project. The 120-page manuscript bore the cryptic, tongue-twisting title "The Extrapolation, Interpolation and Smoothing of Stationary Time Series." When the report reached the head of Section D-2, Warren Weaver, he promptly had the manuscript classified, bound with bright yellow covers, and distributed to a select group of wartime scientists and engineers with the required security clearances. They soon began referring to the imposing document as Wiener's "Yellow Peril."

In its pages, thick with theory and complex equations, Wiener laid down the details of his ingenious new approach to antiaircraft fire prediction. He spelled out his new statistical methods of extrapolation (projecting an object's future position from observations of its past positions), interpolation (estimating an object's position between two known points of observation), and his improved technique for smoothing or filtering a series of observed points that formed a jagged or discontinuous line into a precise, continuous, and far more useful mathematical formulation.

His new fire control theory was a momentous achievement in itself, but Wiener's Yellow Peril went far beyond the fine points of antiaircraft fire control. It redefined the whole concept of "control," a concept essential to engineers in every field, in new scientific terms for the electronic age. Wiener's pyrotechnic paper made plain to his wartime colleagues that a technological revolution, if not yet a full-blown scientific revolution, was close at hand. For the first time, he distinguished the established domain of power engineering—which had been dedicated for centuries to the hard work of controlling heavy machinery by purely mechanical means and, since Edison's day, to generating and harnessing the strong currents of raw electrical power—from the new field of communication engineering. That flourishing domain of fluxing telephone signals, radio waves, and reflected radar beams, Wiener's paper made clear, was the new scientific terrain where the tools for controlling every form of modern technology, from electronic gadgets to gun turrets, would reside forever after.

Wiener's historic step in the Yellow Peril was to identify and unify the disparate domains of communication engineering, whether the diverse ranks of working engineers were ready or not. In a surgical cut, he severed the entire practice of control engineering, which had been the province of power engineers historically, and brought it bodily into the camp of communication. He defined this greater technical science of communication and control in its entirety as "the study of messages and their transmission." Building on his earlier work in the mathematical analysis of electronic signals, Wiener became the first person to define the elementary unit of communication—the message—in the broadest technical terms as an "array of measurable quantities distributed in time," and to provide precise statistical methods for filtering, refining, and reconstructing "a message . . . corrupted by a noise."

He also made the crucial conceptual leap that the communication of messages was not limited to "a conscious human effort for transmission of ideas." The signal used to control an electric motor, a self-regulating "servomechanism"—the technical term for the first automated industrial machines—or any mechanical or electrical control device "is also a message, and . . . belongs to the field of communication engineering," wrote Wiener. In the same stroke, drawing on his long experience with analog computers and his early insights into the new mode of digital computing, he joined the budding enterprise of computation to the larger science of communication. Wiener stated unequivocally that all communication operations, "carried out by electrical or mechanical or other such means, are in no way essentially different from the operations computationally carried out by the . . . computing machine."

Then, after establishing the "fundamental unity of all fields of communication," Wiener went on to make one last, all-important distinction for his incipient science. He described that elusive commodity common to all devices used "for conveying information," and a precise means of measuring the "effectiveness" of the information contained in any message, using the new technical method telephone engineers had begun to employ, which he described, in the most basic terms, as *the mathematical likelihood of that particular message emerging from a larger "measure or probability of possible messages"*—a measure that would eventually come to be designated in "bits." And, summing up two decades of his trailblazing work, he put forth the principle that would guide the labors of communication engineers in every field from that moment on, that *"such information will generally be of a statistical nature."*

Years later, Wiener's Yellow Peril would be declassified, published openly, and properly credited as the founding document of the new technical science of communication, but in that dismal winter of '42, his spectacular work was swallowed whole by the wartime bureaucracy.

"I think it was distributed to about fifty people," Bigelow confirmed. "Copies were sent to various places but all to people in the war effort who were sworn to secrecy."

Among those certified secret-keepers was Warren Weaver's liaison at Section D-2, George R. Stibitz, the inventor of the digital calculator Wiener had seen at Dartmouth in 1940. A week after Wiener delivered his Yellow Peril, Stibitz produced his own report attempting to explicate Wiener's paper, which he then circulated among other project teams in Section D-2, including researchers working on both the Bell Labs M-9 predictor and the Rad Lab's XT-1 automatic tracking radar.

Many bright minds on those teams were baffled by Wiener's new statistical theory and mathematics. Some dismissed it entirely. Ivan Getting, who was then the director of the Rad Lab's XT-1 project, recalled that Wiener "did some work on the statistical approach to fire control which as far as I could make out at that time made no sense." Getting was dismissive of Wiener and his work, and his attitude was shared by many of the Rad Lab's young hires, but Wiener's feelings about the Rad Lab's "big shots" were mutual. While their parallel projects were in progress, Wiener was brought in to the Rad Lab to teach higher mathematics and electronics theory to the young physicists who, according to the Rad Lab's director, "knew little to nothing about the microwave electronics that would be needed to translate the British . . . magnetron into a working radar system."

However, a month after Wiener delivered his Yellow Peril to the NDRC, he resigned from the Rad Lab in frustration over the tight time constraints that were placed on his teaching, a limit he perceived as potentially disastrous to the lab's wartime missions. In a letter written to the NDRC's supervisor at the Rad Lab, Wiener complained about "the highly chaotic and anarchic regime of theoretical work in the radiation laboratory" and declared that "my efforts there have proved unwelcome [and] a waste of time on my part."

The Rad Lab's kid-physicists and older staffers did well enough without Wiener. On April 1, 1942, ten days after Wiener resigned from the Lab, the joint Rad Lab-Bell Labs prototype fire-control apparatus was tested at Army antiaircraft command headquarters at Ft. Monroe. The test was deemed highly successful by those in attendance, and the next day the Army ordered 1,256 units for use in the European and Pacific theaters.

Soon after that test, Wiener and Bigelow made another trip to Ft. Monroe to observe the new apparatus in action. As they saw for themselves, the system's performance was far from flawless. The target airplane's image on the Rad Lab's circular radar screen flitted about in a "herky-jerky" way as radar pulses rebounded unevenly from different parts of the aircraft. The Bell device's linear predictions were not right on target either, but given the Allies' urgent defense needs following the events at Pearl Harbor, Wiener and Bigelow agreed that the system worked well enough to justify its expedited production and deployment.

"It did," Bigelow conceded, "because in the war the projectiles you were firing at the enemy target were altitude-sensitive shells that went up and then burst when they reached a certain height above the ground. They scattered and covered a lot of area, so they didn't have to be all that accurate, although you wasted a lot of shells before you got a hit. I remember we watched the shells go up and burst over the ocean. Afterwards, we could see all the fragments falling into the water."

Bigelow recalled Wiener's muted response to that first field test. "He didn't say a great deal about it. We both knew it was a less sophisticated predictor than the one we were working on."

Wiener and Bigelow pressed on with their project, still confident that their statistical approach ultimately would provide superior targeting capability. At the same time, Rad Lab mathematicians and engineers who had received copies of Wiener's Yellow Peril report devoted considerable attention to understanding his new smoothing and filtering techniques. As their field trials progressed, the Rad Lab team devised a method of smoothing their jerky radar images, a technical improvement that helped the Bell predictor to calculate more accurately where the plane would be when the artillery shell arrived. Bigelow recalled that at least two top members of the Rad Lab team came to him for help in applying Wiener's statistical work to their radar-tracking apparatus. He also discussed Wiener's work with team leaders of the Rad Lab "theory group." "They felt there might be some great theoretical secret here that Wiener had disguised in mathematical operations they didn't understand, because Wiener talked a different language and expressed things in ways that were not clear to the rest of the engineering and circuit-designing world," Bigelow recalled.

Before long Wiener himself was fielding inquiries about his new theory from Rad Lab higher-ups. The volume of requests was so great that Wiener beseeched Warren Weaver for additional staff to handle the inundation. "[Rad Lab] people are coming to us all the time with problems that are right down our alley. I don't see any rest for the wicked," he told Weaver, as he put in for "a small number of young mathematicians to . . . assign to the handling of requests for the Radiation Laboratory."

By the late spring of 1942, Wiener and Bigelow had assembled their prototype predictor. Like Bell Labs' M-9 predictor, their device was a specialized analog computer, bigger than a breadbox but smaller than a baby grand piano. However, Wiener's new statistical methods made the Wiener-Bigelow machine more sophisticated mathematically and electronically. Its circuitry translated the target plane's observed positions, over an observation time of ten to twenty seconds, into a series of electronic signals and computed the point in space where Wiener's statistics predicted the plane would be at a given moment in the future.

When it came time to test their machine, Wiener and Bigelow hooked up their predictor to the red spotlight on their model antiaircraft controller, which by this time had been improved to replicate more accurately the aiming of actual antiaircraft artillery. Bigelow dimmed the lights in their workspace, took charge of the mechanism's control stick, and chased their mock warplane on its irregular flight around the walls of Room 2-244. Their predictor led the warplane and accurately predicted its future path a half second *before* it arrived at the targeted coordinates!

Wiener was elated. "He began puffing on his cigar in a violent way," Bigelow recalled. "The room filled up with smoke. He sort of jumped up and down."

For Wiener it was a major conquest, the first practical demonstration of his new statistical prediction theory and the first evidence in the metal that "his calculations were relevant and serviceable."

In the summer their prototype was ready to show. The team tidied up Room 2-244, and, on July 1, Wiener and Bigelow demonstrated their still nameless, numberless apparatus for Weaver and Stibitz. When the demonstration was over and assessed, as it turned out, Wiener and Bigelow's test results were four times more accurate than Bell Labs' linear predictor and ten times more accurate than the next best method then in development.

Weaver called the Wiener-Bigelow predictor a "miracle," but asked, "was it a useful miracle?" By now their lead time was up to a full second, and Weaver found that, for a one-second lead time, their predictor's performance was "positively uncanny" and "astonishingly close" to the plane's actual position. But to be of practical value in the field its effective prediction time would have to be doubled or tripled, at least, to allow sufficient lead time for the military's high-caliber artillery shells to reach their target altitudes. After the demonstration, Weaver sent a letter to Wiener stating his view that Wiener's theoretical work had been successfully completed and expressing his belief that it would find wide application. His only question was whether their prototype could be translated quickly and practically into a full-scale fire control device.

That was no easy task. As Wiener knew, putting his statistical prediction theory into practice in a working mechanism suitable for mass production and heavy action in combat was a technical feat that posed unprecedented engineering challenges. Like Bell Labs' linear predictor, their statistical predictor would need to be linked to the Rad Lab's radar tracking device and yoked to a battery of antiaircraft guns on a moveable assembly, with each component exchanging control signals and electronic data with the others, and modifying its operations accordingly in response to a continuous flow of feedback about the system's performance. Those tangible engineering practicalities forced Wiener and Bigelow to confront the enigmatic feedback process and the thorny problems Wiener did not tend to in his project with Lee in China. Now, once again, in the push to translate his prediction theory into a viable production prototype, feedback emerged as the focal point and final obstacle in his war work with Bigelow.

The term "feedback" was still new in the engineering argot, although the idea had been around since ancient times. The Greeks invented automatic wine dispensers and water clocks controlled, like modern plumbing, by the feedback action of a float. In 1789, the still nameless principle became an emblem of the industrial revolution, when the Scottish engineer James Watt invented a clever new device he called a "governor," which used a little portion of his new steam engine's output to regulate its speed automatically. The same principle was employed by European shipbuilders in the 1850s to create "slave-motors" or servomechanisms that served as automatic steering engines for oceangoing steamships. Similar devices were used to rotate and stabilize the giant gun turrets on British and French naval vessels; however, those early feedback mechanisms fell prey to a problem Wiener and Bigelow had already encountered in their prototype fire director: often the mechanisms ran amok in spasms of overcompensation, causing the big guns to swing widely around their mark.

In the 1920s, communication engineers found a similar condition convulsing their networks. The wailing, banshee-like noise they called "singing" arose when an amplified sound was picked up by a nearby microphone and re-amplified many times over in a vicious cycle of runaway "positive" feedback. The problem persisted until 1927, when a young electrical engineer at the newly formed Bell Labs discovered that, when a portion of the output of the amplified signal was fed back into the circuit in reversed phase—electronically speaking, as a negative feedback—the ear-splitting singing ceased and the signal came through louder and clearer.

By 1942, engineers were routinely using feedback principles to create machines that automatically regulated their own operations. Thousands of servomechanisms were in commercial use in mechanical and electrical systems. MIT even organized the first "Servomechanisms Laboratory" in the winter of 1940. But, as it had been with electronic circuits for decades after their first practical applications, there was scant theory beneath those diverse feedback inventions, and scarcely an inkling among engineers that one form of feedback had anything to do with another.

Wiener learned the facts about feedback from Bigelow, who told him about a paper he had read in the Bell Labs journal and other articles that were beginning to appear in the engineering literature. "As soon as I described it he understood it," said Bigelow. Wiener instantly grasped the importance of feedback to circuit theory, servomechanism design, and the fledgling field of electronic computing. But he saw beyond the technical details of the feedback circuits and mechanisms they were building into their fire director to the all-important human factors involved in real-world fire control situations, which he and Bigelow had incorporated into the design of their statistical device: the underlying motivations, intentions, and split-second decision-making processes that determined the enemy pilot's zigzagging path, the complex sensorimotor abilities that controlled both the pilot's maneuverings and the artillery gunner's tracking and firing behavior, the inevitable human errors of observation and aiming made by field operators as they turned the cranks on the guns to keep their targets in the crosshairs, and their reflex responses of compensating and, more often, overcompensating for their errors with each hurried turn of the mechanism.

Those basic problems of sensory perception and motor response drew Wiener back to biology, and to his good friend Arturo Rosenblueth of Harvard Medical School. In the summer of '42, Wiener requested permission from Weaver to share information with Rosenblueth about his classified project. Shortly after that Wiener, Bigelow, and Rosenblueth embarked on a three-way collaboration to tease out the tangle of physical and neurophysiological factors involved in antiaircraft fire control.

Wiener and Bigelow told Rosenblueth about the curious problem they had been encountering with their prototype when its mock gun turret would begin to oscillate, swing menacingly from one extreme to the other, and wildly overshoot its mark. Rosenblueth recalled an eerily similar neurological disorder called "purpose tremor" that might cause a person trying to carry out the simplest intentional act—say, picking up a pencil—to swing his arm back and forth uncontrollably and completely miss his objective. Neurophysiologists had traced the disorder to a defect in the internal circuitry of the cerebellum, the brain region that regulates and coordinates muscle movement. The similarity of the two disorders—one technological, one neurological—was striking.

For Wiener, the discovery of feedback was tantamount to the discovery of fire. With Rosenblueth's help, and from his own knowledge of their mentor Walter Cannon's concept of "homeostasis," Wiener drew the connections between feedback in the technical sense, in the physiological sense, and the innumerable feedback loops wired into the living electrical networks of the brain and nervous system. And he made one more link that brought the feedback process fully into focus. As Wiener and Bigelow looked deeper into the human factors they were incorporating into their fire-control apparatus, they found that, like the automated mechanisms they were building to operate on negative feedback principles, the warplane pilots and antiaircraft gunners whose actions they were modeling mathematically also used negative feedback intuitively "to regulate their conduct by observing the errors committed in a certain pattern of behavior and by opposing these errors by actions deliberately tending to reduce them."

With feedback firmly in their sights, Wiener and Bigelow built into their prototype predictor the feedback circuits needed to incorporate the tracking data derived from radar signals reflecting off enemy warplanes, and to feed that stream of new inputs to the computer's targeting apparatus. Rosenblueth's insights helped them to isolate their prototype's mechanical "nervous disorder" and make improvements to minimize its effects. However, despite their best efforts, their prototype continued to display residual tremors and other erratic movements—behaviors Wiener was now coming to accept as inevitable "pathological conditions of very great feedback." Wiener and Bigelow had felt confident that, with Rosenblueth on their team, they could proceed from their crude experimental device to the design of a complete apparatus for anti-aircraft control and prediction. Wiener had solved the feedback problem in principle. Yet, once again, he found that the complexities of feedback would not be so easily resolved in practice.

In October 1942, when Stibitz returned to MIT to assess Wiener and Bigelow's progress, he learned that Bigelow and his crew were having technical problems putting the last points of Wiener's feedback theory in the metal on the timetable they had been given. A month later, in a meeting with Warren Weaver, Bigelow confirmed that their prototype could not be made into a full-scale device in time for use in the war. "It was an enormous job to produce the mechanism and circuitry needed to carry out the very complicated mathematical operations Wiener had written down," he admitted. "It would just take too much time and equipment and everything else to put it all together."

Bigelow was not so blunt with Wiener. "It was clear to me, and to the OSRD, that we couldn't get Wiener's ideas into production, not for this war, but I would avoid such discussions with Wiener." When Bigelow did try ever so gently to inject Wiener with a dose of reality, "He would ignore it and keep right on working. He was unshaken. He wanted to feel he was doing something to help the war effort, and you couldn't undermine that." But that fall, Wiener was forced to scale back on his promise to provide a significant improvement in performance over the Bell Labs predictor. "I think realism came to sink into his thoughts at that time," said Bigelow.

Finally, late in November, Wiener, too, contacted Weaver and reluctantly reported that the best full-scale system he and Bigelow could produce would provide only a ten percent improvement over Bell's M-9, a marginal gain Wiener himself did not consider to be significant. Soon after, their contract was terminated.

———

In July 1943, the joint Rad Lab–Bell Labs fire control system went into mass production, its performance measurably improved during the period after its designers read Wiener's Yellow Peril and consulted directly with Wiener and Bigelow. By February 1944, the first sets reached Europe and saved Allied lives at Anzio south of Rome. In June, thirty-nine units came ashore at Normandy with Allied troops on D-Day. Weeks later, several hundred more gunlaying sets, equipped with new radar-controlled proximity fuses in their artillery shells, reached England in time to thwart the first rocket-propelled V-1 "buzz bomb" attacks on London. In December, more tracker-predictor sets reached France and provided air defense in the Ardennes during the decisive Battle of the Bulge. The Rad Lab-Bell Labs system became one of the American war machine's "greatest success stories."

In the end, Wiener's statistically based fire director never saw combat, but the work of the myopic mathematician—who as a young man in the Great War could not hit "a barn among a flock of barns" with an army rifle—made a significant contribution to the art and science of antiaircraft fire. As the Rad Lab's wartime reports confirmed, Rad Lab theorists drew directly on Wiener's theories and mathematics to solve their radar tracking and smoothing problems. The Bell Labs team, too, used Wiener's work to help them improve the performance of their linear predictor, while other engineers and mathematicians, including Wiener's MIT protégé Norman Levinson and a leading figure at Princeton, were assigned to cull the Yellow Peril for urgent applications to other wartime technical problems.

Above all else, Wiener's war work—like his earlier mathematical work, his innovative electronic circuits and filters, and all his theoretical work—showed engineers the limits of their knowledge and the measure of their ignorance, as he spelled out with statistical certainty and a known margin of error the physical limits on what their projects could hope to achieve. His contributions helped to win the war, and as Rad Lab theorists later acknowledged, his statistical approach to communication and automated control systems ultimately won the day. It became the method of choice in the design of radar systems, servomechanisms, and across the field of communication engineering from that point on.

Wiener's overseers in Section D-2 held his work in high regard, and when a navy captain began having trouble with his ship's newly installed fire control apparatus, he, too, was directed to call upon Wiener for help with his problem. As MIT's wartime records chronicled:

> At the Philadelphia Navy Yard one day late in the war, the automatic fire control of a gun on a warship was behaving erratically. Navy officials hurriedly sent word to the Massachusetts Institute of Technology mathematician who had helped to develop the device, describing the short circuits occurring at certain positions of

the gun muzzle. The professor told them that if they would look into the mechanism at a specific place they would find that a mouse had crawled in there and died. This turned out to be correct.

———

Bigelow tried to account for the discrepancy between Wiener's brilliant war work and the dismissive attitude Wiener himself engendered among many of the war's younger scientists and engineers. He cited Wiener's notorious temperament, and the self-image he projected at age fifty as an eternal prodigy, which did not impress the lab's young physicists and engineers.

"To most people Wiener was this boisterous, preposterous little guy who saw himself as the center of all activity," said Bigelow. "That reputation preceded him everywhere he went. There was a good deal of prejudice against him." But, to Bigelow, Wiener's strengths as a scientific thinker also proved to be as much of a drawback for him as any quirk of his personality.

"Wiener was a generous and lovable guy, extremely bright, not just intuitive but *insightful*. Few people were quicker than he was to see a point, to perceive what the issue was about and what decisions would be most useful in a given application. But our problem during the war was to get something out in the field quickly that would help fire control operations as they actually were encountered, and Wiener was always trying to solve the problem *after* next."

Good minds at MIT and beyond soon realized that Wiener had indeed solved the problem after next. During those years, Wiener midwifed, not one, but two new communication sciences in Room 2-244. In the Yellow Peril, where he laid out his new statistical methods for measuring the "effectiveness" of the information in a message and for "estimating which messages are frequent and which are rare," he made a major contribution to the new technical science that would become famous in its own right as "information theory." Yet, for most of his manuscript's years in captivity, like all Wiener's wartime contributions, his seminal work on the theory of information was known only within the NDRC's small circle of government-approved scientists and engineers.

One member of that inner circle, and a frequent visitor to Room 2-244, was the young mathematician Claude E. Shannon, a recent MIT Ph.D. and a new hire at Bell Labs who was then working on Bell's fire control team and on another team assigned to devise secure coding procedures for wartime communications. He would later publish a seminal paper that would be cited as the founding work in the field of information theory. Early in the war, Shannon came to learn Wiener's new communication theories at the source. Bigelow was present at many meetings between the two men where he watched Wiener give liberally of his ideas to his younger colleague.

"In the time I was associated with Wiener, Shannon would come and talk to Wiener every couple of weeks and spend an hour or two talking with him," Bigelow recalled. "Wiener would exchange ideas with him in a most generous fashion, because Wiener had all the insights of what information theory would be like and he spewed out all these ideas and his comments and suggestions to Shannon."

On several occasions, Bigelow remembered seeing the two men working together at the blackboard on equations that, Bigelow believed, laid the foundation for Shannon's own theorems that would become cornerstones of information-age mathematics and electronics engineering. In fact, Bigelow said that one of his strongest recollections was "seeing Wiener giving Shannon advice, help and ideas over and over again. Shannon would come around and talk with him about communication ideas and how to express them, and Wiener would free associate and free float. I think Wiener was the support for Shannon's ideas and much of his thinking on information theory."

Wiener shared his thoughts unstintingly with Shannon, as he did with every young student who showed talent and a genuine interest in his work. But, after a while, Shannon's visits began to strain Wiener's generosity. Fagi Levinson heard that private response directly from Wiener. "Wiener was very kindly," she said, "but when Claude Shannon came to MIT, Wiener said, 'He's coming to pluck my brains.' He didn't want to see him, avoided him. He worried about people stealing his ideas and getting credit for his work."

Through the later years of the war, Wiener continued to submit ideas to the NDRC and OSRD, but his talents were not put to use in any significant way after his fire control project ended. And, although he ranked among the great mathematical minds of that era, he was never tapped to contribute to the government's all-out mobilization for the Manhattan Project.

One reason for this may have been the dislike for Wiener personally and intolerance of his eccentricities that was felt by younger wartime scientists and many of Wiener's peers. But with little doubt a major reason for Wiener's absence from the atomic bomb effort, and all the war's later projects, was an official concern over security. Wiener objected fiercely to the government's demand for secrecy among scientists involved in war research. He found secrecy ethically objectionable as a member of the international scientific community, and practically impossible for a person of his gregarious nature.

As the war progressed, the government's security concerns were compounded by a deeper concern about Wiener's emotional stability. The war's pressures sat heavy on his hair-trigger psyche. To meet the stiff deadlines he had set for his project, he worked days and nights on end. To stay awake, he took amphetamines without concern for their known physiological effects. Before long the combined stresses, secrecy restrictions, and sheer exhaustion began to take a toll on him. His family and colleagues began to worry about his physical and psychological well-being. At one point, by his own admission, he pushed himself to the brink of a breakdown.

Wiener had other things bothering him during those years, among them, his growing anxieties about his wife's Nazi sympathies and her family ties in enemy territory. Margaret Wiener's relatives in Germany were loyal Nazis "almost to a man." One of her cousins was an administrator in a concentration camp. Her sister's husband managed the routing of trains around Breslau, not far from Auschwitz. Details were sparse, but Wiener felt to his roots each word he heard about the Nazis' genocidal war against Europe's Jews.

As the news grew more dire, Wiener's mood darkened. Bigelow watched him become steadily more depressed, "not over our project, but over a feeling of powerlessness" and his inability to make a greater contribution to the war. By the spring of 1945, Wiener told a colleague he was suffering from "war fatigue." In August, when the first atomic bomb dropped on Japan, Wiener was distraught. He especially "detested the arrogance displayed in the use of the bomb against Asians," Dirk Struik recalled.

By that time, Wiener was no longer associated with any aspect of the national defense program. In May 1944, more than a year after his classified project closed down, he was formally severed from the war effort in a letter from his old friend Vannevar Bush. The blunt boilerplate acknowledged his "assistance in the early phases of the war," described a purely bureaucratic process in which "the list of appointees is periodically reviewed for purposes of removing names of those persons not called on," and thanked him for his "splendid cooperation."

————

And splendid it was, even if his statistical predictor did not go off to war. In the span of only two years, working on that obscure wartime engineering project, Wiener conceived, defined, and quietly announced the coming of a new unified science of communication. He identified the new science's elementary unit, the message, made up of the new statistical substance he called information; he grounded its quintessential process of feedback within a larger conceptual framework; and he linked the fundamental operations of modern telecommunications, computing, and automation to the living communication processes of the human nervous system.

His new statistical theories and communication concepts were portents of the new science in its birth pangs, the first glimpses of a new way of understanding modern technology and taking it in hand, of seeing the world, and ultimately of doing science itself. Yet Wiener's residual experience of the "scientific confusion of wartime" was one of disillusion and hurt, as he watched younger theorists and technicians pluck the fruits of his work while his own wartime theories and writings were still being officially suppressed. Years later he would condemn the "general breakdown of the decencies in science" during the war and claim cryptically, "I found that among those I was trusting were some who could not be held to any trust."

His way of collaboration and cross-disciplinary science was coming to the fore, but he would emerge from the war with deep qualms about what others would do with the new knowledge and technology. And, already, in his early fifties, he was beginning to worry that his own prodigious mental powers were declining. Colleagues spoke of his constant questioning. "Tell me, am I slipping?" he would ask.

But his greatest work still lay ahead.

PART TWO

In the Court of Cybernetics

7

The Knights of Circular Causality

And these knights with the Round Table please me more than right great riches.

—Sir Thomas Malory, *Le Morte d'Arthur*

NORBERT WIENER COULD KEEP A SECRET when he had to, but he could not keep a good idea to himself. The lockdown of his war work frustrated him immensely and forced him to seek alternate outlets for his new communication concepts. And find them he did. In fact, only five months after the United States entered the Second World War, Wiener quietly began lobbing a few bombshells of his own into the stately halls of the established sciences, where they caught fire in the minds of some of America's most distinguished scientists and innovative thinkers.

That turn of events marked the start of a postwar scientific revolution that would change the life of the mind and all the world's societies. Yet, when his new communication ideas burst onto the scientific stage, Wiener was nowhere near the scene and not yet free to claim them as his own.

The first glimmerings of his new thinking lit up a sedate conference room in New York's Beekman Hotel on the morning of May 13, 1942, at a conclave of psychologists, physiologists, and social scientists who had gathered under the auspices of the Josiah Macy Jr. Foundation. The foundation, established a decade earlier by an heir to the prominent New York family of maritime merchants, was a sponsor of eclectic conferences in the new interdisciplinary spirit, and this meeting had been called to debate some timely issues at the junction between psychology and brain science.

The two dozen attendees at the conference were an illustrious group. They included Warren McCulloch, a neurophysiologist from the University of Illinois and one of the world's foremost authorities on the functions and organization of the brain; another leading neurophysiologist, Rafael Lorente de Nó of the Rockefeller Institute in New York; Lawrence Kubie, a neurologist turned Freudian psychoanalyst with a client list of eminent New York artists and play-wrights; and two celebrated anthropologists, Gregory Bateson and Margaret Mead, who were husband and wife at the time and renowned figures interna-tionally for their trailblazing studies of life in remote Pacific Island cultures. Wiener was hard at work back in Cambridge preparing for the debut of his pro-totype antiaircraft fire director, and his contributions that day were in the hands of his good friend and colleague Arturo Rosenblueth, the Harvard neurophysi-ologist, who brought his own considerable talents to the table.

The conferees had barely begun their discussion of minds and brains when they were diverted by Rosenblueth's wholly unexpected presentation. Rosen-blueth, a "burly, vigorous man of middle height, quick in his actions and speech," was an impressive figure. Born in Chihuahua, Mexico, he studied medicine in Mexico, Berlin, and Paris before coming to the United States in 1928. His lin-eage was part Hungarian Jew, part Spaniard, part Mexican Indian, all of which he carried with an aristocratic bearing. With style and sophistication, he laid out the primordial ideas of Wiener's developing communication science, which Wiener had only recently delivered to his government supervisors in his impos-ing "Yellow Peril" manuscript. Without spilling the secrets of Wiener's classified project, Rosenblueth told his audience about messages, feedback, and the sur-prising similarities he and Wiener and the engineer Julian Bigelow were finding in the actions of electronic devices, automatic machines, and human nervous systems. He described the rich insights into human behavior and physiological response the three men had already distilled from their study of man–machine interactions.

Then Rosenblueth let loose a radical idea he and his colleagues were just be-ginning to flesh out in Cambridge. In their work on problems of communica-tion and automatic control, Rosenblueth said, they had identified a new realm of orderly processes observable in nature and the human world. These new com-munication processes were not governed by the traditional logic of linear, cause-and-effect relations that had driven the scientific method since its inception. They were governed by a new logical principle Rosenblueth called "circular causality"—after the circuitous feedback loops Wiener and Bigelow had tapped and harnessed in their device for predicting the future positions of fast-flying airplanes.

This new causality was one in which living things and machines alike be-haved *with purpose*. It was a sizeable leap, from machines that took aim at targets to creatures and machines with aims of their own, and the first formulation in scientific terms of the strange circular logic of feedback that lay at the root of all intelligent behavior.

Rosenblueth's presentation was clearly intended to shake up the astute minds in his audience. He acknowledged that the notion of purpose was heresy among serious scientists in the twentieth century. The very idea of purposeful action defied science's reigning cause-and-effect paradigm, which held that no ultimate purpose or end result could govern an action that preceded it in time. Practitioners of that paradigm had outlawed talk of purpose with respect to inanimate objects and even living things. In their urge to bring greater rigor to the modern mind, reductionist philosophers and scientists—following in the line from Russell and Whitehead in England, to Wittgenstein and his disciples in the influential Vienna Circle of "logical positivists," to the strict behaviorists who dominated American psychology—had banished mention of the word "mind" itself and, with it, the whole subjective realm of human experiences, emotions, purposes, and other disputed "internal states" that could not be readily detected by the senses, described mathematically, and verified experimentally.

But those objections posed only passing obstacles to Wiener and his colleagues. Their daring new logic, mathematics, and experiments conducted in Wiener's wartime laboratory had uncovered a huge class of living things and mechanical devices that carried out purposeful acts along circular paths—acts that were as real and tangible as any cause-and-effect action. Rosenblueth offered his audience one practical example of a class of purposeful devices that was already known to the wartime public: torpedoes with built-in target-seeking mechanisms that homed in on a ship or submarine by following the magnetic pull of its hull or the sound coming from its propellers. He followed with examples of biological activities, readily observable in animals and lower organisms, that displayed the same unmistakable signs of purposeful action "directed toward the attainment of a goal": light- and heat-seeking movements by plants and primitive creatures; homeostatic processes such as the body's internal mechanisms for regulating appetite and temperature; and virtually every form of higher-order animal behavior.

All those purposeful actions were governed by circular communication processes and guided to their goals by error-correcting negative feedback—in Wiener's new communication terminology, by information that looped back continuously to its source to show how far off the mark it was straying and the corrections needed for the system to reach its goal. That fundamental insight raised exciting new possibilities for theory and research in biology, brain science, and all the sciences, Rosenblueth said. Wiener's communication principles and statistical methods provided the theoretical foundation and rigorous mathematics needed to ground and verify those complex living processes, and the technical means to reproduce them in working models; and Rosenblueth proposed to his colleagues that they should begin just such a program of research, based on those theoretical and technical breakthroughs.

Rosenblueth's talk stole the show at the Macy Foundation conference. His words set off depth charges among the psychologists in attendance, some of whom had been at war for years with strict behaviorists over the importance of purpose in animal and human behavior. Among the brain scientists, his notions of feedback and circularity struck a resonant chord. Twelve years before, Lawrence Kubie, the neurophysiologist turned psychoanalyst, had published one of the first papers proposing that circular waves of electrical activity could arise in the brain's networks of connected neurons and "move along pathways which ultimately return them to their starting points." Rafael Lorente de Nó of the Rockefeller Institute had confirmed the existence of those circular neural networks several years later.

Warren McCulloch, the foremost neurophysiologist at the meeting, and one with a strong philosophical bent, seized on Rosenblueth's presentation. McCulloch found the new communication concepts to be deeply intriguing, ideally suited for application to his own laboratory research, and of immediate relevance to the logical theory of brain function he was developing in Chicago with his young colleague Walter Pitts. In his lifelong quest to learn how the mind's higher powers could arise from the brain's incessant chatter of electrical activity, McCulloch had pioneered the use of new electrical techniques to study the brain's functions and physiology. He had been watching developments in the field of electronic computing and pondering the parallels to neurological processes he had observed in the laboratory. For McCulloch, the new communication perspective and research program Rosenblueth was describing raised exciting prospects for novel interdisciplinary projects that could shine new light on age-old questions of brain and mind.

And McCulloch was not alone in his excitement. Among the social scientists at the conference, Gregory Bateson was especially enthusiastic. He saw in Rosenblueth's brief introduction to Wiener's new communication concepts something he had been urgently seeking for years: a rich new resource for theory and research in anthropology and all the social sciences. As a scrutinizer of far-flung cultures, Bateson quickly grasped the human implications of a science of communication and saw in its logically sound, mathematically precise processes and principles practical tools to help sort out the hodgepodge of human relationships that shape the lives of individuals and societies.

Margaret Mead was thunderstruck by what she heard. Years later, she wrote in a memoir, "I did not notice that I had broken one of my teeth until the Conference was over."

Those first knights of circular causality soon found themselves joined in a common cause with a greater purpose: to find and retrieve the defining features of humanity—the rudiments of human intelligence in all its inward and outward expressions—to ground them logically and neurologically, theoretically and experimentally, and, thereby, to provide an alternative to the twentieth century's rampant forces of scientific reductionism. They were the advance guard of a new counterforce in scientific thought that would become known as "the cybernetics group," and Wiener would soon take his place in their front ranks. But their dawning vision and scientific crusade would be deferred until the other war was won.

———

Early the following year, the first inklings of Wiener's new science were put out for public consumption. The short article Wiener, Rosenblueth, and Bigelow published jointly in the distinguished journal *Philosophy of Science*, titled "Behavior, Purpose and Teleology," contained no complicated equations or classified engineering designs. It was instead a broadly philosophical paper that was more profound in many ways than Wiener's brain-blistering Yellow Peril. In six pages, Wiener and his colleagues unfurled their proposition that the complex workings of automatic machines and electronic computers—and living nervous systems, too—could all be studied from a unified viewpoint grounded in the advancing science of communication. Formally, now, they declared that their scientific framework offered a whole new way of looking at the ubiquitous communication and control processes carried out, to varying degrees, by intelligent machines, human beings, and all living things, and that each of those remarkable entities achieved their goals through *purposeful action governed by negative feedback and the logic of circular causality.*

It was the first manifesto of the communication revolution, and it proposed something truly seditious: an alternate system for understanding the new realm of communication and control processes that were purposeful, goal-directed, and *teleological* in nature. Their use of the word "teleology" was even more radical than their use of the word "purpose." The term traced back to ancient Greece—and, for Wiener, to the classic texts he had learned in the kneehole of his father's desk. In Aristotle's *Physics,* beyond the purely physical causes that explained why things existed and behaved as they did, lay their purpose or "final cause" (Gr., *telos,* end). For Aristotle, as for Wiener, the end came first. Purpose had primacy, it was the highest good, as the old sage averred, "for being the purpose means being the best of things and the end of everything else."

The concept of teleology had met its own end during the Dark Ages, when questions about final causes were ceded to religion, where they flourished in endless theological debates about God's grand design for the world. But now Wiener and his teammates were conspiring to revive teleology at the center of a new science of communicating beings and machines. With unwavering resolve, the authors argued that "concepts of purpose and teleology . . . although rather discredited at present, are . . . important." They rejected the conventional wisdom of causality and insisted that their new perspective on purposeful communication processes could provide a greater understanding of human beings and all living things. It could illuminate pressing issues in the design of intelligent machines, and point the way to the building of computers, robots, and other automated devices possessing rudimentary powers of learning and memory.

Their manifesto was a mote in the eye of the reigning orthodoxies. In principle, it provided a rational explanation for the appearance in nature of intentional, intelligent behavior without recourse to metaphysics or divine intervention, and it laid out a systematic program for putting their new principles into practice across a wide range of scientific and technical endeavors.

Wiener was excited by the prospect, and soon so were a lot of other people.

———

In Chicago, Warren McCulloch picked up the banner of Wiener's new ideas and brought his own ample powers to the revolution. Tall and lean, with a craggy face and a bushy beard, McCulloch was a swashbuckling Scottish-American with the courtly air of an old Highland laird. Born in Orange, New Jersey, in 1898, the son of a successful businessman and a devoutly religious Southern belle, he went to a small college in Pennsylvania and began his studies for the ministry, but his inquisitive mind led him into philosophy and far from his ordained calling.

As an undergraduate in the 1920s, as his own legend had it, McCulloch asked the question he would ponder the rest of his life: *"What is a number that a man may know it; and a man, that he may know a number?"* It was a high auguring, and it put him on a trajectory that would run parallel to Wiener's for the next four decades. He transferred to Yale, and plunged into Russell and Whitehead's *Principia Mathematica* in a quest to uncover the mind's logical structure. During graduate work in psychology at Columbia, he became convinced that the sure route to the mind, and to the real foundations of mathematical logic, would be through a systematic study of the brain and nervous system.

In 1927, McCulloch received his M.D. from Columbia and began his descent into the labyrinths of the brain's billions of individual cells, the neurons, and their complex networks of working connections, the synapses. His suspicion was that neurons in the brain's uppermost layer, the cortex, where human reasoning and reckoning were thought to take place, were connected in ways that embodied the logical relations described in the *Principia*. If such a resemblance could be verified experimentally, McCulloch believed, it could explain how human beings made logical deductions and carried out calculations—to wit, how a man could know a number or any other thing—and how ideas generally took form and trafficked in the brain.

McCulloch returned to Yale in 1934 as a research fellow in Yale Medical School's Laboratory of Neurophysiology, and over the next half dozen years he methodically created the first detailed map of the functional anatomy of the cortex. By the early 1940s, McCulloch had found solid evidence that the brain's neural networks fired in patterns that obeyed the formal rules of symbolic logic. He confirmed that neurons were "all-or-nothing" decision makers: They fired or did not depending on whether the sum of incoming signals from neighboring nerve cells added up to a "true" (fire) or "false" (don't fire) statement. Their connected networks performed complex mathematical summations in elaborate sequences of electrochemical "statements," each contingent on the one before, that could be likened to the propositions of an extended logical argument.

McCulloch's grand hypothesis—that the human brain, the physical substrate of the mind, was a little electrical engine for doing mathematical logic—seemed to explain how the brain performed rational processes of logical inference, deduction, reckoning with numbers, basic functions of sensory perception, and the mind's higher-order operations of language, learning, and memory. But along with those logical connections, McCulloch also saw something he had not expected to find in the brain—something totally *il*logical. As his Macy colleagues had observed a decade earlier, McCulloch confirmed in his own laboratory that many of the brain's branching networks contained inexplicable loops of *circular* connections. They were, in effect, wired by nature to make endless circular arguments that contradicted themselves neurologically and, ultimately, went nowhere. Those seemingly irrational arguments posed an intractable puzzle for McCulloch, as their symbolic counterparts had decades earlier for Russell and Whitehead in the infamous paradoxes of the *Principia*.

McCulloch's paradox of "nets with circles" exceeded his grasp. The image of closed circular arguments in the brain, shunting around in endless eddies of neurological jabber, made McCulloch's head spin. They made no sense. They had no . . . *purpose*. But that was before McCulloch ran into Arturo Rosenblueth at the Macy conference. Rosenblueth's revelations about feedback and circular causality seemed to hold the key to the logical puzzle McCulloch was working, but he could not put those final pieces together on his own.

Late in 1941, McCulloch had relocated to Chicago to take a position as director of the new research laboratory of the Neuropsychiatric Institute at the University of Illinois Medical School. By the spring of '42, he had one of the best shops around, with a staff of almost thirty and some of the brightest brain researchers in the field. But none of them could solve his puzzle of paradoxical neural nets.

Then he met a smart young kid who showed him the way.

———

Walter Pitts was an awkward, painfully shy, boy-wonder mathematician whose troubled youth made Norbert Wiener's look idyllic. Born into a working-class family in Detroit in 1923, Pitts ran away from home at thirteen and made his way to Chicago.

Jerome Lettvin, a professor emeritus of electrical engineering and biomedical engineering at MIT, was the instigator of the McCulloch-Pitts collaboration and remains the primary source of information about the little-known Pitts, whose reputation was more mythical than real among historians of twentieth-century science. According to Lettvin, who was a premed student at the University of Chicago when he met Pitts in 1938, "Walter's father was a plumber who beat up on him and Walter finally escaped and was living on the streets. One day he was chased into a library by some bullies and he hid out in the stacks where they kept the mathematics books. He came across the *Principia Mathematica* and he could not put it down. He stayed in the library for a week and went through all three volumes. Then he sat down and wrote a criticism of a long section in the first volume, which he sent off to Russell in England. Russell responded favorably. He sent back a letter inviting Walter to do graduate study at Cambridge— mind you, this was a thirteen-year-old."

Evidently, Russell was more keen on Pitts than he had been on "the infant prodigy named Wiener," but Pitts was in no position to accept Russell's invitation. Two years later, Pitts arrived in Chicago, where he attended a lecture on mathematical logic given by Russell, who was a visiting professor at the University of Chicago in the fall of 1938. According to separate accounts by Lettvin and McCulloch, Russell directed Pitts to study with Rudolf Carnap, the leader of the Vienna Circle of logical positivists, who had recently come to America from Austria and settled at the university. That fall, without even a high school diploma to his credit, Pitts became an unofficial student at the University of Chicago.

Lettvin recalled Pitts's irregular college career. "Walter would attend classes occasionally. He had no money so he couldn't register. He was a homeless waif, living in a shabby room for four dollars a week, but he was quickly recognized as a 'weird kid genius' and given a small stipend from UC." Pitts met Carnap in an encounter reminiscent of his earlier exchange with Russell. "Carnap had just written a book on logic, and Walter brought in his copy with written annotations pointing out what he considered to be some serious defects," said Lettvin, who was there. "He walked into Carnap's office with the book, didn't introduce himself, and started asking questions. Carnap was enthralled." Pitts studied with Carnap just long enough to master the logician's abstruse symbolic notation. Then he became intrigued by the work of another prominent émigré at the university, the Russian mathematical physicist Nicolas Rashevsky, founder of the field of mathematical biophysics, which aimed to rebuild biology on the bedrock of the physical sciences using the new cutting-edge tools of mathematical logic.

The quiet Pitts, with his slight frame and thin face eclipsed by his glasses, barely made a physical impression. But, by 1941, he had consumed the university's libraries and much of its curriculum, leaving a trail of nascent brilliance in his wake. Pitts also had an impish streak he made no effort to suppress. "Walter once came into a science survey class when they were having a true-false final exam," Lettvin remembered. "He sat in the front row tossing up a coin and marking down answers, tossing, marking. He got the highest grade in the class. Of course, the coin tossing was just for show."

Lettvin was in medical school when McCulloch set up shop across town, and he was among the first to learn about the brain research getting under way in McCulloch's bustling laboratory. He met McCulloch and, soon after, introduced him to Pitts. Like Russell, Carnap, and Rashevsky before him, McCulloch instantly recognized Pitts's genius, and he moved quickly to bring Pitts into his project to devise a logical model of the brain and nervous system. McCulloch took a personal interest in all the young people who entered his scientific circle. He often took hard-pressed students into his home, where they were welcomed into the family by McCulloch's wife, Rook, and their four children; and early in 1942, Pitts and Lettvin both moved into McCulloch's rambling residence on the outskirts of Chicago. There, in the quiet of the night, after McCulloch's own children, who were not much younger than Pitts, had been put to bed, McCulloch and Pitts proceeded to crack the logic of the brain's neural networks. In a matter of weeks, they had parsed the flow of signals through the brain's branching pathways in their unprecedented effort, as Lettvin put it, "to understand how [the] brain could so operate as to be the mechanism for mental process."

Their first collaborated paper, "A Logical Calculus of Ideas Immanent in the Nervous System," published in the *Bulletin of Mathematical Biophysics* in 1943, was as profound and revolutionary as Wiener's paper with Rosenblueth and Bigelow that appeared the same year. Using elaborate logical notation, McCulloch and Pitts showed that all the activities "which we are wont to call *mental* are rigorously deducible from present neurophysiology." They explained how simple sensory experiences could be "computed" logically in the brain from signals carried to the organ from sense receptors in the skin. They drew the first schematic model of a logical "net of neurons" and made the case for their radical proposition "that every idea and every sensation is realized by activity within that net." They even showed how higher mental processes such as learning and memory could be computed and lead to the formation of new synaptic connections between neurons.

Then they solved McCulloch's paradox of circular networks. Drawing on the new concept of circular causality McCulloch had gleaned from Rosenblueth at the Macy conference, McCulloch and Pitts set down a series of theorems describing the action in neural "nets with circles" that showed how the brain's looping networks could give rise to self-perpetuating cycles of electrical activity, and how that "activity may be set up in a circuit and continue reverberating around it for an indefinite period of time." Such a phenomenon, they asserted, could create persisting memories through electrical action alone. More important, it could enable the brain's computing networks to "predict future from present activities." In that way, McCulloch and Pitts observed, their model accounted for human "systems which . . . exhibit purposive behavior" and other "organisms . . . known to possess many such systems."

And they made another portentous connection. Their exotic brain calculus described a new order of "immanent" computational activity *innate within the brain* that conformed almost perfectly to the British math icon Alan Turing's definition of a "universal" computer, one that, as McCulloch said, "could compute any logical consequence of its input or, in Turing's phrase, compute any computable number." It was a big step forward for the new field Wiener envisioned as a unified communication science applicable equally to brains and machines.

Yet, for all their insights, McCulloch and Pitts failed to connect with the people they were most eager to reach. Like Wiener's papers in the 1920s propounding a new statistical approach to communication engineering, which were not recognized by communication engineers for two decades, the McCulloch-Pitts paper riddled with hieroglyphs of symbolic logic made barely a ripple among mainstream brain scientists. "The whole field of neurology and neurobiology ignored the structure, the message, and the form of McCulloch and Pitts's theory," Lettvin confirmed. The paper raised even fewer eyebrows among mainstream psychologists and philosophers. But it surged through the networks of mathematicians and engineers working on the theory and design of the new electronic computing machines.

Ultimately their paper would be cited as a breakthrough in the evolution of digital computing, as the founding work in the field of artificial intelligence, and as a crucial catalyst in the postwar project to create the world's first "electronic brain." Wiener would be the other catalyst in that project, and soon he would form an historic alliance with the neurophysiologist whose scientific urges were so closely attuned to his own, and with the boy genius mathematician of the next generation.

By 1943, McCulloch already knew quite a bit about Wiener and his work. The two men had met several years earlier at a dinner arranged by Rosenblueth where Wiener wasted no time engaging McCulloch's incipient ideas about neural networks.

"To me he was a myth before I met the man," McCulloch wrote later. "He told me promptly what I could expect of my own theories of the working of the brain. Time proved him right." By the end of that first evening, McCulloch had made a favorable impression on Wiener, and Wiener had overawed McCulloch. "I was amazed at Norbert's exact knowledge, pointed questions and clear thinking in neurophysiology," he recalled. "He talked also of various kinds of computation and was happy with my notion of brains as, to a first guess, digital computers."

The Cambridge-Chicago axis formed within weeks of the publication of their two papers in 1943. That summer, Lettvin moved to Boston to begin his internship in neurology. His co-intern, a young cousin of Wiener's from New York, invited Lettvin to join him on a visit to Wiener's office at MIT. Wiener received the two young doctors with a lamentation about his own lack of talented doctoral students. Lettvin seized the opportunity to tell Wiener about his precocious friend in Chicago, Walter Pitts. "I said, 'I know a mathematician you would like,'" Lettvin recalled. He described Pitts to Wiener as "a young genius, self-taught, who already knew Sanskrit, Latin and Greek by teaching it to himself." He recounted how Pitts had matched wits with Russell, Carnap, and Rashevsky, and signed on as McCulloch's collaborator before the age of twenty. "Wiener looks at me and says, 'There's no such person,'" Lettvin remembered.

That night, Lettvin placed a long-distance call to Chicago and, with McCulloch's help, bought Pitts a train ticket to Boston. Days later, Lettvin took Pitts to Wiener's office, where the two math whizzes wasted no time getting acquainted.

"Wiener greets us very gruffly, 'Hello, hello.' Then he says, 'Come, I want to show you my proof of the ergodic theorem.' He takes Walter next door to a classroom with long blackboards and starts writing. He gets halfway down one board—you know, Wiener made errors every now and then—and Walter says, 'Wait a minute, Professor, this doesn't make sense.' He asks Wiener to defend it. Wiener thinks, defends it, then as he's going along Walter points out a questionable assumption. Wiener thinks, defends it. By the end of the third blackboard, it was clear that Walter was here to stay."

Wiener saw in Pitts the prospect of working with a brilliant young collaborator whose analytical powers rivaled his own, and a chance to help a gifted youth to become a world-class mathematician. He invited Pitts to come to MIT under his auspices and pledged to get Pitts—who had never graduated from any school or even formally enrolled in a university—a Ph.D. in mathematics. Mc-Culloch gave the deal his blessing. "Pitts has told me of your offer. I'm delighted," McCulloch wrote to Wiener, with a playful caveat. "You know that you are hijacking my bootlegged collaborator. . . . Lacking him, I shall probably turn to you both for help with my own naughtiest notions."

In the autumn of 1943, Pitts moved to Cambridge with his few possessions and registered at MIT as a special student under Wiener's supervision, and Wiener quickly set about tailoring a doctoral program designed to fill in the blanks in Pitts's education. He put Pitts to work learning electronic circuit theory and the mathematics of communication engineering. At the same time, he started an active exchange of ideas with McCulloch.

The three men began traveling between Cambridge and Chicago at regular intervals to coordinate their common goals and plan wider collaborative ventures. In 1944, a third outpost was established when Rosenblueth returned to Mexico, where he had been chosen to head the physiology laboratory at the new Instituto Nacional de Cardiología in Mexico City, and Wiener and his new teammates began shuttling north and south of the border. Since the start of their dinner seminars at Harvard a decade earlier, Wiener and Rosenblueth had talked of establishing an interdisciplinary institute in Cambridge to advance their explorations across the fields of science. Now, with Rosenblueth relocated in Mexico, and Pitts and McCulloch in the mix, their dream turned into something larger and more fluid: a moveable feast of lively interdisciplinary action, cross-country and cross-border communications, working visits, extended sabbaticals, and collaborative projects—all radiating outward from Wiener's home base at MIT.

Soon others joined the party. Lettvin completed his training as a psychiatrist and became a capable brain researcher himself. And another bright knight came to the table: a handsome, young Englishman named Oliver Selfridge, the grandson of the founder of London's fashionable Selfridges department store. The dashing Selfridge, a creative thinker and researcher with a quick wit like Pitts and a mischievous streak of his own, began work on several formidable projects with Wiener and Rosenblueth, both in Cambridge and in Mexico, while he was still an undergraduate engineering student at MIT.

Their dispersed research group was multitalented, the work and travel exciting and constantly surprising, and as their various projects unfolded, Wiener and his men found one another great fun to be around. But it was Pitts who really dazzled Wiener. In a letter to the Guggenheim Foundation recommending Pitts for one of their prized fellowships to support his doctoral work, similar to the one Wiener himself had won as a young postdoc two decades before, Wiener sang the praises of his newest pupil and collaborator. "He is without question the strongest young scientist whom I have ever met. I should be extremely astonished if he does not prove to be one of the two or three most important scientists of his generation, not merely in America but in the world at large. . . . He has as a scientist magnificent equipment."

Like McCulloch, Wiener in his own way adopted "the boys"—Pitts, Lettvin, and Selfridge. Pitts, the troubled boy genius, became the son he never had. Lettvin, with his dark beard and premature paunch, was Wiener's mirror image as a younger, more vigorous man. Selfridge, the self-assured Anglo-American, had the dash and derring-do Wiener at once admired and admonished in his younger English colleagues. The three became frequent visitors to Wiener's home in Belmont. Wiener's daughters enjoyed their company and attention. Margaret welcomed them and served them home-cooked meals side-by-side with Wiener's vegetarian fare. At times, she even joined in a bit of mischief with a hidden motive. "We used to sneak beef stock into the soup and his wife was all for that. She would say, 'He mustn't know,'" Selfridge recalled.

For his part in the war effort, Pitts relocated to New York to do mathematical work for a petrochemicals company that was refining radioactive matter for the atomic bomb. Lettvin moved with him to train at Bellevue Hospital for overseas duty as a psychiatrist, and the two shared an apartment in the city. Wiener dropped in on them during his trips to the city on official matters of his own. "He was wonderful company," Lettvin remembered, but Wiener's roaring apnea made it impossible for Pitts and Lettvin to sleep. "We had only a single room. I slept on the sofa, Walter took one bed and Wiener slept in the other. Wiener would start to snore, then suddenly stop, and Walter and I would go, 'My god, he's dead! What do we do?' Then after a minute or so he would let out a big snore and start all over again. After two nights, we were exhausted."

Wiener shared other adventures with his boys and his girls. One summer in New Hampshire, Lettvin and Selfridge came to visit and accompanied the family to the county fair, where Wiener routinely performed a feat people were still talking about decades later. "At country fairs one of his favorite things was the booth where you take a hammer and ring the bell and get a cigar. He would do this time after time and he would always hit it," Barbara recalled. "Jerry tried to ring the bell and he couldn't hit it," Peggy remembered. "Then Dad rang the bell. Jerry was much younger, much bigger and much stronger, but Dad had the knack."

———

Wiener had high hopes for the work commencing within his expanding circle of friends and colleagues, and his hopes would rise higher still when another giant among mathematicians enlisted in his interdisciplinary enterprise.

At Princeton, the polymath John von Neumann took a special interest in Wiener's incubating ideas about communication, control, and computation. Von Neumann, a Hungarian Jew who emigrated to the United States in the early 1930s and took up residence at the Institute for Advanced Study, was an ex-prodigy like Wiener, but with a very different temperament and social background, and with his own ambitions in the fast-changing world of science and technology. Born in 1903, the son of a wealthy Budapest banker, von Neumann had risen to prominence at Göttingen in the late 1920s, accepting an appointment there just as Wiener was departing after his disastrous term as a visiting lecturer. In only a few years, he single-handedly completed the mathematical framework for quantum theory and, in his spare time, devised a new mathematical "theory of games" that was no frivolous matter.

The jaunty von Neumann, whose wide eyes and high forehead gave him the benign look of "a kindly milquetoast uncle," quickly emerged as a dynamo among the powerhouses that packed the little institute on Princeton's bucolic Olden Lane. His peers hailed von Neumann's brain as "a perfect instrument." He belied the stereotypes of ex-prodigies and absentminded professors. He dressed, not in the disheveled vogue of Einstein but, like his father, in "conservative, bankerlike attire." His English was mellifluous and he "exuded Hungarian charm."

Von Neumann and Wiener could not have been more dissimilar, yet the two ex-prodigies had much in common as mathematicians. They were destined to converge, and soon after von Neumann settled in at Princeton, he and Wiener began a dialogue that would change the way mathematics was done, and the men and machines that did it. The two men exchanged papers and met for extended conversations on mathematical matters. Wiener and Margaret came to Princeton as houseguests of von Neumann and his wife. Von Neumann visited Wiener's home in turn. Barbara recalled the family gently mocking "Gentleman Johnny," though not to his face, for his courtly ways and savoir faire that cut a strange figure beside Wiener.

The outlook for machine-assisted computation was improving year by year, and beginning to excite mathematicians as much as electrical engineers. During that period, the young British mathematician Alan Turing was at Princeton completing his doctoral work at the IAS, and refining the ideas in his seminal paper that introduced the concept of a "universal machine" for doing mathematics. Von Neumann was fascinated by Turing's hypothetical machine, and he invited Turing to stay on at the institute as his assistant and develop his concepts further, but Turing chose to return to England.

As the Second World War bore down on America, von Neumann's attention turned from hypothetical to practical matters, while other American mathematicians took steps to actualize Turing's machine. At Bell Labs, George Stibitz began to assemble his binary "complex number calculator" from a pile of two-position telephone relay switches. At Iowa State University in 1939, John V. Atanasoff unveiled his prototype "ABC" computer, the first electronic device to use vacuum tubes for computing—the idea Wiener had proposed to Vannevar Bush in the late 1920s. Also in 1939, engineers at IBM began work, under the direction of Harvard mathematics professor Howard Aiken, on the IBM Mark I, the first large-scale digital calculator that performed computations in the manner Turing had prescribed, by automatically following logical instructions spelled out in sequence on a long paper tape.

Wiener watched those developments with his own discerning eye. The proposal he submitted to Vannevar Bush in September 1940, for an all-electronic digital computer with a built-in logical program, was put on the shelf while Wiener went to work on the problems of antiaircraft fire control, but his interest in computers never flagged, and he was named as the principal consultant on computation for the joint wartime committee of the American Mathematical Society and the Mathematical Association of America. At the same time, von Neumann moved into the inner sanctums of the wartime bureaucracy. He advised the Army's Ballistic Research Laboratory at Aberdeen Proving Ground and the Navy's Bureau of Ordnance. In 1943, he joined the Manhattan Project and became its principal mathematician, and the theorist who proved the viability of the implosion method used to detonate the second atomic bomb. Also that year, he traveled to England to observe the British war effort firsthand and catch up on Turing's efforts to turn his theoretical machine into a practical device for breaking German codes. The visit revived his interest in mechanized mathematics and, by the spring, von Neumann by his own admission had developed an "obscene interest" in computing.

Back at Aberdeen, Vannevar Bush's new 100-ton analog computer, which had only been completed a year earlier, was already being overwhelmed by the work of calculating firing tables for the new artillery rolling out of the nation's armaments factories. A successor machine, a new digital computer dubbed the ENIAC (Electronic Numerical Integrator and Calculator), was being built nearby, at the University of Pennsylvania's Moore School of Electrical Engineering, by a mixed team of military and civilian personnel led by John Mauchly, a physicist, and his chief engineer, J. Presper Eckert. Von Neumann did not learn about the project until the following summer, while he was in the thick of his theoretical work on the implosion bomb. Then, suddenly, his interest in computing found a new application. As chief mathematician on the bomb project, von Neumann was urgently in need of faster and more powerful computing tools, and ENIAC quickly emerged as his best bet. In August 1944, he joined the ENIAC group as a consultant and immediately set to work devising ways to apply the new computer to the death-defying mathematics of atomic fission. At the same time, he began proposing improvements for ENIAC's successor, an even more powerful machine that was already in the planning stages.

Three months later, a very different approach to computing was born, as the date was set for a meeting on von Neumann's turf at Princeton that would lead to the creation of the first electronic computer built to emulate the workings of the human brain. That historic meeting would also lead to an even greater technological revolution. It was initiated, not by von Neumann, but by Wiener, and it was thoroughly informed by his ideas and scientific ideals.

———

In December 1944, Wiener called together the foremost American mathematicians working in the domain of electronic computing and allied technical fields, and the leading theorists and researchers in the field of neurophysiology. Wiener, von Neumann, and Harvard's Howard Aiken jointly signed the letter inviting the select group to the two-day conference in Princeton to exchange ideas and research that would advance the design of computers and lay the groundwork for a broader scientific undertaking in the postwar era. The letter, written by Wiener and sent out on his letterhead in the MIT Department of Mathematics, made the quiet proclamation that:

> a group of people interested in communication engineering, the engineering of computing machines, the engineering of control devises [*sic*] . . . and the communication and control aspects of the nervous system has come to a tentative conclusion that the relations between these fields of research have developed to a degree of intimacy that makes a get-together meeting between people interested in them highly desirable.

Owing to the war, the letter made clear, their get-together would not be "a completely open meeting," but an opportunity "to summon together a small group of those interested to discuss questions of common interest and make plans for the future development of this field of effort, which as yet is not even named."

The offering letter from the three sponsors went out to a mere seven invitees, including the two leading brain researchers who had attended the Macy conference in New York in 1942, Rafael Lorente de Nó and Warren McCulloch, McCulloch's young collaborator Walter Pitts, who was now working under Wiener's direction at MIT, and four other prominent mathematicians and statisticians. In a second letter, also written by Wiener, the three sponsors proclaimed the new principle of teleology Wiener had framed in his paper with his wartime colleagues to be the defining theme of the coming conference, and of the larger scientific venture they were inaugurating. They proposed that their fledgling group "be known as the Teleological Society" because a major part of their interests were "devoted on the one hand to the study of how purpose is realized in human and animal conduct and on the other hand how purpose can be imitated by mechanical and electrical means." Among the items on their working agenda were: "the name of the science," plans for publishing a journal and organizing a research center, determining "what policy should be adopted about patents and inventions" resulting from the group's efforts, "what measures should be taken to bring our ideas to general scientific attention . . . how to protect the researches of the group from dangerous and sensational publicity," and ideas for soliciting institutional support for their venture.

On that last matter the three co-founders purportedly were of one mind: Their group would not solicit, or be beholden to, the largesse of commercial communications, electronics, or business machine corporations. Wiener's concerns no doubt reflected his earlier experience with the American Telegraph and Telephone Company, when he and Yuk Wing Lee sold the patents on their electronic filtering network to the giant company, which then quashed development of their invention to protect the company's similar devices. Now that unhappy entrepreneurial experience, combined with Wiener's growing aversion to secrecy and control of any kind over scientific matters, raised a new ethical standard he would uphold for the rest of his professional life: Scientific knowledge should not be used for the benefit of private interests but in the public interest for the betterment of humankind.

Wiener's position was seconded by Aiken, who had experienced even greater frustrations in his dealings with IBM. Von Neumann also signed off on the letter, and Wiener believed—or wished to—that von Neumann's views were in harmony with his and Aiken's. In an exhilarated phone call to MIT's Vice President James R. Killian, Jr. while the Princeton meeting was in its planning stages, Wiener expressed his understanding, as Killian's secretary recorded it in a memo, that all three conference sponsors were "unanimous in their feeling that such companies as IBM, RCA, Bell Labs, should not be represented at such a meeting since if they fully appreciated the 'tremendous economic consequences' involved in the development of computing machines as control devices they would undertake research immediately, inspired chiefly by the profit motive."

In that call, Wiener reaffirmed his longstanding goal for his new science, "that as soon as the war stops a center may be established at M.I.T. to push research in this field," and he assured Killian that members of the group "would regard it as a cooperative undertaking with frequent conferences to which representatives from other universities would be invited." Killian's secretary ended her memo commemorating Wiener's call with a brief editorial comment: "Norbert was really on the crest of the wave yesterday."

The tiny Teleological Society met at Princeton's Institute for Advanced Study on January 6 and 7, 1945. Most of the major figures who would lead the communication revolution were there, except for Rosenblueth, who had assumed his new post in Mexico City several months earlier, and Aiken, who was unable to attend the meeting.

The excitement among the participants was palpable. On the first day, von Neumann spoke about the state of computer development, Wiener spoke about his work in communication engineering, and spirited discussions followed. On the second day, McCulloch teamed up with Lorente de Nó, as Wiener recalled, "for a very convincing presentation of the present status of the problem of the organization of the brain" and the potential applications of brain research to the new technology. McCulloch captured the air of the meeting in his inimitable way, as he recorded von Neumann's jousting with Wiener over a classic problem of electronic technology, the "black box" problem, set in a wartime context:

> Lorente de Nó and I . . . were asked to consider . . . two hypothetical black boxes that the Allies had liberated from the Germans. . . . No one knew what they were supposed to do [but both boxes] had inputs and outputs, so labeled. . . . Norbert was snoring at the top of his lungs and his cigar ashes were falling on his stomach. But when Lorente and I had tried to answer, Norbert rose abruptly and said, "You could of course give it all possible sinusoidal frequencies one after the other and record the output, but it would be better to feed it noise—say white noise."
> . . . I caught the sparkle in Johnny von Neumann's eye. . . . He knew that a stimulus for man or machine must be shaped to match . . . his [input] filters, and that white noise would not do. There followed a wonderful duel: Norbert with an enormous club chasing Johnny, and Johnny with a rapier waltzing around Norbert—at the end of which they went to lunch arm in arm.

Wiener was elated by the action. As he recalled years later:

> Very shortly we found that people working in all these fields were beginning to talk the same language, with a vocabulary containing expressions from the communication engineer, the servomechanism man, the computing-machine man, and the neurophysiologist. . . . All of them were interested in the storage of information. . . . All of them found that the term *feedback* . . . was an appropriate way of describing phenomena in the living organism as well as in the machine.

In a letter to Rosenblueth written soon after the conclave ended, Wiener called the Princeton meeting "a great success." He declared his conviction that they could now undertake an ongoing program of research and proceed with their plans to create an institute, at MIT or elsewhere, to serve as a base for their new scientific ideas. Von Neumann also was quite pleased with the meeting. He offered to serve on the society's two working groups on computers, and he appeared to be fully on board Wiener's train as it left Princeton Station.

Through the winter and spring of 1945, Wiener and von Neumann stayed in close contact. Wiener sent von Neumann his thoughts on the new research center at MIT, and von Neumann's reply was positive, but he put some distance between himself and Wiener, suggesting that the two of them could make contributions in several directions. They met again in March and, afterwards, Wiener wrote enthusiastically to von Neumann, praising his postwar plans for practical projects in applied mathematics, as von Neumann had described them, and suggesting that "our little [Teleological Society] fits perfectly into the picture." Both men knew that von Neumann's base of operations at the Institute for Advanced Study was a haven for Nobel Prize–winning physicists and pure mathematicians, but that its ivory tower was off limits to applied engineering projects like those the new society's members were looking to develop, and which MIT had been eagerly cultivating in its 200,000-square-foot Rad Lab.

Then, out of nowhere, Wiener made a bid to woo von Neumann from his prestigious perch at the place Wiener called "the Princetitute." After meeting with MIT officials, Wiener had been authorized to offer von Neumann the chairmanship of the mathematics department if he would move to Tech. He promised von Neumann a laboratory of his own, with minimal administrative requirements, and he pledged that their planned research program "would go. through on wheels."

Wiener, who had no head or family gene for entrepreneurial ventures, valued von Neumann's talents as "a very slick organizer." He was confident that his new research center would flourish with the added oomph of von Neumann's brainpower, prestige, political connections, and well-honed social skills that Wiener knew he himself was lacking. Von Neumann gave him good reasons to believe a major coup was imminent. In several encouraging letters, he advised Wiener on the planning of the research center and implied that he would campaign actively for it in the belief that "the best way to get 'something' done is to propagandize everybody who is a reasonable potential support."

But von Neumann had plans of his own.

———

By June 1945, ENIAC was coming together in glass and metal. It was the largest agglomeration of electronic circuitry ever assembled, but by that time the work ENIAC had been designed to do three years earlier was all but done. The war in Europe was over. The final field test of the atomic bomb was two weeks away. The first "gadget" would fall on Japan three weeks later.

And Gentleman Johnny was at last going to build himself a *real* computer. On June 30, 1945, von Neumann unveiled his grand design for a new computing machine called EDVAC—Electronic Discrete Variable Automatic Computer. His seminal "First Draft of a Report on the EDVAC," distributed within the military's classified channels, proposed some radical innovations in computer design—including a more logical structure than previous machines, a single, centralized calculating unit, and a new "stored program" capability—that would come to be known as the "von Neumann architecture." The same basic structure would still be embodied in almost every digital computer operating at the turn of the twenty-first century.

The EDVAC's design embodied von Neumann's own impeccable logic. The new machine was smaller and far more flexible than any of its predecessors. It would be the first truly universal machine in Turing's sense, or as von Neumann described it, a general-purpose, automatic digital computer, capable of executing any appropriately prescribed sequence of operations provided that "the instructions which govern this operation must be given to the device in absolutely exhaustive detail." In von Neumann's vivid depiction, EDVAC would also possess its own human attributes and "specialized organs": a central processing unit or "arithmetical" organ, a logical "control organ," "considerable memory," and specialized "input and output organs."

Von Neumann cited only one published source in his report: the 1943 McCulloch-Pitts paper, which had been ignored by mainstream neurologists and psychologists. The paper explicating the inner calculus of the brain's living electrical connections had convinced von Neumann that a brain-like general-purpose computer could compute any number or carry out any sequence of logical operations. Moreover, it had provided him with a working schematic showing how basic computing units could be programmed and linked together logically in the manner of a McCulloch-Pitts nerve net. Throughout his report, von Neumann richly employed his explicit "neuron analogy" to justify "a consistent use of the binary system" of arithmetic, which mirrored the "all-or-none" characteristics of neural signaling, "vacuum tube aggregates" that functioned as EDVAC's neural networks and "synaptic delays," his "logical control organ" and computer "memory" that corresponded to "the *associative* neurons in the human nervous system," and "input and output organs" that were "the equivalents of the *sensory* or *afferent* and the *motor* or *efferent* neurons."

Language like this had never been used before to describe the workings of a machine. Ironically, those words and the ideas they expressed established von Neumann's "automatic digital computing system" as the ultimate *analog* computer—a deliberate attempt to model by analogy and reproduce in technology the organic components, structure, and logical operations of the human brain. Von Neumann credited his neuron analogy to the McCulloch-Pitts paper, but his imagery also bore striking similarities to the rich neural analogies in Wiener's wartime paper with Rosenblueth and Bigelow.

EDVAC as von Neumann described it was the first computer to combine in one design all five of the prescient suggestions Wiener had submitted to Vannevar Bush five years earlier: digital operation, all-electronic vacuum-tube computing elements, a binary counting system, an easily recordable and erasable electronic memory, and a fully automatic, internal program of logical commands and calculating instructions (although in Wiener's earlier design the program was hard-wired in the machine, not stored electronically in the computer's memory, a method no one had imagined and that was not realizable in practice in 1940). Von Neumann's digital computer also incorporated key features of Wiener's analog fire director, including built-in error-correction mechanisms that operated on negative feedback principles.

Bush never acted on Wiener's proposal, nor did he circulate Wiener's memorandum, but Wiener probably conveyed all, or nearly all, his ideas about computers and automatic machines to von Neumann at the Princeton conference, and in their many informal exchanges and face-to-face conversations. Indeed, as two technical referees, D. K. Ferry, formerly of the Office of Naval Research, and R. E. Saeks, chairman of the Department of Electrical and Computer Engineering at Arizona State University, concluded,

> Most of the elements of the von Neumann machine, save the stored program, are present in [Wiener's] memorandum. It is interesting then to wonder how much von Neumann and Wiener influenced each other. . . . It is hard to estimate how much cross-fertilization occurred between these two men. Yet, had Bush circulated the Wiener memorandum, we might today be talking about the . . . Wiener-von Neumann, if not the Wiener machine, instead of the von Neumann machine.

Through the summer of 1945, Wiener pressed on in his effort to bring von Neumann to MIT. By early July, he was confident that his bid was a winner. "It really looks to me now as if the appointment and his acceptance were in the bag," he wrote to Rosenblueth. A month later, just days after the atomic bombs fell, Wiener was still upbeat in his next letter to Rosenblueth: "Johnny was down here the last two days. He is almost hooked." At the time, it appears, Wiener had no idea of the extent of von Neumann's secret work on the bomb, or about the new classified project von Neumann had begun for the military to build a far more powerful hydrogen "superbomb" for use in the postwar era. Wiener may not have known the details of von Neumann's EDVAC report, since the project was classified and he was no longer cleared to receive such documents. Nor did Wiener know that, while he was confidently awaiting von Neumann's acceptance of his generous job offer, von Neumann was quietly negotiating a better deal with the Institute for Advanced Study.

Late in November, von Neumann wrote to Wiener and formally declined his offer. He also informed Wiener that, in an unprecedented move, the IAS had agreed to let him build his own EDVAC computer at the institute in cooperation with one of the large electronics corporations that Wiener had vowed just a year earlier to resist—RCA. Von Neumann sent warm thanks to Wiener and yet another request. "I hope that we shall work together in the field just the same. . . . Could you find the time to visit here before the end of this month?"

As the expert gamesman he was, von Neumann did not show all his cards. His stated aim for his EDVAC, which would now become known as the IAS machine, was to build the first general-purpose, stored-program computer at Princeton. But von Neumann had another motive he never revealed to Wiener or anyone else without the highest-level security clearances. He needed the powerful new computer to help him and his Los Alamos colleagues design and build the next generation of atomic superweapons. And von Neumann had other new partners he did not disclose to Wiener, including the Army, the Navy, and, later, the Air Force and the newborn Atomic Energy Commission. However, von Neumann still needed one more crucial player to get this project rolling: a chief engineer.

On Wiener's next visit to Princeton, von Neumann got his man: Wiener's own engineer, Julian Bigelow. On Wiener's recommendation, von Neumann hired Bigelow for the IAS project in March 1946, and he quickly became the key man in the effort to render von Neumann's new computer architecture in a fully functioning machine. The IAS computer's vital organs were regulated by complex feedback mechanisms that shuttled data and logical commands around their tracks until the program was played out. Its results were checked by the fastidious error-correction circuitry von Neumann had prescribed in his design, and Bigelow was the perfect person to implement those archetypal feedback mechanisms.

Von Neumann's aggressive push for funds from the military and federal agencies, and his early alliance with RCA on the project, presaged the strong positions he would take on the desired role of government and industry in the research, development, and mass production of computing machines. Ultimately, he would depart entirely from Wiener's public-interest position. But, in the pattern that traced back to the self-doubts of his childhood, Wiener's response to von Neumann's maneuvers, and to his own disappointments, was not to lash out or strike back—but silence.

Quietly, behind the scenes during those waning months of the Second World War, a new world of postwar science and technology was coming together from many separate points. From the Macy conference in New York to the inaugural meeting of the Teleological Society in Princeton and the diverse projects that followed, Wiener's new concepts of feedback and circular causality had come full circle: from their origins in his quixotic fire control project, into the living world of the human brain and nervous system, and back into the brain-like circuits Bigelow was building into von Neumann's new computing machine. In the process, some of America's best minds had become the adherents of a "great heresy," as McCulloch called it. Beyond their shared technical ambitions, the epic challenge Wiener and his colleagues took upon themselves was an audacious scientific mission: to breach the bulwark of reduction that had held fast in Western science for centuries, and crack the riddle of human intelligence, using Wiener's new logic and communication concepts as the levers and building blocks of a new understanding.

And that, ultimately, was what Wiener's scientific revolution would be about. From the outset, the mission Wiener and his knights had embarked on was not a tinker's dream of making machinery but a higher philosophical and scientific quest. Already, their loose-knit band had solved the first mysteries of the mind's higher powers of reasoning and reckoning, and their revolution would keep rolling across the hard sciences, the life sciences, the human sciences, and the forefronts of philosophy itself.

Before the war was history, the seeds of that new scientific worldview were pushing up through the holes in the old one. And, in the new air of collaboration and camaraderie Wiener had sought since he was a young mathematician, his infant science that had no name was beginning to take on a life of its own.

8

Breakfast at Macy's

> The players all played at once without waiting for turns, quarrelling
> all the while, and fighting for the hedgehogs. . . . "I don't think they
> play at all fairly," Alice began . . . "and they all quarrel so dread-
> fully one can't hear oneself speak—and they don't seem to have any
> rules in particular; at least, if there are, nobody attends to them."
>
> —Lewis Carroll, *Alice's Adventures in Wonderland*

WIENER WALKED NEW YORK'S MAJESTIC Park Avenue on a sunny morning in March 1946 with more bounce than usual in his step. The clouds of war were gone. His new peacetime mission was just beginning, and he had come to New York for the most important scientific meeting of his career, the long-awaited postwar conference of the Josiah Macy Jr. Foundation.

McCulloch had arranged the meeting and would serve as chairman of the conference and nine more that would follow over the next eight years. The undertaking was enthusiastically supported by the Macy Foundation's medical director, Dr. Frank Fremont-Smith, who had been a charter member of Rosenblueth's monthly "supper club" at Harvard Medical School and was now a driving force behind the foundation's interdisciplinary conferences in the same vein.

But no one was more excited about the coming conference than Wiener. "This meeting is going to be a big thing for us and our cause," he wrote to Mc-Culloch three weeks earlier, during a prep session in Princeton. "I am now down with von Neumann discussing plans and I can assure you that his part and mine will be well coordinated. Pitts and I are also getting busy together and so is Rosenblueth. . . . We are impatient for the meeting . . . when we shall see you and talk over many things of common interest."

All the principals were eager to pick up their conversation about brains and computers where it had left off at the Princeton meeting the year before, and their dialogue was about to open wider. During the meeting's planning stages, Gregory Bateson, the anthropologist, who had returned from his wartime chores in the Pacific, pressed McCulloch and Fremont-Smith on their promise to reconvene the original group of Macy conferees and take up the psychological and social implications of the new concepts Rosenblueth had unveiled at the conference in '42. No meeting had ever attempted such an ambitious agenda of intertwined scientific ideas and disciplines, but Bateson prevailed. To accommodate all the parties, and for want of a more succinct name, the conference was torturously titled "The Feedback Mechanisms and Circular Causal Systems in Biology and the Social Sciences Meeting."

The table was set for a melding of minds that would give Wiener the final pieces he needed to assemble his new science, and a contingent of renowned scientists and scholars to take it into the world.

They gathered for two days and two nights around a great circle of tables at the Beekman Hotel on Manhattan's Upper East Side, with their lodging, meals, and the cocktails Wiener would not touch paid for by the Macy Foundation. The twenty conferees included the core group of mathematicians and physiologists from the Princeton conference: Wiener, von Neumann, McCulloch, Pitts, Rafael Lorente de Nó; along with Wiener's colleagues Rosenblueth, who came back from Mexico for the meeting, and Bigelow, who had started work on von Neumann's computer project the day before. The social scientists included the leading figures from the 1942 meeting: the anthropologists Bateson and Mead, the psychiatrist Lawrence Kubie, and the learning theorist Lawrence K. Frank. And some prominent new figures joined the group: Heinrich Klüver, a German émigré and Gestalt psychologist at the University of Chicago; MIT social psychologist Kurt Lewin, another German émigré from the Gestalt school; Columbia University sociologist Paul Lazarsfeld, an Austrian who had conducted pioneering research on the effects of modern mass communication technologies; and British ecologist G. E. Hutchinson, a colleague of Bateson's and a trailblazer in the study of environmental systems.

Von Neumann led off the presentations with a report on electronic computing and its analogies to the brain's neurological computing networks. Nattily attired, as always, in his vested suit, a handkerchief folded perfectly in his pocket, he described the new digital computer he was building at the Institute for Advanced Study and its innovative stored-program design that "could compute any computable number or solve any logical problem." In a cross-disciplinary "duet" with the neurophysiologist Lorente de Nó, von Neumann dazzled his audience with visions of his new computer's brain-like logic, computational powers, and impressive storage and recall capacities that mirrored the dynamic action of "the computing machine of the nervous system." For many scientists in the crowd, it was their first insight into those logical doings inside their own heads, and their first inkling that such a brain-like machine was conceivable, let alone about to be constructed.

Wiener was taking nothing for granted in the new man-machine equation, and he interrupted von Neumann's talk to question his computer's logical prowess. He wondered how von Neumann's machine would react to the kind of logical dilemmas human minds confronted every day, and, deep down, he feared for the contraption's metal health. In a primal puzzle of computation, Wiener predicted that if a computer were programmed to solve one of Russell's famed paradoxes—for example, to determine the truth value of the proposition "this statement is false"—it would go into an oscillating irrational state, "so that if it first decided that something was true it would next decide that it was false and vice versa," without ever reaching a solution. When challenged, von Neumann could not resolve that paradox himself.

After lunch, Wiener and Rosenblueth took the floor and led the conferees beyond computers to the larger mission of the conference. Wiener spelled out the basics of his developing science of communication and gave a precise new meaning to the everyday notions of messages and information, redefining them as elementary units of communication that could be measured mathematically. He described the new concept of feedback and the core principle of "control through negative feedback" that governed all purposeful, goal-seeking activities in machines and living things. He reviewed the history of automatic feedback mechanisms, from the practical gadgets invented by the Greeks to modern servomechanisms, and the new kind of intelligent machine he and Bigelow had built during the war that "took cognizance of the world about it and of its own performance."

Wiener explained how this new breed of man-made mechanisms, equipped with built-in computing devices, looping feedback circuits, and sensitive "receptors and effectors," could perform practical tasks more advanced than any machine built to that time and "so base their action on previous information as to guess the future." Then, as Lorente de Nó had done for von Neumann, Rosenblueth put flesh on Wiener's new conceptions and his concerns, describing the physiology of living feedback processes, and the consequences for living systems and automata alike when those vital communication and control processes break down pathologically.

By dinnertime, it was clear that what Wiener, von Neumann, and the neurophysiologists were offering to their newfound peers was a versatile new language and scientific framework for understanding the mind's cognitive powers, the body's organic control processes, and the new arena of communication processes common to humans and machines that could be fruitful and immensely useful to scientists on both sides of the divide.

That night, Gregory Bateson and Margaret Mead, the husband-and-wife team of anthropologists, stepped onto the court of the mathematicians and physicians. Hawk-nosed and tousle-haired, Bateson at age forty-two was a tower of a man, six foot five, with "more limbs and height than he knew what to do with." Born into Britain's scientific gentry (his father, the biologist William Bateson, coined the term "genetics"), Bateson, like Wiener, had started college as a zoology major, then in the mid-1920s he made his move into anthropology, the study of the origins and social development of people and cultures. The traditional field was in transition, sloughing off its old façade of baseless racial theories and armchair analyses. Bateson brought to it his naturalist's belief in the guiding principles of ecology, the new movement in science that focused on the reciprocal interactions between living things and their environments, which Wiener, too, had embraced in his boyhood reading.

Mead, barely five feet tall, was round-faced and button-nosed, with a soft smile that belied her fierceness. Born in Philadelphia into a Quaker family of academicians, she had trained at New York's Barnard College with Franz Boas, the father of modern anthropology, who stressed the importance of observations in the field. Under his influence, in 1928, Mead published her best-selling book *Coming of Age in Samoa*, a study of the uninhibited sexual behavior of teenage girls in that Pacific Island culture, which established her, at age twenty-six, as one of the most controversial figures in American science. Her presence was powerful and would reign quietly, and sometimes not so quietly, over all the Macy meetings.

In the years since Rosenblueth had awakened them to Wiener's new communication concepts, the pair had come to see the world through new eyes and to speak the new language of communication. They saw how the notions of feedback and circular causality could rationalize and re-energize the outmoded vehicles of traditional social science research, and they made it their mission to convey to their peers what the new communication theories had to offer all the social sciences. Now they talked fluently to the conferees about remote cultures that "achieved stability by inverse feedback." Bateson lit up the serious crowd with his colorful descriptions of the social feedback processes he and Mead had observed in their expeditions to the South Pacific—including a comical transvestite ritual of the Iatmul tribe of New Guinea, which, as Bateson noted, played a stabilizing feedback role when aggressive urges among male tribesmen threatened to break out in internal warfare.

And they went further, showing how the new communication tools could be used to extract patterns of knowledge from primitive and modern societies alike: from the circular flow of messages that pass between individuals and groups, outward into their cultures through diverse customs, ceremonies, and communications media, and then back to their senders, in endless loops of messages and feedback that inform, influence, teach, shape, and continuously reshape each individual's behavior, purpose, and personality, and the life of the larger society. For Bateson and Mead, the new communication concepts seemed almost too good to be true, but there they were—ripe for harvesting and replanting in the fertile fields of the human sciences.

It was a long day, but the historic work of the Macy conferences had begun.

In fits and starts, the motley group began to mesh and go forward as a unit. Some of the newcomers joined in the conversation. Heinrich Klüver, the Gestalt psychologist, wanted to talk about perception, the subjective process of mind that was the essence of Gestalt psychology and the nemesis of the dominant American school of behaviorism. Klüver appealed to the conferees to help him solve the ultimate Gestalt riddle of how the mind perceives *forms*—shapes and patterns of experience, whether a face, a chair, or any other universal figure. He wondered how the human brain could pick out a form or *any* thing from the chaos of signals impinging on it from moment to moment, and whether a brain-like electronic device made of nothing but nets of simple, neuron-like components could do the same?

Here was a topic all the attendees could sink their teeth into: a primary problem for the neurophysiologists and psychologists, an issue with far-reaching cultural implications for the anthropologists, and a challenge to the group's mathematicians and technicians. *Could a "universal" computing machine compute a form—a perception—using logic alone?* Klüver's challenge marked the start of a quest that would become a centerpiece of computer science: to understand the mind's perceptual processes and to emulate them in electronic machines. In the months that followed, Wiener would ponder the problem from a communication perspective. To explain how messages picked up by the eye's millions of sense receptors are transformed into perceptible images in the brain, he postulated a complex organic process akin to the television scanning technology he had been fascinated by since the 1920s. McCulloch and Pitts, too, would take up Klüver's question in a new collaborative project that would prove even more influential than their initial work on the logic of neural nets.

Other giants of social science came to the table. Kurt Lewin, the social psychologist, was a pioneer of human communication research and founder of the modern study of group dynamics. His early discoveries of the dramatic physical and psychological effects produced by people in groups, ranging from accelerated learning and decision-making, to peer pressure, to euphoria and other "electric effects" on group participants and leaders, made a persuasive case for the group's working hypothesis that the power of communication was palpable at the neurological level and a force to be reckoned with in society. Lewin was determined to bring those ephemeral human communication dynamics into the fold of rigorous science, and, like Bateson and Mead, he began at once to incorporate Wiener's feedback concept and fluid new logic of circular causality into his own theories and research methods.

Paul Lazarsfeld, the sociologist, was a pioneer of mass communication research. As director of Columbia's Bureau of Applied Social Research, he conducted the first major study of the effects of radio on American society and groundbreaking research on the formation of public opinion and voting patterns. He was especially interested in Wiener's theories about the statistical nature of communication, and he planned to incorporate Wiener's technical communication concepts and mathematical methods into his advancing work in survey design and statistical analysis.

As the conference's final hour approached, Fremont-Smith had no trouble mustering a quorum for a second conference and a third, and Wiener's fledgling Teleological Society was transformed into a seasonal happening under the auspices of the Macy Foundation. McCulloch was delighted with the response. From the outset, he had sounded the charge for the Macy group's yearning scientific revolutionaries, who had come together in the wake of their wartime assignments, "looking to mathematics and engineers working on communications" for benevolent uses of the war's discoveries and scientific theories. In the years after Hiroshima, physical scientists were turning their talents away from weaponry and migrating to the life sciences, and social scientists were forswearing propaganda and strategic studies of foreign populations for more altruistic pursuits.

The Macy meetings delivered what the foundation's doctors and Wiener, too, had been hoping to provide: a cure for the collective shudder so many scientists were feeling in the shadow of American science's wartime triumphs, and a first step toward assembling a set of peacetime scientific tools for confronting the complexities of the new postwar society. Wiener hailed the positive outcomes of the opening conference in which he had so much of his own conceptual capital at stake, but McCulloch's take on the roundtable, with its colloquial and at times chaotic format, was more jaundiced than the view through Wiener's rosy lenses. "Of our first meeting Norbert wrote that it 'was largely devoted to didactic papers by those of us who had been present at the Princeton meeting, and to a general assessment of the importance of the field by all present,'" McCulloch dutifully recorded, then he set the record straight. "In fact it was, characteristically, without any papers, and everyone who tried to speak was challenged again and again for his obscurity. I can still remember Norbert in a loud voice pleading or commanding; *'May I finish my sentence?'* and hearing his noisy antagonist . . . shouting: *'Don't stop me when I am interrupting.'"*

In the months after the conference, McCulloch and Pitts plunged into Heinrich Klüver's perception question. Von Neumann and Bigelow got started on construction of the IAS computer, and von Neumann commuted to Los Alamos to continue his secret work on the atomic superbomb. Wiener began migrating regularly to Mexico to begin a new phase of his work with Rosenblueth on "the study of the nervous system as a communication apparatus." In their multi-year plan funded in part by the Rockefeller Foundation, Wiener would spend half of every other year at the Instituto Nacional de Cardiología working with Rosenblueth in his lab to extract experimental data on communication processes from living tissue. Rosenblueth would spend half of the alternate years with Wiener at MIT working on theoretical matters and the plans for their long-envisioned interdisciplinary institute.

Wiener was thrilled by the new science and scenery south of the border. The untamed Mexican landscape beguiled him from his first blast of high-desert air. He was enchanted by the adobe dwellings of the *campaña*, by the lush plants and flowers that defied the country's arid climate, and by his first "indications of a new way of living with more gusto in it than belongs to us inhibited North Americans." He became a fluent speaker of Spanish with a convincing Latin accent. He indulged in the country's colorful cuisine and spicy sauces. And, once his red blood count had risen sufficiently to offset Mexico City's 7,000-foot elevation, he embarked on his wanderings through the city's spacious public squares and crowded *paseos* in the same indefatigable way he had conquered the capitals of Europe and the back alleys of Beijing.

He and Rosenblueth were eager to apply Wiener's statistical communication theories to the study of signal transmission in long nerve fibers in the limbs of cats and other creatures; and, in keeping with the mandate of the Instituto, they also turned their attention to the flow of electrical signals through the tight nets and circles of nervous tissue that governed the rhythmic beating of the heart. In those groundbreaking studies in the physiology of communication, Wiener and Rosenblueth ran into an old war buddy: the peculiar problem caused by "pathological conditions of very great feedback." While testing the long motor neurons in the legs of anesthetized cats, they observed that, as they raised the level of electrical stimulation, increasing the "load" on the cats' nerves, the animals' legs went into spastic convulsions and crazy oscillating rhythms that were wilder than a runaway gun turret—a condition known to neurophysiologists as clonus.

Once again, Wiener and Rosenblueth set out to understand the pathology and similar spasms that afflicted the limbs of humans. They sought to crack the code of clonus mathematically and tame its spasms in the laboratory, but this task proved even more difficult than their wartime project and, for Wiener, deeply frustrating. Early in their research, Wiener wrote to Pitts and Lettvin in Cambridge saying, as Lettvin recalled, "I want nothing more to do with the nervous system. To hell with it. It has absolutely no properties that I can analyze." However, despite Wiener's rant, he and Rosenblueth persevered, and by the summer's end they had found something fascinating.

The flow of impulses coursing through the cat's leg was not linear, as they had expected, but *logarithmic*: The output changed exponentially as they increased the input of stimulating electrical current at a steady rate. The strange finding—that in a cat's leg, if not in all nervous tissue, electrical messages flowed through the nerve's living wiring in a way that "is possible only in a logarithmic system"—spoke to Wiener at deep theoretical levels. Their numbers were almost identical to recent measurements of signals and messages obtained by mechanical engineers studying the spastic behavior of oscillating servomechanisms, and by electrical engineers tracking the rate of signal transmission through noisy telephone lines as far back as the 1920s.

Wiener would spell out the significance of that connection, and make more stunning connections of his own, a few months later.

———

The action that fall increased exponentially as well. In September 1946, Paul Lazarsfeld convened a special one-day subsession of the Macy conference series, which he titled "Teleological Mechanisms in Society." There he introduced the new communication concepts to a wider circle of social scientists and brought more prominent scholars into the orbit of the conferences: his Columbia colleague Robert Merton, the leading authority on the impact of social factors on the progress of science; Harvard anthropologist Clyde Kluckhohn, a progressive force in American anthropology close behind Bateson and Mead; and the dean of American sociologists, Talcott Parsons, chairman of Harvard's Department of Social Relations. Parsons especially "became an enthusiast" of the Macy group's work, and he hand-carried Wiener's new concepts of communication, feedback, and circular causality into the highest echelons of American social science.

A month later, the whole Macy group reported the results of their summer researches. McCulloch gave an update on his and Pitts's new project on the perception of forms. They were confident that the human processes of perception and pattern recognition required for the apprehension of universal forms could be replicated by computers equipped with the equivalent logic and electronic

neural networks. Wiener and von Neumann agreed that such a feat would require computing machines to become capable of learning from their past operations and outcomes, as human minds learned from their experiences, to
improve their future performance and to prevent them from making the same
mistakes repeatedly. But, again, Wiener expressed his objection to "attempts to
exclude paradoxes" from their discussion of computers that could be designed
and programmed, now, for both logic and learning.

For Wiener, the problem was becoming a preoccupation, especially if machines that reason and learn might be put to work in society in place of human
decision makers. *Would such machines be able to make complex decisions in critical situations? Would their reasoning be more "rational" than that of human beings faced with
similar decisions—or less? And how would they deal with life's inevitable dilemmas?* Finally, McCulloch grasped Wiener's argument and saw that it was no trifling
philosophical exercise, but not everyone shared their concern, especially von
Neumann, who could not be bothered with paradoxes. "I am afraid . . . that day
suffered . . . from too much hormonal or too little formal discourse," McCulloch said glumly.

There was more breaking news at the second conference. Pitts revealed that
he was preparing to write his doctoral dissertation under Wiener's direction. His
topic marked a leap beyond his work with McCulloch. At Wiener's suggestion,
Pitts proposed to expand the logical calculus of computation from a flat, two-
dimensional schematic into a full-bodied, three-dimensional model bearing a
real physical resemblance to the brain's living communication networks. But
scaling up from two to three dimensions would be no easy task. No one had
ever attempted to devise such a complex logical scheme and work out its mathematics. Moreover, to be truly brain-like, as the best anatomical research at the
time indicated, Pitts's 3-D neural network would have to be, not fixed and utterly predictable, like his models with McCulloch and the logical circuits in von
Neumann's electronic brain, but probabilistic in its computations and riddled
with random factors.

Pitts's project to parse the math and logic of random nerve nets took the entire company of assembled geniuses aback. As McCulloch observed, in a cautious
comment that reflected the magnitude of the challenge Pitts was undertaking,
"the mathematics for the ordering of the [random] net has yet to be evolved."
But Wiener, who had invented the mathematics of random paths, knew Pitts was
fully the equal of the other young mathematicians he had worked with, and he
was confident that, if anyone could blaze that trail, it was Pitts.

Kurt Lewin took the next step into the social dimensions of the Macy group's
mission, and he, too, had done his homework. In only a few months, Lewin had
made broad practical applications of the new communication theories. As Lewin
told his audience, he had found circular processes abounding in "the interplay

between a leader . . . and a group," in the patterns of mass marketing and mass consumption that determined the American diet, and in many other social and economic processes of American culture. He had laid out his findings in a long article that was already in the pipeline for publication, an article that would have lasting impact on the social sciences and on the whole of society.

Before the conferees left town, Lawrence Frank, the learning theorist and former executive vice president of the Macy Foundation, arranged a special session of the conference series on the heretical topic of "Teleological Mechanisms" at the New York Academy of Sciences, and Wiener was asked to be the keynote speaker. The event marked a coming out for the cloistered conferees and their revolutionary concepts, and it marked a major turn in Wiener's thinking about communication and about the curious new stuff he called information.

Frank's opening statement to the Academy assured the scientific establishment that the Macy group's new approach to teleology and purposeful behavior "is not a regressive movement to an earlier stage in the history of ideas, but a forward movement toward a more effective conception of the problems we face today." Yet he admonished his audience: "We should not miss . . . the significance and larger meaning of this conference, held at this time. . . . We are engaged, today, in one of the major transitions or upheavals in the history of ideas, as we recognize that many of our older ideas and assumptions are now obsolescent and strive to develop a new frame of reference." Then Frank introduced Wiener, whose paper with Rosenblueth and Bigelow, he said, "largely initiated these discussions."

Wiener began with a history of the new scientific ideas and their connection to a basic principle of statistical mechanics: the second law of thermodynamics. He described how physical systems generally run downhill from states of higher organization to states of increasing disorder and, eventually, to the state of maximum randomness and disorganization known as entropy. Then he pointed out one important exception to the rule. He explained how all living things defied that fundamental directive of nature, in the act of life itself, by means of purposeful, circular processes that enabled them to beat the law of entropy, progress to higher states of organization, and maintain those exceptional states throughout their lifetimes.

The new automatic machines and electronic computers he and his colleagues were devising performed in similar ways, Wiener observed, and their operations offered valuable insights into the workings of complex physical and biological systems—including the most extraordinary system of organization in nature, the human mind. He identified the universal process that connected the new electronic inventions, and the intricate circuits and networks they were comprised of, with every living thing, with the mind, and with the larger social systems his Macy colleagues had brought into the conversation. It was the same ubiquitous process that had run like a river through his mathematics and technical work for two decades—communication—and it was only now coming into its own on the forefronts of the postwar scientific landscape.

For the first time, Wiener revealed the key element of his communication theory to an audience wider than the Macy group and the few wartime readers of his classified Yellow Peril monograph. "The unifying idea of these divers disciplines is the message," he proclaimed, and he restated his simple definition that "the message, to convey information, must represent a choice from among possible messages." He linked his evolving notion of messages that convey information to the fundamental ideas of probability theory and statistical mechanics. Then he made a third, historic distinction about information:

> The notion involves *entropy* as well.

For Wiener, that esoteric concept from thermodynamics was essential to the science of communication. Drawing on the insights he had derived from his research in Mexico two months before, he offered a new answer to the question posed two decades earlier by the engineers at Bell Laboratories in their efforts to measure the amount of information in a message. In the first statement on record that described information the way it is universally conceived today, Wiener defined information in physical terms as a function of the entropy or degree of randomness in the message. In the plainest language possible, he made clear to his audience that, "Entropy here appears as the *negative* of the amount of information contained in the message. . . . *Thus, essentially, amount of information is the negative of entropy.*"

His new physical conception of information was a revolutionary move for communication engineering, as radical and profound in its own way as his earlier conceptions of feedback, purpose, and circular causality were for understanding the actions of living things and machines. He tied the accelerating flow of messages through electronic circuits, within machines, and between human beings to the most basic physical measures and mathematical relations in the material world, marking a major step toward understanding the intangible stuff of information in tangible terms. But for Wiener it was only logical, as he told the Academy:

> In fact, it is not surprising that entropy and information are negatives of one another. Information measures order and entropy measures disorder. It is indeed possible to conceive all order in terms of message.

Wiener made one more important connection. As he did for digital computing in 1940, he now embraced the binary scale for measuring information using the same logarithmic methods he had adopted in Mexico. When measuring the amount of information in any message, whether that information was being conveyed through telephone lines or nervous tissue, Wiener declared:

The number of digits ... will be the logarithm ... to the base 2.

That measure would become a benchmark for the new technological era and its ubiquitous commodity of information.

Wiener described to his audience how his new communication theories and information concepts could be applied to the practical task of building computing machines with prodigious memories and the ability to search through large quantities of data. He foresaw still more useful ways to harness "the information furnished by such a machine, *i.e.,* its output" to do real work in the world, and to serve as the "central nervous system in future automatic-control machines" and entire automated factories. He saw profound implications of the new communication theories for the fields of medicine and mental health. And, as he had learned from his Macy colleagues, the new science offered even greater understandings about "the coupling of human beings into a larger communication system" that possessed its own infinitely rich capacity to convey messages and meanings among individuals, groups, and entire societies.

Wiener ended his presentation on a high cultural note, as he connected his new technical science of communication to the age-old arts of human communication and paid homage to language, "emotional non-verbal communication," and the great oral and written traditions on which he was weaned, which he saw as no less essential to human progress than all the technology of modern life.

The sizeable contingent of Macy players who had come to show their colors were elated. Others in the group built on the themes in Wiener's presentation. McCulloch offered a "forecast of several extensions" to research in neurophysiology. Bateson laid out an ambitious program for applying the new communication framework across the social sciences. He reeled off a litany of timely social phenomena that were driven by circular feedback processes, "business cycles, armaments races ... systems of checks and balances in government," and closed with the warning that, "in the social sciences we need not less but rather more rigorous thinking than is usual among the physicists [for] the entities with which we deal are much more complex than even their computing machines."

G. E. Hutchinson, the ecologist, provided a few global examples of feedback processes at work in the environment, and of organisms that "may be acted upon by their environment, and ... may react upon it," in colossal circular operations with far-reaching implications for the future: the cycle of photosynthesis that plays out between green plants and the earth's atmosphere; changes in the balance of oxygen and carbon dioxide in the air caused, in part, by "the modern industrial combustion of fuel"; cycles of commodity production, consumption, and "long-term privation owing to an excessive rate of depletion of natural resources" in human populations. He even described the measurable effects of those material cycles historically on the cultural production of art and ideas. His address that day would redefine the field of ecology and environmental science.

But McCulloch was most pleased with Wiener's speech and eager to see it disseminated to the wider scientific community. The summer after the New York Academy meeting, he sent advance copies of Wiener's address as it would appear in the Academy's *Annals* to forty prominent scientists in America and around the world, with his advice that they should "regard it as a personal communication of some notions which we trust you may find fruitful in your own field." Among those on his mailing list were Claude Shannon at Bell Labs and other members of the lab's technical staff, probability theorists Kolmogoroff and Kintchine at the Soviet Academy of Science in Moscow, Nobel prize–winning biochemist Albert Szent-Györgyi at the University of Budapest, and Austrian quantum physicist Erwin Schrödinger. Several years earlier, Schrödinger had made the first link between entropy in physical systems and the negative entropy required for the survival of living things, which had inspired Wiener to make the next connection between negative entropy and information.

The response was encouraging. "The Bell people are fully accepting my thesis concerning the relationship of statistics and communication engineering," Wiener advised McCulloch in May 1947. McCulloch, too, was making liberal use of Wiener's notions in his new writings and anticipating many more requests for reprints of Wiener's address. "You see to what extent your ideas on . . . information are taking hold," he told Wiener.

———

By the third Macy conference in the spring of 1947, debate over Wiener's new science was roiling. Von Neumann was shaken by Wiener's equation linking one of the most important principles of physical science in the modern era—the entropy measure of randomness enshrined in the second law of thermodynamics—with something so ephemeral, *non*random, and quintessentially human as information. That the new stuff of messages and intelligence itself could be derived from a mere statistical probability, and from pure chaos at that, seemed almost too much for von Neumann's logical mind to handle.

He was also jealous that Wiener, and not he, had made the connection. According to Heims, von Neumann "had appreciated the connection between entropy and information even before Wiener made it central . . . and was peeved at the credit that fell to Wiener for exploring that connection and assessing its importance at full value, which von Neumann had failed to do." Von Neumann's biographer went a step further. He claimed that "Johnny had a genuine admiration for Wiener's mind, which he suspected might be intrinsically better than his own," and he suggested that von Neumann's main reason for participating in the Macy conferences was to gain access to Wiener's ideas. Both men were masters of probability theory and statistical mechanics, but Wiener was more comfortable than von Neumann with the uncertainties inherent in communication theory and other applied sciences emerging in the postwar era.

Like Wiener in Mexico the summer before, von Neumann, too, was having problems cracking the complexities of the brain and nervous system. Several months earlier, in a stunning turnabout, von Neumann, the originator and chief proponent of the neural network approach to computing, had sent a long letter to Wiener in which he voiced serious doubts about the wisdom of using the brain as his template for the new digital computer technology. In their effort to create automata that operated like human brains, "we selected . . . the most complicated object under the sun—literally," von Neumann complained. "The complexity of the subject is overawing. . . . I feel that we have to turn to simpler systems." As an alternative, he proposed to model automata on the molecular mechanisms of single-celled organisms, an idea he would develop more fully a few years later.

Von Neumann feared that his "anti-neurological tirade" might provoke a tirade in turn from Wiener, but Wiener remained calm. He had no personal stake in McCulloch and Pitts's neural networks, and his own thinking was already gravitating to deeper realms of biology and the body's chemical and molecular signaling systems. Nevertheless, something happened to the warm relations between Wiener and von Neumann around that time. Von Neumann's letter may have played a part, or Wiener may have had a delayed reaction to von Neumann's rebuff of his offer to join his interdisciplinary venture at MIT, or Wiener may have been distressed personally by von Neumann's work on the bomb and his enthusiasm for building even bigger atomic weapons. Whatever the cause, the chill was already detectable after the first Macy conference. As Heims reported, "At subsequent meetings . . . there was a noticeable coolness and even friction between the two men. When von Neumann spoke, Wiener would ostentatiously doodle or go to sleep very noisily." Von Neumann returned the insult at a mathematics conference a few months later. "When Wiener was lecturing, von Neumann sat in the front row and as ostentatiously and noisily as possible read *The New York Times*, to Wiener's annoyance."

They were strong personalities mixing combustibly in the Macy foundry. Bateson remembered von Neumann expounding pugilistically, "pouring all the stuff out and punching with both fists." Pitts, the quiet one, was no less combative in "the sharpness of his intellectual arguments and show of contempt toward sloppy reasoning by social scientists." Heinz von Foerster, a physicist from Vienna who joined the Macy group a few years later and became the editor of its proceedings, remembered Margaret Mead as another Macy conferee who did not flinch from fisticuffs when the need arose. "She kept the sessions going. I remember one situation where they fell into citations, and two members were tossing Greek proverbs at each other. Margaret suddenly took her fist, knocked it on the table and said, 'We *know* you can speak Greek! Now, why don't you address the problem we are discussing?' She was a very important catalyst."

Many people played leadership roles in the proceedings, but without question Wiener was the "papa" of the group, as von Foerster put it. "His presence was very dominating. Everybody listened to the 'papa's' ideas." Von Foerster's memories of Wiener were not like those of some younger engineers at MIT, or of others who viewed him as a befuddled genius with an outsized ego. "He was very gentle, very clear. My impression was of an extraordinarily modest and humble man who was absolutely in control of everything he said."

Heims, too, perceived Wiener as the dominant figure at the Macy conferences "in his role as brilliant originator of ideas and *enfant terrible.*" He was "irrepressible in his enthusiasm for the scientific ideas presented at the conferences," said Heims, and "evidently enjoyed the meetings and his central role in them: sometimes he got up from his chair and in his ducklike fashion walked around and around the circle of tables, holding forth exuberantly, cigar in hand, apparently unstoppable."

Through the first years of the Macy conferences, the group's core members regrouped in the summer at Cloverly, the country home of Lawrence Frank, which was situated on a small lake in New Hampshire, an hour's drive from the Wieners' place in South Tamworth. Bateson and Mead, Fremont-Smith, Kurt Lewin, and others came with their families. The children played in the woods by day and, by night, sang around the campfire, while the adults convened in "a perpetual conference and in the wash of speculation and good humor that continued in the recesses of the adult sessions." Mead spent a dozen summers at Cloverly holding court on the porch and "floating dreamily in the water with a balsa board or a tire." Wiener was remembered by Mead and Bateson's young daughter, Mary Catherine, as someone who "used to stop by, smoking smelly cigars and pouring out his latest idea to Larry or Margaret, not much interested in listening to their responses."

More festive times transpired at McCulloch's country estate in Old Lyme, Connecticut. During his years at Yale, McCulloch had purchased an old 500-acre farm in the coastal town thirty miles east of New Haven, and he lovingly maintained its great stone house and barn, tiny lake, and working fields, with the help of the large entourage that migrated there each summer from Chicago, Boston, and elsewhere. The summer scene in Old Lyme was footloose and free-wheeling. There was serious scientific talk ongoing at all hours around the big dining table in the old farmhouse's great room, and in lesser forums scattered across the grounds, and McCulloch raised the proceedings to sublime heights with his "enormous sense of delight, fun, joy."

Wiener came often to Old Lyme, where he joined in the dialogue and good times with a freedom he seldom knew in his childhood or his home life as an adult. But, at Old Lyme, as in New York and on all his visitations, he was a mixed gift as a houseguest. "Wiener was a fascinating man, but he was such a child. Everybody took care of him," McCulloch's daughter Taffy remembered. "He used to sleep upstairs and nobody could go to sleep because of his snoring. It was horrible. Then, in the morning, he'd come padding into the wrong room, wandering around, because he couldn't see."

Taffy retained a vivid image of Wiener in the wild. "He was a character. He looked like a frog with bulging eyes. I remember him floating in the lake with his tummy sticking up, just talking away, waving his cigar in the air, and slowly sinking in the water." Swimming in the nude was the norm at Old Lyme. "No one ever wore bathing suits in the lake," she added.

The liberated scene at Old Lyme was a balm for Wiener's psyche. He never threw a tantrum, unleashed an emotional storm, or sank into depression while he was there. But Margaret Wiener did not care for the McCulloch crowd's bohemian ways. She never accompanied Wiener on his summer visits to Old Lyme and did not know the full extent of the freedoms that were on display there. "I am absolutely certain he never told Mother about the skinny dipping," Peggy attested. "Oh, if Mother had got wind of that I can just see her going into orbit."

Other scientists shared bohemian lifestyles during that fervid postwar period, and a generation of younger scientists, scholars—and an entire youth culture— would follow their lead two decades later. But the extended tribe of the Macy group and the nomadic McCulloch clan, with its progressive elders and inspired rapscallions, was on the cusp of its time, and its members' unfettered behavior, irreverent attitudes, and flouting of social conventions leavened all their labors.

———

Heims found that "the elaboration, critique, extension, refinement, and following out the implications" of the ideas introduced at the Macy conferences "took more than a generation, and continues today." He maintained that the conferences "played a significant historical role in the development of the human and the natural sciences in the United States," and the new technological sciences may have benefited even more from the exchange. Yet all was not sweetness and light at those mythic Macy meetings. As McCulloch acknowledged in a memoir, with no attempt to pretty the picture, "The first five meetings were intolerable. . . . The smoke, the noise, the smell of battle were not printable." He fumed:

Nothing that I have ever lived through . . . has ever been like . . . those meetings. . . . You never have heard adult human beings, of such academic stature, use such language to attack each other. I have seen member after member depart in tears, and one never returned. . . . I well remember one of our scientists, and won't say who, shaking his fist in Margaret [Mead]'s face and shouting, "Hell man, if you don't think that the squirrel knows what the blue jay is saying when you go into the woods with a gun, you've never been hunting. Hell, man, you've only been blundering around the woods with a gun." This is a mild sample.

But Margaret Mead saw something greater going on. From her vantage point as an anthropologist, Mead perceived the tiny Macy tribe as a textbook instance of a "unit of cultural micro-evolution . . . a cluster of interacting individuals who . . . make choices which set a direction" for the larger culture. And the Macy group itself was not the most impressive feature of Mead's "evolutionary cluster." In her conception, "the most distinctive characteristic of an evolutionary cluster is the presence in it of at least one irreplaceable individual, someone with such special gifts of imagination and thought that without him the cluster would assume an entirely different character."

For Mead, von Foerster, McCulloch, and many others at the Macy conferences, Wiener was that "one irreplaceable individual." He opened the new world of communication concepts and principles to the conferees, and, without question, the Macy conferences did a world of good for Wiener. They vastly expanded his awareness of the social dimensions of his new communication science, and they alerted him early on to the human consequences of the new technologies that were already beginning to flow from it.

Wiener looked, listened, took it all in, and put the pieces together as only an ex-prodigy of the omnibus type could do. Then, a year after the Macy conferences began, he assembled his grand synthesis of all the new knowledge and technical know-how, along with his deepening human understandings and social concerns, in the little book that marked the big bang of the information age.

9

The Big Bang: *Cybernetics*

"Information is information, not matter or energy."
—Norbert Wiener, *Cybernetics*

ON MAY 20, 1947, WIENER SET SAIL from New York for his first trip to Europe since the war. That morning, just before dawn, a bright comet streaked through the southern sky. Soon after, a total eclipse of the sun preceded Wiener across the Atlantic. Such a cosmic convergence had occurred only once before in recorded history.

After a short stay at the London home of his biologist friend J. B. S. Haldane, Wiener made a quick tour of Britain to meet with the country's key players in the accelerating race to build the first fully programmable, general-purpose, digital computer. He met with teams in Cambridge and Manchester who were developing machines based on von Neumann's architecture. At the National Physical Laboratories in Teddington, he talked with Alan Turing, the great conceptualizer of computation, who was directing the design of another von Neumann–style machine and developing his own logical programming language. Wiener had much to discuss with Turing about his new science of communication and control. He found the British atmosphere "entirely ripe for the assimilation of the new ideas" he was developing and "the engineering work excellent."

But England was only a stopover for Wiener on this trip. He was bound for France to attend an international conference on harmonic analysis at the Collège de France organized by another old friend and collaborator, the Polish-born mathematician Szolem Mandelbrojt. Wiener's work was a centerpiece of the proceedings.

Then, in Paris, after the conference, Wiener struck a lucky match that lit the flame for his new science on the Continent and beyond.

The City of Light was beginning to shine again after the dark days of the war. People were strolling the city's promenades without fear and flocking to its gardens. The bookstalls by the Seine were bursting with ideas, and talk of philosophy could be heard again around the tables of the outdoor cafés. Wiener walked the winding alleys of the Left Bank, with their wafting smells of hot baguettes and exotic cheeses, up the bustling Boulevard St. Michel into the heart of the old city's university district, and made his way to a cluttered little book shop across from the Sorbonne. There a meeting had been arranged by one of his MIT colleagues with the gentleman Wiener lauded as "one of the most interesting men I have ever met," the Mexican Frenchman, former diplomat, and jacques-of-all-trades he would refer to forever after as "the publisher Freymann of the firm of Hermann et Cie."

Freymann won Wiener's respect with stories of his professional exploits and his adroit efforts to secure publication contracts for French learned societies, and he won Wiener's heart with his candid disclosure of the ways he had parlayed those commitments into an intellectual publishing house "as nearly free from the motive of profit as any publishing house can be." Then Freymann made Wiener an offer he couldn't refuse.

The French historian of science Pierre de Latil captured the moment as it passed down from Freymann himself:

"Why don't you write a book on the theories that you are always talking about?"—"The public isn't ripe yet. Maybe in another twenty years. . . ."—"All the same, I think I know of a publisher who might be interested. . . ."—"No publisher would ever take such a risk!"—"Oh, I think he might." The interchange continued thus for a moment, and then Wiener suddenly said, "I get you! You are the publisher." They shook hands on it. "In three months' time I shall hand over my manuscript."

Afterwards, as Wiener remembered, "we sealed the contract over a cup of cocoa in a neighboring *patisserie*." At the time, Wiener did not know that his new publisher had no faith he would ever deliver on his commitment. Latil continued:

When Wiener left, Freymann smiled and said, "Of course he'll never give it another thought"; and in fact no further mention was made of the subject throughout Wiener's stay in Paris.

Wiener did give it another thought. The book he promised the publisher Freymann would pour from his mind in a controlled creative explosion, propelled by his excitement at the prospect of disseminating his ideas to an international audience of his peers and the wider public. But the work of writing the book would not be done in France or even in the United States. Wiener returned to Boston in mid-July, and, five days later, he left for Mexico City and his next extended work session with Rosenblueth at the Instituto Nacional de Cardiología. There, in his spare time, he wrote the small volume that would name his new science and proclaim the scientific revolution he had been scripting for a quarter century.

After so many years of preparation, and so much time spent in the hinterlands of American science, Wiener knew exactly what he wanted to say, but he also knew early on that his task as an author was more than technical. "It became clear to me almost at the very beginning that these new concepts of communication and control involved a new interpretation of man, of man's knowledge of the universe, and of society."

And the work began. He wrote longhand in pencil on lined legal pads, proceeding in a studious but "somewhat manic mood." He worked in the living room of his small apartment near the Instituto, in the garden on the roof that looked out across the city to the snowcapped mountains beyond, and in a quiet work space Rosenblueth had provided in the rooms adjacent to his laboratory. He did his best writing in the morning, on waking up brimming with ideas from his predawn reveries, as his thoughts flowed onto the page in a scribble of oversize script. His new communication theory came together from opposite ends of the scientific universe: engineering and biology, thermodynamics and homeostasis, information and entropy, computing machines and nervous systems.

The mass of higher mathematics Wiener produced to prove his points rolled out in strings of recondite symbols and shorthand notations—there would be time to check his complex equations later. The new concepts from Gestalt psychology, anthropology, and the other social sciences, which he had taken in at the Macy meetings, fit neatly into the larger puzzle he was working: the palpable effects of information and communication on perception and personality, the power of mass communication to shape the lives of individuals and cultures, the interplay of complex communication processes that connect human beings to the living world around them. His "new interpretation of man," the universe, and society came down on the page with clarity and confidence, as Wiener found himself flexing philosophical muscles he had not exercised since his youth with a new wisdom and maturity.

All that remained was for him to find a name for the new science.

Through the war years and after, Wiener had been reaching for just the right term that would capture the essence of the new science of communication and control. Now, with that need and his book deadline bearing down, he dug into his vast store of knowledge in classical languages and the history of science. As he reported later:

> I went to work very hard on this, but the . . . thing that puzzled me was what title to choose for the book and what name for the subject. I first looked for a Greek word signifying "messenger," but the only one I knew was *angelos* . . . meaning "angel," a messenger of God. The word was thus pre-empted and would not give me the right context. Then I looked for an appropriate word from the field of control. The only word I could think of was the Greek word for steersman, *kubernêtês*. I decided that, as the word I was looking for was to be used in English, I ought to take advantage of the English pronunciation of the Greek, and hit on the name *cybernetics*.

The word had a nice ring to it and a proud etymology. The classical texts Wiener recited as a youth were peopled with *kubernêtai*—with steersmen, helmsmen, and pilots of humble birth and high purpose. Steersmen made frequent appearances in Greek epic poems and dramas, where they navigated majestic ships and heroic figures safely across wine-dark seas. Greek philosophers, too, were fascinated by steersmen. In a dialogue on rhetoric that was thick with nautical allusions, Socrates veered off on a rhetorical tangent that foreshadowed Wiener's mission two millennia later:

> The art of the steersman saves the souls of men and their bodies . . . from the extremity of danger . . . yet it has no airs or pretences of doing anything extraordinary. . . . He who is the master of the art . . . walks about on the sea-shore by his ship in an unassuming way [but] he knows . . . which of his passengers he has benefited . . . by not letting them be lost at sea.

The term had survived through the ages. In the nineteenth century, the French physicist André-Marie Ampère, a founder of the science of electricity, used the word *cybernétique* to describe the "art of government," although Wiener would not learn that until after his book was published. Instead, he knew the word "cybernetics" in its Latin form, *gubernātor*—the root of the English word governor. In 1789, Watt first used that term in a technical context to describe the flyball feedback apparatus that controlled the speed of his steam engine. A century later, James Clerk Maxwell secured the term in the scientific literature in his famous paper, "On Governors," published in the *Proceedings of the Royal Society*, but by then the concept already had come stem to stern with the advent of the first automatic, feedback-controlled steering engines for seafaring steamships.

To commemorate the human *kubernêtai* of antiquity and the first practical cybernetic devices of the industrial age, Wiener named his new science cybernetics—because "it was the best word I could find to express the art and science of control over the whole range of fields in which this notion is applicable."

―――――

Late in 1947, three months after Wiener and Freymann had shaken hands on the deal the publisher never thought would come to fruition, as Latil chronicled, "an air-mail package arrived at the Rue de la Sorbonne. Freymann opened it—and there was the manuscript." Freymann put the book into production fifteen minutes later. "There's something to get on with—that manuscript on the table over there," Freymann told a Parisian printer who had come to his shop looking for work. The type was set in eleven days and the galley proofs were rushed to Wiener in Boston, but Freymann was not free and clear. Wiener sent back a cryptic cable: "You'll have to beat American efficiency twice over."

The impending publication of *Cybernetics* had caught Wiener's employers by surprise. When Wiener showed the manuscript to "the MIT authorities," as he called them, "they were much interested . . . and hoped that they might find a way to publish the book in America." Indeed, the folks at Tech moved with alacrity to reclaim the rights to *Cybernetics*. The director of the institute's publishing company, The Technology Press, telephoned Paris and implored Freymann to release Wiener from his contract on the grounds that MIT "could not let the work of one of its own professors be published by another firm." But Freymann was unmoved. After six telephone calls, the transatlantic tug-of-war was peaceably resolved to the satisfaction of both parties—and the greater glory of France. Freymann consented to publish the book jointly with MIT in the United States, but he retained the copyright and the French edition was distributed internationally.

The book's galleys did not fare so well. With his eyesight failing from the strain of the work and cataracts he began to develop, Wiener asked his two most talented doctoral students, Walter Pitts and Oliver Selfridge, to proof and correct one of the two sets of galleys Freymann had provided; but someone slipped up and the uncorrected copy was shipped back to France. Selfridge claimed Wiener had confused the two copies and returned the wrong one to Freymann. Wiener blamed his charges. He lamented later that the book "came out in a rather unsatisfactory form, as the proofreading was done at a period at which I could not use my eyes and the young assistants who were to have helped did not take their responsibility seriously."

McCulloch got a glimpse of the book at the fifth Macy conference in March 1948. He saw that Wiener was apoplectic with worry over the reception it would receive, but McCulloch perceived Wiener's concern as more than childish anxiety or the average author's obsession. "I had only an hour's chance to skim through the book and we discussed its possible circulation which we both underestimated grossly," McCulloch recalled. "I could and did praise it to him then and there, but he remained overanxious and too tense about the book's real value. He was like a prophet with a message that had to be delivered, not merely like a 12-year-old seeking approbation."

Cybernetics: or Control and Communication in the Animal and the Machine was published simultaneously in France and the United States on October 22, 1948. Both editions were printed in English. The hardcover price in America was three dollars. When it appeared, some scientific reviewers dismissed the book as abstruse, badly structured, and riddled with errors in its many complicated mathematical formulas. The first charge was inevitable given the book's highly theoretical subject matter. The second criticism was debatable. Wiener's writing style, like his science, was conversational and multidimensional. He made frequent leaps from his theories to their ramifications and injected colorful examples of his science in action, in the laboratory and in real life, across the wide range of fields to which his cybernetic concepts were applicable. On the third charge there was no dispute: The first edition of the book contained numerous mathematical errors, owing to Wiener's penchant for blunders, the snafu with the galleys, and his cataracts, which prevented him from reading the book in its final stages.

Notwithstanding those flaws, *Cybernetics* was a masterstroke of scientific reasoning and writing. At the turn of the millennium, according to *American Scientist* magazine, it still ranked among the most "memorable and influential" works of twentieth-century science.

The introduction alone was worth the price of the book. In a comprehensive overview written in his lively, personal style, Wiener recapped the short history of the new science and its long gestation in Europe and America. He acknowledged the field's forebears in philosophy, mathematics, and thermodynamics, especially Leibniz—whom Wiener lauded as "a patron saint for cybernetics"—and Maxwell. (He saved Gibbs for special treatment in a later chapter on statistical mechanics.) He credited each of his contemporaries and collaborators who had contributed to the new science and technology, including

Bush, Lee, McCulloch, Pitts, Turing, Aiken, von Neumann, and Shannon. He described at length his war work with Bigelow, and he put on the record the prescient memo on digital computer design he had submitted to Bush in the fall of 1940.

From the outset, however, Wiener ascribed the start of cybernetics not to technology but to biology. He traced its inception to Rosenblueth's supper seminars at Harvard Medical School, and to the conviction he and Rosenblueth shared that the most fertile turf for growth in the sciences lay in the "no-man's land" between the established fields. Now Wiener evoked the full power of the dream he and Rosenblueth had pursued jointly of blazing those "boundary regions" and "blank spaces on the map of science," and their unquenched longing to create an interdisciplinary "institution of independent scientists, working together in one of these backwoods of science . . . joined by the desire, indeed by the spiritual necessity, to understand the region as a whole, and to lend one another the strength of that understanding."

Wiener scrupulously acknowledged all his other allies in no-man's land: the participants in the Macy conferences, the Princeton computer meeting, and everyone involved in the research advancing in Cambridge, Chicago, and Mexico City. The synthesis of cybernetics had emerged from their growing awareness of "the essential unity of the set of problems centering about communication [and] control . . . whether in the machine or in living tissue," and because the group collectively had been "seriously hampered . . . by the absence of any common terminology, or even of a single name for the field." Before long, said Wiener, the jargon of the engineers "became contaminated with the terms of the neurophysiologist and the psychologist," and "it had become clear to all that . . . some attempt should be made to achieve a common vocabulary."

Wiener laid out the new terms and concepts of cybernetics in a grand tour of the new terrain as he had mapped it. His book combined all the communication and control principles he had presented piecemeal in his still-classified Yellow Peril, in his 1943 paper with Rosenblueth and Bigelow, and in his 1946 speech to the New York Academy of Sciences, in a greater synthesis that applied equally now to animals, machines, and human beings. His "fundamental notion of the message" was enlarged to incorporate messages "transmitted by electrical, mechanical, or nervous means." His breakthrough notion of information, expressed vaguely six years earlier as a statistical "measure or probability," was now formulated precisely as "*a statistical theory of the amount of information, in which the unit amount of information was that transmitted as a single decision between equally probable alternatives.*"

As the book unfolded, Wiener distinguished the two major modes of information, "discrete or continuous"—digital or analog—and their diverse applications in communication, electronic computing, and automatic control systems. He supplied pages of equations to prove his important point that information is order, the negative of entropy or disorder in physics, and should be measured logarithmically in binary units, and he put his proofs to work solving practical problems of communication engineering. He laid bare the mathematical and engineering secrets to separating messages (order) from noise (disorder), and he showed how his new statistical techniques could be used to design improved electronic circuits, feedback mechanisms, and automated machines. He made the all-important distinction between physical and biological systems, their different modes of organization, and the different forms of information that could be found in each realm of nature, and he proved mathematically that the engineering principle of feedback was equivalent to the physiological process of homeostasis.

At the book's midpoint, Wiener brought his most important cybernetic terms and concepts together and gave to the world, not just to closed circles of engineers and technicians, the keys to understanding complex systems of all kinds. He proclaimed the essential unity of information processes and showed how the new technical methods of "control by informative feedback" that engineers were beginning to use ubiquitously to control, organize, stabilize, self-regulate, and govern vast communication networks and intelligent automatic machines were, in their essence, the same universal processes that nature long ago selected as its basic operating system for human beings and all living things.

In so doing, in *Cybernetics*, Wiener established the universal principle of feedback as more than merely a good technical idea. He gave concrete, practical examples of both negative and positive feedback at work in mechanical, electrical, and living systems, and he described common disorders caused by too much or too little of each kind of feedback. Negative feedback, as he made clear, was not such a negative thing at all. Its vital error-correcting information gave order and self-control to automatic machines, to the body and the brain, and to people in their daily lives. Positive feedback—information that confirms and reinforces the outcome of a communication or control process—could be equally valuable to cybernetic systems, especially human ones. But it could also be downright devastating, as Wiener learned the hard way in his fire control project. When left unchecked, its reinforcing effects tended to compound a system's errors, leading to runaway processes, wild oscillations, and, ultimately, to "something catastrophic." The effect could be readily observed in people afflicted with debilitating neurological disorders like purpose tremor and Parkinson's disease, and—one of Wiener's favorite examples—in the plight of hapless drivers on icy roads who can veer out of control with one tap on the accelerator, and whose brakes are useless to provide the crucial negative feedback required to stop the car.

To survive and adapt to the world around them, Wiener made clear, all communication systems need a healthy balance of negative and positive feedback. Under his influence, the term would leap into the minds of individuals, the lexicons of cultures, and the conceptual repertoire of the information age.

In *Cybernetics*, Wiener made more enlightening connections between human beings and machines, and he widened the horizons of his new science to encompass the psychological insights that had begun to emerge from the Macy conferences. He introduced his readers to the new genus of electronic computers and showed how their physical components and logical operations mimicked the action in the brains and nervous systems of their human creators. Building on the insights of Gestalt psychologists, and the new work of McCulloch and Pitts, he explained the mind's subjective processes of perception in the new terms of information processing. He proposed a comparable technology for computer perception, using an electronic scanning mechanism operating at the same frequency as brain waves in the visual cortex, more than a decade before the first computer was hooked up to a television camera. And he described the ways in which both types of cybernetic systems—computers and brains—could malfunction similarly and catastrophically.

Wiener believed that society's understanding of mental illness could benefit greatly from the lessons afforded by the new brain-like computing machines. He suggested that many human "functional" (as opposed to organic) mental disorders were "fundamentally diseases of memory, of the circulating information kept by the brain in the active state," and that "even the grosser disorders . . . may produce a large part of their effects not so much by the destruction of tissue . . . as by the secondary disturbances of traffic" in the nervous system—a problem he suspected early on could involve the brain's myriad circulating chemical messengers. He explained how modern problems such as "malignant worry," anxiety attacks, and other classic neurotic disorders might start with a relatively trivial concern and "build itself up into a process totally destructive to the ordinary mental life," just as a computer caught up in a logical paradox "may go into a circular process which there seems to be no way to stop." He compared other pathological states of mind to the new technical problem of information "overload," which may take place, in the human animal as in the machine, "either by an excess in the amount of traffic to be carried" or by an excess of undesirable messages. In both cases, he predicted, "a point will come—quite suddenly—when the normal traffic will not have space enough allotted to it, and we shall have a form of mental breakdown, very possibly amounting to insanity."

In the book's closing chapters, Wiener introduced his readers to the social and cultural dimensions of cybernetics. Unlike a computer or a brain, Wiener observed, the channels of communication in society were formed, not from wires or neural nets, but from the exchange of information between individuals using language and nonverbal communication, from learning and group communication in families and larger social organizations, and from the exchange of knowledge and experience among people of different cultures. Drawing on the ideas of Bateson, Mead, and other social scientists at the Macy conferences, Wiener described the stabilizing "homeostatic processes" found in primitive and developed societies alike, and the myriad forces that may strengthen or undermine those essential processes. He observed that "small, closely knit communities have a very considerable measure of homeostasis; and this, whether they are highly literate communities in a civilized country or villages of primitive savages," but he was not so approving of contemporary mass communication cultures.

Now, for the first time in his professional writing, Wiener injected some social criticism into his scientific perspective. He saw a dearth of healthful homeostatic processes in large developed societies, due to the modern glut of information and its counterpart, the "constriction of the means of communication" by vested social and political interests. He assailed the simplistic theories of free-market ideologists, and he reserved a special antipathy for modern-day "hucksters," media barons, captains of industry, and politicians who sought power over large populations through manipulation of the mass media. "Any organism is held together . . . by the possession of means for the acquisition, use, retention, and transmission of information," wrote Wiener, especially the mass society, which is "too large for the direct contact of its members." He finished with a flourish, and a warning to his readers: "Of all of these anti-homeostatic factors in society, the control of the means of communication is the most effective and most important."

Cybernetics was hard going in places and heavy with higher mathematics, but Wiener's little book had flair and punch. In the first show of a popular writing style he had not used since his days as a beat reporter at the *Boston Herald*, Wiener spiced *Cybernetics* with strong leads, self-assured prose, and philosophical ruminations. His science was revolutionary. His voice was defiant and destined to be controversial. The book was peppered with provocative statements that proved prophetic in hindsight. In one oracular passage, Wiener hailed the primary resource of the newborn information age as a unique force and substance in nature, and he warned those who failed to heed its new physical realities and human imperatives that they would do so at their peril:

Information is information, not matter or energy. No materialism which does not admit this can survive at the present day.

From the start, he expressed high hopes for cybernetics and the new technological era it heralded. However, he also injected a darker theme that would loom large in his writing and professional endeavors from that point onward: his concern for the human consequences of cybernetics and the powerful new knowledge he was handing over to science and society. He foresaw the sweeping advances cybernetics made possible for modern enterprises ranging from weapons production to mass production. He voiced his alarm at the potential impact of the new intelligent technology on working people in every business and industry, and his fear that, like the first industrial revolution, which devalued "the human arm by the competition of machinery," the second, cybernetic, industrial revolution "is similarly bound to devalue the human brain, at least in its simpler and more routine decisions [until] the average human being of mediocre attainments or less has nothing to sell that it is worth anyone's money to buy."

Writing in the first years of the postwar era, with the war's horrors still sharp in his view, Wiener took a fatalistic position on the prospects for his science, and he declared unequivocally the direction in which his personal priorities and responsibilities as a scientist were pointing him:

> Those of us who have contributed to the new science of cybernetics thus stand in a moral position which is, to say the least, not very comfortable. We have contributed to the initiation of a new science which . . . embraces technical developments with great possibilities for good and for evil. We can only hand it over into the world that exists about us, and this is the world of Belsen and Hiroshima. We do not even have the choice of suppressing these new technical developments. They belong to the age. . . . The best we can do is to see that a large public understands the trend and the bearing of the present work, and to confine our personal efforts to those fields . . . most remote from war and exploitation.

Cybernetics hit the postwar world with a bang. Publication of the book, just two years short of the twentieth century's midpoint, put science and societies worldwide on a new trajectory. America's major media helped to usher in the new science and showcase its ex-prodigy founding father. *Scientific American* gave Wiener the leadoff spot in its November 1948 issue, premiering his new work before a wide audience of scientists and the public. *Newsweek* broke the news in a full-page report that depicted Wiener, in the first of many verbal caricatures that would follow in the press, as "a true boy genius" grown into "a bearded, fast-talking professor of mathematics," and accurately portrayed his role in the new interdisciplinary enterprise as "something of a scientific matchmaker."

Within weeks the media were detecting signs of a scientific groundswell. *Time* magazine in its year-end issue reported, "Once in a great while a scientific book is published that sets bells jangling wildly in a dozen different sciences. Such a book is *Cybernetics*." In its signature style, the magazine repositioned Wiener as the advance man for the next phase of the American century. "He looks like a Quiz Kid grown into a Santa Claus—and that's about what he is." But *Time* also grasped the substance of the new synthesis Wiener and his colleagues were forging. "The cyberneticists are like explorers pushing into a new country and finding that nature, by constructing the human brain, pioneered there before them." And it did not refrain from reporting Wiener's anxiety at the consequences he saw coming over the horizon. "Many times throughout his book Dr. Wiener stops in a cold sweat and looks a few years ahead."

The book flew off the shelves. *Cybernetics* went through five printings in its first six months and became the talk of international publishing circles. No one was more surprised by the book's success than Wiener, save, perhaps, his French publisher. "Freymann had not rated the commercial prospects of *Cybernetics* very highly—nor, as a matter of fact, had anybody on either side of the ocean," Wiener acknowledged later. "When it became a scientific best-seller we were all astonished, not least myself."

By February 1949, even staid *Business Week* felt obliged to inform its readers of the phenomenon of *Cybernetics,* which it compared favorably to the year's other scientific best-seller on a much sexier subject, Dr. Alfred Kinsey's report on the sexual behavior of American men. The magazine took note of "a book which it had been expected would appeal to only a small technical audience. . . . *Cybernetics* . . . has a mystifying title; its pages are spiky with mathematical signs and Greek letters; it is wretchedly printed. . . . Yet in . . . one respect Wiener's book resembles *The Kinsey Report*: the public response to it is at least as significant as the content of the book itself." By April, *The New York Times* was tracking Wiener's book that had "amazed dealers by [its] sales" and left them powerless to "understand why [the] volume has the popular appeal it has." A writer for the *Times Book Review* interviewed Wiener and was pleased to discover "a roly-poly little man . . . full of nervous energy" whose "bright-eyed eagerness for new ideas and . . . trigger-quick responses make it easy to see through the beard to the boy genius." The *Times's* reviewer was even more impressed with *Cybernetics*. "Between the equations . . . are pages of sparkling literature and provocative prose. . . . Every sentence is forcefully voiced and rhetorically impeccable." Later, another *Times* critic called the book one of the "seminal books . . . comparable in ultimate importance to . . . Galileo or Malthus or Rousseau or Mill."

Cybernetics caught the wave of a population newly attuned to matters of science and technology in the light of the Allies' wartime technological triumphs and eager to reap the fruits of a new harvest of postwar inventions. It also struck a nerve in a public sensitized to the growing threat of atomic warfare and, now, to the new prospect Wiener broached in his book that machines might soon replace human labor in the factories and offices to which millions of victorious veterans were returning.

Technicians in heavy industry and the electronics industries were primed to receive Wiener's ideas. In the eighty years since Maxwell published his historic paper on governors, the production of commercial feedback devices had grown geometrically, from the first automatic steering engines and thermostats, to the new automatic clothes washers and automobile transmissions rolling off the assembly lines where the term "automation" was coined to describe the use of automatic machines to aid or replace human workers in industrial operations. By the late 1940s, automated devices were widespread in industry and commercial applications, but there was still little theory or hard scientific knowledge underlying their manufacture.

Mechanical engineers were only dimly aware that the fragile electronic feedback circuits being developed at Bell Labs had anything in common with their heavy metal contraptions. In fact, before *Cybernetics* appeared, most servomechanisms and automatic control devices were being "designed, built, and manufactured . . . without any clear understanding of the dynamics . . . of the system to be controlled." Even at MIT, power engineers and communication engineers, including top people at the Rad Lab and the Servomechanisms Laboratory, still did not get the point Wiener had made in his Yellow Peril: that control and communication were two sides of the same universal process that had little to do with power and everything to do with information. As the Rad Lab's Ivan Getting recalled, "There was no connection between the Bell Labs work in communications feedback and the feedback which was an essential element of all servomechanisms."

Then came *Cybernetics* and, hot on its heels, Wiener's elated publishers released a second technical book by Wiener, an updated edition of his Yellow Peril, which Wiener had been working to get declassified since the war's end. The obscure monograph with the large name, already a legend in the inner circles of engineering, finally broke free from its restraints in 1949, when it was published with a long appendix spelling out its wider peacetime applications and an even longer title, *Extrapolation, Interpolation and Smoothing of Stationary Time Series, with Engineering Applications.* Together, the two books, *Cybernetics* and *Time Series*, as it became known, took the postwar engineering world by storm.

Wiener's former student and young cousin from New York, Gordon "Tobey" Raisbeck, who edited the Yellow Peril for publication, and who was employed in the engineering department of Bell Laboratories when *Time Series* appeared, recalled the sudden impact of Wiener's books on his colleagues. "By '49, people began to apply Norbert's work to the coding and detection of signals, and from that point on you *had* to use those methods because they were so much more powerful than what had been used up to that time. That was when Norbert was first recognized in the engineering community outside of MIT and in other reputable schools of electrical engineering." Raisbeck knew Wiener had been greatly frustrated by the lack of recognition of his work among American mathematicians and engineers, but he saw Wiener's triumph from the longer perspective of the acceptance of scientific ideas historically. "I consider twenty years to be pretty fast, really, especially when you consider that from 1940 to 1946, and later in many instances there were severe restrictions on scientific communication."

An historian from a later generation, Kevin Kelly, editor-at-large of *Wired* magazine and an avatar of all the new technologies, tracked the path Wiener's work cut through the postwar engineering community. "Within a year or two of *Cybernetics's* publication, electronic control circuits revolutionized industry. . . . Generations of technicians spent years strenuously perfecting the regulation of [industrial production processes] and more years attempting their synchronization. To no avail . . . until Wiener's brilliant generalization published in *Cybernetics*. Engineers around the world immediately grasped the crucial idea and installed electronic feedback devices in their mills." And the impact was felt far beyond the factory. Under Wiener's influence, the separate streams of servomechanisms and electronic feedback circuits merged into a unified current of technical know-how, and a great river of automated machinery, home appliances, and electronic devices began pouring into the marketplace.

Indeed, the postwar explosion of industrial expansion, economic growth, and technological progress owed much to Wiener's work. His new science brought changes in the research and development of electronic technology, in the methods and costs of mass producing it, in the tasks performed by workers in every field, in the buying habits and lifestyles of consumers—and, more important, in the conscious connections people everywhere were beginning to make between intelligent machines, human beings, and all living things.

The success of *Cybernetics* propelled Wiener beyond his standing as a merely brilliant mathematician to a new status as a certified "genius" and ascending superstar on the American stage and in the rarefied realms of international science. After his debut in *Time* and *Newsweek,* Wiener's portly frame and prognostications were featured regularly in full-page photospreads in *Life.* His work became the subject of long articles in *Fortune, The New Yorker,* and a *Time* cover story on computers. The French newspaper *Le Monde* ran an article on *Cybernetics* and response to the book was especially strong in Sweden. In 1949, Wiener was invited to deliver the Josiah Willard Gibbs Lecture at the American Mathematical Society's annual meeting. In 1950, he addressed the International Congress of Mathematicians, which convened at Harvard that year.

Cybernetics, feedback, and the unwieldy name Norbert Wiener had become household words in America and many other places, yet, for Wiener, the respect of his peers was far more important. His colleagues in the Macy group were overjoyed when *Cybernetics* appeared (although McCulloch protested that Wiener's book was "too popular. I have purchased two copies and both of them have already been stolen by my friends"). Most of the Macy conferees saw Wiener's leap onto the public stage as a boost for their mutual scientific interests, and the mood at the group's post-publication conferences was jubilant.

Heinz von Foerster remembered the air of excitement that engulfed the sixth Macy conference in March 1949, and his own contribution to the proceedings that touched Wiener deeply.

"I arrived in New York from Vienna and, three weeks later, I was invited to participate in the Macy Foundation meetings. Here was Norbert Wiener and the whole group, John von Neumann, Warren McCulloch, Margaret Mead, Gregory Bateson, and my English vocabulary was maybe fifty words! I said, 'My god, I can't even pronounce the *title* of the conference'—which at that time was 'Circular Causal and Feedback Mechanisms in Biological and Social Systems.' I said, 'I propose we should call the conference "Cybernetics,"' because I had just read Wiener's book. All the others accepted this at once with applause and laughter and said, 'Excellent proposal!' Wiener was sitting next to me, and he was so touched that his peers accepted the funny name of cybernetics for their conference. Tears came into his eyes. He had to leave the room to hide his emotions."

The foundation endorsed the conferees' consensus, and the next five conferences and the published volumes of their proceedings that followed were titled simply, "Cybernetics," with the old name consigned to a subtitle.

———

Now others joined Wiener in the front ranks of the communication revolution. Side by side with the publication of *Cybernetics,* in 1948, a long technical article appeared that launched cybernetics' sister science of information theory. The article was written by Wiener's younger colleague from MIT, Claude Shannon, who had become one of the leading theorists at Bell Laboratories.

Shannon, like Wiener, had toyed with electronics since his youth and he had once strung a mean telegraph from his house to a friend's using half a mile of barbed wire. He came to MIT in 1936 as a graduate student in mathematics and electrical engineering, and drew his first assignment as a laboratory assistant to Vannevar Bush, who was then building his next-generation differential analyzer. Shannon's work on the cumbersome analog computer's control unit, consisting of a hundred flip-flop relays, led to his acclaimed master's thesis on the symbolic logic of switching circuits, which Shannon analyzed using the binary mathematics of Boolean algebra. The ten-page thesis, written two years after Turing's landmark theoretical paper on computation, proved that it was possible to perform complex mathematical operations using simple binary circuits, and that computers built in the binary mode could make logical decisions as well as mathematical ones.

Wiener had little contact with Shannon during Shannon's graduate years at MIT, but he saw early on that Shannon had "hit on an idea which even then showed a profound originality." During the war, Shannon in turn became keenly interested in Wiener's ideas and sought his help in recurring visits to Wiener's wartime laboratory, although over time Wiener grew weary of those visits and worried that Shannon was "coming to pluck my brains."

Shannon's two-part paper, "A Mathematical Theory of Communication," published in the *Bell System Technical Journal,* built on the work of Bell Labs' theorists dating back to the 1920s and, equally, on Wiener's work in the Yellow Peril. The paper took a statistical approach to communication theory that derived directly from Wiener's wartime work, and it promoted other ideas Wiener had introduced two years earlier in his presentation to the New York Academy of Sciences: the use of logarithms in base 2 to measure the amount of information in a message in binary units, and the notion that information and entropy were related in physical terms. However, in a divergence that turned on a technical difference in their mathematics, Shannon declared information to be the equivalent of entropy, not negative entropy, as Wiener maintained.

Shannon's paper offered other historic definitions. It introduced the notion of the "binary digit" or "bit," a term proposed by the Princeton mathematician and Bell Labs consultant John W. Tukey, which quickly became the standard unit of information measured in base 2 logarithms. Shannon's twenty-odd theorems gave Bell Labs' engineers precise formulas for calculating the "channel capacity" of a transmission line measured in bits per second and the amount of "redundancy" or repetition required to send a signal faithfully over a noisy channel. And he made one more important distinction for information theory. In his boldest move, Shannon severed his technical definition of information from any consideration of the *meaning* of the messages being sent and received. He declared unequivocally: "These semantic aspects of communication are irrelevant to the engineering problem." That dictum flew in the face of every definition of the word "information" on record since the fourteenth century, but from a purely technical standpoint, Shannon's move made sense and cleared away a lot of semantic confusion that was indeed irrelevant to the technical problems of communication.

The first installment of Shannon's paper, published in July 1948, drew on his war work in cryptography and dealt with his theoretical forte of discrete messages—distinct, disconnected signals, like the dots and dashes of Morse code or the separate letters of an encrypted message, which could be transmitted by the tapping of a telegraph key or in similar electrical pulses. The second installment, published the same month as *Cybernetics*, focused on the continuous messages embodied in voice transmissions by telephone or radio and other electromagnetic wave phenomena. Wiener's work in communication theory had almost nothing to do with the discrete domain and everything to do with the continuous, although their methods of measurement were virtually identical in statistical terms.

Shannon gave Wiener credit for some of the central ideas in his new mathematical theory. He cited Wiener's war work and stated explicitly: "Communication theory is heavily indebted to Wiener for much of its basic philosophy and theory." He confirmed that Wiener's "classic NDRC report" (the Yellow Peril) contained "the first clear-cut formulation of communication theory as a statistical problem [and] is an important collateral reference in connection with the present paper." In a closing acknowledgment, he said forthrightly that Wiener's "elegant solution" of fundamental problems of communication theory had "considerably influenced the writer's thinking in this field."

Shannon's technical article for communication engineers soon caught the eye of people higher up in the postwar scientific establishment, and the following year the paper was reprinted in a slim volume Shannon co-authored with Wiener's wartime supervisor, Warren Weaver. The book included a long essay by Weaver that reinterpreted Shannon's theory for nontechnical readers and hailed its implications for technology and communications across the spectrum of society.

Concurrent with cybernetics, Shannon's mathematical communication theory "exploded with the force of a bomb" on the engineering world. Engineers were in awe of his new theory that, in fact, staked its claim in the engineering literature three months before *Cybernetics* was published. And, despite Shannon's acknowledgments of Wiener's contributions, the credit for information theory accrued largely to Shannon. "It was like a bolt out of the blue. . . . I don't know of any other theory that came in a complete form like that, with very few antecedents or history," said John Pierce, another pioneering theorist at Bell Labs who coined the word "transistor" and would later invent the communications satellite. Oliver Selfridge recorded the shock wave when Shannon's theory swept through Cambridge. "It was a revelation. . . . Around MIT the reaction was, 'Brilliant! Why didn't I think of that?'"

By the time Shannon and Weaver's book appeared, Wiener was world famous for cybernetics, but he chafed at the acclaim Shannon received for his meticulous theorems and elaborate development of ideas Wiener had shared with him in their private sessions. Publicly, Wiener voiced the "highest regard for Dr. Shannon both as to his scientific accomplishment and to his personal integrity" and extolled him in print as "one of the major spirits behind the present age," but he was determined to be acknowledged for his contributions to information theory, as he was for his contributions to computing. In a review of Shannon and Weaver's book in *Physics Today*, Wiener recognized it as "a work whose origins were independent of my own work, but which has been bound from the beginning to my investigations by cross influences spreading in both directions." In his autobiography, he would speak assertively of "the Shannon-Wiener definition of quantity of information (for it belongs to the two of us equally)."

In the end, both men deserved, and received, credit for contributing two complementary concepts of information to the postwar world: Shannon's discrete, digital conception derived from his war work on coding, and Wiener's continuous, analog conception, which he had been developing statistically since the 1920s. Shannon's innovative digital methods appealed to many electronics engineers, the younger ones especially, but Wiener was intent on keeping the two modes of information from becoming mutually exclusive technological faiths. In the computing arena, he lobbied for the digital approach. The laboratory research he was conducting with his colleagues was largely analog, like the life processes they were exploring. Wiener's aim for information theory, as it was for antiaircraft guns, was to keep things from swinging too widely to one side or the other.

However, Shannon's technical approach to information raised some serious new problems of its own. Many communication theorists and researchers were troubled by his stipulation that questions of meaning and the "semantic aspects of communication" were irrelevant to the theory of information. Weaver himself had liberally extended Shannon's technical communication theory to include "all of the procedures by which one mind may affect another . . . not only written and oral speech, but also music, the pictorial arts, the theatre, the ballet, and in fact all human behavior," which only served to confuse the matter further.

When Shannon came as a guest to a later Macy conference, most of the group's social scientists, and even some physical scientists, took issue with his ban on semantics. Heinz von Foerster, the biophysicist from Vienna and editor of the conferences' transactions, who had family ties to Wittgenstein, the century's leading philosopher of language and meaning, argued that information, even in the technical sense, could not be severed from meaning without horrendous consequences for human understanding. "I complained about the use of the word 'information' in situations where there was no information at all, where they were just passing on signals," von Foerster remembered. "I wanted to call the whole of what they called information theory *signal* theory, because information was not yet there. There were *'beep beeps'* but that was all, no information. The moment one transforms that set of signals into other signals our brains can make an understanding of, *then* information is born! It's not in the beeps." Others in the group agreed, but their view was drowned out by the growing influence of the new technology in the postwar society and within the Macy group itself.

In the late 1980s, as his memory was beginning to decline, Shannon and his wife Betty, a numerical analyst he had met at Bell Labs and married in 1949, spoke candidly about his theory and his intentions for it, which like Shannon himself were quite modest and reserved. He defended his decision to detach the technical theory of information from age-old philosophical debates about meaning and modern-day theories of semantics which were irrelevant to the engineering problem, but he made a historical distinction that could have saved a lot of needless confusion and scientific commotion. Shannon denied that he had ever fathered anything called "information theory."

"In the first place, you called it a theory of *communication*," Betty Shannon reminded her husband. "You didn't call it a theory of information."

"Yes, I thought that communication is a matter of getting bits from here to here, whether they're part of the Bible or just which way a coin is tossed," Shannon confirmed. His point at that late date was almost identical to von Foerster's four decades earlier.

Betty testified to Shannon's distress when scientists far removed from the field of communication engineering jumped on the information "bandwagon," as he called it, and Shannon watched his work career out of proportion to the technical theory he had propounded. "It bothered you along the way a few times but by that time it was out of your hands," said Betty.

Shannon reaffirmed the limits he put on his work that confined his part of the theory to something specific and purely technical, without the larger philosophical aspirations and social connections that characterized Wiener's mission and vision for cybernetics. "The theory has to do with just getting bits from here to here," Shannon repeated. "That's the communication part of it, what the communication engineers were trying to do. Information, where you attach meaning to it, that's the next thing, that's a step beyond, and that isn't a concern of the engineer, although it's interesting to talk about."

Wiener's conception of information was greater than Shannon's or Weaver's. Wiener approached information, as he had come to cybernetics, from the vantage points of both engineering and biology. For him, information was not merely discrete or continuous, not strictly linear or even circular, not matter or energy, but something altogether new, extended in space and time—and very often alive. In Wiener's view, information was not just a string of bits to be transmitted or a succession of signals with or without meaning, but *a measure of the degree of organization in a system.*

That key concept for Wiener—organization—took information a step beyond the simple order embodied in linear modes of information transmission. Organization came from biology (Gr., *organon*, organ of the body), and, by the 1930s, a new breed of theoretical biologists and "living systems" thinkers were using the word to explain life processes specifically and, generally, to describe "the harmonious interaction of hierarchies and parts that gave meaning to the whole." Wiener embraced the concept during that period in his interactions with Rosenblueth and Cannon at Harvard, and it gained new meaning for him in the 1940s when Schrödinger linked entropy in physical systems with negative entropy in living things.

It was that new, dynamic quality of organization that Wiener brought to his conception of information, as he did to cybernetics and all of communication science. In that insight, he joined the animate and inanimate worlds, and completed his bridge across the no man's lands of science. Wiener's "essential unity" of analog and digital information processes spanned the whole of nature, society, and human invention—from the minute chemical messengers that conveyed their life-giving messages throughout the body in the distinctive shape of their molecular structure, to the incessant switching and pulsing of the brain's neural networks and electromagnetic waves, to the electric streams of analog and digital data that trafficked through the ether and telephone lines and computer circuits, to the myriad semantic and symbolic communications of modern societies that conveyed so many separate messages through the "body politic," as Wiener called it—and he would range freely through all those domains in the years ahead.

At the time, Wiener's engineering-oriented colleagues failed to apprehend that physical dimension of information and its direct connection to the basic processes of life, but others would soon join him in the new biological domains of information and communication.

———

Back at Princeton, John von Neumann was not oblivious to all these developments. Work on his IAS digital computer was progressing at a painfully slow pace, owing to problems with the design of the computer's memory. The delay was frustrating, and von Neumann was not a man to stand idly by while the foundations of the new technological age, and a whole new era of scientific thinking, were being laid.

At the Macy conferences, as McCulloch recalled, von Neumann "spent hours after the meetings and other long sessions with me, for he had become excited by the possibility of formulating the central problems of cybernetics as he had done for games." As he first told Wiener privately in 1946, von Neumann was becoming increasingly uneasy about the neural network model at the core of his digital computer architecture, and his growing interest in simple organisms like viruses, and their surprisingly complex behavior, had started him thinking about computation from an alternate view. Now von Neumann, too, was shifting his focus from inanimate processes to the complex processes of life itself—organization, growth, adaptation, and the elaborate sequence of operations necessary for reproduction. He wondered how those intricate processes could be carried out successfully by organisms that were quite crude compared to the human brain, and he set out to determine whether machines could be designed and built along the same simple lines to grow and learn from their experiences, and even to reproduce themselves automatically.

Von Neumann mulled over with McCulloch "the possibilities of building reliable computers of unreliable components." He met with other Macy conferees, and with physicists working at the forefront of the new cell biology, to discuss the prospects for creating intelligent machines, modeled on cellular organisms, that could program themselves in a biological manner and solve problems through an analogous electronic process of self-organization, "of things essentially chaotic becoming organized." In September 1948, he presented his first formal thoughts on the subject to a conference on computers and "cerebral mechanisms" sponsored by the Hixon Fund at the California Institute of Technology in Pasadena. McCulloch was there, and eight other Macy conferees, but not Wiener, who was busy in the east preparing for the publication of *Cybernetics*.

Von Neumann's talk at the Hixon symposium provided another dazzling display of his logical prowess, as he set down his own grand synthesis, which he called "The General and Logical Theory of Automata." Like his EDVAC report, his new theory built on the earlier work of Turing and McCulloch-Pitts. This time, however, he delivered a piercing critique of the McCulloch-Pitts neural network model, then he unveiled his alternate vision for the next generation of intelligent technology modeled on simple cellular organisms. Among his proposed innovations was an elaborate, tape-driven, self-reproducing "automaton whose output is other automata" that astutely anticipated the discovery a few years later of the genetic material DNA.

Von Neumann incorporated important ideas from Wiener's work into his general theory of automata. He analyzed the logical properties of systems with circular feedback and echoed Wiener's analog view of information, emphasizing the "serious weakness" of strictly digital computer theories that had "very little contact with the continuous." And, alluding to the information-entropy connection he had kicked himself for missing, von Neumann called for the development of a new system of logic rooted in thermodynamics and the new statistical methods for measuring information. All those ideas, along with his later plan for a new "probabilistic logic" inspired by the new model of random nerve nets that Pitts was developing under Wiener's supervision, would be refined and extended by von Neumann, McCulloch, and others in the years that followed.

The movement of cybernetics was advancing on both sides of the Atlantic. In England, in 1950, J. B. S. Haldane became the first biologist to apply Wiener's theories directly to the field of genetics. The same year, his colleague at the University of London, Hans Kalmus, published a piece in the *Journal of Heredity* titled "A Cybernetical Aspect of Genetics," which put the field on notice "that this new way of looking at life offered unifying principles and a powerful interpretive framework."

Also in 1950, the Austrian biologist Ludwig von Bertalanffy, who had emigrated to Canada the year before, published his first papers in English describing the comprehensive "general system theory" he had been developing in Europe for two decades. His systems approach was a fitting complement to cybernetics. Both shared deep roots in biological principles of homeostasis and organization, and a common mission to find an antidote to reduction in biology and all the sciences, and each developed its own circle of adherents and propagators in the established fields. But Bertalanffy, who had survived a rough passage through the war, was fiercely possessive of his theory. He never entered Wiener's circle and did not participate in the Macy conferences on cybernetics. Eventually he attacked cybernetics outright, in part, as Bertalanffy's biographer suspected, because "cybernetics upstaged . . . systems thought to such an extent that many writers began assuming cybernetics was synonymous with [general system theory]."

Wiener's influence was also being felt in domains far from engineering and biology. Soon after *Cybernetics* was published, Wiener's Macy colleague Gregory Bateson turned his attention away from tribal cultures and toward the human problems of the advancing technological society. In 1949, he joined forces with the Swiss psychiatrist Jurgen Ruesch in a groundbreaking study of the role of communication in psychiatry. Bateson and Ruesch's study, later published as a book, *Communication: The Social Matrix of Psychiatry*, presented the first theory of human communication informed by the new principles of cybernetics. And their social matrix reached into broader realms of communication and culture, as they became the first social scientists to examine the new communication links between humans and computers, between computers and other computers, and their social effects, which were only beginning to unfold.

Wiener was thinking about those social effects, too. In 1950, he published his first popular work on cybernetics for a nontechnical audience. The book, arrestingly titled *The Human Use of Human Beings: Cybernetics and Society*, cut across all the domains in which the new science and technology were taking hold: computers, automation, telecommunications, biology, medicine, psychiatry, economics, mass communication, popular culture, and the arts. The slim, easy-to-read volume appealed to people who were concerned, as he was, about the impact of the new technologies on their jobs and their daily lives, and to others who were simply seeking to understand the fast-changing times. It became Wiener's best-selling work yet, and it would take him deeper into social criticism and activism on a national and international scale.

All that and more would follow from *Cybernetics*. Wiener's big bang was the first work of the new technological era that enabled people to see that new world at once in its technical and human dimensions, not merely in matter and energy terms but, now, in a universe with two added dimensions: information and communication. Many people sensed the turn cybernetics signaled, but none more so than Wiener's colleagues at the Macy conferences. Warren McCulloch marked the milestone in a paean to the new science:

> Cybernetics . . . was born in 1943 [and] christened in 1948. . . . It has been a challenge to logic and to mathematics, an inspiration to neurophysiology and to the theory of automata. . . . Above all, it is ready to officiate at the expiration of philosophical Dualism and Reductionism. . . . Our world is again one, and so are we. Or at least, that is how it looks to one who was never a prodigy and is not a mathematician.

Bateson went further, declaring with unbounded enthusiasm:

> I think that cybernetics is the biggest bite out of the fruit of the Tree of Knowledge that mankind has taken in the past 2000 years.

With surprising speed, Wiener's new science had supplied a new conceptual base and practical foundation for the design and production of automatic machinery and electronic technology, for theory and research in the hard sciences, the life sciences, the social sciences—and for people in their daily lives as they struggled to grasp and master the wondrous ways of the new technological age.

Finally, in his mid-fifties, Wiener had fulfilled his promise as a prodigy and proved his worth to the world at a turning point in humankind's own life story. But the triumph Wiener and his colleagues rejoiced in would bring new pressures to bear on Wiener and his family, and breed a simmering disquiet in Margaret Wiener toward members of her husband's inner circle.

10

Wienerwalks II

I have made all sorts of acquaintances but have not yet found any congenial company. I don't know what there is about me that attracts people; so many like me and become attached to me, and then I am always sorry we can travel only a short way together.

—J. W. von Goethe, *The Sorrows of Young Werther*

WIENER'S NEWFOUND FAME DID NOT CHANGE his routines. Around MIT, "he was a familiar sight, standing splayfoot, his cigar posed in his right hand at the level of his mouth, pouring out on student, janitor, business manager or astounded colleague witticisms or profundities of science with equal gusto." His office in Building 2 was strewn with loose pages of text scribbled with mathematical notes, and kept tidy by a succession of secretaries who looked after him at work as Margaret did at home. His sartorial style, too, remained consistent. On the job, Wiener wore nondescript vested suits, usually of a light-colored tweed. "He had a habit of writing on the blackboard and then leaning against it that ruled out dark colors," Barbara recalled.

His worsening eyesight made just finding his way a problem, but he devised a novel workaround. He directed his optometrist to make him a pair of inverted bifocals, with the reading lenses on the top and the distance lenses on the bottom, so he could see more clearly where he was going. The invention forced him to throw his head back almost perpendicular to the ground when he walked and caused many people he encountered to feel, with good reason, that he was looking down his nose at them.

His formal speaking style, too, struck many listeners as an affectation. But Wiener was the genuine article, the product of an earlier time and his father's tutelage that, for all his modern scientific ideas, made Wiener appear increasingly like a walking anachronism. "He wrote and talked with the clarity of the late nineteenth-century Englishman," Jerry Lettvin observed. "His mode of speech was clear, non-elliptical, very straightforward. There was no cuteness about it."

He continued to teach Tech's mainstay class on harmonic analysis and its engineering applications. Oliver Selfridge, the youngest member of Wiener's inner circle, found his lectures to be "pretty damn incomprehensible." Occasionally, Wiener would stand in for a sick professor. Mildred Siegel, a rare co-ed at Tech during that time, remembered Wiener's abortive attempt to teach the beginning calculus course of a department colleague. "He came in and started writing on the blackboard and *writes writes writes writes* and suddenly got this look on his face and went out and never came back."

A few years later, during her graduate work, Siegel encountered Wiener again—in her kitchen—when he wandered into the Belmont home she shared with her husband, Armand, a physicist with whom Wiener had become acquainted professionally. "I was making dinner, and Norbert came over. I was having trouble with a calculus problem, and I said, 'Norbert, how you do this problem?' He looked at it and said, 'The answer is five.' I said, 'But, Norbert, I don't understand.' He said, 'I'll do it another way.' He looked at it, paused. 'The answer is five.' I never asked him for help again."

Wiener walked tirelessly around Tech during those eventful years, and his unannounced drop-ins evoked a mixed response from his colleagues. Jerome Wiesner, a rising star in the Rad Lab, who worked with Wiener on several postwar projects, remembered Wiener's "daily visits around the Institute from office to office and his conversation that always began with 'How's it going?' He never waited for the answer before sailing into his latest idea." Wiesner counted himself among those who were generally glad when Wiener appeared and eager to hear his ruminations. "Whatever was on his mind, Norbert Wiener's visit was one of the high points of the day at MIT for me and many others."

Some were not so giving of their time or attention. One group of engineers resented Wiener's intrusions and devised an extreme countermeasure they called their "Wiener Early Warning System." Heims reported that "they would contrive to place a man where he could see Wiener coming. He would alert the others, who would then scatter in all directions, even hiding in the men's room." Fagi Levinson knew of one colleague who hid under his desk when he saw Wiener coming. Her husband Norman recalled that those collegial encounters with Wiener—in which his colleagues would be required "to affirm in the strongest terms the great excellence of whatever piece of his research he himself would proceed to describe . . . in the most glowing terms"—could be "an exhausting experience" and an expensive one.

There was an instructor at MIT . . . who would complain bitterly about the expense of a chance meeting with Wiener. The man . . . found that the degree of enthusiasm he had to feign to meet the demands of Wiener left him enervated and without confidence in his own work. Such was his state that he would rush off to a psychiatrist. The fee . . . was much more than he could afford. Whether this played a role in his early departure from MIT he never said.

And Wiener did not always show others the same courtesies he expected. When a new theorem piqued his interest he was a rapt listener, but more often his colleagues complained that Wiener expressed little interest in their ideas. When he became famous, he hounded his faculty colleagues to learn "what others at MIT thought of him." When the talk turned to people at other institutions, "his first question was 'What do they think of my work?'" Tobey Raisbeck relayed one incident that occurred when George Pólya, the Hungarian mathematician whom Wiener had helped to find a job in America before the war, visited MIT. "After their visit, Norbert was driving Pólya to the airport, and Norbert came to a dead stop in the middle of Boston's rush-hour traffic and said, 'And what do they think of my work at Stanford?' All around them horns were honking, but Norbert wouldn't start moving again until Pólya told him. George was sure he was going to miss the plane. 'They think your work is splendid,' he said. Wiener started the car and Pólya made his plane."

Sometimes Wiener exhausted himself from his output, and from his gross weight, which put him to sleep in professional situations with greater frequency as he grew older. His snoring in seminars became such a distraction that, one year, during meetings of the Harvard-MIT Mathematics Colloquium, the presenters waged a competition to see if anyone could keep Wiener awake through one entire seminar. A European mathematician won the contest by injecting Wiener's name into his presentation whenever Wiener appeared to doze off, but his attempt to catch Wiener napping backfired. "One time he said Norbert's name and it was purely gratuitous," an audience member recalled. "He made his point, then added, 'and this has nothing whatsoever to do with Wiener's work on the ergodic theorem.' Norbert jumped up, thought very hard for a minute, and said, 'Oh, yes it does!' and proceeded to build a conceptual bridge."

Steven Burns, a young Rad Lab researcher, saw Wiener as a mentor and a fire hazard. "He used to come to the physics department colloquia with a copy of *The New York Times*. He would sit in the front row, and when the lecturer began he would light his cigar and open the paper and start reading, then he would nod off. The ash would get longer and longer until the entire audience was focused on Wiener's cigar and wondering would the ash drop on the newspaper and catch him alight?" Joseph L. Doob, a prominent figure in probability theory, recalled a near disaster that occurred when Wiener visited him at the University of Illinois in 1949. "He was in Urbana at the dedication of our electrical engineering building, and they were showing slides. The lights were out, and all I could see besides the slides was a spark in the dark. This was Norbert's cigar. Then all of a sudden there was a tremendous crash and sparks flew! He had fallen asleep and fallen out of his chair."

Tales of Wiener's eccentricities grew year by year. When guests came to dinner in his home, "he'd get up from the table as soon as he had had enough to eat and say, 'Well, I'm going up to have a nap' and disappear," Tobey Raisbeck recalled. "I don't think the concept of being rude was something he understood." His friends and detractors alike testified to Wiener's repeat offenses in his social and professional behavior, which they described as "immature," "petulant," and, at his most extreme, "infantile." Margaret, too, bemoaned the travails of caring for her forgetful, impolite, overly excitable husband, which she had computed to an exact degree of difficulty. "It was like caring for triplets," she told other faculty wives at MIT coffee klatsches, and her complaint was not confined to occasional meetings of her fellow *frau-professors*. "No, that she said every day," Fagi Levinson confirmed.

Yet Wiener was not always as helpless as Margaret portrayed him to be. "Margaret had a big stake in having Norbert seen as someone who required her constant attention and management, but I was with him for months at a time when he was clearly able to cope with everyday forces," Tobey Raisbeck maintained. He recalled that, during the war, when MIT students and faculty were garrisoned in Cambridge for the summer to work on their military projects, he and Wiener were assigned to neighboring rooming houses. Margaret directed Raisbeck to watch over Wiener and tend to his needs. "I was to see that he had a reasonable dinner every night, take his clothes to the laundry, she reeled off a long list." But no such vigil was required. "She thought he would be absolutely unable to survive unless somebody looked after him. As a matter of fact, he got along just fine." Jerry Lettvin, the psychiatrist, who was around him through much of the 1940s, concurred that when Wiener was not in the throes of a depression, "he was quite competent."

Indeed, there were good times during those years. The preferred family pastime was Wiener's passion, going to the movies. When his moods were good he took in a film weekly with his family and assorted guests. Wiener, the mystery buff, was a big Hitchcock fan. Like the famed English director whose silhouette was similar to his own, Wiener was a formidable presence in the cinema—and sometimes a nuisance. "His principal objective at the mystery movie was to announce who the murderer was in a loud voice long before the plot was resolved," Barbara remembered.

Wiener's other passion was women with pierced ears. During a stay in New York with Pitts and Lettvin, the threesome walked the city's bustling streets one evening, and Wiener shared his predilection. "A woman passed us with big earrings dangling from her ears, and Wiener told us that one of the things he found sexually exciting was women who had pierced ears," Lettvin recalled. "Walter and I said, 'Yeah, all right.' We thought it was amusing."

His fondness for pierced ears, and his urge to talk about it, fit the pattern of his boyish responses. "It was just one of these innocent fetishes that led to no action," observed Lettvin. But it was not viewed as such by Wiener's wife, who took all his libidinous remarks as proof of perversion and an affront to her Old World sensibilities. By now, Wiener's daughters were used to their father's ribald banter. Around the dinner table, in his ebullient moods, Wiener kept up a patter of puns punctuated with occasional sexual innuendoes. Barbara had perfected her response to his single and double entendres. "I would pretend to be too stupid to understand them, or to not respond at all," she remembered. Peggy was bothered less by her father's remarks than by her mother's reactions to them. She recalled Margaret's expressions of shock and disgust "even to a comment from our father that we were growing up. If Dad made comments, he was not salacious, but Mother would interpret them as being something he shouldn't have said."

———

Margaret was increasingly at odds with her husband and her children. Relations within the family had begun to deteriorate well before *Cybernetics* made Wiener famous, and on several occasions the family's internal problems had caused trouble outside the Wiener home.

At the height of the war, Barbara, just entering her teens, had recited passages by heart from the book she found on her mother's dresser, the English-language edition of Hitler's *Mein Kampf*, in her history class at school. She was briefly suspended from school and the incident set off a flurry of gossip about the Wiener family in the highbrow Belmont community. Soon after, Margaret began to worry that her headstrong daughter might disclose more intimate secrets of her life at home—such as Wiener's emotional storms, his frequent talk of suicide, and Margaret's own extreme beliefs—and do serious harm to Wiener's reputation and the family's rising social position. To head off any potential threat to the family from within or without, Margaret began a multi-pronged effort to keep her elder daughter silent about her home life in social circles beyond the family.

Margaret threatened her daughter with a grave reprisal if she spilled the family's secrets. "Look what happened to Uncle Fritz. We'll put you away!" she vowed when Wiener was not present. She fanned out through the town, pulling people aside, relatives, family friends, Barbara's teachers, babysitters, camp counselors—even the minister of the Unitarian Church the family attended—and warned them to beware of her daughter, who made a habit of "telling lies that ruined men's reputations."

As a result, Barbara developed a growing sense of ostracism that made her feel she would have to move to another environment entirely to continue her schooling. She persuaded her parents to send her to a private boarding school in Canada, but Margaret proved to have a long reach. Early in the school year, Barbara became ill with severe pains in her stomach; the school nurse dismissed her complaints with the terse excuse, "Oh, we know all about you. You make things up." When she returned home for Christmas, still in pain, Wiener took the situation in hand and called the family doctor. Barbara was hospitalized and, within hours, her ruptured appendix was removed.

Later, Margaret made another, more bizarre accusation. While riding the train home from Canada for her spring break, Barbara was approached by a young soldier, who took the seat next to her and started an aggressive flirtation. She rejected his advances, but when the train arrived in Boston he reached for her suitcase in an uninvited act of gallantry. The two travelers stepped off the train together and came face-to-face with Wiener and his wife. Margaret saw the suitcase in the young man's hand and leaped to a wholly unwarranted conclusion. "The next day was Sunday. I went to church and when I came home there was the doctor waiting to examine me," Barbara recalled. "My mother was convinced that I couldn't have had an encounter like this without 'giving in.'" Wiener, knowing his wife's sexual attitudes, defended his daughter in the dispute. "After looking at my face, my father sent the doctor home and said, 'No one's ever going to bring this up to you again.'"

Peggy confirmed Margaret's "paranoia" over sexual matters. "Mother seemed to assume sexual involvements and implications where they did not exist," she remembered. "Anything to do with sex upset the hell out of her. She was so hung up about it. It was just something that made her very, very uncomfortable."

And her discomfort escalated. A year later, during the family's migration to Mexico for Wiener's summer research with Rosenblueth, McCulloch's teenage daughter Taffy came down to Mexico with a girlfriend and paid a visit to the Wieners' apartment. Taffy was several years older than Wiener's girls and considerably more worldly. Charmed by the abundant silver jewelry for sale in Mexico City, Taffy and her friend decided to get their ears pierced, and they invited Barbara and Peggy to join them. When all the girls returned to the family's apartment with small gold rings in their ears, Margaret became furious at her daughters and, now, worked Wiener's expressed fondness for pierced ears into the charges she had leveled against Barbara since she was a small child.

"She accused me of trying to 'seduce' my father by doing something that would excite him," Barbara recalled. "She misinterpreted grossly," said Peggy. "We just wanted to be able to wear nice earrings without losing them, but Mother went straight up and turned left!" For McCulloch's daughter, who watched the family drama unfold, the episode revealed less about Wiener's penchants and more about Margaret's "prurient view of everything." Wiener had no such problem with the piercings. In a note to Arturo, he praised his daughters' adventures in Mexico as "very valuable to the girls in enlarging their outlook and in giving them a more mature view of life."

─────

Wiener and Rosenblueth made good progress during their extended sessions in Mexico, but their work was hindered by needless delays in the north. Wiener had arranged for Pitts to help them with their laboratory studies of nerve fibers and heart muscles, two research areas in which his analytical skills would be of value, and that also would put some grounding under the theories Pitts was devising for his dissertation. With Wiener's help, Pitts had won the Guggenheim grant he applied for to support his doctoral project, but Wiener soon learned that Pitts was plagued by two flaws Wiener himself never suffered as a prodigy or as an adult: an incorrigible habit of procrastination and a terror of being judged, which Pitts masked with bravado.

Heims observed that Pitts, in his rebellion against authority and seeming indifference to the trappings of postwar society, and even to his own career, had much in common with his contemporaries in the Beat Generation. He also noted that Pitts, "like the Beats, spent a good bit of time 'on the road'" with his pals. In the summer and fall of 1946, Pitts, Selfridge, and Lettvin set out from Boston on a series of rambling summer road trips that took them, by roundabout routes, to Chicago, Colorado, California, and, eventually, to Mexico, where Wiener and Rosenblueth were hard at work and waiting for Pitts to appear.

Wiener was annoyed by Pitts's peregrinations, but he was more distressed by Pitts's failure to perform his assigned laboratory research, and to produce the rigorous mathematical analysis of random nerve nets that he and Rosenblueth and the Guggenheim Foundation were awaiting. "Walter is such a migratory person that we are not quite certain how to reach him," Wiener complained to Mc-Culloch from Mexico. Five months later, back at MIT, Wiener's patience was wearing thin. When Pitts finally showed up in Mexico in January 1947, Wiener cracked the whip and authorized Rosenblueth to put an end to Pitts's dilly-dallying. "He must get [his] paper on the statistical mechanics of the nervous network . . . ready for publication within six weeks. . . . There should be no more monkey business about a long auto trip. . . . Everything should be done to discourage, or even prevent, him from any [more travel] until he has completed his . . . work."

But, that spring, Pitts caused another problem for Wiener that provoked a crisis in the group's interactions and, inadvertently, in the life of Wiener's elder daughter as well.

Barbara graduated from her boarding school at age sixteen and was accepted at Radcliffe College. She rode the bus into Cambridge with her father each day, then met Wiener at MIT and rode the bus back home with him. After a lonely year as a nonresident student at Radcliffe, she transferred to MIT and was ushered into the inner circle of Wiener's faculty colleagues and his boys, whom she had come to know during their visits to the Wiener home and their summer trips to New Hampshire and Mexico. "I clung to the boys, who seemed to me the first real friends I had ever had," she said. "They offered an opportunity for conversation and joking with boys my age that didn't involve any hidden demand for sex," a subject about which Barbara, more than most teenage girls of that time, was by her own admission "extraordinarily inexperienced."

Early in 1947, Warren McCulloch offered Barbara, who was contemplating a career in biology, a training position in his laboratory at the University of Illinois and safe haven for a term at his home in suburban Chicago. She was eager for some independence and for a chance to break free of the emotional burdens imposed by her parents. After receiving McCulloch's assurance that the job was not make-work, Wiener approved the plan, and in February Barbara headed to Chicago to begin her apprenticeship—and her unexpected adventure as the newest resident in the McCulloch home.

Barbara recalled her entry onto the exotic turf of the McCulloch clan, which resided in an old farmhouse outside the city, its lively crowd of houseguests, and the constant stream of company from McCulloch's coterie at the university. "The laird of the castle appeared and disappeared at will, talking grandly and holding court. The household was proudly radical. The phonograph in the living room kept spinning out songs of the Spanish Civil War and union songs to which everyone knew the words but me. Total sexual freedom was proclaimed by rule of the laird, and the company seemed to be incredibly sophisticated and knowledgeable."

The free-spirited environment was confusing for this refugee from Margaret Wiener's rigid domestic world. Much of the difference was due to McCulloch's wife, Rook, a modern woman from a prominent, politically liberal, New York Jewish family. Rook had her own career as a social worker, but she was no less committed to caring for her husband and children and the young researchers the family took under its roof. "Rook was a saint, sometimes a little *too* saintly," said Jerry Lettvin, alluding only dimly to McCulloch's cavortings, which were common knowledge in Chicago's scientific circles.

Barbara worked diligently at her apprentice chores in McCulloch's lab, abstracting articles for publication and assisting in studies of animal neurophysiology in the lab's operating room. She found new freedom as an independent woman and she even had her first love affair with a charming young medical student she met through McCulloch's daughter. During those months, while Barbara was growing up, Pitts and Selfridge were settling down and industriously pursuing their assigned research projects with Wiener and Rosenblueth. At the same time, Lettvin began studying higher mathematics at MIT and conducting his own experimental research in neurophysiology. Then, in April 1947, a sudden storm swept through their network, threatening to disrupt the pioneering communication research that was still only beginning among the three senior scientists and their disciples.

Word flashed through the group that Wiener was furious about something and would have nothing further to do with any of the boys, or with McCulloch, or with the projects they were engaged in individually and jointly. No one in the group knew what had set Wiener off, but their suspicions centered on Barbara, who was having her own difficulties navigating her newly liberated social environment in Chicago. Her first love affair had played itself out. A second relationship had put her at odds with the close-knit group around McCulloch, who began to worry that Wiener and his wife would hold them accountable if word of their daughter's love affairs and the McCulloch crowd's liberal lifestyle filtered back to Cambridge. Concerned about what Barbara was telling the folks back home, McCulloch and Lettvin, who was visiting in Chicago at the time, confronted Barbara one night in the living room of the McCulloch home.

"I had just eaten supper with Rook and the children when Warren and Jerry stormed into the house," Barbara recalled. "We were clearly in the middle of a major crisis, and I was at the center of the fury. It seemed I had done something horrible but no one would tell me what. I did grasp that it had something to do with a letter they accused me of writing to my father containing accusations that made him turn on them and break up their association." Barbara had done nothing to invoke her father's wrath, but her denials only made matters worse. "The next morning I was out—of a job, out of the house. Nobody would talk to me."

Days later the real reason for Wiener's rage became clear. A draft manuscript he had given to Pitts to review had been missing for months. Wiener had just learned of Pitts's negligence and recovered the paper, but not before he had "lost priority on some important work," as he told McCulloch. Wiener did not say what that work was, but it turned out to be a paper he was writing setting forth the first formal statement in the technical literature of his new theories of communication and information—work that, if those months had not been lost, would have established Wiener as the leading thinker in the field well before the publication of *Cybernetics*, and preceded Shannon's paper on information theory by a year.

In a volley of incrimination, Wiener accused Pitts, McCulloch, and the entire group of joining in a "conspiracy of silence" to keep him from learning of the loss of his work in progress. In a letter to McCulloch, he threatened to end their joint scientific effort and to "completely withdraw from the Macy meetings." But there was no conspiracy to deceive Wiener. In fact, Pitts had mislaid the manuscript in the course of his wanderings. A few more letters set matters straight and Wiener's anger abated. He apologized to McCulloch, but he would not give any of the boys a free pass. Wiener moved to further separate Pitts from Lettvin and Selfridge, and he banned "the three boys together, or any two of them," from assembling in Mexico where Pitts was working.

In the end, Lettvin took the blame—and the fall—for what was, in fact, Pitts's dereliction. Lettvin recalled, "I saw that the damage to Walter would be considerable, so I said, 'Why don't you blame it on me?' Wiener wouldn't talk to me for a while. I was under a cloud."

That first blowup passed, but it left scars on everyone involved. Lettvin quit MIT soon after and went to work at a state mental hospital in Illinois. Pitts returned to Mexico with Selfridge that summer, and the two young wanderers finally settled down and worked productively with Wiener and Rosenblueth in their laboratory studies. Along with *Cybernetics*, Wiener's Mexican sojourn that year produced the only paper he co-authored with Pitts, a study of signal flow in nerve fibers, and Selfridge solved a mathematical problem that had stumped both Wiener and Pitts and turned out his first published paper. But relations between them remained strained, and Selfridge left MIT the following year without earning his Ph.D. Wiener's daughter came back to Boston in an unsettled state, traumatized by her alleged role in a crisis for which no one ever bothered to absolve her.

And, unbeknownst to Barbara and everyone else in the Wiener-McCulloch circle, her short stay in Chicago would become the catalyst in a far more grievous blowup that would rip through the court of cybernetics a few years later.

———

Writing *Cybernetics* had proved to be the final insult to Wiener's overburdened eyes. He underwent surgery to remove his cataracts. During his recovery, Margaret and the girls had to do his reading aloud to him and, from that point on, Margaret and his secretaries played more active roles in his writing, taking dictation and, on some matters, composing his letters for him. Now, along with his usual worries—and his growing concerns about the impact of his new science on the postwar world—Wiener feared he would lose his sight completely. To prepare himself, one MIT colleague observed, "he practiced being blind by burying his face in a book and walking the halls by following along with his finger. If he reached the open door of a classroom, he would simply forge ahead and circumnavigate the room while the entire class stared." The exercise was replicated several years later when Wiener and a young MIT engineer built a prototype robot on wheels equipped with a light-sensitive "eyestalk." As Wiener had done, the squat creature rolled down Tech's long corridor and turned into every open doorway, guided not by touch but by the light beams pouring from the brighter offices and classrooms. Wiener nicknamed the machine "the automatic barfly."

Wiener had no shortage of companions during those years. When Rosenblueth moved back to Mexico, his monthly supper seminars at Harvard Medical School had ended; and in the spring of '48, Wiener revived them on a weekly basis at his favorite Chinese restaurant near the MIT campus. Jerome Wiesner remembered the serious science and high spirits those suppers nurtured among his generation of scientists and engineers at Tech, much as the Macy conferences were doing for their elect group of theorists and researchers in New York. "The first meeting reminded me of the tower of Babel," Wiesner wrote in a memoir that echoed the frustration McCulloch chronicled at the initial Macy conferences. But soon the excitement of their multidisciplinary venture—and Wiener's paternal presence—brought the group's effort into focus. "After the first meeting, one of us would take the lead each time, giving a brief summary of their research, usually accompanied by a running commentary by Wiener to set the stage for the evening's discussion. As time went on, we came to understand each other's lingo and to understand and even believe in Wiener's view of the universal role of communications in the universe."

Wiener held court in the restaurants near MIT, but he continued to entertain his closest friends at home in Belmont and at the family's summer place in South Tamworth. The ample dining rooms, screened-in porches, and secluded studies were big enough to accommodate all Wiener's moods and distinguished visitors who came to call. J. B. S. Haldane stayed with the family in Belmont on his first visit to America after the war. His reputation as Britain's most controversial biologist was at its peak, and he continued to outrage the British establishment with his communist views, but Haldane's politics did not bother Wiener or even interest him much. He and Haldane were joined by a deeper set of scientific values and by age-old scholarly traditions that sang in their bones—and aloud at every opportunity.

Peggy recalled them "sitting at the table after one of my mother's excellent meals singing *De Contemptu Mundi* by Bernard of Cluny." The two infidels crooned the medieval Latin hymn in a rousing antiphonal chorus:

Urbs Sion aurea,
patria lactea. . . .
Jerusalem the Golden,
with milk and honey blessed. . . .

and the poem's sober conclusion:

The world is very evil,
the times are waxing late,
Be sober and keep vigil,
the Judge is at the gate.

Wiener would later paraphrase the poem in his warning to the postwar world about the dangers of the new automated technologies.

Other positive forces were gathering around Wiener at MIT. His good friend and collaborator Yuk Wing Lee returned to Tech following his long entrapment in China. To bring Lee up to speed on developments in American science and technology, Wiener gave him research assignments on topics at the forefront of cybernetics and challenging problems whose solutions Wiener had sketched but not worked out in detail. At Tech, Lee taught basic courses on Wiener's statistical methods and the specialized knowledge required to build automated machinery and factories—a new technical field, which, as Wiener knew, "went well beyond the ambit of a theoretical man like me." Lee also wrote technical papers and books explicating Wiener's theories and their applications that quickly established him as Wiener's "interpreter to the engineering public" and emissary to the corporations Wiener continued to shun.

———

Wiener welcomed the accolades he received for *Cybernetics*, but fame was no panacea for his deep-rooted insecurities and persisting manic-depressive states. Peggy believed the universal acclaim that greeted *Cybernetics* brought Wiener the first peace of mind he had known in his quest to prove himself as an adult. "I think it was the first time in his life he felt he really got the recognition he deserved." Yet all the praise did not take much pressure off Wiener or his family. "It should have, but did it? I don't know," said Peggy. In many ways, fame only added to Wiener's distractions and to the deeper responsibilities he was beginning to feel personally and professionally. "After *Cybernetics* came out, a reporter from *Life* magazine came to our house," she recalled. "Alfred Eisenstadt, the famous photographer, took pictures of the two of us playing chess. I was a rotten player but Dad's mind was on other things. I mated him in three moves."

Shown above:
Bertha and Leo Wiener, ca. 1900.

Courtesy of the Wiener Family

Shown left:
Norbert Wiener, age 2, ca. 1896.
By then he knew the alphabet.

Courtesy of the Wiener Family

Shown right:
Norbert Wiener, Ph.D.,
age 18, 1913.

Courtesy of the Wiener Family

Shown below:
Pvt. Wiener (far right)
with the 21st Recruit
Company, Aberdeen
Proving Ground,
October 1918.

Courtesy of MIT Museum

Shown left:
Wiener with his wife Margaret and young daughters Barbara (left) and Peggy, 1931.

Courtesy of the Wiener Family

Shown below:
Wiener with his daughters in New Hampshire, ca. 1933.

Courtesy of the Wiener Family

Shown facing page, above:
Wiener mountaineering in the Alps with an American colleague (not shown) and his wife, Switzerland, 1925.

Courtesy of the Wiener Family

Shown facing page, below:
Wiener with physicist Max Born, MIT, 1925.

Photo by George H. Davis, Jr.
Courtesy of MIT Museum

Shown left:
Wiener clowning with a colleague
at the International Congress of
Mathematics, Zurich, 1932.

Courtesy of the Wiener Family

Shown right:
Wiener and Margaret,
Tsing Hua University, Beijing, 1936.

Courtesy of the Wiener Family

Shown right:
Wiener on one
of his *Wienerwegs*
around MIT, 1940s.

Courtesy of MIT Museum

Shown left:
Wiener and
colleague Arturo
Rosenblueth at
the Instituto
Nacional de
Cardiología,
Mexico City,
1945.

*Courtesy of Instituto
Nacional de
Cardiología/Institute
Archives and Special
Collections, MIT Libraries,
Cambridge, Massachusetts*

Shown left:
Walter Pitts soon after his arrival at MIT, 1943.

Courtesy of Jerry and Maggie Lettvin

Shown below:
Walter Pitts (left) and Oliver Selfridge on the road, 1946.

Courtesy of Jerry and Maggie Lettvin

Shown above:
Warren and Rook McCulloch, on their farm in Old Lyme, Connecticut, 1944.

Courtesy of the McCulloch family

Shown right:
Warren McCulloch
(left) and Jerry Lettvin
picnicking at Old
Lyme, 1947.

*Courtesy of Jerry and
Maggie Lettvin*

Shown left:
Margaret Mead and
Gregory Bateson shortly
before World War II.

*Library of Congress/Courtesy of the
Institute for Intercultural Studies, Inc.,
New York*

Shown bottom (inset):
Heinz von Foerster,
editor of the
Transactions of the
Macy Conferences on
Cybernetics, ca 1950.

Courtesy of Tom von Foerster

Shown above:
Walter Pitts and Warren McCulloch
in Scotland, 1953, a year after the
break with Wiener.

Courtesy of the McCulloch family

Shown top:
Wiener, RLE Associate Director Jerome Wiesner (center), and Yuk Wing Lee
with the autocorrelator machine, MIT Research Laboratory of Electronics, 1949.

Courtesy of MIT Museum

Shown bottom:
Chief Engineer Julian Bigelow (left), Project Director Herman Goldstine,
Institute for Advanced Study Director J. Robert Oppenheimer, and John von
Neumann at the dedication ceremony for the IAS computer, June 1952.

Photo by Alan W. Richards. Courtesy of the Institute for Advanced Study Archives, Historical Studies—Social Science Library

Shown above:
Y. W. Lee (left), Amar G. Bose, and Wiener at MIT, 1957.

Courtesy of MIT Museum

Shown above:
Wiener lecturing at MIT, 1957.

Courtesy of MIT Museum

Shown right:
Wiener, official MIT portrait, 1950s.

Courtesy of MIT Museum

Shown below:
Wiener's friend Swami Sarvagatananda of the MIT Chaplain's Office and the Ramakrishna Vedanta Society of Boston.

Courtesy of MIT Museum

Shown above:
Warren McCulloch, ca. 1960.

Courtesy of the McCulloch family

Shown above:
Wiener, wife Margaret, daughters Peggy (center), Barbara, and Barbara's husband Gordon "Tobey" Raisbeck, 1960.

Shown above:
Wiener (far left) at First International Congress on Control and Automation, Moscow, Summer 1960.

Courtesy of MIT Museum

Shown left:
Wiener at his retirement dinner with MIT President Julius A. Stratton (left) and information theorist Claude Shannon, Spring 1960.

Courtesy of MIT Museum

Shown right: Wiener and his grandson, Michael Norbert Raisbeck, New Hampshire, 1950s.

Courtesy of the Wiener Family

Shown above:

Wiener receiving the National Medal of Science from President Lyndon B. Johnson at the White House, with Presidential Science Advisor Jerome Wiesner (left) and fellow medal recipients, Bell Labs engineer John R. Pierce, Vannevar Bush, biologist Cornelius B. van Niel, and physicist Luis W. Alvarez, January 13, 1964.

Associated Press/Wide World Photos. Courtesy of MIT Museum

Shown above:

Frank Fremont-Smith of the Josiah Macy Jr. Foundation (left), John J. Ford, CIA Soviet science and technology analyst, and Warren McCulloch, at a symposium of the American Society for Cybernetics, 1968.

Courtesy of J. Patrick Ford

Margaret took pride and personal satisfaction in Wiener's newfound stature, which she accepted as partial compensation for her own sacrifices. "She was his *frau-professor* and my father's fame, when he had some, she took as a credit to her," Barbara observed, but Wiener's daughters had a harder time with their father's renown. He acknowledged the new tensions his celebrity status forced on the travails of their late teens and the ongoing burdens they had carried throughout their childhood. In his autobiography, he reported that "Peggy said, on more occasions than one, 'I'm tired of being Norbert Wiener's daughter; I want to be Peggy Wiener.'" And he apologized to his daughters, as he had to his wife, for the added hardships his temperament and, now, his very existence imposed on their lives. "I made no attempt to force the children into my frame, but the mere fact of being what I am inevitably subjected them to a sort of pressure with which my will had nothing to do."

His manic depression continued to ambush him. His temper would fire without warning, and his new feelings of self-worth would give way to attacks on his peers, competitors, and students under his tutelage. What many of his MIT colleagues had a hard time accepting, even in enlightened Boston-Cambridge, was that Wiener's infamous outbursts were neither vicious nor wholly capricious but impelled by internal forces and, more often than not, triggered by real events, like the blowup over his missing manuscript.

Peggy defended her father on that count. "Dad was not malicious," she said. "He might provoke people in repartee, but he did not bait them. He always had a reason when he took someone on, or at least he believed he did, and he was direct. He didn't go around whispering behind people's backs. That wasn't his style at all." To be sure, Wiener's eruptions had a style all their own. Barbara, who witnessed many of Wiener's outbursts during her time at MIT, found them at once painful to watch and highly entertaining. "My father's vocabulary was enormous, so when he got mad at somebody he could go on for a long time. He spoke in rolling sentences of Victorian English. Sometimes it was just worth listening to as a literary product." But she also saw the deeper effects of Wiener's combustible moods on his colleagues and students. "If he was in a good mood he was on top of the world and he would generate that feeling in the people around him. Then he'd go down, and when he did everybody got frightened, because a lot of their belief in themselves came from their belief in him."

Indeed, during his up times Wiener was incandescent, but there were dark moments during those triumphant years. Successive MIT presidents, provosts, and department heads came to accept the fact of Wiener barging into their offices unannounced to spume about some matter that had set him off. "Sometimes he was disturbed: someone had challenged one of his ideas; or he had become convinced that some foolish action of the President or Secretary of State was going to catapult the world into oblivion," Jerome Wiesner remembered.

It was on those blue days, as Wiesner called them, almost invariably, that Wiener fired off furious letters of resignation from the institute or one professional group or another. The math department bore the brunt of his wrath by virtue of its proximity—one witness said he resigned from the department fifty times—however, Tech's faculty and administrators had a time-tested plan for dealing with Wiener's ritual. "They knew what was happening and would come running to bolster him up and make sure he didn't go off the rails," Tobey Raisbeck recalled.

Others became attuned to the rhythms of Wiener's resignations. Warren McCulloch scrambled to save the Macy conferences when Wiener threatened to resign after the first meeting and, again, during the mess over his lost manuscript that erupted a month after the third Macy meeting. However, one organization that did not roll with Wiener's punches was the National Academy of Sciences. Wiener had been elected as a fellow in the early 1930s, but a decade later he gruffly resigned from the group he now called "a self-perpetuating and . . . irresponsible body of men." The dispute arose over the Academy's awarding of medals and prizes in ceremonial practices Wiener considered to be corrupted by organizational politics—a charge other Academy members upheld and documented in later investigations. However, in his resignation letter, Wiener let out a litany of secondary complaints about the Academy's "bad catering . . . tedious and expensive dinners" and "the general atmosphere of . . . pomposity which has hung over the meetings." The Academy's officers did not beg him to reconsider, and Wiener never allied himself with any other scientific society.

Some of Wiener's peers turned against him for those actions, most were merely perplexed by them, and others found them quietly amusing. Wiener often justified his resignations on moral and ethical grounds, and some of them were well-grounded, but others were clearly by-products of his manic-depressive states. And, by the late 1940s, Wiener's closest colleagues had detected another pattern in his depressions. "It seemed to us whenever we saw Wiener elsewhere than in Massachusetts he was jovial, cheerful, and got along very well with everybody," Jerry Lettvin remembered. "In fact, it was curious, when his wife went away for any length of time, he never got depressed."

Others confirmed the picture Wiener's daughters had watched since they were small children. Indeed, far from freeing Wiener from his inner torments, Margaret Wiener in her own emotional deafness often exacerbated her husband's mood swings. Tobey Raisbeck saw the pattern on his weekly visits to Wiener's home. "She unbalanced him. She would induce his depressions," he said. Jerry Lettvin went a step further. He saw Margaret's hand stoking Wiener's emotional storms and the professional tantrums they precipitated, and he believed that she sometimes deliberately triggered them. Lettvin recalled one incident he witnessed during that period. "Wiener received a letter from a math student at Worcester Polytech that said, 'We don't have any money, but would you consider coming to give us a lecture?' He was delighted. He always said, 'Any student of mathematics is a friend of mine.' So he wrote back and told them he would come. Then, that evening his wife said, 'How dare you go to some little place for free. It's as if you're worth nothing.' She triggered him into writing a horrible letter severing all relations with Worcester Polytech. The students never understood what caused it."

By then the pattern was unmistakable. Margaret had concluded from her own travails with Wiener that he was more manageable domestically, more acceptable socially, and more in conformity with her image of a man of stature when he was depressed than when he was euphoric; and she acted on those assumptions, consciously or not, in Wiener's interest as she perceived it, bursting his balloon by force of habit and bringing him down to earth by any means at her disposal. And the observant minds who gathered around him had detected one more pattern in Wiener's mood swings that would have a bearing on all their lives. As Lettvin observed, "Wiener had this problem that whatever decisions he made in a depression he held to, even after he came out of it."

His moods peaked and plummeted. He might sulk for days at a stretch or slump in a chair muttering, "I'm no good. I'm no good," or his German standard, *"Ich bin muede"* (I am tired), then spring back into action with a sanguine salute and his other stock phrase, "I'm just doing the best I can." He did not wallow in his mood disorder nor did he flaunt it, but he remained "a little sheepish, a little embarrassed" about it, as Barbara recalled. He even warned his closest friends and colleagues to steer clear of him when he was heading into a depression. "Then you stayed out of his way," Lettvin recalled. "You tried not to be noticed by him during a depression."

With cold comfort at home, Wiener began seeing a psychiatrist again. On his trips to New York, he met discreetly with the eminent Dr. Janet Rioch, a progressive therapist who gave Wiener the outlet he needed to vent his feelings and fend off the greater breakdown he still feared he might suffer. In contrast to his earlier experiences with strict Freudian psychoanalysts, Wiener welcomed those sessions with Rioch, who as he said did not go "by the dream book" or "make so much of a fetish of the ritual of the couch," but "who made a far greater effort to get in rapport with me as a human being."

———

Wiener's girls seized their freedom by degrees. After her disastrous stay in Chicago, Barbara transferred from MIT to Boston University, and Peggy followed her father's path to Tufts. As his daughters moved into their own lives, Margaret tried to keep the world around Wiener stable and supportive, but she was no match for the enormity of attention, personal pressures, and professional demands coming at him. Instead, she stepped up her efforts to control her husband and others she perceived as potential threats to his prominence around MIT.

Margaret continued obsessing over earrings, and one particular pair prompted her to act with compulsion. As the 1940s came to a close, Margot Zemurray, the math department secretary, who had typed much of Wiener's voluminous correspondence after *Cybernetics* was published, and who dutifully filed away his irate resignation letters, became engaged. The young secretary, a striking woman, had her ears pierced in preparation for her wedding. There was no inkling of an untoward response by Wiener, but there were numerous witnesses to his wife's reaction. Jerry Lettvin recalled the incident that became the talk of MIT.

"A few days later, Mrs. Wiener showed up in the math department. She wanted that secretary fired 'immediately' and, at that time, the department went along. She was fired and never knew why. Wiener didn't know why. Just because her ears were pierced and Wiener had reported that to his wife, that was it. She engineered the discharge of the secretary."

Margaret may have had her hands full caring for Wiener, but there was a growing desperation to her care and her machinations that was not warranted by Wiener's temperament or even his oscillating brain chemistry. She continued to make her elder daughter a scapegoat for her own fears and frustrations. Several years earlier, when Barbara's friendship with her second cousin Tobey Raisbeck began to blossom into a full-fledged courtship, Margaret took steps to curtail it, fearful that Barbara might leach more family secrets into the grapevine at MIT.

Tobey remembered Margaret's calculated effort to bring an end to their romance. "Not very many weeks after we started behaving as though we *might* be dating, Barbara's mother invited me to the house, very formally. 'Would you come to tea on such and such a date.'" It was a time when the other family members were not at home. "She sat me down and gave me a heart-to-heart lecture. She said, 'You must be very careful about getting too close to Barbara.'" Raisbeck ticked off Margaret's arguments against her daughter one by one and weighed his own evidence for or against them. "She said, for one thing, 'Barbara is a pathological liar. You can never rely on what she says.' Well, that was contrary to my observations because I had gone out with her casually, and I never heard her say anything which later turned out to be untrue. The second thing she said was, 'Barbara is an uncontrollable alcoholic.' I knew that was untrue because once or twice I had tried to pour a couple of drinks down her, as every teenage boy will, and I saw that, like Norbert, she got sick pretty fast. The third thing Margaret said was, 'Barbara is a promiscuous nymphomaniac.' Once again, this did not jibe with my experience at all."

Raisbeck pursued Barbara in defiance of Margaret, and the bond between them grew. When their union became imminent, Wiener was giddy with fatherly pride. "He kept asking me for weeks ahead of time, 'Has he asked you yet? Has he asked you yet?'" Barbara recalled. "And I said, 'He'll *never* ask me if you keep saying that.'"

The couple were married in December 1948. Raisbeck, who by now had his electrical engineering degree, a Rhodes scholarship, and a tour of duty in the Navy to his credit, began his career at Bell Labs, and the newlyweds moved to New Jersey to make their life far from the family pressures both of them had endured and were determined to put behind them.

———

In the fall of 1950, Wiener was again invited to France by his friend Szolem Mandelbrojt to spend a term at the Collège de France as a visiting lecturer. He was also invited to address an international congress on computers and automation in Paris early in the new year. He sailed alone to France that December, freshly funded with a Fulbright teaching fellowship, and spent several weeks in Paris preparing for the congress and his college lectures. When the congress ended, he made a quick trip to England to visit Haldane, and Margaret and Peggy sailed over to meet him. Wiener had arranged for Peggy, who was studying biochemistry at Tufts, to work for a term as an assistant to Haldane in his genetics laboratory at University College in London, as Barbara had done with McCulloch in Chicago. Wiener and Margaret returned to Paris, and Wiener commenced his teaching duties. Then, a few weeks into the term, Peggy flew to Paris for a visit. Her mother met her at the plane, where yet another of Margaret Wiener's strange encounters with her daughters took place just steps from the tarmac.

"On the flight over I had bad pressure in my ears and my seatmate was concerned about me," Peggy remembered. "When we got off the plane together, Mother looked at this man looking at me with concern and, immediately, before I had said anything to her, she accused me of coming on to him sexually. I was stunned. It was so out of left field I didn't know what to say." Years later, Peggy was still trying to understand her mother's motivations. "I'd seen Mother out of control like that before and it was always about sex. I don't know how she got that way, I know her generation was more apt to be, but she took it to extremes," she said, with a caustic twist. "How she had children I'll never know. Probably she just closed her eyes and thought of Germany."

In the meantime, Wiener, oblivious to his wife's latest attack on his daughters' virtue, roamed the city with his French friends, played chess in the Bar Select on Boulevard Montparnasse, and learned "to know a little bit about the good restaurants and cafés of Paris." Mandelbrojt's nephew, Benoit Mandelbrot, a budding Polish-born French mathematician, who would later trailblaze the field of fractal geometry known as chaos theory, was assigned by his uncle to introduce Wiener to the city's finer bistros. Mandelbrot remembered his night on the town with Wiener.

"My uncle spoke of Wiener in extraordinarily glowing terms. I remember, even before the war, I heard his name when I was thirteen, and he became one of my role models. When he came to Paris, I met him at the Gare de Lyon. He cut a very strange figure. I was amazed at his bulk, his shortness. I took him to a fine restaurant and the headwaiter saw Wiener with his enormous girth and was sure he must be a very prominent American gourmet. Everyone was waiting to see what he would order. The waiter came and handed him the menu, he looked at it—he couldn't read it because of his myopia—and he closed the menu and said, 'I'll have the vegetables.' The waiter looked at him and said, 'Which vegetables?' He said, *'All* the vegetables.' I told the waiter, *'Monsieur est un vegetarian et un myope.* Please make him a selection of your finest vegetables.' They brought him a plate of vegetables and he was happy, but they were very upset that he did not order some fancy gourmet meal."

11

Breach and Betrayal

Infirm of purpose!
Give me the daggers. . . .
'tis the eye of childhood
That fears a painted devil. If he do bleed,
I'll gild the faces of the grooms withal;
For it must seem their guilt.

—**William Shakespeare,** *Macbeth*

THROUGH THE END OF THE 1940S and into the 1950s, cybernetics continued its march on industry, technology, and the public imagination. Sales of Wiener's books multiplied worldwide, and requests began pouring in for his services as a visiting professor and guest lecturer.

Late in the spring of 1951, when his lectures in Paris on *la cybernétique* were concluded, Wiener and his wife moved on to Madrid, where he had been invited to lecture on *la cibernética* to Spanish mathematicians and engineers. Before accepting the invitation, Wiener had warned his host that government authorities in Spain, the only dictatorship in Western Europe that remained in power after the war, "might not like my views when they found out what they were." When he arrived, prepared to speak in Spanish, he was informed that his ideas indeed had been deemed "dangerously liberal," and he was instructed to address his Spanish audiences in French, so that he would not be so widely understood.

After this constrained stay in Spain, Wiener and Margaret returned to France. There Wiener turned his full attention to a project he had been working on piecemeal for months, the first volume of his autobiography, which would begin with a narrative account of his shaping childhood as a prodigy and his early adult years. By now, Wiener believed, his writings on cybernetics and its human consequences, and his growing international reputation as a scientific genius, had made his story worth telling, and one that would be of interest both to his technological disciples and to readers curious about the fate of prodigies generally.

Through the early part of the summer, he dictated his memoir to Margaret in the seaside resort of St. Jean-de-Luz on the Côte Basque, back in Paris in a little hotel on the Left Bank, and in a mountain retreat in the Savoy on the French shore of Lake Geneva. But the process of reliving his pressured childhood proved to be overwhelming even from the pinnacle of his adulthood. He found the re-immersion in his past, which his new psychiatrist had recommended as an ideal form of therapy, to be instead a cause of intense emotional strain. By the end of his European sabbatical, as he acknowledged later, "my overexertion in lecturing and in writing sent me to bed with a racking headache" that required him to be hospitalized in a Swiss hospital across the lake in Geneva.

Wiener alluded only vaguely to that episode, which appeared to be more serious than the bad headache he had described. Upon his release from the hospital, he and Margaret made their way to the port at Genoa for their return trip to the States. However, even that short passage proved too taxing, and upon boarding Wiener was placed in the care of the ship's surgeon, who continued the treatment prescribed on shore. By the time he reached home, he was "a reasonably well man but dead tired," as he wrote later, using his standard euphemism for those times when he was in the throes of, or just emerging from, a severe depression.

After returning to Boston, he left immediately for Mexico City, where the University of Mexico, the oldest in the hemisphere, was celebrating its four hundredth anniversary. For two weeks the city was ablaze in ceremonies and fiestas. Wiener received an honorary degree, but the revelry, for all its delights, exhausted him all over again. His distress, however, was now about to grow by several orders of magnitude.

With the explosion of cybernetics, Wiener could no longer meet the demand for cybernetic insights and education on his own, nor did he need to. Warren McCulloch, the charismatic founder and chairman of the Macy Conferences on Cybernetics, had become an expert evangelist of cybernetic principles and perspectives. McCulloch had promoted Wiener's theories and ideas with almost as much enthusiasm as Wiener himself. He, too, had traveled the country and crossed the ocean to spread the word of cybernetics in his field of neurophysiology and throughout the sciences, and early in 1951 Wiener returned the favor.

Before Wiener departed MIT for his sabbatical in Europe and Mexico, Jerome Wiesner, who was now associate director of the Rad Lab, with Wiener's blessing, invited McCulloch to come to Cambridge to head up a major new research effort on the brain and its cybernetic connections. McCulloch, an Easterner by birth and at heart, gladly gave up his full professorship and perquisites at the University of Illinois for the humble position of research associate at the Rad Lab.

The lab was in a new ferment. In 1946, the secret wartime radar laboratory was rechartered with an expanded scientific and technological mission and rechristened the Research Laboratory of Electronics—although it would always retain its cryptic nickname. After an eventful transition, in 1951, the RLE's main radar projects for the military, and its director, were transferred to MIT's newly established Lincoln Laboratory, and Wiesner took over as director of the RLE. Under his enlightened leadership, the lab's mandate grew to encompass the entire domain of communication theory and research. In addition to its basic work in electronics, Wiesner launched new laboratories within the RLE devoted to scientific study of the human components of communication, and to developing new mechanical and electronic devices to aid and enhance human communication. Eclectic interdisciplinary projects were initiated to study speech, hearing, and language, and to develop electronic prostheses for the speech and hearing impaired. The RLE's language laboratory was established and would eventually lure away from Harvard the renowned Russian linguist Roman Jakobson, his prize pupil Morris Halle, and the young superstar linguist Noam Chomsky. Among the pioneering efforts Wiesner oversaw in the RLE was an early project to produce an automatic language-translation machine to bridge the world's babble of tongues, and to help American diplomats and intelligence analysts with their burgeoning communication chores in the new Cold War era.

Establishing a first-rate brain research laboratory within the RLE was one of Wiesner's highest priorities, and McCulloch was the obvious choice for the job. Through the spring and summer of 1951, while Wiener was away, Wiesner finalized the arrangements with McCulloch to set up his new lab in Building 20, the sprawling, "temporary" wartime structure that had housed the Rad Lab and would serve as the RLE's home for the next fifty years. Wiesner also arranged to bring two more top-notch brain researchers to MIT from Chicago: Jerry Lettvin, the polymorphous psychiatrist, who had made amends with Wiener and had gone on to become a formidable neurophysiologist with his own lab at Manteno State Hospital in Illinois; and Patrick Wall, a young British neurophysiologist with credentials from Oxford, Yale, and the University of Chicago, who had worked with Lettvin at Manteno State. Also that year, Wiener's other prodigal son, Oliver Selfridge, returned to MIT, after two years in the wilderness serving with the Army Signal Corps. Selfridge made his move, not to the RLE, but to the Lincoln Laboratory, where he set to work developing programs for the lab's new Whirlwind digital computer, although he remained a close friend and associate of McCulloch and Lettvin and their RLE teammates.

Wiener's number one boy, Walter Pitts, was already at MIT and had still not completed his dissertation, but Wiener was hopeful that McCulloch's added paternal influence, and the new energy of a state-of-the-art brain research laboratory on the premises, would give Pitts the renewed impetus and discipline he needed to finish his languishing theory of randomly connected neural nets.

Lettvin remembered the vision that he, McCulloch, Pitts, and Wall were charged with implementing at the RLE, and the central role in the laboratory's scientific mission that Wiener and his new science were slated to play. "When Jerry Wiesner invited us to RLE, he had in mind a center to create the science which was at the time always in the offing. . . . Wiener was an essential part of the dream we all shared. [He] was to serve as a critical colleague against whom to sharpen ideas." Pat Wall confirmed the McCulloch group's mission and its direct connection to Wiener's work. "We'd all been brought in together with the idea that the brain was a supercommunicator. . . . The intellectual background was in fact partly the mathematics of cybernetics and control, and partly this whole communication question."

And Wiener shared a larger philosophical goal with McCulloch and his team. Ultimately, the aim of McCulloch's new lab was to take the next steps in the group's collective quest to understand the brain's higher-order cognitive operations in the new terms of information and communication, using the best tools of modern science and the latest electronic technology. "The McCulloch laboratory, more than any other, cared about the relations between cellular events and the processes of mind," Robert Gesteland, a young brain researcher who came later to the lab, recalled. "The spirit of the laboratory was well characterized by the sign on the door. It read '*Experimental Epistemology*'"—a phrase coined by Lettvin and Selfridge.

No subject was too far afield for study at the new RLE. The place was thoroughly interdisciplinary, and, by consensus, Wiener was the lab's guiding light. Much of the RLE's work "was inspired by . . . Wiener and his exciting ideas about communications and feedback in man and machines," Wiesner wrote in a memoir. Wiener's presence in the RLE's labs and hallways was indispensable to those who sought him out and to as many who fled him. As one historian reported, he made his "daily rounds of the laboratory to investigate everyone else's research and to hold forth on whatever new idea had happened to seize his restless mind." Wiener also conducted much of his own postwar communication research and technical tinkering in collaboration with the young researchers and engineers who inhabited the "plywood palace" at the northern edge of the MIT campus.

The prospect of working with his closest colleagues and cybernetics' biggest boosters in Building 20, where Rosenblueth, too, could join in the action during his biennial terms at Tech, put Wiener on the threshold of achieving his longtime dream: to have an interdisciplinary center for research in both the technical and biological dimensions of his new science right at MIT. Wiesner, too, was confident that, by uniting Wiener's proven genius and overarching view of communication science with McCulloch's special expertise in neurophysiology and the philosophy of mind, and with Pitts's quirky genius and uncanny skill at parsing the math and logic of the brain's neural networks, a stream of innovative experiments and basic scientific insights would surge and bring forth important new understandings of human perception, cognition, and the nature of intelligence itself. Everyone assumed that, down the road, the synergy of their joint venture would produce crossover insights and valuable applications to the new communication and computer technologies, and spur the design of electronic prosthetics and other new devices to help people, which was Wiener's personal priority for his new science. Together, at the RLE, the triple-threat Wiener-McCulloch-Pitts team and their supporting group would be the vanguard of the cybernetics revolution, literally, a new brain trust for the emerging technical and human communication sciences.

But before their great adventure at MIT got started, the Wiener-McCulloch alliance detonated in one of the most tumultuous and tragic episodes of the early information age. Late in 1951, four years after their first, short-lived feud, Wiener abruptly broke with McCulloch and severed all relations with the talented young members of his extended research group—including Pitts and Lettvin, their steadfast friend Selfridge, even Jerome Wiesner himself. The epic rift grew into an irreparable breach that would change the course of Wiener's scientific revolution and the infant information age.

———

The sad tale of the Wiener-McCulloch split, the events that precipitated it, and those that ensued, has been a mystery even to the main players whose careers and lives were most profoundly affected by it. Speculation ran rampant around MIT and in cybernetics circles as to why Wiener would break his decade-long friendship with the scientist who had done more than any other, except Wiener himself, to disseminate cybernetics. Equally bewildering was the question of why Wiener would turn his back on a promising new enterprise that seemed poised to fulfill his dream. But Wiener's motivations remained obscure. Different accounts seeped out into the Macy cybernetics group and wider scientific channels, and swirled in the shadows of late twentieth-century scientific history and hearsay.

Wiener's colleague Pesi Masani, who edited his *Collected Works* and wrote a scholarly biography of Wiener, made short shrift of the split, which he claimed "arose from exacerbation of some silly dispute." Heims, the leading chronicler of the period, reported charily that, "The blow-up arose in part from the two men's differing temperaments, outlooks, and life styles, but was the specific result of some personal matters involving McCulloch and members of Wiener's family." Jerry Lettvin's only published comment, in an essay written forty years later for McCulloch's *Collected Works*, referred obliquely to "a personal misunderstanding that was manipulated by the Wiener family . . . into a violent rift."

But the truth as it emerged drew a picture of the events that went far beyond what anyone had suspected.

Wiener was not a happy man in the autumn of 1951. Morris Chafetz, a young psychiatrist from Massachusetts who met him that fall in Mexico City, saw Wiener's turbulent state of mind as he struggled to complete his memoir. "We arranged to have lunch at a local restaurant," Chafetz recalled. "Here I am sitting at the feet of this great man, and do you know what that whole lunch consisted of? His crying—actual crying. He was crying because I was a psychiatrist and he wanted to sit at *my* feet while he was working out the emotions of his relationship with his father in the book."

Midway through the fall, Wiener finished the manuscript, which he self-effacingly titled "The Bent Twig." The work contained intimate revelations and piercing criticisms of Wiener's mother and some of his early mentors and colleagues. The manuscript struck some who read it as being most embarrassing to Wiener and defamatory to his targets. Now a new round of torments began. Late into the fall, a steady rain of rejection letters from publishers poured into Wiener's apartment in Mexico, each one more upsetting than the one before, explaining why their house was declining to publish his blunt and, as several publishers perceived it, self-serving book. The trade publisher Houghton Mifflin, which only a year before had published Wiener's popular work, *The Human Use of Human Beings*, was the first to pass on his autobiography. Two British publishers rejected "The Bent Twig" for lack of "sufficient enthusiasm" and as "a book of almost wholly American interest." Stung by those rejections in the popular book trade, early in November, Wiener wrote to the director of MIT's Technology Press, which had fought so hard for a piece of the action on *Cybernetics*, seeking his advice and, no doubt, a bid for "The Bent Twig."

Then, in the midst of the rebuffs, Wiener received a letter from Pitts and Lettvin in Cambridge overflowing with excitement about their joint venture in the making at the RLE. The letter, written to both Wiener and Rosenblueth on the RLE's letterhead, extolled McCulloch's new laboratory and cutting-edge research apparatus. The boys' playful greeting was couched in the mock medieval language and lore that infected everyone in McCulloch's orb.

Dear Arturo and Norbert:

Know, o most noble, magnanimous and puissant lords, that we be now established in a winsome laboratory—a savage place, As holy and enchanted, As e'er
beneath a waning moon was haunted By woman wailing for her demon lover—
in . . . the R.L.E. All our wants are instantly supplied by an attentive and obedient Jinn . . . pre-amplifiers . . . cameras . . . stimulators and a synchronizer to
control them . . . three oscilloscopes, togagher [sic] with all appurtenances to
delect the experimenter and to dilate his heart.

Pitts and Lettvin went on to describe in technical detail the innovative neurophysiological research they would soon begin, including their experimental
protocols, instrument settings, working hypotheses and anticipated findings.
Then they made one last pitch to persuade Rosenblueth to join them in their
new laboratory. "Anything that you might want can be bought or built in a
week, since we can draw freely on the great resources of Tech," the boys
bragged. They flaunted MIT's unmatched analytical equipment—its wide array
of analog devices and their access to the Lincoln Lab's new general-purpose digital computer, the Whirlwind—along with the RLE's plentiful staff of research
assistants. They ended their letter with a "threat," albeit an idle one, and a farewell
in the same vein:

If you (Arturo) don't come, so that we can all work together, we have every intention of stealing Norbert from you. . . . No doubt his admirable loyalty will
enable him to hold out for a month or two; but when we are here every day
with delicious problems involving the calculation of information in . . . infinite
random point-patterns, and you are reduced to sending letters from Mexico, can
you doubt the eventual outcome? Clearly, you should come too.

We are, Sirs,
your most humble and obedient servants,
Walter and Jerry

Wiener was in no mood for levity, but his boys back in Cambridge had no
idea he was hurting or how badly. His reaction to their letter was swift and un-
merciful. He dispatched an angry telegram to Wiesner: "IMPERTINENT LETTER
RECEIVED FROM PITTS AND LETTVIN. PLEASE INFORM THEM THAT ALL CONNEC-
TION BETWEEN ME AND YOUR PROJECTS IS PERMANENTLY ABOLISHED. THEY ARE
YOUR PROBLEM. WIENER." He did not give Wiesner, Pitts, Lettvin, or McCulloch
an opportunity to respond—or another word of explanation.

But there was an explanation for Wiener's punishing reaction. Just the day before he wrote his telegram to Wiesner, he had received his most distressing rejection letter yet. Frederick G. Fassett, Jr., the publisher of MIT's Technology Press, advised Wiener in no uncertain terms not to publish his overwrought coming-of-age story. "You are now in the midflight of a brilliant life," Fassett wrote. "I think it is a fair question whether the publication of *The Bent Twig* now will contribute to furthering your life work. Indeed, I question whether the publication of it at a later date would do so. . . . I should counsel you . . . to hold it for the time possibly ten or fifteen years from now when you may wish to write a reminiscent volume covering the full span."

It was a devastating blow, but Wiener did not express his disappointment to Fassett. Instead, he lashed out at his boys Pitts and Lettvin and his RLE colleagues McCulloch and Wiesner. He sent his curt telegram to Wiesner and, the same day, wrote a long letter to James R. Killian, Jr., who was now MIT's President

Wiener prefaced his complaint to Killian with the startling claim that he had "for a long time been distracted and worried about the status of cybernetics at MIT" and that he felt "it my duty to call your attention to a grave and dangerous situation which has developed." He reminded Killian of his position as the originator of cybernetics and noted that he had made no attempt to advance his science at the expense of MIT or its principal patron, the U.S. government—a matter of principle for Wiener which he adhered to scrupulously with respect to all his postwar activities. He spelled out his position on the funding of his research and reaffirmed that "I have not accepted one cent of government money on behalf of cybernetics or of any other project since the end of the last war."

Then he opened fire on McCulloch, Wiesner, and the boys. In a hail of vitriol, he called McCulloch "a picturesque and swashbuckling figure in science whose attractiveness is considerably in advance of his reliability." He suggested that Wiesner's motive for bringing McCulloch to MIT was to obtain government money to fund the RLE's neurophysiology laboratory, and he claimed that McCulloch had previously hired Pitts and Lettvin "for a government project . . . at exorbitant and unreasonable rates." He reprised all his earlier complaints about the boys, asserting that he knew "by long and hard experience that neither Mr. Pitts nor Dr. Lettvin are persons to be entrusted with the disbursement and accounting for sums of money as well as equipment," and that Pitts had "systematically failed to perform for me his duties as assistant for which he was receiving his salary." He pounded Pitts for his continuing failure to deliver his dissertation, and for shirking his responsibilities to MIT and the Guggenheim Foundation to complete his protracted doctoral program.

Finally, Wiener turned to the Pitts-Lettvin letter, a copy of which he appended "to show how impertinent it was." He huffed, "I am certainly not going to tolerate pressure from those two young irresponsibles to tell me what to do next with my work, nor can I bring myself to relish the air of gloating with which they report on their success in getting something put over on me behind my back." Wiener closed his letter to MIT's president with a thinly veiled threat of his own:

> I cannot resume the willingness which I have readily shown in the past to put my services at the disposal of other Technology projects until and unless my time and patience are no longer abused by those who wish a further outlet for their administrative desires. . . . It is probably impossible for me to prevent my field of work being made into a rat race by every eager beaver that wishes to appropriate it, but I can definitely say that the present atmosphere is not one in which I can continue my work in cybernetics.

Very sincerely yours,
Norbert Wiener

Killian was no stranger to Wiener's impulses. He had his own file of Wiener's resignation letters in his drawer, but he was dumbfounded by Wiener's broadside attack on Tech's rising star Jerome Wiesner and the RLE's newest research team. A week later, he sent a solicitous reply. He countered Wiener's upset with the admission that Wiener's letter had disturbed him deeply in turn because he was unaware of any conflict between Wiener and Wiesner over the plan for a new neurophysiology lab. Citing Wiesner's expressed "feeling of discipleship," Killian said, "I find it very difficult to accept any conclusion that he has knowingly followed any course unfriendly to you." He defended Wiesner's actions, and the McCulloch group's convergence at MIT, as part of the RLE's greater research mission, which was a direct extension of Wiener's work and in everyone's interest—Wiener's most of all:

> I had assumed that the activities in the Research Laboratory of Electronics, such as those involving Professor McCulloch, represented a wholly laudable and proper investigation of some of the practical implications of your fundamental concepts. . . . Is not the fact that it is undertaken one of the best tributes to the importance and validity of your work?

Killian advised Wiener that the funding for McCulloch's lab had come, not from the government but in a grant from Bell Laboratories (a source that scored no points with Wiener), and he assured him that MIT had made no permanent commitment to McCulloch. He affirmed the obvious truth everyone at MIT had witnessed in Wiener's interactions with McCulloch, Pitts, Lettvin, Wiesner, and all the researchers in the RLE:

> I want to reassure you that the group in the Electronics Laboratory are heart and soul back of you, that they have moved ahead in the program there in response to the leadership and stimulation of yourself. I hope that what they do will further extend the influence of your basic work.

He closed his letter with just the kind of reassurance Wiener so frequently needed to hear:

> No one can ever deny you the credit and the honor which properly belong to you for your creative scholarship in cybernetics. . . . Your contribution has been so great and your scholarly position in cybernetics is so impregnably secure that you can continue as you constantly have to cherish the luxury of a magnanimous attitude toward those who work on the fringes of cybernetics.
>
> All of this adds up to my conviction that you have really nothing to be worried about. I can assure you of the undiminished support of your friends and colleagues here at Tech and of our desire to further your own program and recognition in every way available to us.
>
> *Yours cordially,*
> J. R. Killian, Jr., President

Killian was not being patronizing. He meant what he said about Wiener and his work. And, like all those at MIT who were familiar with Wiener's changeable moods, Killian assumed Wiener's rage would abate soon enough. But that was not to be.

―――――

Jerry Lettvin, the sole survivor of McCulloch's original group at the RLE, was not buying Wiener's bill of particulars. "Walter and I meant nothing but a wonderful enthusiasm in that letter, but Wiener's misconstruction really flabbergasted me," he said, when he first saw a copy of Wiener's diatribe to MIT's president long after the fact. Lettvin was at a loss to understand Wiener's thinking and had more than enough facts to refute Wiener's claims.

"He talked about us taking official funds to pay ourselves high salaries. It's not true. Warren took a big cut in salary to leave his professorship at the University of Illinois. Pat Wall took an equally severe cut. I was living on $20,000 a year in Manteno, and when I came to MIT my salary was $3,000—and this was for someone with a wife and two kids. As for federal funds, that also was crazy because RLE was given a blanket grant which was to be spread over everybody. Sure, federal money paid us, and later Bell Labs subsidized us, but never in such a way that we lived palatially. Warren and Rook lived in a bedraggled apartment in Kendall Square for forty or fifty dollars a month. Pat Wall and his wife lived with us on the second floor of the house we rented for $75 a month. And Walter lived in what he could find. So all of that was pure nonsense. It wasn't a regal existence, but it was worth it because of what we were doing."

Moreover, as Lettvin pointed out, Wiener was fully aware of the RLE's government funding. Wiener himself had worked in the RLE on research projects with Wiesner, Lee, and others, and he continued to work there with new collaborators—just a few doors down from McCulloch's lab—for years afterwards.

"That's the point. He *never* was concerned," said Lettvin. "It's disingenuous."

No one, not Killian, Wiesner, McCulloch, or anyone on McCulloch's team, had the faintest idea what really had incited Wiener's rage. As it turned out, his action had nothing to do with government money, or Pitts and Lettvin's impertinence, or any of the other complaints he lodged with Killian, or even with his rejected autobiography. But they would not learn that for nearly a decade.

Lettvin was still mulling those events half a century later. Through all that time, he had declined to surrender details of the incident that had come into his sole possession. He was almost eighty, but he still felt sadness over the breach, and anger. The story came out in stages.

"When they had the international conference in Genoa in memory of Wiener, they insisted that I come and talk about his contributions," he recalled. "I prepared a very careful, very adulatory talk and, afterwards, Mrs. Wiener came up to congratulate me and offered her hand. I shook her hand—you know, she was a slight woman—but I really wanted to hit her as hard as I could, because I knew that she had contrived the break. She hated McCulloch with a passion that was unbelievable, and she wanted to engineer a break as strongly as she could."

Margaret's motive for contriving a break between Wiener and McCulloch— if that was her motive and her doing—was unclear. From all accounts, she had only one uneventful interaction with McCulloch, the evening he came to dinner at the Wiener home in the mid-1940s. Lettvin was unable to explain the antipathy he perceived. "I don't know why she hated him. It was a *tropism*," he said, pondering his own visceral response. "Oh, Warren was flamboyant. He would talk in the most obscure ways possible, sometimes with the air of high inspiration, but that flamboyance violated Mrs. Wiener's notion of propriety. She felt that Warren was a corrupter of Wiener and she was going to engineer the break, however. She hated Warren and she hated everybody associated with Warren."

Margaret Wiener disliked Warren McCulloch for many reasons: for his charisma and his colorful mannerisms; for his Orphic aphorisms, which escaped her, for his liberal lifestyle and the social values he and his wife and many of their contemporaries shared; and, surely, for his thirst for alcohol, which was widely observed in the social circles of cybernetics. "I'm sure she disapproved of him," said Taffy McCulloch. Dr. Pauline Cooke, a research fellow at the University of Illinois Medical School and a member of McCulloch's retinue who later also moved to Cambridge, got the same impression. "It would be impossible for a person like Margaret Wiener to like Warren McCulloch. He was the antithesis of everything a middle-class German believes in. They were social worlds apart."

Wiener's daughter Peggy, speaking from her own experience, could only imagine her mother's reaction to the liberated McCulloch crowd's free-wheeling, free-loving lifestyle. "If this was the kind of household they had, Mother would not have understood it. If she had even a suspicion of 'loose-living' at the McCullochs' she would have thrown a fit. She would have been extremely upset and regarded them as morally depraved."

Beyond her moral judgments, Margaret also may have feared for her husband's well-being among those unbridled bohemians. She alone among Wiener's contemporaries knew how deeply depressed and even suicidal he could become, and how distraught he had been for much of the year leading up to the split. She could defend her claim that she was looking out for Wiener's interests and psychological stability in proximity to a social scene she perceived as unwholesome and fraught with emotional perils. But that defense had holes of its own. By all accounts, Margaret was "emotionally deaf" and didn't have a clue about her husband's feelings, her children's, or anyone else's. She herself often triggered Wiener's depressions and was powerless to assuage them. In the opinion of both her daughters, she worried as much about Wiener's stature as a scientist as she did about his stability.

With little doubt, Margaret perceived McCulloch's move to MIT as an encroachment on Wiener's dominion and a threat to his prominence, and, as she had demonstrated repeatedly, she did not hesitate to assert herself in Wiener's professional affairs when she believed his interests were at stake. Early in 1951, before she joined Wiener in Europe, and before the new neurophysiology lab was established, Margaret made a private visit to Jerome Wiesner at the RLE in an effort to dissuade him from bringing McCulloch and his entourage to Tech. However, unlike her earlier campaign, which succeeded in getting an MIT employee fired for having her ears pierced, Wiesner took no action in response to Margaret's visit.

This time, Wiener was aware of his wife's actions, but only after the fact. Months later, when he wrote his enraged letter to President Killian, in a surprising turnaround, he claimed he had opposed Wiesner's plan for McCulloch's laboratory from the outset and that the situation had been very distressing to him, but that he had said nothing because he was "loth to be put in a position where

I had to damn a colleague." He defended Margaret's initiative in meeting with Wiesner "on her own behalf" because she "was so disturbed by my worries about the situation." However, Wiener's testimony, and his entire letter, may have been heavily influenced by Margaret's version of events. At the time, Wiener was still recovering from his two cataract operations, and he acknowledged that his wife "collaborated" in his writing activities, which were conducted almost entirely by dictation.

Peggy saw further into Margaret's motivations. She perceived her mother's urgency to intervene in Wiener's professional interactions, and to protect his stature as a scientist, as evidence of another deep-rooted desire on Margaret's part. "For years I've thought that the real motivating force in Mother's life was protecting her social position," Peggy said, "and I think Mother saw a threat to Dad as a threat to her social position."

In her urge to preserve her dominant place in Wiener's life and career, Margaret needed to keep Wiener's activities at levels where she could function without interference as his closest companion and primary advisor. After her failed attempt with Wiesner, those final weeks in Mexico were the last moments she had in which to prevent the Wiener-McCulloch collaboration from going forward. She had to act before Wiener returned to Cambridge in January 1952, when McCulloch would complete his move from Chicago and the two men would at last come together at the RLE, and to do that in a way that would preclude any argument from Wiener. She knew instinctively that her case against McCulloch and his team would have to be personal—and irrefutable.

As Wiener's return date came near, she made one last, desperate attempt to kill the Wiener-McCulloch collaboration in its crib.

———

"I didn't know what happened until Arturo told me ten years later," said Lettvin. Early in the 1960s, Lettvin and his wife Maggie paid a visit to Rosenblueth and his wife Virginia in Mexico. At dinner in a posh Mexico City restaurant, Lettvin asked Rosenblueth if he had any insight into Wiener's actions so many years ago and his continuing silence with the entire McCulloch group.

"Don't you know?" asked Rosenblueth, incredulous, then he proceeded to replay the events of that fateful night in December 1951. He recounted Wiener's torment over the writing of his autobiography and his progressive despair at the flood of rejection letters. He confirmed the bad timing of Pitts and Lettvin's jocular letter and the last, crushing rejection Wiener received from the director of Technology Press on the first Saturday in December 1951. It was at that low point, at a glum dinner the Wieners and Rosenblueths were sharing just hours after Wiener had received the rejection, that Margaret told Wiener something appalling and far more disturbing. She said that the boys in McCulloch's group—Wiener's boys—had seduced his elder daughter during her stay at the McCulloch home in Chicago four years earlier. In her damning indictment, as Lettvin recalled Rosenblueth's words, Margaret alleged that not one but "more than one" of the boys had seduced the chaste nineteen-year-old during her first foray away from home and the protected environment of her boarding school.

Wiener paled. His mind staggered at the thought of the boys defiling his "pure universal" in McCulloch's lair of iniquity. The next morning, he sent his angry telegram to Jerome Wiesner and wrote his long letter to President Killian formally severing all relations with McCulloch and his team at the RLE.

It was a gangbuster revelation, except for one important detail: The whole story was a fabrication. "She made it all up!" said Lettvin. "There was nothing to it." At a moment when she felt herself to be in danger of losing control over Wiener's territory and affairs, consciously or not, she took a page from her favorite book and chose to propagate a big lie where no little lie, or the truth by any stretch, would suffice. After two decades spent discrediting her elder daughter and falsely blaming her for "spreading lies that ruined men's reputations," Margaret Wiener did just that—and used her daughter, yet again, as her scapegoat.

A less vulnerable and more even-tempered man might have sought to validate those serious accusations before he acted on them, but Wiener did not question Margaret's version of the events. He did not discuss her allegations with Barbara, who had never made any such claims to her mother. Nor did he give his accused colleagues a chance to respond to Margaret's charges, or even to know they had been charged.

Margaret's plan worked perfectly. By charging McCulloch and his team *in toto* with a grave and intimately personal offense, she insured the secrecy of her scheme. Wiener would not divulge his daughter's supposed shame to anyone in those discreet days of the 1950s. Better yet, by framing the whole matter in a sexual context, and then assisting Wiener in his efforts to paper it over with purely professional complaints, Margaret all but guaranteed that no one in any official position at MIT would even know about her allegations and give the truth a chance to work its way out.

For Lettvin, all Margaret's machinations about sex and debauchery in the McCulloch group were only a cover for the deeper fear and paranoia she imparted to Wiener directly and indirectly. "Margaret wanted Wiener to think Warren was stealing cybernetics from him. When I saw that in his letter my hair stood up. I never heard anything so ridiculous. Nobody was trying to take it away from him." Whatever the source, Wiener did come to believe that. In a letter to a colleague in London written after the split, he expressed his concern that McCulloch was trying to eclipse him as the principle proponent of cybernetics, although only a few years before Wiener had enthusiastically supported McCulloch's efforts to carry the torch of cybernetics to England. Wiener vowed once again to quit the Macy conferences—a threat he now carried out. His stated reasons were laced with paranoia and patently absurd. "Professor McCulloch . . . has taken such measures to aggrandize his role in cybernetics at the Macy meetings and elsewhere, that I feel that I am gradually being elbowed out of public participation in them."

McCulloch's colleague Pauline Cooke also saw that paranoid fear as the deeper motive behind Margaret's move against McCulloch and all Wiener's subsequent actions. Cooke likened the high drama in the court of cybernetics to a famous tragedy in another medieval court, as she pictured the Scottish laird McCulloch being done in by the hidden hand of Margaret's Lady Macbeth, who sought to frame his knaves for the crime and secure her husband's throne. "There she was, feeding poison into the king's ears like in Shakespeare, lapping it in. *'Look what they're doing to you. Before you know it, you'll be in a back building and they'll be in the front office.'*"

Wiener's daughters were not entirely surprised to learn the truth about the breach. "It is the sort of story that would spring to Mother's mind," Peggy said of Margaret's seduction charge. She agreed with Lettvin that Wiener's action had nothing to do with the money, as Wiener claimed, "unless that had been planted in his mind. He was not an ungenerous person." She weighed the evidence. "McCulloch and the others were delighted at the prospect of the work to come. If Dad had not been poisoned against them, he surely would have seen that." Still, Peggy wondered about Margaret's ultimate intent. "The question is, did Mother deliberately make up this story to separate Dad from rivals and preserve her position, or did she really believe what she said about the McCulloch group? Was she malicious and self-serving or was she obsessed? Neither prospect is pleasing." Barbara, who was the object of all the fuss in the first place, was further embittered toward her mother, but not entirely forgiving of her father, whom she viewed as a passive and unquestioning partner in Margaret's scheme.

Outwardly, Wiener appeared unfazed by the fiasco, but it affected him deeply. The breakup consumed him emotionally for months and took a physical toll. In a letter he sent to Morris Chafetz in Mexico in March 1952, Wiener told his young doctor friend about a new health problem he was experiencing. "The affair with 'the boys' is still rather nasty," Wiener wrote. "The worry about it gave me a minor angina attack." A week later he was still thumping mad. "The Wiesner-McCulloch-Pitts-Lettvin imbroglio stinks more than I thought it would," he wrote to Rosenblueth. Again he complained of a "very minor anginal attack," but he reassured his cardiologist colleague that he had been "under observation."

In April, Wiener's prognosis was positive, but just barely. "The McCulloch thing is settling itself essentially in my favor, but in a rough and not too immediately satisfactory way," he told Chafetz. "I am very tired at present, but I think I have come out of it all without a seriously damaged heart." Yet by July, Wiener still had not recovered. He was seen by another colleague to be "in a rather nervous state" and plagued by his continuing fear that McCulloch and others were out to steal cybernetics from him. Nearly a year later, he still would not work with Wiesner on the ambitious project they had started several years before to build an automated machine that would enable the deaf to "hear" by touch. "You will realize that neither now nor in the future is any collaboration possible between us," he reminded Wiesner. His heart problems, exhaustion, and emotional doldrums caused him to curtail his research and public speaking activities for much of that year and the next.

For McCulloch and the boys, the full impact of Wiener's action was delayed. At the outset, all were confident that the storm would blow over, like the brief blowup of '47. "Walter and I didn't believe it for a long time," said Lettvin. "We thought it was a petulant outburst that could be remedied. Wiener's colleagues and others went to bat for us. They all came back and said, 'It's impossible to talk to him. We don't know what's happened, but there's absolutely no way.'" For months after the split, Wiener would not even acknowledge the existence of his former friends. When they encountered one another in the RLE, at MIT faculty meetings, or in Tech's dining halls, McCulloch and his teammates would greet Wiener amiably, but he would not reply. "He wouldn't talk and we didn't know why," said Lettvin, recalling the group's ongoing state of confusion.

McCulloch was shell-shocked by the whole affair. At first, he did not know what to make of Wiener's actions, then gradually it began to dawn on him that Wiener's verdict was final. When the implications for McCulloch, his team, and their new research venture hit home, McCulloch could barely grasp the enormity of it all. "Warren kept on going, but it was clear that he was devastated," said Lettvin. Two months into the breach, during a speech in Chicago, McCulloch unleashed a rant against his longtime anathema, the Freudian school of psychoanalysis, that was so splenetic it led some of his colleagues to gossip that McCulloch himself "was going through a disturbed episode." During a later speech at Yale, observers noted that "something about his talk or behavior was erratic." McCulloch's psychiatrist friend Lawrence Kubie told a colleague he was certain that, "One force which may be relevant to [McCulloch's] upset is the fantastic state of megalomanic and paranoid rage in which Norbert Wiener seems to have been ever since Warren came to MIT."

But it was Pitts who took the hardest hit. Lettvin witnessed Pitts's extreme reaction to their group's mass eviction from Wiener's good graces. "Walter suffered monstrously. For him the world stopped," said Lettvin. "Walter loved Wiener enormously. Wiener had become the father Walter never had and when Walter lost Wiener he lost his *raison d'être*." For Lettvin, seeing his friend suffer was more wrenching than the break with Wiener itself. "I think if it hadn't had so strong an effect on Walter everything else would have been all right," said

Lettvin, "but for Walter it was fatal." In the months after the split, the shy Pitts became a near-total recluse. He began drinking heavily. And his scientific work that had been so promising, including his much anticipated doctoral dissertation, suffered a worse fate. "He burned everything he'd ever written," said Lettvin, "the whole manuscript he was working on for his dissertation and everything else. And from that point on he did nothing."

McCulloch struggled to regain his bearings and his perspective. His epitaph for his severed partnership with Wiener at MIT, typically anachronistic, evoked the split between the Puritans and Royalists in England's civil war. *"Remember, mutatis mutandi, he was a Roundhead; I, a cavalier!"* McCulloch lamented.

But this rift was not merely a family squabble. As Heims confirmed, the breach was permanent, profound, and "destined to have an effect on the history of cybernetics."

————

Work at the RLE's new neurophysiology lab went forward, but without Wiener and often without McCulloch. In the lab's first years of operation, Lettvin, Pat Wall, Pitts on occasion, and other keen young MIT minds equipped with the best electronic instruments available—many of which were designed and built within the RLE's ramshackle walls—conceived the bold theories and hypotheses that defined the new field of "neuroscience," devised ingenious laboratory experiments to test them, and let fly other ideas that were years ahead of their time. For the first time, those state-of-the-art weak-signal generators, detectors, amplifiers, synchronizers, cameras, data recorders, oscilloscopes, and the latest analog and digital computers that Pitts and Lettvin had celebrated in their ill-fated letter to Wiener were applied to the study of the living information-processing activities transpiring in the brains and nervous systems of animals and humans.

Then, early in the 1950s, almost simultaneously with the Wiener-McCulloch split, a rift of another kind occurred: a break in the underlying framework of the new communication sciences that would rip through every domain of modern thought and technology. The entire realm of communication theory, research, and engineering practice, which had only come together as a unified field a few years earlier, began to subdivide. Like one living cell mitosing into two, the communication revolution pulled apart at its poles and branched into two increasingly divergent worlds: the world of continuous, analog communication and control processes, which Wiener had codified in *Cybernetics* and all his earlier work; and the new world of logical, discrete, digital information processes, which Shannon had first codified in his Bell Labs paper, "A Mathematical Theory of Communication." The analog mode had characterized human communication historically, from gestures and speech to the first symbol systems for counting and writing, to the first electronic media, automated technologies, and modern day computers. But the digital mode was rapidly being incorporated into the logical design and circuitry of all the newest electronic technologies, and it would soon come to symbolize the very essence of the information age.

That technical split marked a sea change in the lives of the postwar generation of communication engineers; and scientists in allied fields, and in many of the older, established fields as well, began to feel the growing impulse to pick one paradigm or the other. For thinkers and practitioners across the spectrum of postwar society—including those at the forefronts of the technical communication revolution, the organic communication revolution in neuroscience and biology, and the human communication revolution in psychology and the social sciences—two roads diverged in the information age, and the multitudes, by and large, took the glimmering new digital highway.

Jerry Lettvin, who was en route to the RLE with McCulloch's group when the road forked before them, felt that great divide in theory and practice as a "cleavage" of historic proportions. The cleaving sent highly logical minds like McCulloch, Shannon, and von Neumann traveling down the digital road, highly *ana*logical thinkers like Wiener, Bateson, and others in the Macy group traveling down the other—and everyone scrambling to find the road map to the future.

During the years when Pitts and Lettvin were pursuing their apprenticeships under McCulloch and Wiener, Lettvin had watched the two technical domains develop side by side. A common bond joined digital and analog operations, Lettvin said, because "it was obvious that they were both founded on the same basic things in the world." Those basic things, as Wiener and his peers discerned them, included information, communication, feedback control processes found throughout nature, and the ubiquitous acts of computation that comprised the universal language of nature and the logical underpinnings of the mind. Lettvin remembered the flexible attitudes and open architectures Wiener, McCulloch, Pitts, and others brought back from the Macy conferences, the Princeton computer conference, and later meetings, where analog and digital models mixed freely in discussions aimed at finding the best method, or combination of methods, to achieve a desired technical, biological, or social scientific purpose.

"There were meetings to determine whether you could design a machine so it would be both analog *and* digital, not a bad idea," said Lettvin. He remembered the fluid use of devices of both types at the RLE, and the exchange of data between the sensitive analog signal detectors and wave analyzers essential to modern neuroscience and MIT's Whirlwind digital computer. "The analog devices were far more complex and intriguing than logical machines. There was a certain *niceness* about analog operations," said Lettvin, with a touch of nostalgia, "but analog computation is hell on earth. The devices have a life of their own.

So once it became clear that they were going to be able to tame logical devices first, and that you could compute complex, continuous functions on a logical machine, there was no longer a need for many of the old analog machines."

Lettvin traced the high point of the unified approach, and, ironically, the moment that split the technical world into two counterposed conceptual camps, back to the legendary Hixon Symposium at Caltech in the fall of 1948. There John von Neumann first voiced the dissident view that the human brain was hopelessly complex and the wrong real-world model to use for designing electronic computers and other intelligent machines. Von Neumann made the astonishing assertion that the McCulloch-Pitts neural network model, on which his own digital computer architecture was based, was fundamentally flawed as a model for computing machines and even for understanding the workings of the brain itself. His "General and Logical Theory of Automata" maintained that, in the brain and nervous system, "elements of both procedures, digital and analogy [sic], are discernible. . . . The nerve impulse seems in the main to be an all-or-none affair, comparable to a binary digit . . . but it is equally evident that this is not the entire story."

Von Neumann restated the point Wiener had made in his New York Academy address and, again, in *Cybernetics*: that neurochemical and hormonal processes of an analog nature play critical roles in the brain's computational activities. He insisted that "even the neuron is not exactly a digital organ," and he cautioned that the brain's "chemical and other processes . . . may even be more important than the electrical phenomena." Von Neumann declared, "In fact, the 'digital method' used in computing may be entirely alien to the nervous system." At Hixon and in his later addresses, von Neumann went further and suggested that logic itself would have to undergo a complete rethinking and restructuring in the light of present and future discoveries in brain science.

Von Neumann's bombshell began a debate in the upper echelons of computer theory that his disciples would continue for decades, but, at the time, it left a sizeable hole in the theoretical foundation of the developing digital computer industry. The first commercial "electronic brains" had just begun to roll out of major electronics corporations in 1951, the year McCulloch and his crew set up their neurophysiology laboratory at MIT. And here was von Neumann, the master architect of digital computing, whose Hixon paper only appeared in print that same year, proclaiming that the new computers did not work like human brains after all! That discrepancy did not stop commercial computer companies, or mathematicians in universities and government computer laboratories, in their race to build faster and more proficient digital computers of the von Neumann type, and von Neumann was not about to renounce his own architecture. But that basic design flaw in his blueprint gnawed away at all the new machines' progenitors—von Neumann, McCulloch, Pitts, and Wiener, too—and each in his own way took steps to keep the two poles of analog and digital thinking from flying apart irretrievably.

For brain scientists in the early 1950s, the two modes of communication and computation posed major dilemmas. "There's nothing that you deal with in ordinary operations where digital and analog are in combat until you come to the nervous system," said Lettvin. "Neurons are curious devices. The messages are discrete but the processing is analog. Warren and Walter both realized that their model was not really sufficient, but the work they would have had to do to model it correctly was beyond belief."

From the Hixon symposium in '48, through the break with Wiener in '51, and for years afterward, McCulloch tried repeatedly to revise his original brain schematic. But instead of heeding the call to develop a more truly brain-like theory based on a mixed analog–digital model, McCulloch became a partisan for the logical cause. He plunged into the alchemy of transmuting logic into neurology and vice versa, devising arcane systems of "multi-valued logic" and "probabilistic logic," as he struggled in vain to build on statistical ideas developed over the years by Wiener, von Neumann, and Pitts. But without Wiener on his team, and effectively without Pitts, his own vision foundered.

For some time, Wiener had been thinking about the brain's analog communication channels. In *Cybernetics,* he described the brain's alternate system of communication that operated, not by all-or-nothing mechanisms and yes–no logical decisions, but by a more subtle and random chemical signaling system made up of analog "messages which go out generally into the nervous system, to all elements which are in a state to receive them." Those generalized, "to whom it may concern" messages, as Wiener called them, were transmitted by versatile chemical messengers which scientists would soon recognize as a broad class of specialized "neurotransmitters," but after his break with McCulloch, Wiener had no one with whom to pursue his speculations—least of all Walter Pitts.

Pitts's doctoral work on random nets was moving in a parallel direction, and pushing deep into the analog domain. "Walter was ahead of his time," Lettvin said of Pitts's doctoral project. "He envisioned a layered device, a three-dimensional net, and he was exploring the properties such a system might display with respect to the information put into it. He did it analog, using continuous mathematics. Wiener was very much taken with this. Nobody else was tackling it. Walter had done a huge amount of work and gotten some very interesting results. He had already written between two and three hundred pages of serious stuff—a generalized theory of data processing in a three-dimensional system—but when the blow fell with Wiener, Walter simply stopped."

———

No one could say for sure what the breach cost Wiener and the others personally, what it cost cybernetics in theory and in practice, or what would have resulted if the split had not occurred; but some markers point down the road not taken.

Had the breach not happened, quite possibly, the brain model that dominated neuroscience for a generation, the all-or-nothing neural network theory of the brain's function and working organization, might have been modified long ago to include those hellish analog processes that preoccupied Wiener throughout his career—processes that have yet to be fully integrated into science's theoretical models of brain and mind. On one technical forefront, in the domain of electronic computing, if those flaws in the McCulloch-Pitts model had been corrected, von Neumann's vaunted architecture might have been improved to incorporate the truer and more lifelike computing methods for which von Neumann himself was reaching. But that work was cut short at a crucial moment when a concerted team effort still might have influenced the advancing architecture of computing and all the technologies of the information age.

Lettvin did not dare to speculate about the what-ifs. "I haven't the vaguest idea what would have happened," he said. Pitts never spoke about the split, nor did McCulloch conjecture or bemoan what might have been. Oliver Selfridge, who witnessed the debacle at the RLE from his safe haven in the Lincoln Laboratory, was the only survivor of the original Wiener-McCulloch group who was willing to hazard an opinion. "I think one thing that might have occurred if Norbert and Warren had been talking is that there would have been more attention by those great guys to teleology, the science of purposes," said Selfridge, who made that very subject the focus of his later work in artificial intelligence and computer theory. He mourned the loss for the players, Pitts especially. "Walter was my age, there's no reason why he couldn't have gone into computing, too. There was no reason with all these wonderful new things happening, that he could not take charge of a new field like this and know it all at once. To me, that wasn't just a little tragedy."

However, in Selfridge's view, the greatest consequence of the breach was for Wiener's new science itself. "It really fucked up cybernetics," he said bluntly, "because here you've got the guy who invented the term and invented the idea right there with you, but there was no interaction at all with Norbert, which was a crying shame. The real tragedy was that we were taking cybernetics seriously and looking at it in just the way in which Norbert should have thrived, so this breakup really did a hell of a lot of damage."

Lettvin looked back longingly to the triumphs of that time and its broken promise. "They were halcyon days. They were unbelievable," said Lettvin. He missed his fallen comrades, and perhaps Wiener most of all. "Oh, I thought the world of him. We all did."

They had made primary contributions to cybernetics and the evolution of Wiener's thinking, but now Wiener's vital connection to McCulloch, Pitts, and their teammates was gone, cut off by his own rash actions and his wife's hidden hand. The human factors they supplied to Wiener and his revolution were the ones Margaret Wiener could not grasp or feel: the spirited camaraderie, the feelings of acceptance, the surcease from worry he found, at least momentarily, when he was with McCulloch and the boys at their best.

By Margaret Wiener's deeds, not theirs, the forward progress of cybernetics was brought to a halt on its home turf, but that fact was unknown to anyone outside their small group in the cold Cambridge winter of 1952, and its full effects would not become clear for years to come. The upbeat decade of unparalleled technical progress would go forward. Wiener's revolution would continue its sweep over the sciences and societies worldwide.

And Wiener himself would reconceive his mission and purpose as a scientist and head off in new directions.

PART THREE

Aftermath

12

A Scientist Rebels

One dark evening he came back to all his dear families, and he coiled up his trunk and said, "How do you do?" They were very glad to see him and immediately said, "Come here and be spanked for your 'satiable curtiosity."

"Pooh," said the Elephant's Child. "I don't think you people know anything about spanking; but I do, and I'll show you.". . .

Then that bad Elephant's Child spanked all his dear families for a long time, till they were very warm and greatly astonished.

—**Rudyard Kipling,** *Just So Stories*

LONG BEFORE HE BROKE WITH MCCULLOCH, before the birth of cybernetics and the big bang of the information age, Wiener was becoming deeply concerned about problems he saw coming over the horizons of the new technological era. He had been a rebel in his own right, at times openly defiant of his parents, his mentors, and his peers. Now, in his fifties, he would become a formidable foe on the new battlefields of postwar science and society, and a fierce fighter for human beings.

The atomic bomb had laid waste to his faith in the forward motion of science and technology and confirmed his worst fears about the dangers of placing scientific knowledge in the hands of society's elites. The day after Hiroshima, on his rounds around MIT, Wiener told Dirk Struik that the decision to drop the bomb on a civilian population "signified the beginning of a new and terrifying period in human history." To him, the bomb's salient fact was not that it had brought a swift end to the war, but that it marked the coming of "a new world . . . with which we should have to live ever after," one in which "for the first time in history, it has become possible for a limited group of a few thousand people to threaten the absolute destruction of millions . . . without any . . . immediate risk to themselves."

He saw "the practical certainty that other people will follow where we have already gone and that we shall be exposed to the same perils to which we have exposed others." He questioned the subtle seduction of scientists and engineers by the new weaponry they were unleashing. He questioned the motives of "the lords of the present science," the government officials, military commanders, and administrators of nuclear research who had dispensed billions of dollars in research funding and sooner or later would be impelled to justify their expenditures. More worrying still was a deeper impulse he perceived in the scientists and technocrats of the newborn atomic age. "Behind all this I sensed the desires of the gadgeteer to see the wheels go round."

While he had played no part in the Manhattan Project, he knew the intelligent technology he had helped to develop for the war effort would be incorporated in the next generation of atomic weapons, already on the drawing boards, for use in a projected war between the United States and its former ally, the Soviet Union. At the time, many people in and out of the government felt that war might be only a few years away. In October 1945, two months after the Second World War ended, Wiener vowed to remove himself and his knowledge from any connection to the next war—even if that meant quitting mathematics and science altogether.

Wiener had other reservations about the direction of postwar science. He cursed the continuing secrecy imposed on basic science by military and government funders. He deplored the emerging alliance of science, industry, and the military, with its proprietary controls over the new technologies the government and its corporate partners were jointly developing and putting into production. Scientists had accepted strict measures of secrecy and tight restraints on their choice of work, and on their movements at home and abroad, for the sake of the war. "We had expected that after this war—as after all before—we should return to the free spirit of communication, intranational and international, which is the very life of science," he lamented. But now he found that "whether we wished it or not, we were to be the custodians of secrets on which the whole national life might depend. At no time in the foreseeable future could we again do our research as free men. Those who had gained rank and power over us during the war were most loath to relinquish any part of the prestige they had obtained."

His concerns were well founded. With its stunning victories in two theaters, the United States had emerged from the war as the world's dominant military, industrial, and economic power, and much of that credit was owed to its scientists. The nation's science administrators, led by Wiener's MIT colleague Vannevar Bush, director of the wartime Office of Scientific Research and Development, were eager to maintain and increase America's scientific output, and to make its fruits available to the military and to industry; and after the war the national science project was retooled for service in the new Cold War and its technological trenches.

Under the master plan Bush set forth in a report to President Truman in 1945, ties among government agencies, universities, and industry were strengthened and funds were slated to flow through a new civilian controlled national research foundation. At the same time, the military itself, through the newly established Office of Naval Research and its counterparts in the other armed services, also became a primary funder of basic and applied science. By 1950, when the new National Science Foundation was chartered by Congress, the many-headed funding structure found the government paying more money to large corporations to conduct scientific research than it paid to scientists at universities and not-for-profit research institutes.

American science had turned a corner and would never go back. The great universities and technical institutes where scientists were trained and sheltered were forging permanent ties to government and the military and making their own lucrative arrangements with the private sector, but Wiener was leery of the new national science leviathan. Above all, he could not bear the thought that his new science of communication and control might be applied wholesale by the military-industrial machine, behind impenetrable layers of secrecy, to the contemptible business of making bigger and more deadly atomic weapons.

The new prospect of "push-button warfare" that was being promoted by the nation's top brass—war waged with atomic weapons that could be fired automatically by computerized "command and control" technologies—was horrifying to Wiener. "The whole idea . . . has an enormous temptation for those who are confident of their power of invention and have a deep distrust of human beings," he said. He felt compelled to actively oppose the military enterprise that, he believed, threatened the souls of scientists and posed grave perils to the world.

"I thus decided that I would have to turn from a position of the greatest secrecy to a position of the greatest publicity, and bring to the attention of all the possibilities and dangers of the new developments."

His first opportunity presented itself late in 1946, when he received a letter from a researcher at the Boeing Aircraft Company in Seattle requesting a copy of his Yellow Peril manuscript. The letter made clear that Wiener's wartime insights and later technical work were wanted for use in the company's project to design guided missiles to deliver the new weapons in America's Cold War arsenal. Wiener's response was terse, as he refused with undisguised satisfaction to provide a copy of his still classified monograph or any of his newer work.

Since the termination of the war I have highly regretted the large percentage of scientific effort in this country which is being put into the preparation of the next calamity. I therefore am much gratified to find that my [papers are] no longer available to those who construct controlled missiles. I can, of course, furnish you with no advice as to where to find them.

His next move was more audacious: He sent an expanded version of his reply to the *Atlantic Monthly* in Boston. The magazine's editors printed it in its entirety in the *Atlantic*'s January 1947 issue, under the defiant headline, "A Scientist Rebels," with a brief introduction implicitly endorsing Wiener's position. The *Atlantic* described Wiener as "one of the world's foremost mathematical analysts" whose "ideas played a significant part in the development of the theories of communication and control which were essential in winning the war." It portrayed his indignation at being asked to join in a new postwar arms buildup as a feeling shared by many scientists who had served loyally in the war effort. Then, without mentioning the company by name, the magazine unleashed Wiener's letter "to a research scientist of a great aircraft corporation." The letter ended with an oath and a call to the conscience of his fellow scientists:

I do not expect to publish any future work . . . which may do damage in the hands of irresponsible militarists. I am taking the liberty of calling this letter to the attention of other people in scientific work. I believe it is only proper that they should know of it in order to make their own independent decisions, if similar situations should confront them.

He was not the first prominent scientist to forswear atomic weapons and warfare. Years earlier, Albert Einstein, the premier pacifist among twentieth-century scientists, had urged "every thoughtful, well-meaning and conscientious human being . . . not to participate in any war, for any reason, or to lend support of any kind, whether direct or indirect," and many physicists followed his lead after the war. But Wiener was the first scientist associated with the new communication and control technologies who publicly refused to cooperate with the government and its agents on a project ostensibly in the nation's defense.

More than a year before *Cybernetics* was published, Wiener's fiery letter in the *Atlantic Monthly* propelled him to national attention. He would not change his position, but his letter would change the course of his career—in some ways he would come to regret. "If I had thought out fully how I was thus subjecting myself to a deep moral commitment . . . I might well have hesitated," he confessed later, "although I probably would put this hesitation behind me as an act of cowardice. The moral consequences of my act were soon to follow."

At that moment, the brass ring in the new digital computing domain was still up for grabs. Von Neumann's project with Bigelow to build his EDVAC machine at Princeton had only begun, and von Neumann's design had not yet emerged as the dominant computer architecture. Howard Aiken's Mark I electromechanical computer had been operating for two years; his all-electronic Mark II was approaching completion in Harvard's new Computation Laboratory. And Mauchley and Eckert had just left the University of Pennsylvania to build their Universal Automatic Computer, the first commercial machine, which would become known as the UNIVAC.

Late in 1946, following the second Macy conference and Wiener's keynote address to the New York Academy of Sciences, and just weeks after Wiener wrote his fateful letter to his admirer at Boeing, he received an invitation from a colleague at the University of California in Los Angeles to direct a "semimilitary project on mechanical computation" in California. The offer was tempting. At last, Wiener was being handed a chance to build an operational model of the all-electronic digital computer he had proposed to Vannevar Bush at the OSRD six years earlier, in the memorandum Bush never circulated and which also remained classified. But Wiener was caught in an ethical dilemma of his own making.

He determined more precisely the exact nature of the "semimilitary" project. "It was in fact to be under the Bureau of Standards, but it was quite clear that all the facilities engineered by the project . . . would be pre-empted for years by the military services," he wrote later. "My acquaintance had tried to commit me not only to work whose objective was distasteful to me but to work which would involve conditions of secrecy, of a police examination of my opinions, and of the confinements of administrative responsibility. These I could not accept. . . . When the invitation was passed on to me, I considered my *Atlantic Monthly* letter, and I had no alternative but to say no. . . . My hand was forced."

Early in January 1947, his hand was forced again. Just days after "A Scientist Rebels" appeared in print, Aiken convened a conference on automatic computing at Harvard, cosponsored by his longtime patron, the U.S. Navy's Bureau of Ordnance. Wiener had agreed months earlier to deliver a paper. "I went to Aiken and tried to explain the situation," he recalled. "I pointed out to him that the California offer had made it necessary for me to take a definite stand on my war work and that I could not accept one sort of a military association and reject another. I therefore asked to be released from my promise to give a talk."

Wiener described the public relations disaster that followed:

I gathered from Aiken that there would be time to take my name off the list of speakers for the meeting. However, when the meeting came, I found that Aiken had done this by merely running a line through my name on the printed programs which had been issued to . . . the press. . . . The newspapermen came to me and asked whether this striking out of my name had anything to do with the letter that had appeared in the *Atlantic Monthly.* I said that it had, and I tried to explain to them the circumstances . . . and that I was not taking this step out of pique or personal animosity. . . . [Aiken] assumed that I had been involved in some deep plot to discredit him and to turn the meeting into a public scandal. In fact . . . [t]he matter would have had no publicity if it had not been for the . . . way he scratched out my name.

The press had a field day with the fiasco. The first major scientific conference on computing boasted 157 attendees from universities, 103 from the government, seventy-five from the electronics industry, and scientists from Great Britain, Belgium, and Sweden—but one no-show got all the attention. "M.I.T. SCIENTIST 'REBELS' AT WAR RESEARCH TALK: CITES MORAL ISSUE, HITS 'IRRE-SPONSIBLE MILITARISTS' " the *Boston Traveler* blared the day after the conference opened, in a front-page headline that tied Wiener's action to his statements in the *Atlantic Monthly.* The story was carried by *The New York Times* a day later and picked up around the world, and it received a second boost after Wiener's new policy of scientific noncooperation was endorsed by Einstein at Princeton. "I greatly admire and approve the attitude of Professor Wiener," Einstein told the press, adding, "I believe that a similar attitude on the part of all the prominent scientists in this country would contribute much toward solving the urgent problem of national security."

The incident was humiliating for Aiken and wholly overshadowed the historic work of his symposium. But it proved more damaging to Wiener, who, by the most inopportune timing, was forced to remove himself completely from the field of electronic computing that he had helped to nurture since the 1920s. True to his word, but ruefully, Wiener wrote to McCulloch the day the *Boston Traveler* story appeared. "I am . . . giving up all work on the computing machine because it is too closely associated with the guided missiles project." Several months later, in an interview in *The New York Herald Tribune,* Wiener reaffirmed his position and went beyond it, declaring that he would undertake no further research that was connected to the U.S. government. The following year, when *Cybernetics* appeared, he vowed again that he would "not work on any project that might mean the ultimate death of innocent people."

He never worked in the field of computing again or took another cent for his work from the military or any agency of the U.S. government.

Wiener's outspoken statements and rebellious acts isolated him from his MIT colleagues far more than his volatile temperament and eccentric mannerisms. By the end of the war, MIT's contracts with the Army, the Navy, and the Air Force had made the institute "the nation's largest non-industrial defense contractor"— a title it would hold proudly for decades—and its ties to industry were even tighter. Tech's faculty and administrators, who worked closely with government and the private sector to develop the latest electronic technologies and their military applications, were embarrassed by Wiener's pronouncements, but they were powerless to mitigate them.

His vow of noncooperation touched off heated debate in the inner circles of American science, but Wiener would not call off his rebellion. Two years later, he stepped up his protest in a tough follow-up piece in *The Bulletin of the Atomic Scientists*. His new statement removed any doubt that his earlier action was taken from a firm moral conviction, and not in the throes of an emotional storm. "In the first place, it is clear that the degradation of the position of the scientist as an independent worker and thinker to that of a morally irresponsible stooge in a science-factory has proceeded even more rapidly and devastatingly than I had expected," he wrote. "This subordination of those who ought to think to those who have the administrative power is ruinous for the morale of the scientist, and quite to the same extent it is ruinous to the quality of the objective scientific output of the country."

He renewed his pledge and strengthened it:

> In view of this, I still see no reason to turn over to any person, whether he be an army officer or the kept scientist of a great corporation, any results which I obtain if I think they are not going to be used for the best interests of science and of humanity.

Beyond the perils of atomic weapons and "megabuck science," as he called it, Wiener's moral concerns were spreading to other arenas of industry and society. The prospect of automation in industry presented him with another professional dilemma he felt personally. "I wondered whether I had not got into a moral situation in which my first duty might be to speak to others concerning material which could be socially harmful." He weighed the pros and cons of coming changes he could only envision dimly:

> The automatic factory could not fail to raise new social problems [as it] threatens to replace [human workers] completely by mechanical agencies. . . . On the other hand, it creates a new demand for the highly skillful professional man who can organize the order of operations. . . . If these changes . . . come upon us in a haphazard and ill-organized way, we may well be in for the greatest period of

unemployment we have yet seen. It seemed . . . quite possible that we could avoid a catastrophe of this sort, but if so, it would only be by much thinking, and not by waiting supinely until the catastrophe is upon us.

Unlike his stance toward the military, Wiener did not wholly oppose the application of his ideas to industry, but he knew a major mobilization would be needed to prepare the world's societies for the changes that were coming. He opened a new front in his scientific activism, beginning the necessary process of public education, in his forthright way, by seeking maximum publicity for his concerns. And he took steps to lead that social mobilization himself, by taking his message directly to those people who he knew would be the most profoundly affected.

Early in the postwar period, Wiener began an active outreach to organized labor. He made contact with two union leaders, one a labor counselor, the other a high official of the typographers' union, which was part of the larger Congress of Industrial Organizations representing millions of factory workers. They gave him an attentive, sympathetic hearing, but neither Wiener nor his listeners could impress the upper reaches of union officialdom with the seriousness of the challenges posed by automation. The experience left him frustrated and strongly suspecting that labor leaders had a limited view of the coming realities of automation, and few tools for dealing with his larger questions about the future of labor itself.

By the time *Cybernetics* appeared in 1948, Wiener's growing concerns about automation had convinced him that, side by side with the dangers of atomic weapons . . .

> We were here in the presence of another social potentiality of unheard-of importance for good and for evil. The automatic factory and the assembly line without human agents . . . makes the metaphorical dominance of the machines . . . a most immediate and non-metaphorical problem. It gives the human race a new and most effective collection of mechanical slaves to perform its labor. Such mechanical labor has most of the economic properties of slave labor. . . . However, *any labor that accepts the conditions of competition with slave labor accepts the conditions of slave labor, and is essentially slave labor.*

In his first public statement on the approaching automated society, Wiener tried to be even-handed. "It may very well be a good thing for humanity to have the machine remove from it the need of menial and disagreeable tasks, or it may not. I do not know." But he did know that "it cannot be good for these new potentialities to be assessed in the terms of the market, of the money they save." To head off the human impact he saw coming, Wiener proposed a simple, commonsense solution that would become his hallmark: *"The answer, of course, is to have a society based on human values other than buying or selling."* That was imperative; the alternative, for Wiener, was unthinkable. "To arrive at this society, we need a good deal of planning and a good deal of struggle," he said, and he proceeded to promote both the planning and the struggle.

In the flush of media interviews that followed *Cybernetics*, he warned of the perils posed by new cybernetic technologies and the press picked up the story, reporting his conviction that "machines that think, pass judgments—and even have nervous breakdowns—will be replacing human labor on the industrial assembly line, making the unskilled human being 'obsolete.'" Wiener did not mince his words. "This thing will come like an earthquake," he predicted, as he linked the social consequences of intelligent technology to the other concern already weighing on people's minds. "The impact of the thinking machine will be a shock certainly of comparable order to that of the atomic bomb." He described industry's eagerness to automate, with profit as its only motive, as "a very dangerous thing socially. If we are going to sell man down the river and replace him, he's going to be a very angry man and an angry man is a dangerous man." He reserved a special wrath for the trade group that served industry's interests, the National Association of Manufacturers (or, as he called the group in private, "the Natural Association of Malefactors"), but that enmity was not mutual.

Invitations began arriving from leading industrial corporations to advise them on the design and implementation of the automated factory. In the spring of 1949, Wiener was approached by executives at General Electric, the country's leading developer of automated technology. The company wanted Wiener to advise managers in its Industrial Controls Department on automation matters, and to teach automation methods to engineers at GE's giant production plant in Lynn, Massachusetts, where he had worked briefly during the First World War—but he refused both requests. In fact, GE's repeated overtures, and its influential position in American industry, prompted Wiener to extend his boycott beyond government and the military, and to abstain from any work that might hasten the deployment of automated machines and the displacement of human workers. It also spurred him on in his efforts to give workers some leverage of their own.

A few months later, Wiener made his second outreach to organized labor. In August 1949, he wrote a long letter to Walter Reuther, president of the United Automobile Workers, the country's largest and most powerful union, to alert him to the technical prospects of automation and its consequences for labor. Wiener told Reuther of his consulting offer from industry and implored him to "show a sufficient interest in the very pressing menace of the large-scale replacement of labor by machine." He urged Reuther "to steal a march upon the existing industrial corporations . . . and while taking a part in the production of such machines to secure the profits in them to an organization dedicated to the benefit of labor." And he offered his services, free of charge, to the labor king-pin. "I am willing to back you loyally, and without any demand or request for personal returns. . . . I do not wish to contribute in any way to selling labor down the river."

Reuther replied by telegram four days later: "DEEPLY INTERESTED IN YOUR LETTER. WOULD LIKE TO DISCUSS IT WITH YOU AT EARLIEST OPPORTUNITY." However, Reuther was busy with tense contract negotiations with the Ford Motor Company over the union's opposition to a speed-up of the company's assembly lines, and preparing for a strike against Chrysler, and the two men did not begin their discussion for seven months. Finally, in March 1950, Reuther came to Boston, and he and Wiener held a private breakfast meeting at Reuther's hotel.

The scrappy union chief and the bearded mathematics professor stared across the starched white tablecloth at one another like emissaries from opposite ends of the world, but their concerns were in close alignment. Wiener briefed Reuther on the new technology and its implications for the factory. Reuther listened intently and asked intelligent questions. Then the two men resolved ad hoc to form a pragmatic new interdisciplinary team, a joint "labor-science council" dedicated to anticipating, and acting on, the changes bearing down on America's factory workers.

Wiener was impressed by Reuther's grasp of his scientific ideas and social concerns, by Reuther's willingness to engage the larger, long-range changes coming for workers everywhere, and by his international perspective on labor's problems, which mirrored his own global approach to the new technology and its human consequences. He rejoiced at having "found in Mr. Reuther and the men about him exactly that more universal union statesmanship which I had missed in my first sporadic attempts to make union contacts." Neither Wiener nor Reuther had any desire to repeat the bitter Luddite rebellion of textile workers in England in the nineteenth century, who smashed the automatic looms that threatened their craft and their livelihood. Both these modern men wanted to work cooperatively with industry's owners and engineers to ease the inevitable transition to automated machinery in the workplace, and to cushion the impact on workers by retraining and upgrading their skills to enable people to work compatibly with the new machines as supervisors and troubleshooters on the shop floor.

But, on that forefront, too, Wiener's initiatives set him squarely in opposition to many of his fellow scientists and their new patrons in government and indus-try. The early postwar years had seen a record number of strikes by organized labor, and for many industrialists the new age of the automatic factory and "machines without men" could not come fast enough. Anti-labor sentiment in industry was shared by powerful political interests, and by high-ranking officials in the military, particularly in the Air Force, which was leading the way in the campaign to bring automation to the aircraft industry and the new domain of guided missiles.

Wiener and Reuther vehemently opposed those efforts that paid no heed to the social consequences of automation, but their joint venture in defense of workers was delayed repeatedly by their busy schedules. They kept up their discussions in writing while Wiener traveled the world spreading the word about cybernetics, and while Reuther was immersed in hard bargaining for UAW workers and the mobilization of American industry for the Korean War, which broke out in the summer of 1950.

Two years would pass before they would attempt to meet again and resume their work.

———

Wiener's social concerns played leapfrog with his scientific work during those years, and gave him a welcome respite from his personal problems and stormy professional relations, as he applied his talents and energies to meaningful projects for the greater good. It was in that eventful summer of 1950, during the off-year in his Mexican migrations, and before he plunged into his childhood memoir, that he published his first book written for a nonscientific audience.

The idea of writing a popular work on cybernetics and its social implications was first suggested to him by the esteemed New York publisher Alfred A. Knopf early in 1949. At the time, Wiener was overwhelmed with offers in response to *Cybernetics* and he rejected Knopf's proposal, but the idea took root in his mind. Months later, Paul Brooks, an editor at Boston's Houghton Mifflin publishers, prodded Wiener again, knowing he had been approached unsuccessfully by others and that he was "mercurial," "unpredictable," and "touchy." To Brooks's surprise, Wiener was interested, and now he "did have a message for the general public: a warning against allowing modern technology to take over our lives."

Wiener retreated to South Tamworth and began to dictate the book to his secretary Margot Zemurray (who had not yet incurred Margaret Wiener's fury for having her ears pierced and was still gainfully employed by MIT), but when Brooks received the first chapter of the book he was floored. It "had nothing to do with the subject we had discussed. . . . It was quite impossible," Brooks chronicled in his own memoir. He telephoned Margot in New Hampshire to discuss "that dreadful first chapter and what we could do about it." In the middle of the call, to Brooks's horror, Wiener took the phone. *"Wiener here! We'll scrap the book!"* Brooks begged him to reconsider and rushed him a long memorandum of editorial suggestions. The next day, Brooks got a call-back from South Tamworth. *"Wiener here. Have your memorandum. Think I can do it!"* "And he did," Brooks confirmed.

The book's title took some doing as well. Wiener proposed two cryptic classical titles, *Pandora* and *Cassandra*—named, respectively, for the fabled daughter of the gods who opened a forbidden box and unleashed all the world's evils on humankind, and for the clear-eyed prophetess of Troy whose warnings no one heeded until disaster was upon them. Brooks roundly rejected both titles, which, he said, "would, in the opinion of everyone here, kill the book dead." He drew the winning entry from Wiener's own words, extracting a passage from the manuscript in which Wiener described the encroaching exploitations of the second industrial revolution and dedicated his efforts to "protest against this inhuman use of human beings." Wiener liked the idea and the finished work went to press.

"What a book!" Brooks wrote approvingly to his author. "Most people who [are] concerned with human values shudder slightly at the whole idea of the part that mathematical machines and the theory behind them are taking in modern life. They assume automatically that the scientist is *not* concerned with humanity. It will be news to them that the leader in this whole branch of science is most concerned with exactly that."

The Human Use of Human Beings: Cybernetics and Society marked the next step in Wiener's rebellion, and in his drive to redirect his energies away from endeavors requiring the greatest secrecy to those affording the greatest publicity. Writing in simple, nontechnical language, Wiener focused his audience on the big picture of the cybernetics revolution and its "real change of point of view." Speaking more as an educator and philosopher than merely as a mathematician, he declared the ultimate purpose of his book: to help people understand the impact of this change "in working science, and . . . in our attitude to life in general."

Wiener stressed the social dimensions of cybernetics, in his conviction that "society can only be understood through a study of the messages and the communication facilities which belong to it." He described communication as the "cement" of society "which binds its fabric together," and he offered his lay audience a novel definition of the new stuff of information that was beginning to appear everywhere in society. For Wiener, information was not a string of digitized bits, as Shannon defined it, or anything and everything, as Weaver portrayed it, but a process with a purpose:

> The process of receiving and of using information is the process of our adjusting to the [circumstances] of [our] environment, and of our living effectively within that environment. The needs and the complexity of modern life make greater demands on this process of information than ever before, and our press . . . our sci-

entific laboratories, our universities, our libraries . . . are obliged to meet the needs of this process or fail in their purpose.

Wiener's definition of information was revolutionary in the scientific sense, but more revolutionary were his thoughts on the economic value of information, and on its nature as a commodity which differed greatly from conventional, matter-and-energy commodities. He questioned the wisdom of applying traditional market values to the realm of information and communication, as many new technology industries were doing. Speaking directly to his American audience, he believed it was important to point out that the mechanism of the market alone "does not represent a universal basis of human values." His problem with the market approach was not a criticism of any social or economic theory but, to the contrary, the recognition of a new economic reality unique to the nature of information itself, and he was convinced that the market approach inevitably "leads to the misunderstanding and the mistreatment of information and its associated concepts."

Indeed, Wiener was the first information-age forebear to consider information, not as a tangible good to be bought and sold, but as "content"—whether that content was an ephemeral commodity like the news, a body of scientific knowledge, or the living substance of everyday experience human beings extracted from the world around them. To his mind, the value of that information, or any technology associated with it, was tied to its value for human survival, and to its real potential to inform and improve the lives of people and societies. While not averse to commercial enterprise, he viewed the exploitation of information to the detriment of those human values as a threat to the wealth of nations, to their security, and to their very survival, and he called for the "unhampered exchange" of knowledge and information in every form.

Wiener urged societies to put those human values first in every arena. The prospect of people having to compete in a coming world of automated factories and tireless robot slaves, on purely economic terms, threatened to devalue work and human workers on a scale the country—and the world—had not seen:

> It is perfectly clear that this will produce an unemployment situation, in comparison with which the . . . depression of the thirties will seem a pleasant joke. This depression will ruin many industries—possibly even the industries which have taken advantage of the new potentialities. . . . Thus the new industrial revolution is a two-edged sword. It may be used for the benefit of humanity. . . . It may also be used to destroy humanity, and if it is not used intelligently it can go very far in that direction.

His message went beyond the loss of jobs and the economics of information. Near the end of the book, he spoke of another peril inherent in his science of communication and control, as he warned of the potential effects of cybernetic technology on society over the long-term. In one of the book's most poignant passages, he wrote:

> The [automatic machine] is not frightening because of any danger that it may achieve autonomous control over humanity. . . . Its real danger . . . is the quite different one that such machines, though helpless by themselves, may be used by a human being or a block of human beings to increase their control over the rest of the human race . . . by means not of machines themselves but through . . . techniques as narrow and indifferent to human possibility as if they had, in fact, been conceived mechanically. . . . In order to avoid the manifold dangers of this, both external and internal . . . we must know . . . what man's nature is and what his built-in purposes are.

Drawing parallels to the legend of the Greek god Prometheus, who gave fire to humankind then suffered the wrath of the gods for daring to give mere mortals so much power, he spoke of the impending "sense of tragedy" he felt hovering around the new technology, and of his concern that people would surrender to machines their human powers of choice and control. He recoiled at the prospect that any person would "calmly transfer to the machine . . . the responsibility for his choice of good and evil, without continuing to accept a full responsibility for that choice." For Wiener, those conscious acts of human choice that were the essence of information itself were at risk from the logical powers inherent in the new computing devices, from their unbeatable speed and efficiency as decisionmakers, and from the next generation of intelligent machines that were being wired to remember and learn from their past decisions, outcomes, and errors.

Drawing with equal aplomb on the Old Testament of the Jews and the Arabian *1001 Nights*, Wiener delivered a prophetic warning to the human race about the new technology he had helped to fashion in its image:

> Any machine constructed for the purpose of making decisions, if it does not possess the power of learning, will be completely literal-minded. Woe to us if we let it decide our conduct, unless we have previously examined the laws of its action, and know fully that its conduct will be carried out on principles acceptable to us. On the other hand, the machine like the djinee [in the bottle] will in no way be obliged to make such decisions as we should have made, or will be acceptable to us. For the man who is not aware of this, to throw the problem of his responsibil-

ity on the machine . . . is to cast his responsibility to the winds, and to find it coming back seated on the whirlwind.

As he looked out over the landscape at the promises of his new science, its beguiling enticements, and the dangers on the threshold of the new technological era, he evoked the song he had sung around his dinner table with J. B. S. Haldane—the malediction of the medieval monk Bernard of Cluny—and exhorted his readers to heed his new warnings to the modern world:

The hour is very late, and the choice of good and evil knocks at our door.

———

The Human Use of Human Beings was open to all the charges that were leveled against *Cybernetics* and then some. It was loosely organized, and far more outspoken in its social criticism, yet those seeming defects struck a nerve in the anxious public of the early 1950s, and the book sold briskly. Readers and reviewers alike seemed quietly thrilled that a scientist of Wiener's rank would speak out so forcefully on matters of urgent and universal import. Even Wiener's cranky old mentor Bertrand Russell, now pushing eighty, who the year before had won the Nobel Prize in Literature, gave his headstrong student high marks and a hearty endorsement in a popular British magazine. In an article titled "Are Human Beings Necessary?" Russell hailed Wiener's work as "a book of enormous importance." He credited Wiener with setting "forth in grave tones the perils facing the human race as the result of his labours and those of some other ingenious people," and vouched that "if this new [industrial] revolution is not to cause vast and unexampled misery, we shall have to change some of the fundamental assumptions upon which the world has been run ever since civilisation began."

The warm public reception suggested that Wiener had succeeded in his mission to take his new science and social concerns directly to the people, and to give cybernetics a fuller role as a science of communication in all its technical and human dimensions—and one with explicit moral imperatives as well. His stand against weapons work was joined by scientists who shared his concerns about the immorality of atomic warfare. His colleagues in the Macy Conferences on Cybernetics, Gregory Bateson, Margaret Mead, and others, were outspoken on postwar problems of mental health and international conflict. However, on issues involving the new cybernetic technology and its human consequences, Wiener led the way alone and, sometimes, in glaring opposition to his peers.

No scientist in the postwar era stood in sharper contrast to Wiener than his fellow ex-prodigy John von Neumann. As a grateful émigré, von Neumann was an ardent believer in America's national mission, eager to prove the superiority of the American system and the supremacy of American technology. As the famed scientist/war hero whose atomic bomb design had delivered the coup de grâce in the Pacific, he was a staunch proponent of the military-industrial establishment and its flourishing mechanisms for technological research and development. When the Cold War commenced, von Neumann enlisted as a consultant to nearly two dozen government and industrial organizations. In the months after the Soviet Union detonated its first atomic bomb, he advocated a "preventive" atomic attack against the communist nation. He embraced official secrecy and abhorred publicity and social activism of any kind.

Under his influence, the mathematical game theory von Neumann developed for achieving desired outcomes among competitors in economic transactions became the centerpiece in America's arsenal for plotting unthinkable conflicts between the world's counterposed atomic powers. At the RAND Corporation in Los Angeles, the prototype civilian "think tank" established by the Air Force in 1946, von Neumann became a major player in the development of the nation's weapons policies that culminated a decade later in the nuclear balance-of-terror strategy that became known as MAD—mutual assured destruction.

Wiener never criticized von Neumann personally for his aggressive activities that violated everything he stood for as a scientist, but he could not let von Neumann's theory go unchallenged. Wiener had thought through the problems of strategy from a cybernetic standpoint. He questioned the primary assumption on which von Neumann had based his highly conjectural theory. "Von Neumann's picture of the player as a completely intelligent, completely ruthless person is an abstraction and a perversion of the facts," Wiener wrote in *Cybernetics*. He knew that people made decisions for all kinds of rational and irrational reasons. They acted on emotion and often were uninformed and misinformed. Moreover, many contests involved multiple players who formed coalitions that terminated "in a welter of betrayal, turncoatism, and deception." Wiener threw pointed counterinstances at von Neumann's theory from the playing fields of everyday life. He described the common marketing and propaganda strategies that influenced the public to buy certain products and vote for particular candidates, and the mix of technical and monetary temptations that drew young scientists into the development of atomic weapons.

For Wiener, human survival was not a game. He conceded that artful forms of game playing held sway among "hucksters" and military strategists, and that they offered "only too true a picture of the higher business life, or the closely related lives of politics, diplomacy, and war." But his understanding of the dynamics of complex cybernetic systems—and of humanity's nature and purpose—had shown him the depravity and ultimate futility of von Neumann's cynical gaming approach. "In the long run, even the most brilliant and unprincipled huckster must expect ruin. . . . There is no homeostasis whatever. We are involved in the business cycles of boom and failure, in the successions of dictatorship and revolution, in the wars which everyone loses, which are so real a feature of modern times." He flatly rejected von Neumann's presumption that "people are selfish and treacherous as . . . laws of nature." "No man is either all fool or all knave," wrote Wiener. "The average man is quite reasonably intelligent concerning subjects which come to his direct attention and quite reasonably altruistic in matters of public benefit or private suffering which are brought before his own eyes."

Many people came around to Wiener's view, including some of von Neumann's greatest admirers. At RAND in 1950, the brash young math genius John Forbes Nash, Jr. arrived from Princeton and developed his alternate theory of "cooperative games" and a new concept of mixed-strategy games that progressed cybernetically in "a logical circle" toward an equilibrium point that optimized the outcome for all the players. By the early 1950s, Nash's mixed approach was winning favor with military strategists, and after a roundabout journey of his own, Nash would win a Nobel Prize for his corrections and extensions of von Neumann's theory.

———

Through those busy years of the late 1940s and early 1950s, Wiener and his new ally Walter Reuther continued to correspond and plan for their joint Council of Labor and Science, but Reuther's heavy schedule of labor negotiations was out of phase with Wiener's itinerary of domestic and international travels. Finally, in January 1952, Reuther invited Wiener to address a national UAW-CIO convention in Cleveland. It was just the kind of high-profile forum Wiener had been seeking to raise public concern for the millions of workers in American factories and offices who would soon be displaced by automated machines, but Wiener passed up the opportunity to meet labor on its main stage. In a curt letter, he turned down Reuther's invitation, citing his exhausting year in Europe and Mexico and his doctor's orders to rest. Coming only weeks after his break with McCulloch and his young colleagues at MIT, his words suggested yet again that Wiener's ailment was not his usual "fatigue" but the consuming depression and accompanying heart problems that immobilized him for months after the emotionally devastating split.

The invitation he had pursued with such vigor, and the chance to take the spotlight with Reuther and advance the cause of American workers against the onrush of automation, never came again.

After the publication of *The Human Use of Human Beings* and the wave of publicity that followed it, Wiener was looking forward to a little respite. His acts of public education and scientific statesmanship had attracted more attention than he had ever anticipated and won him many acolytes and grateful admirers. Yet, inevitably, his obstreperous statements and actions also earned Wiener his share of critics, skeptics, and, unbeknown to him, some secret detractors who would dog him for the rest of his days.

13

A Government Reacts

When a true genius appears in this world, you may know him by this sign, that the dunces are all in confederacy against him.
—Jonathan Swift

WIENER'S REBELLION DID NOT GO unnoticed by government officials.

Soon after his angry letter renouncing military research appeared in the *Atlantic Monthly*, and the day after his withdrawal from the Navy's computer conference at Harvard was reported in *The New York Times*, officials in the Office of Naval Intelligence at the Boston Navy Yard decided that the absent professor posed a threat to the republic. Ninety miles to the west, at Westover Field, near Springfield, the regional commanders of section G–2, the Army's military intelligence division, read their papers and reached a similar conclusion.

On January 10, 1947, the Boston Field Office of the Federal Bureau of Investigation received a complaint about Wiener from an official at one of those military intelligence agencies. Hours later, in a drab federal office overlooking Post Office Square in downtown Boston, an investigative file was opened on "NORBERT WIENER a.k.a. NORBERT WEINER" under the FBI's "Security Matter – C" classification designating "persons suspected of subversive activities against the Government of the United States."

Once again, Wiener had made himself too dangerous to be ignored, as he had in academic circles since the 1920s, and his rebellion sent a shudder through the government's advance guard. To those ranking military and civilian officials, Wiener's public refusal to cooperate with the development of guided missiles, or to participate in any government-funded scientific program, was plainly subversive; and there was fear that his rebellion at the Harvard conference, and at the government's "key facility" for military research at MIT, could ignite a wider insurrection and spread through the ranks of American scientists. It was a threat that brought all the government's agents and agencies together with a common purpose.

The Special Agent in Charge of the FBI's Boston Field Office—or as he was known in the Bureau's argot, SAC Boston—went to work gathering intelligence on his new security subject. A week later, he sent his first report to the FBI's director, J. Edgar Hoover, in Washington.

As it turned out, the FBI had been keeping track of Wiener's activities and affiliations since 1940, when he was first screened for civilian service in the scientific war effort. During the war, a number of confidential informants had supplied information to the FBI about Wiener, his colleagues, his family members, and his social acquaintances. One informant reported on Wiener's ethnic background and paid high praise to his professional achievements, advising the government that "subject . . . was of Russian-Jewish extraction, and is allegedly second only to Einstein as a mathematical analyst." That endorsement was corroborated by a copy of the *Boston Traveler* article, which dispassionately informed the director, in testimony taken verbatim from the *Atlantic Monthly*, that "Dr. Wiener is a scientist whose ideas are credited with playing a significant part in the development of the theories of communication and control which were essential in winning the war."

SAC Boston had some other tidbits. He copied the entry on Leo Wiener in the *Encyclopedia Americana*, duly noting that "subject's father was . . . born in Russia . . . educated at Minsk," and that "his greatest editorial production was the 24-volume compilation of Count Tolstoy's Works." And he relayed verbatim the report of another wartime informant who advised the Bureau in 1941 that:

> There were two very strong "Communists" whom informant thought were possibly doing work for Germany prior to the Russian entry into the war. . . . These individuals . . . were NORBERT WIENER . . . and DIRK VAN DER STUCK [*sic*], whom informant stated was a Russian Communist who had entered this country from Czechoslovakia.

The charge that Wiener was a "very strong Communist" was groundless, although that claim was accurate with regard to Dirk Struik. Like many European and American intellectuals in the first half of the twentieth century, Struik was inspired by the Russian Revolution and embraced communism during the economic upheavals of the Great Depression, and he remained an ardent adherent of the Dutch Marxist school. The Bureau's informant was woefully misinformed about Struik's national origins and party affiliations, but the false report gave added credence and urgency to the allegations against both Struik and Wiener.

A string of confidential sources had kept the government apprised of Wiener's extracurricular activities throughout the war, including his participation in professional, humanitarian, and civil liberties organizations that worked to aid stricken peoples and oppose authoritarian actions at home and abroad. Those groups, all of which the FBI had deemed to be "subversive," were the American Association of Scientific Workers, the National Federation for Constitutional Liberties, Russian War Relief, the American Friends of the Chinese People, and the Joint Anti-Fascist Refugee Committee.

Other tenuous indictments and insinuations followed after the war, as the FBI went fishing for new evidence against Wiener and his associates. SAC Boston learned from one informant that a group of Russian astronomers who were in Cambridge touring the Harvard Observatory "had expressed a desire to talk to Wiener" and have a look at MIT's "calculating machines," and that "an appointment was made for Wiener to be visited . . . at MIT." Another informant reported that "a member of the Communist Political Association in Cambridge . . . was a protégé of . . . WIENER." A third informant brought back some information that appeared to be far more damning. That source, who was present at the local Communist Party headquarters several weeks before, reported that the notorious British communist and biologist J. B. S. Haldane had been in Boston to deliver a series of lectures on "The Relation of Marxism and Science," and that Haldane had stayed as "a guest of Professor . . . WEINER [sic]." The FBI never learned that Wiener and his old Cambridge chum had committed nothing worse than the radical act of singing at the dinner table.

During those years a virulent anticommunist fervor spread across the country. As the Cold War with the Soviet Union intensified, ranking members of the House Un-American Activities Committee in Washington took steps to root out alleged communists and their "fellow travelers" from government, academia, and the entertainment industry. Members of Congress called for federal prosecution of Albert Einstein, who had been a subject of concern to the FBI since 1932 for his pacifism and other ostensibly subversive political beliefs. Many respected scientists, including the director of the National Bureau of Standards, the head of the Harvard Observatory, and the editor of *The Bulletin of the Atomic Scientists*, were stripped of their security clearances, banned from government research projects, and even fired from their jobs or pressured to resign.

The fever of red-baiting spread to state and local governments, and, in April 1947, the Massachusetts Legislature proposed to hold its own hearings on "subversive activities within the Commonwealth," and to bar members of subversive organizations from working for the state. Wiener was appalled by the witch hunts, and he signed his name to a statement of protest cosponsored by the state's Civil Liberties Union, the Council of Churches, the League of Women

Voters, and "70 of the most distinguished citizens of Massachusetts." That act of protest, too, was duly noted in Wiener's expanding FBI file.

But the Bureau was more urgently concerned about Wiener's continuing condemnations of atomic weapons and America's military-industrial establishment. For two years after his outcry in the *Atlantic Monthly*, a heated debate over his policy of noncooperation ensued in *The Bulletin of the Atomic Scientists*. In August 1947, when physicist Louis Ridenour, a leader of the Rad Lab's wartime radar projects, blasted Wiener's stance in a forum in the *Bulletin*, Wiener fired back a shot that made the FBI see red. He wrote:

> The Armed Services are not fit almoners for education and science. They are run by men whose chief purpose in life is war, and to whom the absence of war, even though a war is almost certain to engulf them personally, is a frustration and a denial of the purpose of their existence. . . . To say, as Dr. Ridenour does, that "the moral issue does not involve the technique or equipment with which a war is fought," is completely specious. . . . A weapon which [absolves scientists] of the responsibility for needless and indiscriminate slaughter; a weapon which blackens our name in the eyes of the world; a weapon which gives the control of the entire destinies of our countries into the hands of a few people who know no control; a weapon which has no riposte other than the counter-slaughter of our civilian population; such a weapon is a bad weapon—technically, strategically and morally. . . .
>
> To give the atomic bomb to those who profess to be able to come to a decision as to whether to use it in no more than two weeks after its first trial, does not differ in its wisdom from putting a razor in the hands of a five-year-old child!

More reports of suspicious activity fluttered into the FBI's file. An informant overheard a woman with "a distinct New York Jewish accent" arrange for Wiener to meet a visiting scientist from South America. Madame Irène Joliot-Curie, the Nobel prize-winning French chemist and communist sympathizer, whose research was important in the discovery of uranium fission and who had become an outspoken advocate for the abolition of atomic weapons, came to America to raise funds for the Joint Anti-Fascist Refugee Committee, and "Prof. NORMAN WEINER" [sic] was invited to sit at her table and speak at dinners in her honor in Boston and New York. Then, in November 1948, just weeks after the publication of *Cybernetics*, Dr. Hewlett Johnson, the notorious "Red Dean" of England's Canterbury Cathedral, who initially had been refused a visa to enter the U.S. because his invitation came from the National Council of American-Soviet Friendship, was finally granted a visa and "Prof. ROBERT WEINER" [sic] joined his welcoming committee.

Each new screed and public petition he signed went into the FBI's file, along with the informants' reports on each controversial gathering he attended. But Wiener was not the only person at Tech who had thrown the feds into a tizzy. The institute's faculty, students, and even its president were under suspicion. The math department in particular was alleged to be a hotbed of communists, and with some justification. Along with Dirk Struik, Wiener's prized student Norman Levinson, then the department's vice-chairman, and the department's chairman, William Ted Martin, had been active in local communist groups during the war, though both had cut their ties to the movement when the war ended.

Federal agents on campus did not openly pressure anyone to give testimony against a friend or colleague. They didn't have to: The FBI had spies all over the place. Dozens of local citizens, communist party infiltrators, and at least five MIT professors, administrators, and secretaries fed information to the FBI about Struik, Levinson, Martin—and Wiener. The Bureau's informants had no evidence that Wiener was a communist, or that he had committed any illegal act against the government. Instead, they relayed the many unflattering descriptions of Wiener that were circulating around MIT: that he was "extremely erratic," "naïve politically," "a complete egotist," and "a screwball." One unidentified informant suggested that the Bureau should be "worried about Wiener, as 'he is known to be nuts.'"

Another informant made more serious accusations. On October 4, 1948, two weeks before the publication of *Cybernetics*, the FBI received a memorandum from the Chief of Naval Intelligence reporting that, while attending a social gathering in Worcester, Massachusetts, Wiener had made statements "indicating a disloyal attitude toward the United States." The Navy's intelligence chief conveyed the details described by his eyewitness on the scene, who charged Wiener with openly expressing distrust of the armed services, and with the "vilification of all scientific personnel who have prostituted themselves . . . to increase the military potency of the nation." Wiener also accused the government of "desecrating humanity" by its use of the atomic bomb, and of promoting "definite militaristic and fascistic national tendencies." The source cited as an example Wiener's "unfounded allegation that under the Paperclip program alien scientists who were Nazis were cleared for entry with greater facility and in larger numbers than those who were not Nazis."

As an American, Wiener was entitled to his opinions and, as it turned out, he had his facts right. Somehow, he had learned about the government's postwar Project Paperclip, in which federal agents were compulsively cleansing the dossiers of dubious European scientists. Over the next decade, the secret intelligence operation would grant U.S. citizenship to hundreds of German scientists and others who were known to have been loyal Nazis, and help them to find positions in the American scientific community, in the hope of gaining valuable intelligence and technical advantages over the Soviet Union in the Cold War.

The Navy chief spread word of Wiener's statements to the other military intelligence services and to the nation's newly formed civilian intelligence service, the Central Intelligence Agency. His memo emphasized that Wiener had refused to cooperate with the government's guided missile program. And it contained one more unnerving piece of intelligence:

> The same source reported that WIENER had told him he would commit suicide the first day of the next war, as he realized that "his attitude would not be tolerated."

Word of Wiener's outburst passed down through the government's chain of command and was brought to the attention of MIT's president at that time, Karl T. Compton. Compton assured the government's agencies that he was well aware of Wiener's emotional instability but that "he felt that Dr. WIENER is completely harmless," and that his statements in Worcester were the result of his ongoing reaction to the military's use of the Bomb during the war. He promised to speak to Wiener "at the first appropriate occasion and point out to him that utterances which might be interpreted as subversive reflect in a derogatory manner upon MIT."

Satisfied that the threat to the nation had been dealt with and neutralized, late in October 1948, the FBI closed its Security Matter—C investigation of Wiener. But the case would not stay closed for long.

———

The Cold War was escalating abroad, and America's pursuit of its elusive "enemy within" was only beginning. In August 1949, the Soviet Union detonated its first atomic bomb years ahead of the American military's projections. In January 1950, the German-born British physicist Klaus Fuchs, who had worked at Los Alamos during the war, was arrested in England and charged with supplying atomic secrets to the Soviets. A week later, Senator Joseph McCarthy of Wisconsin began his crusade to purge the country of suspected communists in government and throughout the culture. Soon after, Congress passed an Internal Security Act that required members of the "world Communist movement" to register with the government, barred entry into the United States of communists or members of any "totalitarian organization," and permitted the imprisonment of anyone who was considered to pose a danger to "the territory and people of the United States."

As McCarthy's campaign unfolded, J. Edgar Hoover and his G-men redoubled their efforts. Julius and Ethel Rosenberg were arrested in New York City and charged with passing atomic secrets to Soviet agents during the war. The government also stepped up its campaign against Albert Einstein. FBI agents illegally opened his mail, tapped his telephone, and solicited testimony against him from mental patients and Nazi sympathizers; while the Immigration and Naturalization Service, working hand-in-hand with Hoover, took steps to have Einstein stripped of his naturalization and deported.

The thickening air of suspicion dropped a new cloud over Wiener. Suddenly, his name became a hot topic again within the FBI, and it grew hotter when he unleashed a new tirade against the military and decried the government's tightening grip on the lives of American scientists. In a speech to a national scientific conference that was reported in *The New York Times* and promptly inserted into Wiener's FBI file, he called on American scientists "to resist the inroads of 'tyranny' in this country which seeks to make them 'the milk cows of power.'" Wiener was not cowed by Congress, the military, or the FBI, or by the controversy his remarks sparked in the media. He went about his business undaunted. Then he made a move that really rattled J. Edgar Hoover.

In the fall of 1950, soon after *The Human Use of Human Beings* appeared, Wiener embarked on his long sabbatical to lecture in Paris and Madrid, extending the reach of his science and social concerns into Europe. During his time on the Continent, he had made plans to see Haldane in London. With so many spy cases against American scientists prosecuted and pending, word of Wiener's planned departure with his family set off alarms in official channels on both sides of the Atlantic.

The fuss began when the FBI received a report from one of its informants who watched over the ports in New York City stating that "WIENER or a member of his family will shortly be going to England to visit with the well-known British Communist, J. B. HALDANE." The report also revived two wholly refuted charges that "WIENER, the . . . famous scientist, will freely admit that he is a Communist and has been over a long period of years," and that "This man . . . has access to some of the top secrets in the country." When Hoover received the news, he sent a new order to SAC Boston "TO REOPEN THIS [MATTER] AND CONDUCT THE NECESSARY INVESTIGATION." The same day, Hoover dispatched the first in a series of secret cables to the Justice Department's attaché at the U.S. Embassy in London firmly recommending that "British Intelligence Authorities" be advised of Wiener's pending visit to Haldane.

Soon after, Hoover's man in London made contact with the British security service, MI-5. At the U.S. Embassy's request, His Majesty's Secret Service tracked Wiener's movements around London, starting the moment he arrived from France in mid-January 1951 and met up with his wife and daughter, through his visits with Haldane at the University of London in Tavistock Square, until Wiener returned to France with Margaret three weeks later. But all the king's men were unable to piece together the purpose of Wiener's meetings with his learned colleague Haldane, who was always a communist but never a spy. Neither MI-5 nor the FBI had the foggiest notion about the long-running conversation between the two men on matters of cybernetics and genetics, or just how low the communist struggle for world domination ranked on their scale of chat and good cheer.

MI-5 had more luck solving the mystery of Peggy Wiener's dealings with Haldane in London. Two months after Wiener left his daughter in Haldane's care, MI-5 learned from British immigration officials that Peggy had been admitted to England on a student visa to study genetics in Haldane's laboratory. Finally, in April 1951, after scouring their sources clean, MI-5 informed the U.S. Embassy's attaché, who in turn informed Hoover, that "None of the WIENER family have come to our notice in connection with communist activities up to the present date."

———

Despite that reassuring report, Hoover continued to press the British government for intelligence on Wiener's movements in Europe. In June 1951, when Wiener left France for his politically repressed lecture tour of Spain, MI-5's offices in London, Paris, and Madrid were mobilized, separate from the surveillance the Spanish police were conducting, to track the man "reported to be one of the world's foremost mathematicians [who] has refused to conduct research for the Army and Navy . . . and . . . has access to some of the top secrets in the United States."

Then, just when the FBI was feeling confident that they had all Wiener's moves in check, he vanished. During those nomadic summer months, while he was holed up on the French seashore and in the countryside, agonizing over his childhood memoir and suffering from the "racking headache" that required him to be hospitalized in Switzerland, Hoover's agents, attachés, and foreign associates combed Europe and America searching for him to no avail. They had no idea that Wiener, depressed and exhausted, had returned to the United States with his wife on a steamship from Genoa and promptly headed south for his biennial semester of research with Arturo Rosenblueth in Mexico.

Through the summer and into the fall, while Wiener was dining out in Mexico City restaurants popular with the "foreign colony" and, on at least one occasion, observed wailing uncontrollably at lunch with his young psychiatrist friend, Morris Chafetz, the FBI puzzled over Wiener's disappearance and worried about his suspected subversive activities. Then, late in October, word reached Boston from an FBI outpost on the border in Galveston, Texas, that a reliable informant had encountered Wiener and Rosenblueth in Mexico and that "they displayed anti-American tendencies."

At last, Hoover had found Wiener's trail. He sent new orders to the State Department, and to the CIA's assistant director for special operations, to advise him of any information concerning Wiener's activities in Mexico and his likely date of return to the United States. The same day, a long memo was dispatched from Washington to other government offices providing Wiener's biography and travel records since 1914, his physical description ("Marks: Scar on left forefinger"), and a blow-by-blow account of his alleged subversive activities and disloyal statements beyond America's borders.

Once again, Hoover's panic and paranoia were misplaced, but the FBI did not learn that until March 1952, when its informants at MIT confirmed that Wiener had returned from Mexico without incident two months earlier. During those months, and for half a year before, the FBI was oblivious to Wiener's real activities: to his slow recovery from the depression he had fallen into in Europe, to the spate of painful letters he had received from publishers rejecting his autobiography, to Margaret's intrigues and the torrent of emotions that swept over him in Mexico and precipitated the split with McCulloch and his boys at MIT—and to one other act on Wiener's part that no doubt would have been of great interest to the FBI.

On September 12, 1951, Wiener's friend Dirk Struik was indicted in Boston under Massachusetts' new sedition law for conspiring "to advocate, advise, counsel, and incite the overthrow by force and violence of the Commonwealth." When he got the news in Mexico City, Wiener sent a tough letter to Karl Compton's successor, MIT's President James R. Killian, supporting Struik and threatening drastic action if MIT did not stand by him. Wiener wrote:

> I know Struik to be a person of the highest character and honesty. . . . He has neither the personality nor the intentions of a conspirator. . . . If . . . his relations with M.I.T. suffer, unless there is far more damning testimony against him . . . I shall regretfully be forced to submit to you my resignation from M.I.T.

Wiener's fire this time was not the spinoff of a moodswing or emotional storm but a deeply felt act of conscience, and his most serious threat to MIT to date, as he laid his own job on the line in Struik's defense. He spoke warmly of MIT to Killian as "a place which I honor and love. . . . Nevertheless, where I consider there to be unjust pressure, there must be counter pressure, and the fact that this involves possible loss to the Institute and the termination of my own career does not constitute a sufficiently countervailing factor to enable me to do otherwise."

Struik saw the witch hunt coming. Two years earlier, he had refused to testify before the House Un-American Activities Committee. When the indictment came, he was not surprised or even all that upset. "Wiener was far more upset than I was," he recalled. And Killian heeded Wiener's threat. The MIT administration decided to suspend Struik from teaching but to pay him his full salary until the case was resolved. Wiener never told Struik what he had done—he learned about it after the fact—but Struik believed Wiener's letter saved his job and his livelihood during those treacherous years.

Wiener was not home free himself. In the summer of 1950, even before Struik was indicted, he wrote to his psychiatrist Janet Rioch in New York, describing his anxiety over "the current political situation, in which the informer seems to be running wild throughout the whole structure of our society." He worried about the heat he might take from the views he was expressing so forcefully in *The Human Use of Human Beings*, especially his diatribes against American capitalism and corporations. "You know how much red pepper it contains," he punned, "and how . . . I shall find myself simmering above a brisk fire. . . . I wonder what the McCarthys are going to do?"

His worries were justified. While Wiener was in Europe, the red-baiting fever rose by several degrees. In March 1951, the Rosenbergs went on trial and, a month later, they were convicted of espionage and sentenced to death. The House Un-American Activities Committee opened its second round of hearings, and McCarthy served subpoenas on some of the nation's most respected writers and performing artists demanding that they "name names" of other reputed communists. Hundreds were consigned to industry-wide blacklists that prevented them from working in their fields. The deteriorating domestic situation took down more scientists and scholars who were named to a blacklist of another sort: the government's secret "Security Index" of thousands of known and alleged communists. They, too, faced dismissal from their jobs, exclusion from government employment—including work at universities that received government funds—and arrest and detention in the event of any loosely defined national emergency.

Wiener was afraid he might be next, and those fears compounded the stresses with which he was already contending. Before he returned to the United States after his hospital stay in Switzerland, he had sent a troubled telegram to the Dean of Students at MIT: "RETURNING NYC . . . GREATLY FATIGUED NERVOUS OVER WORK ALARMED AMERICAN WITCH-HUNT PLEASE WIRELESS ESTIMATING MY RISK." President Killian dispatched the Institute's assurances personally: "after consultation we feel you have no need for anxiety and welcome you home."

A month later, once he had returned to Boston and made his way safely to Mexico, Wiener wrote another dark letter to Killian. "These are extremely difficult times," he said. "One wonders what the outcome will be and how soon the state of mind of the world will return to sanity. In the meantime one feels as if one had entered into an enormous ramifying cavern like the Mammoth Cave without any clue or light to guide one."

The hysteria grew year by year. In January 1953, soon after his inauguration, President Dwight D. Eisenhower publicly questioned whether communists should be allowed to teach American students. Hoover ordered the FBI's field offices to investigate and, in April, he informed the Congress that he had found widespread evidence of communist infiltration in higher education. In the inquest that followed, MIT's math department came under withering fire. The department's chairman and vice-chairman, William Ted Martin and Norman Levinson, were called before the House Un-American Activities Committee to confess their membership in the Communist Party a decade earlier, and to account for the "abnormally large percentage of communists at MIT."

"MIT was turned topsy-turvy," Fagi Levinson remembered. "There was strong pressure to name names." Martin "gave a pathetic, frightened performance" before the House committee, but Levinson held his ground. The ordeal left Martin "shattered and deeply depressed," while Levinson's teenage daughter suffered a nervous breakdown caused—at least partly, her parents claimed—by the FBI's harassments.

Wiener watched the whole torturous episode in his own private agony and gave strength to his cherished colleague and protégé. "Wiener was incredibly loyal," Fagi Levinson remembered. "When Norman was called before the committee and refused to name names everyone thought he was a damn fool, but Norbert understood that he couldn't live with himself." Other scientists aggressively opposed the congressional inquest. Einstein implored subpoenaed witnesses to refuse to testify and to choose "the revolutionary way of non-cooperation," but Wiener did not feel that secure—and now with good reason. McCarthy's witch hunt was turning in his direction.

In June 1952, months before his MIT colleagues were subpoenaed, McCarthy and Hoover put Wiener in their crosshairs. Hoover's request to SAC Boston for new evidence that would justify Wiener's inclusion on the Security Index turned up nothing. SAC Boston, knowing Wiener's unpredictable personality and penchant for public outcries, cautioned Hoover about a different kind of danger Wiener could pose to the offices of the inquisition. "This individual has been described as a genius in the field of mathematics but a person who is extremely erratic. . . . The establishment of a security index card on the subject is apparently not warranted . . . and because of Wiener's personality, an interview at this time does not appear practicable."

But that did not deter Hoover or McCarthy. In December 1953, just as MIT's new Lincoln Laboratory was beginning its top-secret project for the Air Force to build the nationwide SAGE automated air defense system for tracking incoming Soviet bombers, Wiener's name was put on McCarthy's new hot list of suspected subversives at Tech who might be called to testify before the Senate Committee on Government Operations, which he chaired. In preparation, Hoover sent yet another urgent message to SAC Boston directing him to review Wiener's file one more time. He gave the Bureau's local "Security Squad" little room to maneuver and two weeks to "report . . . and if justified, recommend either Security Index card or an interview." At the same time, Hoover alerted the Air Force's Office of Special Investigations, the Army's section G-2, and "other interested intelligence agencies" that Wiener was again under suspicion.

SAC Boston left no dossier undisturbed. Files on dozens of individuals and groups were searched for any hint of new subversive actions on Wiener's part, including the file the Bureau had opened on Wiener's daughter when she went to England to study with Haldane. Within hours, SAC Boston's Security Squad confirmed the well-known fact that Wiener had "no association" with "Project Lincoln" or "with any U.S. government contract research." But Hoover still was not satisfied. Early in January 1954, days before the McCarthy committee's deadline, Hoover squeezed SAC Boston again: "In view of the possible interest of Senator McCarthy . . . and the possibility that Wiener may be the subject of a subpoena to testify concerning 'Project Lincoln' . . . consideration should be given to either recommending a Security Index card for the subject or to requesting authority for an interview with him."

Three days later, SAC Boston filed his final report. There were scant pickings for the McCarthy committee. One confidential informant had caught Wiener attending a party gathering at Dirk Struik's home the year before—a Christmas party. The report detailed the futile efforts of local communists for more than a decade to interest Wiener in their activities. A Boston informant stated that "on numerous occasions in the late 1930's and early 1940's he had attempted to interest Wiener in Communism and . . . recruit him into the Communist Party, but was completely unsuccessful." In an advance telegram rushed to Hoover's attention and an accompanying cover memo, SAC Boston took pains to impress upon Hoover: "no unusual subversive activity noted on part of Wiener since last report," and he reaffirmed his earlier assessment that, "Because of Wiener's temperament and attitude, it appears . . . an interview with him would be unwise at this time."

Then, as if none of that mattered, SAC Boston told Hoover the words he longed to hear:

> After further consideration of the material available on WIENER, it has been decided to prepare a summary report on him and recommend him for inclusion on the security index, which will be done in the immediate future.

As before, Wiener had no idea that a new review of his loyalty and activities was in progress, or that he was a candidate for a subpoena to appear before the McCarthy committee. And, again, even at that moment, the FBI had no idea where Wiener was. In fact, he was 7,000 miles away from MIT appearing before the All-India Science Congress in Hyderabad, which he had been invited to attend as an honored guest of the Indian government. At that gala international conference, as luck would have it, Wiener found himself consorting with a large delegation of prominent scientists from the Soviet Union.

The Soviet delegation had come to show the Indians "the cream of Soviet science." Their core group was accompanied by a staff of aides and translators from the Soviet embassy in New Delhi who were, in fact, agents of the Soviet secret police assigned to insulate the Soviet scientists from Western European and American influences, and to keep them from disclosing any Soviet scientific secrets. In the accommodations his hosts had arranged, Wiener was lodged with the Soviets and dined with them at the same table. Well aware of the tensions between their two countries, and in keeping with the spirit of international science, Wiener tried to create a cordial atmosphere and he found that the Soviet scientists responded in kind, conversing with him freely in English, and making no effort to propagandize for Soviet government interests or to press him for information on sensitive matters of American science. However, the Soviet scientists were not so free around their government chaperones, who forced them to speak only Russian, and who put a damper on the entire mood of the congress. Frustrated by the abrupt turnaround, Wiener took to "needling" and "ribbing" his Soviet colleagues for their sudden restraint, and for their reluctance to mix more freely with other scientists and their Indian hosts.

But Wiener was not feeling all that free himself. Back home, in the months before his departure, more scientists had been subpoenaed; the Rosenbergs were executed. And, just days before he left for India, the renowned physicist J. Robert Oppenheimer, who had directed the Manhattan Project and was then serving as chairman of the advisory board of the Atomic Energy Commission, was labeled "a hardened Communist" and "more probably than not [a Soviet] espionage agent" and called before a special security hearing at the AEC on charges of treason. With that perilous political climate awaiting him, Wiener suddenly became fearful for his good standing with his own government and afraid that his interactions with the Soviets would be judged harshly by the witch-hunters. Just how fearful he had become could be seen in his highly uncharacteristic decision to make a preemptive move to reaffirm his allegiance and unwavering loyalty to the United States.

On January 11, 1954, three days before the FBI recommended him for a spot on its Security Index, Wiener traveled from Hyderabad to Bombay, where he telephoned the U.S. Consul General's office and requested a meeting to explain his interactions with the Soviet delegation at the international science congress. The next day, he made his way through Bombay's teeming streets to the American Consulate, where he spoke at length with two senior consular officials. Two weeks later, the FBI and CIA received their copies of the Consul General's report to the State Department in Washington.

The report described a nervous, contrite Wiener testifying to innocuous offenses at Hyderabad and playful gibes aimed at the Soviet security police—whom Wiener referred to as the "pug-uglies":

> The professor said . . . representatives from the Soviet Embassy . . . appeared upon the scene shortly after the conference was convened. It was clear to everyone, said the professor, that these "pug-uglies" as he termed them, came . . . for the obvious purpose of policing the Russian delegation. . . . The professor . . . decided early during the Conference to try and needle the Russian delegates whenever an appropriate occasion arose. For instance . . . Professor Wiener, who speaks Chinese, told [one] Soviet delegate, "You should really learn Chinese, you know. It is the language of a very important people." Professor Wiener jokingly referred to a woman Soviet delegate who wore a particularly drab and unattractive hat. [He] remarked that . . . the hat sat level on her head and that it was at least a proper hat politically speaking having neither a "right" nor a "left" deviation.
>
> Professor Wiener was very obviously concerned lest his having spoken with the Russian scientists and having, to a certain degree, fraternized with them, be misconstrued. From time to time during the interview he paused and asked, "Don't you believe I did right?" or "Was not that the proper course of action?"

Word of Wiener's walk-in statement at the American Consulate in Bombay did not reach Washington before the FBI made its decision to place him on its Security Index—a decision he did not know was pending—but it accrued to his credit with the Bureau. Two months later, after he returned from India, the FBI took a fresh look at Wiener's file and abruptly reversed its earlier recommendation. Several new pieces of information scored points with Hoover and SAC Boston: a passage from *Cybernetics* the Bureau happened upon in which "WIENER mentions how he hopes to perfect artificial limbs for amputees," a worthy goal that in the Bureau's view gave Wiener some redeeming social value; testimony from Wiener's childhood memoir, which had finally found a publisher the year before, in which "the subject mentioned . . . how his father, LEO WIENER . . . had been desperately opposed to the Communists from the beginning"; and the report of Wiener's act of contrition in Bombay, which had finally filtered through the Bureau's pipeline.

One stray datum of information may have been the clincher that exposed the absurdity of the government's case against Wiener and confirmed the view of MIT's president six years earlier that Wiener was "completely harmless." Another informant of some importance had come forward and "indicated that he also felt that Wiener was harmless. . . . He had heard a story . . . that on one occasion Wiener drove to Pittsburgh to attend a meeting and forgetting his car he returned on the train. Not finding his car at home, he then had reported it to be stolen." That new twist on the old tale of Wiener's wayward road-and-train trip to Providence, along with the other new items the FBI had amassed, proved conclusive for SAC Boston:

> In view of the recent data developed, it does not seem justified to place [WIENER] on the Security Index and accordingly, unless advised to the contrary by the Bureau, no further action will be taken in relation to him.

The fever began to break in December 1954, when the U.S. Senate censured McCarthy for his conduct and his committee's witch hunts crumbled, but not before more damage was done to scientists across America. Two months after Wiener's reprieve, following half a year of hearings at the AEC, J. Robert Oppenheimer was found not guilty of treason but nevertheless had his security clearance revoked and was forced to resign from his post at the Commission. Albert Einstein was never charged for his pacifist activities or subpoenaed by any congressional committee, but he remained under investigation by the FBI until his death in 1955. Dirk Struik fought his criminal sedition charge in Massachusetts for four years, until the U.S. Supreme Court ruled in another case that all state sedition laws were unconstitutional, and the indictment against him was quashed. That time was "half reminiscent of Nazi Germany, half of *Alice in Wonderland*," Struik said, of his personal ordeal and the McCarthy era overall.

Perhaps the greatest damage was done to American science itself. Emboldened by the near monopoly of government and military funding for basic research in the early postwar years, the witch-hunters ran roughshod over the intellect and infrastructure of science. Leading scientific organizations buckled under the pressure and became silent in the face of the political hysteria. The National Academy of Sciences and other associations were so thoroughly integrated into the government's Cold War enterprise, and fearful of being labeled subversive organizations themselves, that they did little to defend their members against the McCarthyite siege. Many upstanding scientists and scientific groups recanted their public positions in closed-door negotiations that only encouraged the inquisition, while others eagerly served the military-industrial complex.

Wiener was one of the exceptions. As he made clear in the most public forums, he had no interest in obtaining government contracts or funding. He had no secret information to disclose, no security clearances to lose, and the witch-hunters had no business hunting him in the first place. As his colleagues had known for decades, Wiener was not a joiner, and he made only tenuous alliances even to advance the new science he founded. But Wiener was no passive witness to the political passion play going on in his midst, or to America's emerging role as a global superpower given to its own ideological extremes.

In the summer of 1950, as the Cold War fervor was building, Wiener gathered his thoughts on the subject in a letter to his new friend Walter Reuther. The letter demonstrated his keen understanding of the science and politics of that precarious new mode of warfare, in which hostilities were waged by both sides using strategic games played with unprecedented weapons of mass destruction, sweeping propaganda campaigns, and ruthless suppression of internal political debate.

> I do not think that the average American has much idea of the difference between the present conflict and all others in which we have been engaged. . . . In all previous foreign wars . . . we have not been the main object of the hostility of our enemies. . . . This was . . . true of the European part of the second World War [and] Japan . . . was too far removed from us . . . to threaten our homeland seriously at any time. [But] the good will of the . . . world is not indifferent to us. . . . Let us do nothing to hurry on Armageddon. . . . Even though we can look to the democratic countries only for . . . limited . . . military help, it is far from a matter of indifference to cultivate their good will. . . . Certainly there is nothing to be said for the sabotage which we have applied to all countries which . . . have accepted any form of socialism. We must also avoid the backing of discredited and discreditable regimes. . . . We must show enough cooperative interest in the problems of economically backward . . . countries . . . to make [them] feel that we can give them a more promising and more secure future. . . .
>
> If we do these things . . . we shall . . . have a good fighting chance to survive in a world fit for us to live in. . . . If however we fail to realize that we can win the world only by accepting as ours the interests of the world, moral as well as material, then we shall perish, as we shall deserve to perish.

The witch hunt of American scientists was just one of many grim scenarios Wiener had envisioned from the moment the postwar marriage of government and science was consummated. Unlike his peers who were bound by contract to the government and those who came compliantly before Congress, Wiener's willingness to speak out, forcefully and eloquently, about the perils of the government's oppressions, the hazards of megabuck science, and the immorality and inhumanity of atomic weapons made him an adversary the demagogues in government had every reason to fear.

In the end, Wiener was not too dangerous to be ignored—the distinction he once strived for and prided himself on achieving. Now he was too dangerous *not* to ignore and, for once, too dangerous to engage. His unyielding independence of mind served him well when the McCarthy era descended. But it would make his walk in the world a solitary one in the years that followed.

14

Wienerwalks III

Werther often slowed his brisk pace and often, stopping short,
seemed to consider going back; but again and again he directed
his steps forward.

—J. W. von Goethe, *The Sorrows of Young Werther*

Through the 1950s, Wiener kept up his counteroffensive against the government as he walked his daily rounds along Tech's corridors. Although no one at MIT joined his boycott of government-funded research, everyone knew about it or soon learned when Wiener approached their perimeter.

The clash of wills made for some unlikely scenes on Wiener's wegs. "Ulam was here at the time," Jerry Lettvin recalled, speaking of the Polish-born mathematician, Stanislaw Ulam, who had worked with von Neumann at Los Alamos and who devised the statistical sampling method the two theorists used to model atomic explosions on computers. "Wiener liked to talk to Ulam, but he refused to go into his office because, he said, it was 'federal territory.' So their conversations were conducted with Ulam seated in his office and Wiener standing at the threshold holding forth."

His brisk pace was slowed by his increasing age and weight, and by a conspicuous decline in his spirit that set in around the time of the break with McCulloch and his boys. The loss of his most creative collaborators, and Margaret's accusations against them, had hurt Wiener deeply. At the same time, his growing notoriety as a moral force on the public stage had left him feeling even more alone on his home turf. Many of his colleagues at Tech viewed him not just as an eccentric but as an embarrassment, and as a remote but increasingly unpredictable threat to their government and corporate contracts.

For two years after he returned to MIT from Mexico in January 1952, Wiener was distressed by the breakup of his inner circle, and for much of that time, he was deeply depressed.

Benoit Mandelbrot, the young French mathematician, saw Wiener in that state in the summer of 1952. After earning his doctorate from the Sorbonne, Mandelbrot came to the United States, where he hoped to work with Wiener, who had long been one of his role models. But the man Mandelbrot met at MIT was not the buoyant mind and blithe gourmand of legumes he had dined with in Paris only a year before. "He was not focused at all and he was extraordinarily unhappy," Mandelbrot recalled. "He would come to class and say a few things and then forget himself. When I arrived he said, 'Let's work together.' He was very enthusiastic. We worked on the blackboard for a while, and then suddenly he said, 'Well, I'm sorry, I'm very tired,' and he disappeared."

At the time, Wiener was in the worst of his distress, but Mandelbrot had no way of knowing what was afflicting him. He tried to make their arrangement work, but he saw that Wiener's troubled state was becoming a burden to many people at MIT. "It was difficult. He was very depressed, and it got worse," said Mandelbrot. "At one point, I was having lunch at the faculty club and Wiener entered and my host turned his head away. Everybody turned their heads away. They were avoiding him. I realized how isolated he was and how difficult it was for him to work with anybody. I stayed there for a month until I saw that nothing could come of it. It was disappointing. I would have liked to have done something with Wiener but I was not able to."

Wiener curtailed his speaking schedule for more than a year after the break and scarcely traveled beyond Boston and the quiet of his country home in New Hampshire. His professional output plummeted. In 1952, for the first time since he was freed from the restrictions of wartime censorship, he did not publish one mathematical or scientific paper.

Instead of doing science, Wiener took refuge in his memoirs. Soon after he returned from Mexico, he had received an offer to publish his anguished autobiography, *The Bent Twig*. The bid came straight from Henry W. Simon, of New York's Simon and Schuster publishers, and it was contingent on Wiener's rewriting the manuscript extensively to make it more accessible and appealing to the general public. Months of exhaustive interactions followed, as Simon picked over Wiener's pages, noting the usual omissions, inconsistencies, and more than a few "animadversions" and indignant comments about figures from Wiener's early years that Simon deemed pointless and wholly "dispensable."

The finished work, refocused and retitled, *Ex-Prodigy: My Childhood and Youth*, was a testament to Wiener's prominence and the human interest he had long felt his rarefied boy wonder story warranted. The book was well reviewed, featured in *Time* and *Newsweek*, and in March 1953 it earned Wiener a guest spot on the fledgling NBC television network's new morning program, the *Today* show, as he ventured out on his first public appearances in America in nearly three years. *"Magnificent!"* Simon wrote exuberantly to Wiener of another interview he tuned in a few days later. "We are all proud of the real trooper we have uncovered on our list of authors."

Not everyone was thrilled with Wiener's latest literary achievement. His caustic account of his intellectual upbringing under his father's iron-fisted tutelage, and of the spiritual turmoil caused by his mother's decision to raise her Jewish children as gentiles, was Wiener's final act of rebellion against his parents—and one royal spanking of his elders by an Elephant's Child who never forgot. The memoir was especially hurtful to his mother, who was still alive when the book was published. But Wiener's persisting need to exorcise his childhood demons, even as he approached the age of sixty, took precedence over family diplomacy and social niceties. And Wiener had a higher motive for his memoir: He hoped to explain to his admirers and detractors alike the confluence of personal and social forces that had made him such an extravagantly bent twig, and to bear witness to the undreamed-of pressures the world imposes on all prodigies.

Margaret Mead got the message of Wiener's coming-of-age tale. "The story is shot through with pain," she wrote in an academic journal, yet Wiener "achieves an almost Olympian objectivity and gentleness." It was his favorite review, and the first of many that acknowledged Wiener's heartfelt plea for greater understanding of, and compassion for, nature's "fabulous monsters."

The success of *Ex-Prodigy* brought Wiener some relief from the loss of his closest colleagues, but it brought no reprieve for McCulloch and his boys in the RLE. In time, they came to accept Wiener's banishment as final. Their work went on, but without Wiener to light the way, and a cloud hung over the neurophysiology lab and everyone in it.

After the breach, McCulloch stopped all laboratory experimentation, effectively ending his thirty-year reign as a pioneer of modern neuroscience research, and, instead, immersed himself in the discrete logic of digital models and abstract brain mechanisms. His sponsors at MIT stood by him, but they never endowed him with the title or perquisites he had anticipated. His office in a building adjacent to the RLE was a cramped basement space with no windows. A later disciple recalled, "There were four desks pushed together in the center of the room. . . . Filing cases and bookcases lined . . . the walls, except for a small gap which contained a packing crate on which people would stand to chalk their ideas on a three-foot square blackboard."

After he lost contact with Wiener, Walter Pitts faded from view, materializing and dematerializing at odd hours around MIT, the RLE, and Cambridge's working-class bars, leaving behind only faint sparks from his scintillating mind and stray items of clothing. "Pieces of him appeared, a hat, a jacket, a scarf," his best friend Lettvin recalled. By the mid-fifties, Pitts was in a steep decline accelerated by alcohol and, later, barbiturates. Lettvin and others began to fear for his mental and physical well-being. "After '52 it was straight downhill, there was no way of pulling him back," he remembered. "I used to spend whole nights looking for him all over the city, always worried where he was." Lettvin's wife, Maggie, who watched it all, shared her husband's impression that Pitts's decline was a progressive effect of the break with Wiener. "Walter was so totally devastated by what Wiener did, cutting him off that way, and from that point on it really was a slow suicide. He just found every way he could to self-destruct."

Yet, beyond Cambridge, Pitts's legend continued to grow. In June 1954, he was one of ten young scientists who were hailed by *Fortune* magazine as the rising stars of American science, alongside the physicists Richard Feynman and Julian Schwinger, DNA co-discoverer James Watson, and the geneticist Joshua Lederberg. Their careers would flourish but Pitts's would not. Within a dozen years, while Pitts ran from his colleagues and his genius, five of the young scientists *Fortune* smiled on would win Nobel prizes in their fields.

Along with McCulloch and his team at MIT, as he had vowed, Wiener also parted company with the Macy Conferences on Cybernetics. The conference series had been evolving in sync with Wiener's scientific interests. Leading neurophysiologists and psychologists brought new evidence that "the brain functions more 'analogically' than 'digitally'"—findings that upheld his continuous approach to communication in animals and machines. A parade of new participants came to the conferences: Shannon from Bell Labs, the Russian-born Harvard linguist Roman Jakobson, the German-born physicist-turned-geneticist Max Delbrück from the new biological research center at Cold Spring Harbor on Long Island. But that was not enough to keep Wiener at the table.

The seventh Macy conference in March 1950 was the last one he attended, and his sudden, unexplained departure blew a hole in the proceedings. The conferences continued, and McCulloch stayed on as the group's chairman, but much of his old philosophical flair was gone. Pitts was there bodily but in a detached state of "psychological deterioration," and von Neumann did not attend any more Macy conferences after Wiener departed. Gregory Bateson and Margaret Mead stuck by the group and held fast to the social commitments they had shared with Wiener from the outset, but without Wiener the character of the Macy conferences changed.

The group met three more times. Other distinguished scientists from around the world were brought in to invigorate the proceedings, but no amount of auxiliary brainpower could compensate for Wiener's towering absence. Without his guiding vision and focus on the big picture of cybernetics, the Macy conferences bogged down in minutiae and technical presentations. The powerful theoretical and philosophical concepts he brought to the conversation—such as the notions of purpose, teleology, and circular causality, which had been magnets for many attendees at the early meetings—dropped off the group's agenda. In the political chill of the McCarthy era, discussions of pressing social and cultural issues were studiously avoided, and there was little talk about what the Macy scientists could do concretely "to improve the human condition or alleviate and prevent misery," which was Wiener's highest priority during those years.

By the end, McCulloch himself was restrained and fatalistic about what the series had accomplished. In a summary statement presented at the final conference in May 1953, he conceded the group's lack of focus and persisting "babel of laboratory slangs and technical jargons." He allowed only that, "Our most notable agreement is that we have learned to know one another a bit better, and to fight fair in our shirt sleeves." Mournfully, McCulloch acknowledged the man on everyone's mind, who was not in the room. "Our meetings began chiefly because Norbert Wiener and his friends in mathematics, communication engineering, and physiology . . . were the presiding genii. . . . For all our sakes I wish Wiener were still with us."

With the close of the Macy conferences and that prolific period of Wiener's collaborative work, his new science set off on its own wanderings, as cybernetics entered what Lettvin called its "developmental" stage.

It wandered through psychology and the social sciences, where Bateson led a revolution of his own. Like Wiener after the war, Bateson, too, vowed never to use his knowledge to the detriment of human beings, and he set out to put the new communication tools into the hands of clinicians and people directly, to further their self-knowledge and social relations. After his divorce from Margaret Mead in 1950, Bateson relocated to San Francisco and took action immediately on Wiener's injunction that traditional psychoanalytic theory should be "rewritten in terms of information, communication, feedback and systems." His first clinical breakthrough was his discovery of the "double bind," a pathological communication process that Bateson viewed as a factor in the onset of schizophrenia. His double-bind theory described the disruptive psychological effects of illogical and contradictory messages which, Bateson and his clinical colleagues found, could precipitate confusion, panic, rage, disturbing voices in the head, and more extreme paranoid states. His theory traced back to Wiener's concerns, expressed at the Macy conferences and in *Cybernetics*, about the dangers to people and computers of paradoxical states that could trigger an unstoppable circular process leading to similar forms of mental and machine breakdowns. Bateson's theory won wide acceptance in psychiatry and was the reigning model for more than two decades, until newer genetic and neurochemical explanations began to eclipse the communication approach.

Other social scientists in the 1950s, following on the work of the Macy conferences, applied cybernetic models and methods to the study of communication in small groups and large organizations. With help from Bateson and Kurt Lewin's disciples in group dynamics, and from other European psychologists who came to America after the war, the new "humanistic" psychologists Carl Rogers, Abraham Maslow, and Rollo May pushed beyond the Freudian school and developed effective new communication-based methods of individual and group therapy that would transform the field of mental health and influence the whole of American culture. At the same time, political scientists, beginning with Wiener's colleague Karl Deutsch at MIT, applied cybernetic principles to the art and science of government—the first social enterprise to which such notions were applied historically. Another new enthusiast, University of Michigan economist Kenneth E. Boulding, asked Wiener personally to aid in his "missionary" effort to buck up the dismal science with fresh ideas from cybernetics, which Boulding went on to promote, along with Bertalanffy's general system theory, as essential tools for solving other complex problems of modern technological societies.

And just over the American border, in Canada, in the summer of 1950, Donald Theall, a young American graduate student at the University of Toronto, introduced his English professor, Marshall McLuhan, to Wiener's work and to the new thinking of the cybernetics group. Theall handed McLuhan copies of *Cybernetics* and *The Human Use of Human Beings* and witnessed McLuhan's reaction. "The relevance of Wiener in McLuhan's mind had to do with Wiener's image of the communications network as the contemporary symbol for the 'age of communication and control,'" Theall recalled. Wiener's ideas stimulated McLuhan's thinking and spurred him on to build "a foundation for a contemporary theory of artistic communication" that became a conduit for the flow of cybernetic ideas into art, literature, and the whole of popular culture.

In 1953, McLuhan launched his celebrated seminars on "culture and communication" at the University of Toronto. A decade later, he would use Wiener's ideas liberally, but without attribution, in his own watershed work, *Understanding Media*, which dissected the effects of television and every other medium of communication on human consciousness and culture. The book's subtitle, *The Extensions of Man*, and its oracular pronouncements that "the medium is the message" and that electronic media had turned the world into a "global village," echoed Wiener's words in *The Human Use of Human Beings* that "the transportation of messages serves to forward an extension of man's senses . . . from one end of the world to another," and that "society can only be understood through a study of the messages and the communication facilities which belong to it."

As always, Wiener found some of the most enthusiastic support for his work outside the United States, and by the early 1950s cybernetics was rapidly establishing itself as an international scientific movement. In 1950, French scientists formed the first scientific association for cybernetics, fittingly called the Cercle d'Etudes Cybernétiques—Circle of Cybernetical Studies. In Italy, Nobel physicist Enrico Fermi promoted a seminar on cybernetics at the University of Rome in 1954 and, in 1957, a Divisione di Cibernetica was established at the Institute of Theoretical Physics in Naples. The same year, the first International Association for Cybernetics was established in Belgium.

Some of the most significant advances in cybernetics were taking place in Great Britain. Late in 1949, a small, interdisciplinary dinner club in the American tradition was organized in London by a group of young physiologists, mathematicians, and engineers who had become passionately interested in the new science. The Ratio Club, as it became known for recondite reasons, had some illustrious members, including Alan Turing, the neurophysiologist W. Grey Walter, a pioneer in the study of brain waves who built the first mobile robot, and another neurologist and kitchen-table tinker, W. Ross Ashby, who would become the leading developer of cybernetic theory after Wiener. In the mid-1950s, two other British cyberneticists would put Wiener's ideas into practice on a grand scale. Gordon Pask, an intense Englishman from the Midlands who prosecuted his science in an Edwardian cape, built the first electronic teaching machines based on cybernetic principles and devised a new cybernetic theory of learning. Stafford Beer, a burly Londoner, applied cybernetic principles to organizations. Beer's science of "management cybernetics" was practiced widely in business and government in Britain and, later, inspired ambitious projects to manage government programs in Canada, Mexico, and Uruguay—and, for a time, the entire economy of Chile.

Meanwhile, at Cambridge, two young biologists, one American, one British, reinvented biology in the light of cybernetics. James D. Watson and Francis H. C. Crick enlisted the new tools of cybernetics and information theory in their quest to discover the molecular structure of the genetic material DNA. In a letter to the journal *Nature* in 1953, Watson went beyond Haldane's initial injections of cybernetic ideas into genetics and gave a hint of "the possible future importance of cybernetics at the bacterial level." Weeks later, Watson and Crick unveiled their model of the DNA molecule as a double helix composed of two interlocking strings of biochemical messages and, then, cracked the genetic code that governs the life processes of every organism. Soon after, according to the leading historian of that period in biology, Crick "formalized information as a fundamental property of biological systems" and spelled out the new facts of life in a model that was "remarkably similar to that of Norbert Wiener's a decade earlier."

The "cybernetic groundswell" rippled through genetics and molecular biology in a wave of new communication models and vivid technological metaphors, driven by Wiener's concepts, Shannon's formulas, and von Neumann's speculations about gene-like "cellular automata." Soon physiology, immunology, endocrinology, embryology, and evolutionary biology were brimming with ideas from cybernetics and information theory. At the Pasteur Institute in Paris, the philosophically minded cell biologists Jacques Monod and François Jacob embodied Wiener's principles in a sweeping *Cybernétique Enzymatique* that redefined the organism and life itself as "a cybernetic system governing and controlling the chemical activity at numerous points." They also recast Wiener's revived philosophical notion of teleology as a new organic principle they called *teleonomy,* a quality common to all systems "endowed with a purpose or project," which Monod described as "one of the fundamental characteristics common to all living things without exception."

During that period, cybernetics and information theory helped to lay the foundations for two other new fields that would have profound impact on science and technology: the expanded interdisciplinary venture that became known as "cognitive science" and the specialized technical undertaking that aspired to the name of "artificial intelligence."

The new cognitive science was born in 1956 at a symposium at MIT, where psychologists, brain scientists, and computer theorists formed a new alliance to cross-fertilize their fast-evolving fields. Soon after, Harvard psychologist George Miller, who had made intriguing discoveries about the physical limits of human information-processing capacities, joined forces with neurophysiologist Karl Pribram and psychologist-mathematician Eugene Galanter in a groundbreaking book, *Plans and the Structure of Behavior,* that built on Wiener's 1943 manifesto with Rosenblueth and Bigelow. The three theorists took a new "cybernetic approach to behavior in terms of actions, feedback loops, and readjustments of action in the light of feedback," as the leading chronicler of the new field described it. From that starting point, cognitive science progressed rapidly, using the new communication concepts and emerging tools of computer analysis and modeling to shape a more sophisticated scientific understanding of the mind's subjective processes.

The cognitive perspective attracted a younger generation of social scientists, linguists, and philosophers eager to use the new communication concepts and investigative tools to explore the ways human consciousness itself is shaped and maintained, from moment to moment and place to place, by information and communication, and the combined influences of experience, language, and culture.

The two communication-based approaches to psychology that began in the 1950s, the new science of mind taking shape in the laboratory and the new "third force" of humanistic psychology that was transforming the practices of psychotherapy, came at their subjects from different vantage points. But, together, in the years that followed, they would finally help psychology to break free of its entrenched Freudian and behavioral models.

The new subdiscipline of computer science its founders called artificial intelligence, or simply AI, was another natural outgrowth of cybernetics and information theory, and more proof of the rapid progress being made in the field of digital computing. With the triumph of von Neumann's architecture, as the number of digital computers in use in universities, corporations, and government laboratories began to grow geometrically, the action at the edge of computer science shifted from making hardware to writing the programs needed to make the machines run. To meet the need, a new crop of computer theorists and programmers came forward, led by John McCarthy, a young math instructor from Princeton, Marvin Minsky, a rising star in mathematics at Harvard, and Wiener's former student Oliver Selfridge.

Selfridge turned out to be a pivotal figure in the evolution of AI. After dropping out of MIT in 1949, he had come roaring back to Tech's Lincoln Laboratory two years later, where he infused the lab's Whirlwind computer team with some dazzling ideas he had learned as Wiener's student, and he quickly emerged as the Whirlwind's programming whiz. In the mid-fifties, he unveiled one of the first artificial intelligence programs, his fiendish "Pandemonium" pattern recognition program that enabled the Whirlwind to identify letters and geometric shapes. Pandemonium was a sophisticated software version of McCulloch and Pitts's neural network scheme for recognizing universal forms. Even more impressive was the program's ability to learn from its successes, weed out its weaknesses, and evolve as an ever more accurate pattern recognizer and decision maker.

Now Selfridge, the youngest member of Wiener's inner circle, was leading the way, while his mentors steered clear of the new field their ideas had inspired. McCulloch was supportive of AI but kept his distance. Pitts was indifferent to AI and digital computers generally. And Wiener would not set foot in the new field, which was tied tightly to military projects at MIT and elsewhere.

Marvin Minsky, who came to MIT in 1957 to work in the Lincoln Laboratory and, a year later, started a full-fledged Artificial Intelligence Laboratory in the RLE, remembered seeking out Wiener for whatever insights he could offer the new enterprise, but to little avail. "Wiener didn't think about artificial intelligence much," Minsky said. But the two mathematicians were on different trajectories. Minsky was carving out another new domain of digital computer science, while Wiener was pursuing his own pet projects in the analog mode, including two mobile robots he built with engineers in the RLE and a little help from Grey Walter, a light-seeking "moth" (his "automatic barfly" that rolled into every open doorway) and its opposite number, a light-shunning "bedbug."

AI owed more to digital computing and information theory than to cybernetics. Shannon and von Neumann helped McCarthy, Minsky, and Selfridge to launch the field, writing papers and promoting conferences that gave the new movement momentum, and its rapid advance was yet another reflection of the extent to which digital thinking and technology took hold at Tech and everywhere in the years after Wiener's core group of cyberneticians was sundered. The digital shift was underscored by Claude Shannon's return to MIT in 1958, by the advance of digital technology in the nation's military command and control systems, and by the turn in the development of industrial automation technology from the analog to the digital mode.

By that point, it was clear that the automatic factory was going forward with or without help from Wiener. Early in the 1950s, General Electric had begun testing its ingenious "record-playback" system for automating manufacturing operations. The strictly analog system captured the intricate movements of a skilled machinist's handiwork on punched paper or magnetic tape, and fed them to a workerless machine that could reproduce the machinist's movements exactly, on one machine or hundreds of workerless drones, at a fraction of the cost of human labor. However, another system, developed by the U.S. Air Force with help from engineers at MIT's Servomechanisms Laboratory, became the method of choice for manufacturing jet plane and guided missile parts "by the numbers"—using purely mathematical programming methods and digital computer technology—and it soon swept the field of industrial automation in America.

––––––

Wiener scrupulously avoided any involvement with the work in progress at the Lincoln Lab, the Servo Lab, and MIT's other military-industrial ventures, but he kept up his grand rounds of the RLE. Although he was no longer collaborating with McCulloch and his troupe, or even speaking to them, and barely on civil terms with Wiesner, who led the RLE throughout the 1950s, he continued to inspire innovative research in both the technical and biological domains of cybernetics, and he remained the RLE's catalyst, "though catalyst is a lukewarm description of his role," as Wiesner affirmed years later.

Among the diverse RLE projects Wiener fired up in the fifties were studies by the lab's Information Processing and Transmission Group, its Sensory Communication Group, and one he oversaw personally in the RLE's Communications Biophysics Laboratory, a new multidisciplinary venture organized by Walter Rosenblith, an Austrian-born engineer who came to MIT from Harvard in 1951. After attending seminars led by Wiener and gaining entry into his supper club, Rosenblith formed his group to study biological communication processes "with the aid of up-to-date electronics and Wienerian analytical techniques." The process Wiener was the most eager to analyze was the one to which Grey Walter had introduced him on his travels to England: the mystifying phenomenon of brain waves. Wiener believed those faint bursts of electromagnetic energy the human brain broadcasts day and night opened a window into the mind's eye, and into the brain's innate cybernetic processes, and his young RLE colleagues agreed.

Spurred on by Wiener's enthusiasm and curiosity, Rosenblith and his team, working in tandem with neurophysiologists at Boston's Massachusetts General Hospital, developed a special-purpose computer they called a "correlator" to record and analyze the brain waves of human subjects. The computer ran in the analog mode and operated on the principles of generalized harmonic analysis Wiener had worked out a quarter century earlier.

The correlator was Wiener's baby, the first analog computer to apply his statistical methods automatically to the analysis of brain function. He nursed the device through its design and gestation in the RLE's Machine Shop and watched the finished machine spring to life. "The apparatus made an enormous clatter," John Barlow, a young doctor in the group, recalled. The two dozen relays in the rapid "repeater," the central processing unit that correlated one wave form with another, "chattered incessantly among themselves." The magnetic memory drum that turned 40,000 times a second "added its own tune, producing every sound conceivable and inconceivable." The clanking kept up for twenty minutes for each correlation, until the machine spit out its completed "correlogram" and came to a thumping, self-satisfied halt. Wiener loved the racket. "He was captivated. He was absolutely fixated on the printout as it was coming out," Barlow remembered.

Wiener's goal for the project was nothing less than to "find the Rosetta Stone for the script of brain waves." He believed those seemingly random, unintelligible signals held the key to perception, cognition, and the nature of intelligence itself. After the machine had cut its teeth on simple correlations of everyday brain waves, the team turned to studies of the brain's exotic "alpha" waves, the slowly undulating waves the brain emitted in its most composed and contemplative waking states, which Wiener and others considered to be a tangible measure of the mind's powers of insight. Curious about the underpinnings of his own intelligence, Wiener asked to be the correlator's guinea pig, but getting a clean reading of Wiener's alpha waves posed a challenge to the fellow researchers, who were unaware of his chronic apnea.

"I remember him lying on the table, and we said, 'Well now, Norbert, *relax!*' and Wiener promptly went to sleep and of course his alpha activity disappeared," said Barlow. Finally, the team got a measurable recording, but the correlator's assessment of Wiener's alpha was not the off-the-charts finding he had been expecting. "During the waking state Wiener's alpha activity was not, shall I say, first class," Barlow revealed. "It was more irregular, but well within normal limits, and I could sense Wiener was disappointed that he didn't have all that handsome an alpha." It could have been worse. As Barlow recalled, "Grey Walter had no alpha at all until after the second cocktail."

The Communications Biophysics Laboratory's brain-wave studies were a noteworthy success and a feather in the cap of the RLE. Wiener and his young colleagues published their findings widely, and Wiener wrote his own account of the research for the second edition of *Cybernetics* in which he explained for the first time how the brain's billions of separate neural firings self-organize spontaneously into coherent electromagnetic currents detectable across the whole of the brain's surface. The studies provided hard data on the brain's analog information processes, and Wiener's explanation offered deeper insights into the way the human brain, and all living things, swim upstream with surprising success against nature's inexorable tide of entropy and disorganization. Wiener considered the application of his statistical methods to brain-wave research to be the "most significant" work "of all the things I have done in physiology" and the consummation of the research he and Arturo Rosenblueth had begun a decade earlier.

———

His new team in the RLE made up in part for the loss of McCulloch and his crew, and by the mid-1950s a new crowd of bright young grad students and postdocs was gathering around Wiener at Tech. Wiener's new boys were not a tight-knit tribe like the knights of the cybernetics circle but a miscellany of able minds with whom he would do good work in a smattering of fields, and with whom he would form close friendships that would give his spirit succor in his later years.

And Wiener made another friendship in the fifties that would give him immense satisfaction. In the fall of 1953, Wiener met the young electrical engineer Amar Bose, who was working on his doctorate in the engineering department. Later, Bose would begin his research on sound reproduction in MIT's Acoustics Laboratory that would lead to the audio electronics enterprise that bears his name. Their pairing was not serendipitous but a marriage made in the RLE's front office. "I was drafted by Jerry Wiesner to work with Wiener," Bose remembered. At the time, he knew nothing about the split between Wiener and the McCulloch group, and between Wiener and Wiesner himself, which had occurred only the year before, or that Wiesner was anxiously looking for ways to keep Wiener's ideas flowing into the RLE. "I was pretty annoyed at first, but it was one of the best decisions of my life. It led to ten years where Wiener and I met almost every day."

Wiener came to count on Bose as he counted on so many others he towered over. "When he gave a talk and I was present, he'd come to me afterwards, 'Bose, how did I do? How did I do?' He was talking at a level way above me, but he had to know." Bose gave Wiener the feedback he needed desperately and something more. Bose liked Wiener and met him on his own terms, without prejudice toward his eccentricities or his deficiencies; and he saw strengths in Wiener where others saw only idiosyncrasies. He even discovered the deeper purpose behind the seemingly random walks Wiener navigated around MIT. "They weren't random," Bose insisted. "What he did was pick out somebody in engineering, somebody in political science, somebody in philosophy. He visited them daily and got the essence of what was going on in all their fields just by talking to them for fifteen minutes." It was Wiener's interdisciplinary method in motion. For Bose those *Wienerwegs* were the epitome of Wiener's faith in the free exchange of knowledge across all the domains of science and scholarship, which he practiced indefatigably as he wandered in search of the most useful insights along his path.

Bose saw Wiener's lighter side, too. One day, after the two lunched together in MIT's faculty club, Wiener could not find his London Fog raincoat among the dozens of lookalikes in the coat room. "He said, 'Bose, I have an idea. Let's go into the lobby and discuss things and then everybody will leave and mine will be the only coat left!' We spent an hour-and-a-half outside the cloakroom discussing everything under the sun, until only a few coats were left and he could find his."

Another time Wiener asked Bose to be his stand-in for an important speaking date his advancing age and expressed public positions precluded him from accepting, at the Air Force's main research center for new technology at Wright-Patterson Air Force Base in Ohio. "He came to me and said, 'Bose, I've been invited by some general to come out to Wright Field to give a talk on my theories. You go and speak for me.'" Weeks went by without another word about the event. Eventually, Bose learned that Wiener had lost the letter and forgotten the name of the general who sent it. "He came in and said, 'There's been a fumble in Wright Field.'" Days later, another doctoral student saw Wiener in the math department mailroom. "There was a big bin in there, and there was the derriere of Norbert Wiener sticking up in the air, and letters were flying out of the box and going all over the floor." Soon after, Bose got a call from Wiener's secretary. "She said, 'He just dictated a letter, and the address he wants on it is: To Whomever Contacted Norbert Wiener, Wright-Patterson Air Force Base, Dayton, Ohio. What do I do?'" Bose told her, "'Did you ever send a letter to Santa Claus when you were young? Just do it.' We never got an answer, of course, and the talk never was given."

According to Bose, the power Wiener imparted to the minions who gathered around him came from his ability to "see way over the fence." The phrase was Wiener's. As Bose recalled, "The mathematicians in the department used to hover around his office like bees around a flower, you know, to get a little bit of an idea to give them something they could work on for the next five years. This one mathematician was very good but Wiener said, 'The trouble is he can't see over the fence.'" For Bose the image captured Wiener's ability to envision what was possible, to see the implications of his mathematics and all his ideas straight through to their realization in science and technology, and their impact on society far into the future. "He literally could see the end result of any effort," said Bose. "He knew what could be done and what couldn't be done, and he didn't worry about how the heck you were going to climb over the fence. He just knew you could get there."

Wiener's ability to see over the fence led to one extraordinary team effort. His pioneering analysis of brain waves had prompted Wiener to take up a tough problem in applied mathematics: the analysis of nonlinear random processes, a dizzying class of phenomena, encountered in neurobiology, electrical engineering, physics, economics, and many other fields, that did not yield to simple linear analysis or more conventional methods of statistical analysis. Wiener asked Bose, whose dissertation he was supervising, to work with him on the project.

"He wrote it over two or three years on my blackboard every day," Bose recalled. "He would go along with the chalk in one hand and the eraser in the other. He would write and erase as he went, working out all the equations in his head. He never put two lines on the blackboard. Then one day he came in and put the last result down and said, '*Okay, Bose, that's it! Write it up!*'" Bose was dumbfounded. "I said, 'What do you mean, write it up? There's one line on the board. You erased everything.'"

Bose turned for help to Wiener's longtime collaborator, Y. W. Lee, the senior figure in the electrical engineering department, and the two men proposed a novel solution to Wiener's own nonlinear creative process. They persuaded Wiener to reconstruct his new mathematics, from the beginning, on the blackboard so they could help him prepare it for publication. Wiener reprised his work in a series of lectures to a select group of MIT graduate students. This time, they were ready. "We got him into a classroom and he put it all down again. We photographed it and recorded it, then the doctoral students and I went home and transcribed it." Soon Bose and the others discovered the dilemma of working with Wiener that all his earlier collaborators had learned. "We'd come in every day and say, 'Oh, we caught Wiener! He made a mistake between here and here!' We were so proud of ourselves. He made all these little mistakes. But, the amazing thing was, in the entire book there was never one end result that was wrong."

Eventually Lee and Bose's recordings and photographs of Wiener's lectures on *Nonlinear Problems in Random Theory*, and their painstaking transcriptions and corrections of his serpentine equations, became the first book in English on the subject, and was published simultaneously in the United States and Britain in 1958.

His new friends and projects brought him out of the funk he had fallen into after his break with McCulloch, and by the mid-1950s Wiener was back in motion on all fronts. His visit to the All-India Science Congress in Hyderabad in January 1954, followed by an arduous seven-week lecture tour of the subcontinent, was one of his few overseas journeys without Margaret. Several months before, Peggy had informed her parents that she too was going to marry a young engineer from Tech, a Boston blueblood named John Blake, and Margaret stayed behind to help plan her younger daughter's wedding. Wiener was as pleased with Peggy's marriage as he had been with Barbara's, and offered it as further proof of the inexplicable heredity of famous mathematicians. "Thus, I have in my own family exemplified that peculiar genetics . . . that mathematical ability goes down from father-in-law to son-in-law," he boasted of both his new family members.

That year was a turnaround year for Wiener. His time in India was so successful that he was invited back the following year. On his way home, he made a lecture tour of Japan and taught a summer course on cybernetics at UCLA. He taught at UCLA for two more summers, taught another summer course in Varenna, Italy, and then spent a semester as a visiting professor in the new cybernetics program at the University of Naples. His family could barely keep up with his itinerary or with his growing collection of local dialects and patois. "He could say, 'What is your cheapest cigar?' in any language, usually with a native accent," Barbara remembered. But all that mileage and roadwear took its toll on his body and his psyche. "He would just get completely worn out," Mildred Siegel recalled. "He would go places and they would pick his brains and he just worked and worked, then he would go home and go to bed for a few days. Margaret tried to get him to go for shorter times, but he wouldn't stop because this is what he lived for, this kind of interaction."

The new bonds he was building with Italian and Indian scientists were opening new forefronts for cybernetics in Europe and Asia. But Wiener was not nearly so happy on American soil. Real-world events—the escalating Cold War and atomic arms race, the new alliance of the military and industry working together to replace people with machines, and his increasingly bleak vision of the future—laid him low. "I remember seeing Wiener when he was so depressed, he would sit in a chair and the tears would roll down his face," Fagi Levinson recalled. "He would say, 'There's nothing else for me to do' and make this gesture." She pulled a finger across her throat. Wiener's doctor friend John Barlow saw his distress and the action he took privately to get through his deepening waves of depression. "When the international scene would become unsettled, he'd ask me, 'Do you think there will be a war?' He didn't joke about it, he worried about such things a good deal. Every now and then he would check himself into the MIT infirmary for a few days, and that was one of the things on his mind."

Margaret seemed strangely sanguine during that period. McCulloch and the young wags she had perceived as threats to Wiener's prominence were out of the picture, along with both her daughters, who had moved on to their married lives far from Boston. The couple's relations with both young women were distant emotionally as well, even after Barbara gave birth to Wiener's first grandchildren. Yet everyone around them noticed the marked improvement in Margaret's disposition, including Wiener himself. In a letter to Rosenblueth in Mexico months after the split with McCulloch, he observed in passing that "Margaret is well and happy." Jerry Lettvin saw her around that time and thought she seemed "so cheerful." Barbara believed that her mother finally had won her "last ditch fight for total control" over Wiener's "life and career." Her cheerfulness stood out in sharp contrast to Wiener's unsettled states, as she arranged his social life in the company of his new friends at Tech and their wives, and other nonthreatening younger couples.

Wiener agonized over the fate of cybernetics. He tried to protect his science from misuse but, as he knew better than anyone, cybernetics was out of his hands. "One of the problems cybernetics had in the early days was that it was splashed all over, and this troubled Wiener," John Barlow confirmed. He gave no aid or comfort to those who would exploit cybernetics for low or high purpose. "He was even skeptical of those who were making *cybernetics* a cause," Dirk Struik attested. "He was very worried at the time that there were people who saw in it some kind of universal panacea. He used to say to me, 'I'm not a *Wienerian.*'"

In the summer of 1952, his patience ran out. He lambasted several admirers, and he laid into one young writer with a fury that was wholly unwarranted. An editor at Charles Scribners Sons publishers in New York sent a letter soliciting a comment from Wiener on the first novel written by a twenty-nine-year-old war veteran named Kurt Vonnegut, Jr. Vonnegut had just quit his day job at the General Electric Research Laboratory in Schenectady to write his futuristic tale about a society dominated by automated machines and an elite new class of engineers and technocrats—a place much like the brave new world he saw taking shape in GE's industrial laboratories.

Wiener and Vonnegut were natural allies in the fight for the future—both were former employees of GE—but Wiener was not feeling charitable toward anyone that summer. Vonnegut's novel, *Player Piano*, lauded Wiener as a prophet of the human perils posed by automation. It also contained a fictional character named von Neumann, whom Vonnegut cast as one of the masterminds of a modern-day Luddite rebellion, and when Wiener read that he boiled over. In a letter to Vonnegut's editor, he urged her to advise her rookie author that "he cannot with impunity . . . play fast and loose with the names of living people." He assailed the young practitioners of the "new cult" of postwar science fiction, in which he lumped Vonnegut and accused him of using his future scenario to avoid direct criticism of present-day scientific practices. Days later, Vonnegut wrote to Wiener personally and apologized for any offense Wiener had taken, but he defended his book as a self-evident "indictment of science as it is being run today."

Wiener was doing some dabbling of his own in those creative domains. For years, he had quietly pursued his private passion of writing detective stories and science fiction. His stories, published under the see-through pseudonym W. Norbert, first appeared in MIT's *Tech Engineering News* and were picked up by mainstream publications, including the bible of the new sci-fi cult, *Fantasy and Science Fiction* magazine.

His fiction was highly literate like his conversation, unfailingly accurate in its scientific details, and sprinkled with whimsical asides. It was also cathartic, a purgative for his foul humors and a secondary outlet for his social concerns. And his creative excursions were not confined to print. During that troubled stretch of '52, Wiener developed his own dark story idea for film, and he set out to sell it to his cinematic hero, Alfred Hitchcock. In a query letter, he made his pitch for a story based in a scientific laboratory in Mexico, "where I have run into a combination of characters and . . . situations lending themselves ideally to a suspense and horror movie of the type in which you are expert." Hitchcock never replied, but Wiener was already hard at work on his next writing projects.

With the success of his childhood memoir, Wiener commenced the second volume of his autobiography, covering his adult life and career from the early 1920s through the mid-1950s. In the spring of 1954, his new editor at Doubleday, a young turk named Jason Epstein, asked him to write a small nontechnical book for a general audience on his "philosophy of invention," which Epstein planned to publish in the new paperback format he was promoting in the American market. The handy, inexpensive mass medium offered Wiener a new forum in which to express his growing concerns about science and invention in the postwar era, and he accepted Epstein's offer. Within weeks, he had dictated the first draft of a book he called *Invention: The Care and Feeding of Ideas*, a free-ranging retrospective of scientific ideas and inventions from the Greeks to his day drawn mostly from his immense storehouse of memory.

For three years, he tinkered with the nearly completed manuscript. In the interim, he published the second volume of his autobiography, *I Am a Mathematician*, in which he spoke more personally and passionately about the imperiled state of American science. As in *Ex-Prodigy*, where he condemned "the American lust for standardization" that had driven "the American scientist . . . the way of American cheese," he decried the new wish he perceived among the lords of American science "to decerebrate the scientist, even as the Byzantine State emasculated its civil servants." Again, Wiener seized the moment to settle old scores with his colleagues and competitors, although, in the pattern of silent censure that signaled his most deeply felt emotions, he made no mention of Mc-Culloch, Pitts, Lettvin, or Selfridge, and their prolific interactions during the critical years when cybernetics was being formulated.

In August 1957, Wiener wrote to his editor and announced his intention to abandon his book on invention. He had changed his mind about the project, he told Epstein, because one episode in the book had seized his imagination: the tale of the eccentric English physicist, mathematician, and telegrapher Oliver Heaviside and his inspired conception of the signal-boosting "loading coil" that made long-distance telephoning possible. Heaviside, the creator of the unwieldy operational calculus that comprised virtually the whole of communication engineering theory before Wiener, published his loading coil concept in 1887, but, in his oblivious, unenterprising way, he never bothered to patent it. In 1900, in one of the most disputed episodes in the annals of early communication technology, Columbia University engineering professor Michael Pupin, who had come to the United States as a poor immigrant boy from Serbia, patented Heaviside's loading coil concept himself, without a nod of recognition to Heaviside, and sold the rights to the American Telegraph & Telephone Company for the then astronomical sum of $750,000.

Industry insiders murmured that the deal was prearranged to enable AT&T to corner the market on Heaviside's unprotected idea, but their criticism faded. Pupin went on to win fame, more fortune, and a Pulitzer Prize for his rags-to-riches saga; while the hapless Heaviside died broke, embittered, and half mad in a remote seaside cottage in England in 1925.

Wiener learned the truth about the swindle from older scientists he met at Bell Labs years earlier and others who were still troubled by Pupin's and AT&T's behavior. The steamy tale of scientific invention and corporate intrigue, with its timely morals for the new technological age, appealed to Wiener far more than writing "a purely expository book on invention," and he told Epstein that he wanted to tell the story in fictionalized form. Epstein, who was making his own career move from Doubleday to Random House, reluctantly surrendered *Invention* and agreed to take Wiener's unnamed novel with him.

With Epstein's help, the novel was published in October 1959 as *The Tempter*, Wiener's personal turn on Goethe's timeless Faustian tale recast as a parable for the new technological age. *The Tempter* was no masterpiece, but it received surprisingly good reviews in literary and scientific circles. As Wiener intended, the book gave new impetus to the drive to set the historical record straight on Heaviside and Pupin, and it delivered a stern warning to the postwar generation of scientists and engineers. Yet, like every first-time novelist, Wiener fretted over the response to his artful literary effort. He went to the MIT bookstore daily to see how many copies had been sold. When he asked a colleague in the math department whether he had read the novel and he said he had, Wiener returned, "Then tell me what happens in the section called '1908.'"

Dirk Struik gave it to him straight. "He came to me with a review that was not too favorable. He was very disappointed. I said, 'But Norbert, even Einstein can't be also a Dostoyevsky.'"

———

Wiener was proud of his literary excursions and enjoyed the diversion, but he was not writing for his own amusement. In his fiction, as in all his popular writing, he was pursuing his search for truth in domains beyond science and communicating his defiant ideas directly to people. Yet, behind the scenes, his personal search was taking him in a new direction.

In the late 1950s medical science was at last beginning to decode the myriad neurochemicals that affect the brain in all its states, and their intimate interplay with human emotions and the mind's higher cognitive powers. The new discoveries confirmed some of Wiener's earliest speculations about the impact of those enigmatic "to-whom-it-may-concern" messages—the diverse hormones that flow through the bloodstream into the brain, and the brain's own *neuro*hormones and chemical neurotransmitters—which he first wrote about in *Cybernetics* as likely causes of psychopathology. That new understanding of mental illness began to play on Wiener's mind, spurring new speculations about the chemical underpinnings of his recurring manias, depressions, and emotional storms.

When the first medications to treat schizophrenia were approved, Wiener's younger brother, Fritz, now in his fifties, was released from the Massachusetts mental hospital that had been his home for three decades. Fritz was never far from Wiener's thoughts, or from his deepest fears that he, too, might plunge into full-blown mental illness, and Fritz's reemergence was encouraging. The new drugs gave him tremors, but they made his illness manageable, and he found gainful employment sweeping the floor in a pickle factory in Greenfield, ninety miles west of Boston. Occasionally, Fritz would travel to Belmont to visit Wiener and his wife, who treated him kindly and respected his presence as a reminder of the two brothers' similar dispositions and wildly different fortunes.

Wiener came face to face with mental illness again in the 1950s, and, again, he saw up close the new neurochemical reality he had predicted a decade earlier. He met the dashing, twenty-three-year-old mathematician John Forbes

Nash, Jr., when he arrived at MIT in the spring of 1951, fresh from his initial triumphs in game theory at Princeton and the RAND Corporation. Wiener took a shine to Nash from the start. According to Nash's biographer, he "embraced him enthusiastically and encouraged Nash's growing interest in the subject . . . that led Nash to his most important work," the same subject that had started Wiener himself on the road to greatness—the statistical analysis of fluid turbulence. The complex subject would help Nash prove his case for equilibrium states in game theory and, later, help him earn his Nobel Prize in economics, but first Nash had to conquer his own turbulence.

In the winter of 1959, Nash succumbed to a fever of grandiosity and paranoid delusions that had been building for years, and he was involuntarily committed to Boston's McLean Hospital for treatment of acute schizophrenia. Wiener was one of the few faculty members who understood instantly what Nash was going through and the ordeal that lay ahead for him. Nine months later, after Nash had checked himself out of McLean and fled to Paris in a frenzy of worsening hallucinations, he reached out to Wiener at MIT in a rambling letter festooned with silver foil. "I feel that writing to you there I am writing to the source of a ray of light from within a pit of semi-darkness." The two mathematicians stayed in touch and, four years later, after Nash had returned to America and, finally, began to respond to a combination of the new medications and the new modes of progressive psychotherapy, he wrote to Wiener again from a private mental hospital near Princeton and voiced his hard-won understanding of his illness. "My problems seem to be essentially problems of communications," said Nash.

His celebrated case, retold in the book *A Beautiful Mind* and the popular film of the same name, offered more support for the new theory of mental illness Wiener had postulated in *Cybernetics*. At last, Nash's obsessive thoughts, paranoid delusions, and other schizophrenic symptoms shared by those similarly afflicted, came into view as functional disorders of "the circulating information kept by the brain in the active state" and other "secondary disturbances of traffic," which could be treated effectively with the new brain-regulating medications and communication-based therapeutic methods.

Around that time, another breakthrough in brain science defied the digital computer models that were becoming the rage in cognitive science, artificial intelligence, and many other fields. The discovery, based on a whole new approach to neuroscience theory and laboratory research, confirmed that the human brain did not process information in the mode and manner of a digital machine. The new findings would go beyond all earlier speculations and everyone's imagination—including Wiener's—and, ironically, the news came straight from McCulloch's neurophysiology lab at MIT. However, this time it was not McCulloch or Pitts but Jerry Lettvin who took brain science in a new direction.

With expert support from the RLE's technicians, Lettvin and his English colleague Pat Wall developed probing microelectrodes that could pick up the weak signals generated by the faintest firings of the smallest cells and fibers in the brain and nervous system. The pair launched a series of innovative investigations and, along with their small crew of talented younger researchers, built a formidable reputation for McCulloch's lab as a forefront for studies of the spinal cord, sensory pathways, and brain mechanisms underlying vision and olfaction. While McCulloch, drained of his vital humors, shunned his own laboratory, and Pitts took part torpidly in the lab's dissections, Lettvin kept his colleagues' preeminence aloft by adding their names to important papers reporting on research projects in which they had participated only minimally.

Then, late in the decade, Lettvin and his crew made their finding that turned half a century of neuroscience on its head. Their study of vision in frogs revealed that the brain's most basic information-processing operations were carried out by analog means to an extent never before considered possible. They discovered that the neurons in a lowly frog's eye were capable of sophisticated activities of image detection and analysis, and routinely performed complex tasks, such as determining the size, shape, and motion of objects in the frog's field of view, through analog processes innate in the structure and communication operations of each individual neural cell. Their paper, which Lettvin co-authored with Humberto R. Maturana, a young Chilean researcher in the lab, published in November 1959 with the tantalizing title "What the Frog's Eye Tells the Frog's Brain," forced a sudden rethinking of everything that was known and assumed about sensory perception and the brain's cognitive operations.

The new findings posed a major dilemma for Wiener. The "Frog's Eye" paper upheld the analog mode of information processing he had always favored. It provided indisputable evidence of that action at bedrock levels of the brain and nervous system. It strengthened the biological foundations of cybernetics and unveiled a new order of living information processes in the brain, at a time when those analog processes, and the organic aspects of cybernetics generally, were being stripped from the field's technical applications. From afar, Wiener watched his science being taken forward by the very people he had left behind.

Yet he made no move toward McCulloch or Pitts or Lettvin. Instead, he continued to travel the world as an ambassador for cybernetics, giving talks to overflow audiences on the promises and perils of the new technological age. On occasion, he would even soften his stand against consorting with corporations and meet with their CEOs, without charge, when they came calling at Tech, although he "was not always predictable and could snub an important visitor from the military-industrial establishment or insult him in scathing language."

Fresh accounts of Wiener's footfalls, feats of intellect, and fluency in many languages coursed through the MIT campus. At the faculty club, on being introduced to a visiting Greek dignitary, he broke into a rousing chorus of the Greek national anthem. During a lunch at his favorite Chinese restaurant, Joyce Chen's, up the road from MIT, he gave his order to the waiter in his best dinner-table Mandarin. In the mid-1950s, he moved his weekly supper seminars to Chen's new restaurant on Memorial Drive, just steps from his office in Building 2, where the gentle proprietress personally oversaw the preparation of his ultimate indulgence: bean curd with mushrooms in brown sauce.

But most days Wiener held to his humble routine. He took his usual midday meal, a carton of milk and a bag of potato chips, at the luncheonette in Walker Memorial. He mingled with the MIT students and sanctioned a few jokes on himself, as some of the more intrepid Tech students "made it a habit to regale him with the latest story about him, which he would usually enjoy goodhumoredly." And he would be on his way, popping peanuts as he ambled.

Dirk Struik allowed that Wiener did not always take himself as seriously as most of his disciples and all his critics did, and that on many occasions "his tongue was never far from his cheek." Mildred Siegel saw a deeper method in Wiener's fine madness, and in his renowned absence of mind that harked back to the lessons he learned from the wily dons of Trinity during his formative years. "Was he absentminded? Sure he was absentminded, but he was absentminded *like a fox*."

————

Late in the 1950s, Wiener pulled back from the crowds, as he reluctantly accepted the fact that "if I want to contribute anything more to science, and . . . keep in reasonably good health, I must conserve my energies." But those who still resonated with him continued to experience something extraordinary in his presence. He maintained a close personal relationship with Morris Chafetz, the young psychiatrist he met in Mexico, who had returned to the United States and joined the staff at Massachusetts General Hospital. Chafetz and his wife, Marion, recalled how their entire worldview shifted in the course of their interactions with Wiener. "It's difficult to describe," said Chafetz, as he strained to convey Wiener's impact. "He took society and brought it beyond itself, and he did that for us. And it wasn't just your mind. He took your emotions, he took your self-perception, and broke down barriers, just like he did technologically but in a personal sense, and he set you free. He had a global view. He had full sight." Marion felt Wiener's power coming from the greater scheme of nature. "It was something inside him that was just there, deep, ancient," she said. "He tapped reserves in our minds which we didn't know we had." John Barlow experienced something similar during his private "séances" with Wiener at MIT and Mass General. "I think I was functioning at a different level in his presence, not a mathematical level, an intuitive level," he said.

Amar Bose had many of those ineffable experiences during his ten years of close interactions with Wiener. "He'd come into my office when I was working on my doctorate, but we never talked about my thesis. He had all these ideas and he would talk about those things. Then after a while I noticed a strange thing. After he'd leave I could turn back to whatever I was stuck on and just *go!* It was amazing. At first I thought it was a coincidence, but it happened again and again and again. Your thinking went up to a different plane just trying to envision with him."

For Bose, one final memory summed up the secret of Wiener's singular mind and the force that drove his perambulating genius. "It was one of the last times I saw him. We had lunch in the faculty club. Afterwards, I walked him to his car and, as we were parting, I asked him, 'Professor, how have you been able to make all the incredible contributions you have made to mathematics and science?' He looked at me and said just two words: 'Insatiable curiosity.'"

15

Homage to the Elephant's Child:
The 'Satiable Soul of
Norbert Wiener

So oft in theologic wars,
The disputants, I ween,
Rail on in utter ignorance
Of what each other mean,
And prate about an Elephant
Not one of them has seen!

—John Godfrey Saxe, *The Blind Men and the Elephant*

WIENER'S FOCUS SHIFTED PROFOUNDLY IN the 1950s, and his travels to India were a major factor in that change. He left Boston in December 1953 on his most ambitious mission yet as ambassador of cybernetics. The Indian government had been courting him for several years to come to their ancient land to enlighten its scientists and civic leaders on the country's potential for "rapid industrial growth under the regime of the automatic factory." He was well aware of the long period of internal strife and political turmoil from which the newly independent nation was emerging. He was eager to see the new nation and to get to know its people and culture. He also believed strongly that "we need the Orient more and more to supplement a West which is showing the intellectual and moral enfeeblement following two World Wars."

That trip would mark a turning point for Indian science and technology, and for Wiener as well.

He was welcomed in Bombay with courtesies befitting a foreign dignitary. He was met at the plane by the head of the Indian Atomic Energy Commission, who whisked him through customs, got him settled in at the resplendent Taj

Mahal Hotel, a shimmering Victorian palace fronting on the harbor in the heart of the old section of the city, then took him out to the beach to have tea under the palms.

India's leaders stepped up their VIP treatment as they sought Wiener's wisdom and cybernetic short cuts for modernizing their nation with the aid of Western science and technology. At the new Atomic Energy Institute in Bombay, Wiener was greeted by India's Prime Minister Jawaharlal Nehru. Days later, in Hyderabad, Nehru opened the science congress in the company of the Nizam of Hyderabad, the world's richest man at the time, leaving no doubt about the priority of science in the nation's blueprint for its future. After his good time at the congress, where he "needled" the Russian delegation and their watchdogs from the Soviet secret police, Wiener proceeded on his lecture tour of India's disparate states and scientific institutions. He lectured at the National Chemical Laboratory in Poona and at the Indian Academy of Sciences in Bangalore. He spent a week at the Indian Statistical Institute in Calcutta, and another at the Tata Institute for Fundamental Research in Bombay.

For nearly a month, he taught, lectured, and learned his way across the subcontinent. He stayed in the homes of Indian scientists and feasted with their families. He got a taste of India's abundant history, a feel for the many peoples woven into its colorful new cloth of nationhood, and his own sense of their yearnings and aspirations. Then he moved on to Delhi, the nation's capital and the hub of the country's unfolding plans.

In Delhi, Wiener lectured extensively on the main topic of his trip: "the significance of the automatic factory for the future of India." On his tour of the country, he had reflected on India's potentials and the place it was seeking in the wider scientific and industrial world. He had seen firsthand that "Indian scientists are the intellectual equals of those in any country," but, as he did on his earlier travels to China and Japan, he also witnessed "the special problems of countries combining great intellectual ability with great poverty and now just beginning to enter upon the stage of a truly international scientific life." Based on those initial insights, Wiener made his recommendations to his gracious hosts. He advised India to forgo the old machinery of industrialization and mass factory labor, which had been the engine of progress in the West since the dawn of England's "dark satanic mills," and, instead, to push ahead at its own pace and on its own terms toward a future based on newer and more intelligent technologies.

He acknowledged that the old model was a tempting one for any nation seeking to industrialize broadly, and that it offered India "a chance to capitalize on its unquestioned asset of mass population." But he questioned whether that course was "worth the price in human misery" that threatened to turn the nation's fabled cities, with their gleaming temples and soaring relics of India's imperial past, "into an unlovely hybrid of Indian famine and Manchester drabness."

Wiener was hopeful that a modernizing India could avoid that fate from the outset through the use of automated technology in the new factories that were planned throughout the country, but he was not entirely convinced. He knew that Nehru and other Indian leaders were eager to pursue a cybernetic path to industrialization, and he confirmed Nehru's belief that the automatic factory, with its obvious technological and economic advantages and its resulting increase in production capacities, could expand India's industrial base quickly and "might well be an easier avenue towards a prosperous and effective industrialized country than any of its alternatives." However, from the signs he was seeing already in the United States, he felt compelled to caution his hosts that the "hothouse atmosphere of . . . the automatic factory may conceivably foster evils greater than any which it can alleviate."

Wiener saw another alternative coming over India's horizon. The country's challenge, he said, was not only to produce enough automated factories to make India's industries prosper, or even to produce more world-class scientists in the country's renowned research institutes, but to build "the class of skilled technicians"—or, as he called them, "the non-commissioned officers of science and technology"—to create an entirely new kind of technological society. Because its base of industry and skilled factory workers was so meager to begin with, in India, far more than in the United States, the automatic factory would make "its demands on human efforts not at the bottom but at the very high level of the scientist-engineer," and at the next level of "highly skilled trouble shooters and maintenance workers," and he believed it was "quite in the cards that India can supply both of these within a matter of decades."

Wiener was happy to be a part of India's technological adventure. He returned to the United States in February 1954 greatly reinvigorated, his spirit revived by his travels and reaffirmed by India's thirst for his science and the new technologies it made possible. His new friends invited him to continue his teaching, and he returned to the subcontinent for the academic year 1955–1956 as a visiting professor at the Indian Statistical Institute in Calcutta, where he delivered more than sixty lectures and worked one-on-one with some of the country's foremost scientists and mathematicians.

In the years after Wiener's visits, under Nehru's direction, India established a nationwide system of technical institutes and set to work training a new generation of scientists and engineers, and they soon began exporting them to America and other countries, where they quickly established themselves as some of the world's brightest minds and most sought-after technical workers. Indians became a formidable force in the development of America's computer and telecommunications industries and the predominant population of foreign workers in California's Silicon Valley—by the 1990s, nearly half of all Silicon Valley companies were founded by Indian engineers and entrepreneurs.

And Wiener's vision in the 1950s of the need for a new class of "non-commissioned officers of science and technology" gave India's leaders something even more valuable for the future: their first glimpse of the new technological service economy and its throngs of well-educated, highly skilled young technology workers that would lead their nation, and the global information economy, into the twenty-first century.

————

His time in India struck chords deep in Wiener's psyche. As he did in every culture, he walked the airy esplanades of the nation's majestic cities and wandered through their back alleys and teeming marketplaces. In Madras, on India's eastern coast, he stayed for several days with his old friend T. Vijayaraghavan, a former student of Hardy's at Cambridge who had gone on to become president of the Indian Mathematical Society. In the 1930s, as a young man, Vijayaraghavan had lodged with Wiener and his family in Belmont, where he became the favorite of Wiener's young daughters, who played with the tidy white turban that covered his topknot—the mark of his status as a Brahmin, a member of India's highest caste. Now Vijayaraghavan welcomed Wiener into his home and proudly displayed his urbane new topless look that came with the abolition of the caste system early in the 1950s. His family received Wiener warmly and Wiener was gratified, especially, as he wrote, "when one realizes that I am by Hindu standards a *mlechchha*, an outcast, and that a generation ago any Brahmin would have considered himself polluted by my very presence."

Wiener and Vijayaraghavan tramped to the beach at the Bay of Bengal and went for an early morning swim in the rolling surf of the Indian Ocean. Vijayaraghavan brought along his young grandson, who was close in age to Wiener's grandchildren. The venerable mathematicians walked along the sands, talking about science and more worldly matters, and, as Wiener recalled, "we speculated much on the lives that our grandchildren might live and whether they might not find a better world in which religious and racial prejudices would have abated and in which all the peoples could meet for all purposes in an atmosphere of universal humanity."

Wiener had known many Indian mathematicians in England and the United States, and he was mindful of the central role of religion in Indian life, but his travels on the subcontinent opened his eyes to the sweep of the spiritual in Indian history and across the culture. On his official tours and private wanderings, he beheld the ancient cave temples at Ajanta and Ellora, with their great stone sculptures depicting scenes from the life of the Buddha and the birth of Hinduism. In his visits with his Indian colleagues, he saw the proud Hindu traditions that reigned among the patriarchs of their families. He took notice of those wizened figures and the part they played historically in Indian society, as they withdrew from active involvement in worldly affairs and life's day-to-day struggles to contemplate the divine and assume the role of the *sunnyasi,* or wandering religious ascetic.

Wiener knew that the purely contemplative life would not provide the needed solutions to India's centuries-old cycles of poverty and suffering, and he was impressed by the ease with which the nation's venerable elders were now redirecting their religious impulses to serve their communities and pass on their wisdom without any personal motive or intent of material gain. He marveled at the new role those old working sages were assuming in India's transformation, and at the fact that the Indians themselves had drawn from their own scriptures support for that more worldly and practical yet equally spiritual calling of the *sunnyasi.* In Wiener's eyes, India's elders were providing an invaluable service to the country's scientists, secular leaders and all its people, for "no country can make adequate use of motives and modes of action which are merely passed on to it from the outside, but must find somewhere in its own body of tradition and in its own soul the moral sanction for the developments which are necessary to meet new problems."

In all the other countries he had visited, he had respected their age-old cultural traditions, but Wiener put his chips on the "modern people"—the scientists, mathematicians, engineers, and eager young students he had come to help along the road to the future. Now, for the first time, he stood in awe of those graceful old men he saw everywhere, both those who dressed in Western clothing and, even more, those who retained the trappings of their culture, as he observed "how aristocratic a simple wool shawl can look when it falls over the shoulders of a beautiful and gentle old sage."

He had never been a fan of the religious life, yet there was something about India and its wandering sages that connected with his subterranean spiritual urges, with his lifelong vegetarianism and innate asceticism, his incessant inner questioning—and with his own maturing role as an elder statesman of the new technological age. He even perceived, in the thirst for knowledge and other modernizing impulses he felt from so many Indians he met, an awakening akin to the modernization of Jewish culture that began in Europe when Moses Mendelssohn and other reformers led their people into the world of secular learning and the sciences—an awakening some of Wiener's own forefathers had spearheaded. At a symposium a few years later, he remarked on the two kindred cultural phenomena:

> In India . . . the same groups and even the same villages which in times gone by have furnished great vedic scholars [and] metaphysicians . . . now-a-days are turning out a surprising sequence of natural scientists, mathematicians, economists, and philosophers in the modern tradition. There is the same transfer among these peo-

ple . . . as there was among the Jews a century ago, of the respect for the traditional scholar . . . to the more modern fields of investigation. There is among these people the same drive élan, the same *invincible curiosity* . . . the same conviction . . . that the career of learning is challenging and worthy of a man's best effort.

For a while in the mid-1950s, Wiener was uncharacteristically optimistic about the prospects for the new technological societies he saw blossoming from America to India, and he still held out a qualified vision of the new sciences leading to a technological utopia. "If we accept the primacy of man over his means of production," he assured more than one audience, "there is no reason why the age of the machine may not be one of the greatest flourishing of human prosperity and culture." However, within a few years, he was not feeling nearly so upbeat about America's technological society. The new thrust of automation he saw building at GE and MIT, with its bias against human labor, the quantum leap in the reach and power of atomic weapons, and the growing hunger for government and corporate contracts he perceived among his peers and younger scientists and engineers darkened his outlook by degrees. By the end of the decade, he was denouncing more vehemently than ever the whole megabuck era of postwar science that, he believed, was having "an evil effect on scientific research all down the line."

Even before he left for India, he had begun to refocus his mission. After three decades of mathematical and scientific conquests, the philosopher in him was rising up, scrutinizing the new technological society through the long lens of history, and pondering its new ethical dilemmas and their human consequences. In one of several articles he wrote for the *St. Louis Post-Dispatch*, he took aim at an emerging class of technophiles who, in his view, were falling "into the childish error of worshiping the new gadgets which are our own creation as if they were our masters." When he returned from his first tour of India, he opened fire verbally on the growing faction of Americans he now perceived as a "cult of gadget-worshippers." Using more robust language and imagery, he voiced his increasing displeasure over the direction in which certain "eager beavers" in technology and business circles were pushing the fledgling science of automation, and he expressed his feeling of moral responsibility for the future of a field he had helped to found. His attacks were not personal any longer but the beginning of a larger battle for the soul of his science, and for humankind's survival in the new technological age.

Now Wiener donned the mantle of a sage himself. He began to address his fellow scientists explicitly in the manner of an elder statesman with no vested interests, one who had consistently spurned the lucrative consulting fees and bids for public endorsements that continued to come his way, as his activism shifted to the philosophical plane of his new science and its ethical dimensions.

In a speech at Columbia University in 1957, he drew on his professional interactions with Indian scientists to deliver a pointed message to the next generation of American scientists and engineers. He made his comparison of modern Indian scientists to enlightened nineteenth-century European Jewish scholars, and he implored young Americans to embrace the same conviction that their careers in science were challenging and worthy of their best effort, not for profit or any personal gain but as an "example of devotion and an inner call."

In a speech to the American Association for the Advancement of Science in 1959, he spelled out "Some Moral and Technical Consequences of Automation." Looking back on the first decade of applied cybernetic control processes and their social impact, he observed that many people had dismissed the dangers of the new cybernetic technologies with little understanding of their true risks. He warned of the perils inherent in the speed of the new automated machines, and in the swiftness with which they were proliferating in critical systems of the society. He offered evidence, dating back to his earliest wartime research, of the built-in biological lag in all human information-processing activities, and he voiced his growing fear that "by the very slowness of our human actions, our effective control of our machines may be nullified."

Wiener's gravest concern during that period, like that of many other scientists, was that the automated systems and "fail-safe" mechanisms of the military's atomic arsenal, by virtue of their electronic speed and automatic response strategies, would remove the ultimate decision for a nuclear war from the hands of human decision makers. But his concern extended to every realm of society. Automated mechanisms and strategies based on von Neumann's heralded game theory were seeping into the nation's military, governmental, and economic systems, from the highest reaches of national policy to the lowliest bureaucratic levels. The same idealized logic and calculating principles were being programmed into a new generation of high-speed digital computers equipped with the power to learn from their mistakes and successes in accordance with instructions spelled out in the new artificial intelligence programs that were beginning to pour out of the nation's AI laboratories.

Wiener believed those strategic theories and programs were fundamentally flawed in their most basic assumptions about human reasoning and decision making, and that people and societies that put their trust in electronic systems predicated on such principles were putting themselves at enormous risk. But he was not able to iron out those crucial differences with his fellow ex-prodigy. Von Neumann died early in 1957, at age fifty-three, from bone cancer that was most likely caused by his exposure to radiation from atomic weapons tests. His concepts had infused the fields of digital computing and automata theory. His belief in the benevolent power of learning and self-reproducing machines had burrowed deep into the soil of AI, and into the minds of computer scientists, economists, and military planners.

Now many devotees of the new automata theories and artificial intelligence programs were the same people Wiener decried as cogs in the machinery of the new military-industrial bureaucracy, and the very ones who he feared had cast their responsibility to the winds and would find it coming back seated on the whirlwind. He dreaded the prospect that a master computer running the nation's security apparatus, or any lesser computer programmed on principles derived from flawed theories of games and brains and human motivations, would leave its operators without any clear-cut understanding of the underlying basis of its operations. He feared that when such a machine became capable of responding to its incoming data at a pace no human could match it might act in damaging ways before people could override its decisions.

Wiener worried more about another prospect von Neumann's automata theories made possible. Advances in AI theory and programming methods had made it feasible to create *self*-programming automated machines. The new programs were the last ingredient needed to make fully functioning automata—self-programming, self-reproducing, automated machines, as von Neumann and his disciples had envisioned them, modeled on the cellular operations of living organisms.

The specter of machines endlessly replicating and mutating their artificially smart computer programs was Wiener's ultimate nightmare scenario for the new technological age. The prospect flooded his mind's eye with scenes of uncontrolled destruction, production, and *re*production that raised complex technical questions, and more sobering moral and philosophical ones. As he told his peers in the nation's largest scientific association, and the wider public when his speech was reprinted in the journal *Science* in the spring of 1960: "If we use, to achieve our purposes, a mechanical agency with whose operation we cannot efficiently interfere once we have started it, because the action is so fast and irrevocable that we have not the [means] to intervene before the action is complete, then we had better be quite sure that the purpose put into the machine is the purpose which we really desire and not merely a colorful imitation of it."

His check and balance on the system was the mind of every scientist and engineer, the people he saw as the principal agents of change and control over the world's advancing technological societies. He implored his colleagues to exert more feed*forward* in their programming efforts and in all professional endeavors, to view their contributions to the scientific revolution of their time on the grand time scale of science and to act in every instance, not in their own self-interest, or in the interest of their employers and institutions, but in the greater interest of humankind.

————

His speeches and popular writings were becoming more historical in their scope. His language and imagery had always been spirited and muscular, but now he was reaching to communicate with his peers and the public on a higher plane, as he did with his best students, to convey his growing alarm at the dehumanizing trends he perceived in the rush of technology, and to press for the new ethical actions that were needed to face down the threat. He warned more than one audience:

> If we want to live with the machine. . . we must not worship the machine. We must make a great many changes in the way we live with other people. . . . We must turn the great leaders of business, of industry, of politics, into a state of mind in which they will consider . . . people as their business and not as something to be passed off as none of their business.

Wiener came to rely increasingly on parables, fables, and folk tales to convey his urgent messages to the world. He repeatedly invoked Goethe's parable of the hapless sorcerer's apprentice who started a "devilment" he was unable to stop, and the tale of the *djinnee* in the bottle from the *Arabian Nights,* which taught that "if you are given three wishes, you must be very careful what you wish for"—and another dark metaphor for the dangers of the new technology, the English writer W. W. Jacobs's short story "The Monkey's Paw." In Jacobs's ghastly story, written in 1902, a poor, working-class Englishman was given an enchanted monkey's paw from India that bestowed three wishes on its bearer. As Wiener told the tale, he "wished for a hundred pounds, only to find at his door the agent of the company for which his son works, tendering him one hundred pounds as a consolation for his son's death at the factory." For Wiener, Jacobs's paw was a grim reminder that "Any machine constructed for the purpose of making decisions . . . will be completely literal-minded" and "will in no way be obliged to make such decisions as we should have made, or will be acceptable to us."

Wiener used those tales throughout the 1950s to caution scientists and society against "the worship of the machine as a new brazen calf." Implicit in all his parables were eternal truths and pertinent lessons about the immense power innate in the new technologies and the potent communication processes and principles they embodied, about the new ethical standards they imposed on their inventors, and about the responsibilities that now fell on every citizen, government agency, and corporation to use the new knowledge and technology wisely for the benefit of humankind. He saw those ethical and moral responsibilities, not as altruistic options, but as imperatives—as universal precepts, inseparable from the new technical sciences themselves.

Early in the 1960s, Wiener gathered together all the sensitive questions of ethics and morality that had begun to churn at the center of the cybernetic revolution in engineering, biology, and society. His concerns would form the subject matter of his final popular work, a brief, pensive book titled *God & Golem, Inc.: A Comment on Certain Points Where Cybernetics Impinges on Religion*.

The book evolved from a series of lectures Wiener delivered at Yale in January 1962, the long-running Terry Lectures on "Religion in the Light of Science and Philosophy," and from his presentation on "The Man and the Machine" at a colloquium on the philosophy of science convened by the Société Philosophique de Royaumont, outside Paris, that summer. Wiener's purpose in his new work was to discuss "not religion and science as a whole but certain points in . . . the communication and control sciences" that *impinged* on the domain of religion, and that contained "some of the most important moral traps into which the present generation of human beings is likely to fall."

He refused to become entangled in traditional theological debates and the logical paradoxes that arise when real-world problems of "knowledge" and "power" are conceived only in terms of omniscient, omnipotent beings, and the notion of "worship only in terms of the One Godhead." He drew a line in the intellectual sand, as a prerequisite for his polemic, insisting that the moral dilemmas of knowledge and power and worship posed by the new cybernetic technologies were facts "subject to human investigation quite apart from an accepted theology." And he called on "the scientist . . . the intelligent and honest man of letters and . . . the intelligent and honest clergyman as well" to strip away their personal prejudices and society's taboos and look some of the new era's "unpleasant realities and dangerous comparisons in the face."

"Squeamishness is out of place here," he insisted, "it is even a blasphemy."

The most dangerous comparison on Wiener's mind concerned the shift in humankind's historical role as God's most intelligent creations, "made in His own image," to their new role as creators themselves and—more dangerous still—as creators of intelligent machines that "are very well able to make other machines in *their* own image." That new role raised serious practical and moral dilemmas, including those posed by learning machines that could outwit their creators in simple games and deadly real-world contests. Wiener feared such machines could develop "an uncanny canniness" and "unexpected intelligence [not] built into it by its designer or programmer," a scenario he compared to the celestial conflicts inscribed in *Paradise Lost*, the Book of Job, and other religious texts in which the gods matched wits with their own rebellious creatures, devils, and demiurges, and were stymied by their cunning and unexpected cruelty.

Now he pulled out his newest fable from the Old World: the medieval tale of the Golem, the inanimate creature of clay into which the Rabbi of Prague breathed life by supernatural incantations. In the sixteenth century, as the legend goes, the giant Golem saved the Jews of Prague from a murderous band of marauders, but then it ran amok and began to slaughter the good people of Prague, until the chastened rabbi uttered the secret words that turned the monster back into mud. For Wiener, the hoary Golem was the most timely, relevant metaphor for the new age of self-acting cybernetic technology. As he maintained, "The machine . . . is the modern counterpart of the Golem of the Rabbi of Prague," and, in their latest incarnation, as Wiener retold the tale, those new Golems had innumerable opportunities to run amok in the modern world. Without perfect programming and tireless human oversight, Golems in the military could bypass their fail-safe mechanisms and trigger a nuclear reign of mass destruction. Without adequate planning and attention to their social and economic consequences, Golems in industry could unleash a devastating bounty of uncontrolled mass production and a tide of mass unemployment.

Such technology, warned Wiener, "is a two-edged sword, and sooner or later it will cut you deep," and he left no doubt about where he stood in the contest with the Golems of the new technological age:

> No, the future offers very little hope for those who expect that our new mechanical slaves will offer us a world in which we may rest from thinking. Help us they may, but at the cost of supreme demands upon our honesty and our intelligence. The world of the future will be an ever more demanding struggle against the limitations of our intelligence, not a comfortable hammock in which we can lie down to be waited upon by our robot slaves.

———

Wiener's lectures on "God & Golem" turned the technical questions of his new science into a larger philosophical argument with far-reaching moral, ethical, and spiritual implications, as he stepped fully into his new role as a sage and tribal elder of the dawning information age. That public turn also marked the next step in his private odyssey, and it brought him to a personal reckoning with his own long-suppressed spiritual yearnings.

He had harbored conflicting feelings about Judaism since his childhood. Like his father and his father before him, he had no use for any organized religion, especially the middle-of-the-road Unitarian church Margaret and the girls attended. "He used to say, 'They blow neither hot nor cold and I spit them out of my mouth,'" Barbara remembered, affirming Wiener's knowledge nonetheless of the Book of Revelation. But his travels across India tapped something in his soul that connected with the country's centuries-old spiritual traditions. During his stay at the Indian Statistical Institute in Calcutta, he went more than once to a nearby Hindu temple "to think over my scientific work," as he said, and he was ushered into the inner sanctums of the temple, which had only recently been opened to non-Hindus. Wiener said nothing more about those fleeting encounters, but, along with his soul-stirring introduction to India's venerable *sunnyasis*, he was deeply impressed by the spiritual life of the subcontinent, and he acted on his experiences in surprising ways when he returned to America.

Back at MIT, in the mid-1950s, he formed a warm friendship with Rabbi Herman Pollack, the Jewish chaplain at Tech. Like Wiener, Pollack was an outspoken man of principle and a fighter for social justice, and the two men ventured out together on diverse social, political, and spiritual excursions. During that period, other influential figures at MIT were turning their attention to ethical and spiritual matters. MIT's President Killian took steps to stimulate discussion among Tech's students and faculty of the serious issues raised by the advent of atomic weapons and all the new technologies of the postwar era, and to address forthrightly the ethical and moral challenges they posed for scientists and engineers. Mainline religions and their chaplains were given a higher profile in campus life, and MIT retained the Finnish architect Eero Saarinen to design a chapel that could serve as a gathering place and forum for people of all faiths or no faith. As one campus minister recalled, Killian and the other instigators of the project wanted "a new type of chapel, a chapel which should not look like a church."

Saarinen's design fulfilled their vision. MIT's unlikely new elliptical, windowless, red brick chapel was a cross between the turret of a medieval Finnish castle and the cooling tower of a nuclear power plant. The chapel had an elegant bell on top and an upswept aluminum spire, and the whole structure was surrounded by a moat. Saarinen designed the castle keep to convey "a self-contained, inward-feeling" and, at the same time, an atmosphere of "spiritual unworldliness." MIT celebrated it as "a serene island of contemplation, a sanctuary free of distractions from the urban bustle."

Wiener was bemused by the odd-shaped building that rose up behind an old dorm across from Tech's main entrance on Mass Ave. "When they built it, he said, 'It will put the name of the Lord on everyone's tongue, because they will look at it and say, *Oh, my god!*'" Mildred Siegel recalled. But he visited the chapel and found it to be, as Saarinen intended, a serene island of contemplation whose stillness was broken only by the play of sunlight on the moat as it glimmered through the low arches at the turret's base. He went there often with Rabbi Pollack to attend programs on ethical issues of science and technology sponsored by MIT's newly invigorated ministries.

On one of those visits, Wiener rose from his seat in the rear of the chapel and took a leap of faith.

Swami Sarvagatananda, a minister at the Boston Ramakrishna Vedanta Society who had signed on as MIT's new Hindu chaplain in December 1954, was presiding. His talk that day was on reincarnation. When he concluded, as the Swami recalled it, Wiener walked up to him, shook both his hands, and said, "Swami, what you have said about reincarnation, I accept it. I know you are right."

Wiener's expression of belief in reincarnation—the Hindu doctrine that the soul is reborn in a new body in recurring cycles of death and rebirth governed by *karma*, a sort of divine feedback determined by the moral behavior the individual displays from one life to the next—did not surprise the Swami. But it was a stunning confession by a lifelong agnostic and self-described "disinherited Jew" with family ties to celebrated rabbis and reformists of Eastern Europe and, purportedly, to the greatest Hebrew sage of all time. It was no less shocking to hear that resounding declaration of mystical belief coming from a founding father of the new scientific and technological age. Yet it was not such a stretch for a man who had studied nature's loops and cycles all his life.

The idea of reincarnation was not just a passing fancy for Wiener. The Swami came to MIT on Fridays, and Wiener paid a visit to him in the chapel's counseling office almost every week. In the late 1990s, the elderly holy man explained how Wiener in his early sixties came to embrace the main tenet of the world's oldest continuously practiced religion. He recalled one interchange between them that got to the root of Wiener's extraordinary belief about the origins of his omnibus genius.

"When he came to my office, I asked him, 'Why are you interested in reincarnation?' He said, 'I believe in it.' 'Why?' He told me, 'You know, when I was doing mathematics in my earlier days, I knew all these things. I recalled them when my teachers examined me because I knew these things were so in my past life. In my past life, I was a good mathematician, therefore I come now to fulfill it.' He said, 'I had forgotten and then I remembered and it was all there, in my *storehouse.*' He really believed in it."

That primordial experience of having lived, and learned, before may have been stashed away in Wiener's storehouse since childhood. Now, in his later years, it seemed to give him a quiet new understanding of his precocious mind, helping him put to rest the long torment of his father's self-serving claims and, at the same time, providing him with a tantalizing new source of internal support for his scientific insight.

In fact, a belief in reincarnation was not such an outlandish creed among prodigies and ex-prodigies. David Henry Feldman, the Tufts University psychologist who surveyed many prodigies, found that "a sense of being connected to generations and times past run[s] deep with prodigies and . . . the notion of reincarnation has surfaced more than once as an explanation for [their] astounding achievements." Feldman, who knew Wiener's celebrated case but had no inkling of what the Swami revealed, observed repeated "mystical, metaphysical or otherwise . . . odd situations in the lives of prodigies." His subjects reported "various influences of souls who lived in earlier times . . . along with spontaneous descriptions of . . . past lives experiences that faded as the child matured."

Wiener never said a word about reincarnation in any of his scientific or autobiographical writings, but he described his mystical experience vividly to the Swami decades before social scientists discerned the pattern in other prodigies.

"Oh, we discussed it many times. One day he came to the counseling office and asked me, 'Swami, would you explain this theory of reincarnation, how it happens? How do you go around?' I told him, 'Look here, when we pass away, our *psychic residue* remains, all your knowledge, your experience and your potentials, and in the next life you push it further.'"

Wiener did not seem to be uncomfortable with that explanation.

The Swami remembered other intriguing exchanges during his private discussions with Wiener, and he recalled an impish twist in Wiener's weekly visits. "He would tell the secretary, 'Tell the Swami his bearded student has come.' And whenever he came, he took the whole day, because he was so interested in talking about so many things."

But after his playful entrance, Wiener was often somber and filled with foreboding.

"We talked about philosophy, human progress, the potentials of automation. I tell you, he was not very happy about the world situation," said the Swami. "Many things he was pessimistic about, the trends of American culture, world culture. He said, 'The most important thing is people fail to understand the needs of humanity, the human use of human beings.' This was so important to him. And now I can see everything he predicted."

The old master spoke approvingly about Wiener's new kind of knowledge, but he was quick to relay Wiener's warnings. "In the late fifties, he told me one day, 'Swami, these computers will ruin the brains of all people. A few people will program them and the whole public will just mechanically follow.'" He recalled Wiener's concern that the shift to computerized systems and decision making, over time, would erode the ability of human beings to think and make choices for themselves. "You know, he was very honest about it. He said, 'What you do not use, you lose. These computers have so much potential, but they will ruin people's brains.' He said, 'Swami, you will live to see it in the next century. I will not be here.'"

His conversations with Wiener kept up for a decade and they pondered weighty questions. But, despite their solemn discussions of reincarnation, the Swami flatly rejected reports that he had been Wiener's "religious counsellor" and spiritual advisor. "I advised him of nothing. We were friends. We talked," he said. Wiener never confided in the Swami about his inner turmoil, his emotional storms, or the tensions in his family that had caused him so much anguish. Nor did Wiener confer with him about his fallings-out with his colleagues at MIT. When he met with the Swami, Wiener's mind was attuned to a higher plane. However, the Swami was well aware of the mocking and derogatory comments that whirled around Wiener as he made his way through Tech—and through life. He saw those criticisms, not as a response to Wiener's renowned eccentricities but as testimony to his strengths, to his honesty and outspokenness, his disdain for pretension, and his refusal to engage in petty politicking.

"I'll tell you, he was blunt, very blunt," said the Swami. "I followed him to faculty meetings. People used to avoid him. They were afraid of him because he called a spade a spade."

The Swami also had an answer to the charge that Wiener was chronically childish and, at times, even infantile in his emotions and his actions. "Not infantile, I tell you. *Honest* people are like that. I can say he was childlike, honest, guileless. He cared for people without any kind of motive. He saw things that children only can see."

Coming from his own creed that preached the harmony of all the world's religions, the Swami respected Wiener's personal philosophy and living faith in human beings that he called his universal humanism. "You know, in his spirit Norbert was a man of broad attitude towards life. There was no narrow, dogmatic attitude. I saw in him a great human being, a civilized human being. Norbert was great because he cared for people, not just for this person or that but *all* people. To me that is the meaning of spirituality, to be a decent human being."

His perspective afforded another distinctive view of a man many people claimed to know but whom everyone depicted from a different angle. The old Indian tale about the blind men trying to describe an elephant seemed to fit the Elephant's Child.

"That was Norbert *exactly!*" said the Swami. "He was a pure soul, I know it."

Invoking that wooly parable, he urged people to keep Wiener's work and warnings in view as the information age went forward.

"Ah, that is the most important thing, how he warned us. He warned us not to become blind."

————

The Swami's insights into Wiener's most private beliefs only bolstered the argument that Wiener was indeed one of those rare omnibus prodigies who appeared at a pivotal moment in humankind's history with "something to tell us that might help tip the balance toward our continued existence on this earth." But Wiener's spiritual beliefs did not come from nowhere. His journey to the East had been half a century in the making, beginning with the discovery of his ancestral tie to Maimonides, when he concluded that "even deeper than our simple Jewishness, in a sense the Orient was part of our own family tradition." As a man whose most famous ancestor had lived a respected life in a Moslem community, Wiener asked, "Who was I . . . to identify myself exclusively with the West against the East?"

He had vigorously pursued the scientific marriage of the West and the East. His affinity for teleology and circular causality reflected a mode of thinking more compatible with Eastern thought than with the reigning schools of Western logic. Yet none of Wiener's descendants or colleagues could imagine how this prodigal Jew who blew neither hot nor cold over any religion could have come to seriously entertain a belief in the central tenet of the Hindu faith—reincarnation. "I can't believe that," said his daughter Barbara, "unless he was living a life that was completely outside anything I know." Dirk Struik, too, was deeply skeptical. "I never saw that," he said. "We talked about philosophy. If he had a spiritual side I certainly would have known it." Nevertheless, it was Struik himself, the lifelong communist, who once stunned a professor of divinity at MIT who asked him why Wiener was "so outstanding." "I told him that Wiener was a little closer to God than the rest of us."

Wiener's young colleague Amar Bose, who was of Indian ancestry himself, knew Swami Sarvagatananda well, but he was taken aback by the news that Wiener had been the Swami's "bearded student" during the same years when he and Wiener were meeting together almost daily. In hindsight, he could see some flickers of a spiritual side emerging among Wiener's many facets, but he missed them entirely at the time.

"There were maybe small glimpses, but I did not see them," said Bose. "He may have been afraid people would laugh at him if he talked about such things."

Wiener kept his soul matters to himself, but he gave his philosophy of life to the world. In his later writings, he took on the challenge of elucidating anew for the new technological age just "what man's nature is and what his built-in purposes are." He sketched a revised image of man in the concrete terms of his new scientific concept of information, building on the fundamental biological connection between life and organization, or negative entropy as Schrödinger called it, to reach the highest form of organization in nature—human nature. Three years before the discovery of the structure of life's information-bearing DNA molecules, Wiener described in cybernetic terms the leap of nature that took place with the appearance in the random, physical universe of those unprecedented patterns of "organized complexity" from which all life was assembled and human beings ultimately emerged. That pattern of organization, said Wiener, "is the touchstone of our personal identity. Our tissues change as we live: the food we eat and the air we breathe become flesh of our flesh and bone of our bone. . . . *We are but whirlpools in a river of ever-flowing water. We are not stuff that abides, but patterns that perpetuate themselves.*"

A few years later, near the end of the second volume of his autobiography, *I Am a Mathematician,* he made a deeper and more personal statement about the meaning of life. In a few succinct paragraphs, he identified the process of communication as the organizing force of a chaotic universe, essential to the life of individuals and societies. He offered his new cybernetic perspective as one answer to the widespread pessimism and psychic malaise that had followed the horrors of the war, and to the fatalism embodied in the bleak view of the fifties' existentialist philosophers.

> We are swimming upstream against a great torrent of disorganization. . . . In this, our main obligation is to establish arbitrary enclaves of order and system. . . . It is the greatest possible victory to be, to continue to be, and to have been. No defeat can deprive us of the success of having existed for some moment of time in a universe that seems indifferent to us.
>
> This is no defeatism. . . . The declaration of our own nature and the attempt to build up an enclave of organization in the face of nature's overwhelming tendency to disorder is an insolence against the gods and the iron necessity that they impose. Here lies tragedy, but here lies glory too. . . .
>
> All this represents the manner in which I believe I have been able to add something positive to the pessimism of . . . the existentialists. I have not replaced the gloom of existence by a philosophy which is optimistic in any Pollyanna sense, but . . . with a positive attitude toward the universe and toward our life in it.

As the last formal expression of his personal philosophy and new interpretation of man—coming from one man who had been plagued by paralyzing states of depression all his life—it was a ringing affirmation.

Wiener closed *I Am a Mathematician* on a hopeful note, looking ahead to his future work and already anticipating the third installment of his life story, which he knew he could only call *Aftermath.* "At the age of sixty, I do not find myself at the end of my scientific interests nor, I hope, of my achievements. . . . How many years may be granted me . . . I do not know; but even now I can feel reasonably sure that my scientific career, though it began early, is lasting late."

16

Childhood's End

A part of us lives after us, diffused through all humanity—
more or less—and through all nature. This is the immortality of
the soul. There are large souls . . . stupendously big. Such men live
the best part of their lives after they are dead. . . . His soul will
live and grow for long to come, and hundreds of years hence
will shine as one of the bright stars of the past, whose light takes
ages to reach us.

—Oliver Heaviside (writing on the death of Maxwell)

WIENER'S FIRST GRANDCHILD WAS BORN in the summer of 1949 on Leo Wiener's birthday.

"He was absolutely delighted when our son was born, and he was thrilled when I named him Michael Norbert," Barbara recalled. "I think he really wanted an heir."

The two Norberts would not see much of one another. Young Michael Norbert Raisbeck spent the first dozen years of his life in New Jersey. Wiener spent much of that time in other countries. But Barbara and Tobey brought their children to New Hampshire in the summers when Wiener was there, and Barbara made an extra effort to help Wiener to get to know his namesake. "I thought it was important for both of them," she said.

Wiener's summer games had not changed since the 1930s, when he swam in Bear Camp Pond with his young daughters and the other children of South Tamworth and took them on vigorous tramps through the lesser White Mountains. As he tramped into his sixties, his pace fell off, but his heart was full as he passed on to his grandson his wanderlust and love for the gentle mountains he had roamed since his youth. "He and I would go walking," Michael remembered. "He walked very slowly, and as I got older his physical condition got worse. He was carrying a great deal of weight and a great deal of blood pressure. He would simply go very, very slowly, just a slow, deliberate, easy pace. He loved it, and I did, too."

One summer in the 1950s, Wiener returned from a trip abroad with a mesmerizing present for Michael: a toy steam engine that burned little pellets of real fuel. It had a little boiler, a little smokestack, and a little governor on top that twirled around just like the original device James Watt invented at the dawn of the Industrial Age. Wiener bent over the low table on the back porch and fired up the engine for his grandson. At age four, he could not fully appreciate the brilliance of that archetypal cybernetic device, but five decades later the memory was indelible. "I remember that steam engine!" said Michael. "You put the fuel in and it made a little 'clack' and the governor regulated the speed. That thing was a blast."

As the years went by, Wiener saw less and less of Michael and his three other grandchildren. For a decade after he split with the McCulloch group, Barbara's relations with her parents were severely strained. Early in the 1960s, Tobey Raisbeck took a job with a technology consulting firm and the family moved back to Boston. The move only increased the tension between Barbara and her mother, who continued to make her daughter the scapegoat in her efforts to control Wiener and the people around him—at one point she even denounced Barbara to her own children. Relations with her father deteriorated as well.

Wiener had been fascinated by the "Frog's Eye" paper that came out of McCulloch's lab at the RLE under Jerry Lettvin's direction. He was greatly impressed with the paper's stunning findings that described the analog action of the brain's cognitive information-processing operations at the cellular level, which added a whole new order of action to the analog functions Wiener had long postulated in the brain and nervous system. He knew that the minds in McCulloch's group were unmatched in their grasp of cybernetics' biological and neurological applications, and that they could still play an important part in his master plan to take his science beyond engineering and technology. And now he was prepared to bury the hatchet.

Wiener ran the radical idea by Barbara: After a decade of estrangement, he was thinking about reconciling with McCulloch and restarting their joint research venture at the RLE. Believing as he did that the group's alleged assault on her virtue had been punished and that his forgiveness was now warranted—in the interest of science, if nothing else—he sought her opinion of his plan and her blessing to proceed with it. The discreet father-daughter powwow lent further support to Lettvin's assertion that the breach never was about MIT's money or any patron's tainted research funding, since their breakthrough paper plainly stipulated that the project had been financed by the whole rogues' gallery of government and corporate interests Wiener reviled, including the Army, the Air Force, the Navy, and his bête noire, Bell Labs.

Wiener's reconciliation plan sent a shiver down Barbara's spine. She, too, missed her friendships with Wiener's colleagues, with the gentle Walter Pitts especially. She longed for a reconciliation of her own and had taken steps to reestablish relations with members of McCulloch's group, but her overtures were rejected. Now she feared another round of accusations and personal attacks if those old wounds were reopened, and that "once again I would be the fall guy." Wiener said she was "worrying about nothing." He still did not know that Margaret had made up the charges against McCulloch and the boys, and Barbara still had no idea that her mother had used her as a sacrificial lamb in her intrigue. Later, Wiener told others of his plan and received more mixed reactions, and in the end he never reconciled with his outcast colleagues.

And Wiener continued on his solitary wanderings through the corridors of MIT and the dark, dank RLE. After decades of poking into every portal and laboratory at Tech, he had been appointed to the esteemed post of Institute Professor, a position in which he would be fully interdepartmental and no longer solely affiliated with the math department. His status as an omnibus was official, and he tooted his horn up and down Tech's halls, proudly declaring to his colleagues that he was now "legitimate." "When he came in the office, I asked him, 'What does this mean?'" Amar Bose recalled. "He said, 'It means I can do any damn thing I please, and that's what I've been doing for the last forty years.'"

In the spring of 1960, just after his sixty-fifth birthday, MIT's policies required him to retire, but that was just a formality for Wiener. The institute threw him a gala retirement dinner, and his title was changed to Institute Professor, *Emeritus*, but he remained active at MIT and in his role as cybernetics' ambassador to the world. He was invited to return to Italy to spend the fall term at the University of Naples, which had taken the lead from MIT as the foremost center for cybernetic theory and research. As he had with India's scientists a few years before, he welcomed the opportunity to work with Italy's scientists and with others throughout Europe in whom he hoped to instill "a broad curiosity and the integrity to follow the new problems wherever they may lead."

Then, in the summer of 1960, that positive purpose in Wiener's work took on a new international dimension.

————

At first, the Soviet Union did not look kindly on cybernetics.

In keeping with the strict ideological controls Soviet authorities imposed on science during the opening years of the Cold War, Wiener's science was condemned by the party's organs as a "bourgeois perversion" promulgated by the corrupt West "to transform workers into an extension of the machine" and a "weapon of imperialistic reaction." At one point, the Soviet government even attacked Wiener personally in the pages of the party newspaper, *Pravda*, calling him "a capitalist warmonger" and "a cigar-smoking slave of the industrialists." Wiener loved the two libels—he liked to boast that the Soviet caricature of him as a cigar-smoking capitalist was "half right." The slashing attacks were part of a long tradition of official rejection of Western science and scientists, which began early in the Stalinist era and peaked in 1948 in the farce of Lysenkoism, the Soviets' ideological school of heredity that jettisoned the whole of genetics.

With the death of Stalin in the spring of 1953, the ideological controls over Soviet science were loosened. The windows of international exchange were opened, and cybernetics was one of the first breaths of fresh air that poured into the vacuum. Within a year, Soviet scientists began vigorously propounding cybernetic principles in their institutes, journals, and before the party's Central Committee, which was grappling with serious problems in the Soviet economy and throughout the society. In the spring of 1958, a Scientific Council on Cybernetics was formed.

The new science flourished in the nutrient-starved Soviet soil. Engineers initiated crash programs to produce computers and automated machinery. Physiologists and physicians devised innovative *bio*cybernetics applications. Social scientists applied cybernetic principles to the problems of public administration and to reforming the country's disheveled social structures. Physicists welcomed cybernetics as "a new science providing the key to literally every form of the existence of matter." Philosophers formulated new cybernetic perspectives on the dynamic forces of history and the destiny of socialism. By the early sixties, according to Loren Graham, the leading Western historian of that period in Soviet science, "the range of cybernetics loomed so great that the discipline seemed to some Soviet scholars to be a possible rival to Marxism." Graham tracked the wave as it swept through the ranks of Soviet scientists and the public, concluding, "One can find no other moment in Soviet history when a particular development in science caught the imagination of Soviet writers to the degree to which cybernetics did. . . . In the more popular articles the . . . utilization of cybernetics was equated with the advent of communism and the fulfillment of the Revolution."

And Wiener, the father of the new science, was raised up as a hero of the Soviet cybernetics revolution. By 1960, his technical works had been published in Russian, Czech, and Polish. His autobiographies and *The Human Use of Human Beings*—with Wiener's own railing criticism of America's military and corporations—appeared soon after in Yugoslavia, Romania, and Hungary under the imprint of the state publishing apparatus. Party organs that once ridiculed the cigar-smoking American mathematician now "virtually stated that it was criminal of Soviet philosophers to denounce the founder of cybernetics."

Inevitably, an invitation arrived. In June 1960, a month after he was elevated to *emeritus* status at MIT, and with the fifties' witch hunts formally concluded, Wiener agreed to go to Moscow to address the First International Congress on Control and Automation, before traveling on to his fall teaching assignment at the University of Naples. When he landed in the USSR, he received a reception that, by Soviet standards, was on a par with the welcome British rock bands would receive in the USA a few years later.

During his month-long stay in Russia, he attended conferences and delivered lectures in Moscow, Kiev, and Leningrad. He received the star treatment from his former foes at *Pravda* and other instruments of the party press. He also received the first batch of royalties for his books published in the Eastern bloc. (His accrued earnings, payable in Russian rubles, were worthless in the West, so he accepted payment in cheap Russian caviar and champagne, which he consigned to the cellar of his home in New Hampshire and never so much as tasted.)

Now all the state-approved practitioners of Soviet science clasped Wiener to the bosom of Mother Russia, but Wiener was no tool of the capitalists or the communists at the Moscow conference. Attuned to the Cold War raging on both sides of the Iron Curtain, he crafted his remarks with the care and balance of an Old World watchmaker. He championed the new sciences and their technologies as he always did, hailing their potentials and forewarning of their perils; and he held the powers that be in both blocs equally accountable for their consequences.

Wiener thundered in the East, as he did in the West, that "science must be free from the narrow restraints of political ideology." In his plea to scientists and bureaucrats in both Cold War camps, he warned of the dangers to human survival posed by gadget worshipers and "devoted priests of power," both those in the West, with their "slogans of free enterprise and the profit-motive economy," and their mirror images in "that through-the-looking-glass world where the slogans are the dictatorship of the proletariat and Marxism and communism." His sharp words did not slow the advance of his science in the East. The year after Wiener's visit, at the 22nd Party Congress, the Central Committee endorsed cybernetics as "one of the major tools of the creation of a communist society," and Soviet Premier Nikita Khrushchev personally declared, "It is *imperative* to organize wider application of cybernetics . . . in production, research . . . planning, accounting, statistics and state management."

"Wiener is the only man I know who conquered Russia, and single-handed at that," Wiener's Dutch communist friend Dirk Struik remarked after Wiener's march on Moscow.

Western scientists were stunned by the scope of the Soviet cybernetics program. On their return from the 1960 congress, a group of British scientists admitted that they were "flabbergasted" by Soviet achievements in automation and that the Soviet program surpassed anything in the West. The Soviets' newfound affection for Wiener did not escape the notice of the U.S. government either. The FBI's informants at MIT were watching when a visiting Soviet scientist gave Wiener his first copy of *Кибернетика*—*Cybernetics* translated into Russian. And by the early 1960s the Soviets' love affair with cybernetics had attracted the attention of the Central Intelligence Agency. While Wiener was in Moscow, the CIA's foreign intelligence divisions were tracking the snowballing phenomenon Agency officials referred to as "Soviet cybernetics."

For some time, a team of CIA intelligence analysts, led by John J. Ford, a Soviet expert working in the Agency's Office of Scientific Intelligence, had been monitoring the explosion of cybernetics in the Eastern bloc. Twice the West had been caught off guard by the progress of Soviet science—the Soviets' first atomic bomb detonation in 1949 and then the surprise launch of the first earth-orbiting satellite, Sputnik, in 1957—and the U.S. government did not want to be caught napping again. The year Sputnik went up, Ford began cultivating his sources and collecting intelligence on developments in cybernetics in the Soviet Union. By the time the Kennedy administration came to power in Washington, when the Cold War entered its "hour of maximum danger" over the proliferation of atomic weapons, Ford had already begun circulating a series of intragovernmental reports and classified intelligence memoranda on "The Meaning of Cybernetics in the USSR."

Ford's voluminous database, obtained from the Agency's assets inside the Eastern bloc, from Soviet government publications and scientific and technical journals, and from unnamed sources "in industry, government, labor, finance, and the academic world," told a surprising story. Ford learned that the Soviets' conception of cybernetics was much broader than the prevailing American sense.

According to Ford, the Soviet Cybernetics Council had outlined a massive experimental program to train the "new Communist man" using cybernetic methods. The Council's seventeen technical sections and thousands of subsections had assigned one hundred research and development facilities to cybernetics projects. And while the Soviets were making their biggest strides in industrial automation, they were also developing cybernetic "technology for the optimal control of the economy" and planning for the day "when 'thinking cybernatons' will revolutionize . . . service technology."

Ford recognized the West's commanding lead in computer technology, but he saw signs that the Soviets were moving to narrow the gap. He even uncovered a plan to develop an archetypal, Internet-style, national information network, or "Unified Information Net," with projects already under way to coordinate industrial and economic activities across the USSR. Ford was not convinced that cybernetics would cure all the Soviets' economic and social ills, but as new intelligence came to his attention his early confidence gave way to a growing sense of concern.

A year later, in another classified report, Ford and his team sounded a silent alarm throughout the U.S. government about the expanding scope of Soviet cybernetics. The 126-page report revealed that the Soviets were rapidly operationalizing their master plan for cybernetics and extending their efforts into domains as varied as defense, space vehicle guidance, and urban planning. The Soviets were on their way to achieving something that Wiener had only dreamed of: unifying the many spinoffs and subdisciplines of cybernetics, and all the new communication and control sciences, under one scientific umbrella comprised in the single term "cybernetics." The greater Soviet program combined the production of automated machinery and nationwide telecommunications networks with an ambitious research and development program in the new field of artificial intelligence—or, as the Soviets called it, "autointelligence."

The report also revealed that Soviet cyberneticists had begun research in the embryonic sciences of chaos and complexity theory, and that plans had been laid for the design of von Neumann–style cellular automata. And it uncovered a surge of activity in a field of intense interest to Wiener—medicine—including development of a new generation of cybernetically enhanced prosthetic devices, which served a persisting need among the Eastern bloc's millions of surviving wartime casualties. Prototypes were already on display for automated reading machines for the blind, prosthetic arms and legs, and entire cybernetic organs, or "cyborgs," as they had become known. The Soviets were even addressing Wiener's greatest social concern, the human impact of cybernetic technology, in studies of man-machine interactions and the psychological and social effects of automation.

The long-range goal of the Soviet program, as the CIA assessed it, was to use cybernetic principles and technology to take the Soviet system to "a higher state . . . of social evolution" that extended to every country within the Soviets' sphere of influence, including Cuba, Vietnam, East Germany, and North Korea. By such methods, the head of the Soviet Cybernetics Council proclaimed, "cybernetic methodology . . . will lead to a fuller realization of the fundamental advantages of the socialist over the capitalist system."

Ford and his colleagues warned that Soviet progress in "'military cybernetics' or command and control" was particularly worrisome. However, in Ford's view, the most threatening advances in Soviet cybernetics were not strategic. Soviet progress in cybernetic theory—a subject that had been largely neglected by American scientists in the years since the founding work of Wiener, McCulloch and Pitts, and von Neumann—indicated that the Soviets could soon make new leaps in applying cybernetics to practical problems, and it raised "a distinct possibility that the Soviet Union will gradually assume supremacy in certain areas . . . connected with applications . . . and cut down the temporary Western lead in computation technology."

Soviet philosophers were developing an entire cybernetic worldview, and practitioners were already turning out prototypes of "the New Soviet Man of the day after tomorrow" in special boarding schools throughout the USSR that had set a goal of producing two million young computer programmers. Ford's team confirmed that the Soviets had begun production of a new line of digital computers for economic and military applications, and that Soviet artificial intelligence researchers were embarked on "the path to solution of a whole series of . . . questions . . . which US investigators have left unanswered."

Ford acknowledged that, in most practical applications, the Soviets were still well behind the United States, but he was quick to warn against the dangers of Western complacency in the contest he saw as an escalating cybernetics race between the superpowers with global implications. He emphasized that many Soviets believed their cybernetics program would prove to be the decisive factor in the East-West contest, and he shared their view that it could serve as a model for the world's emerging nations to follow and "influence their paths of future development along lines inimical . . . to US policy."

Ford was no Chicken Little or anticommunist crusader, but he was becoming deeply concerned that his CIA superiors were not taking his team's findings seriously. Early in the Kennedy Administration, he began meeting periodically with the president's brother, Attorney General Robert F. Kennedy, and members of his inner circle to discuss Soviet cybernetics, and to consider steps American policy makers might take to close the widening East-West cybernetics gap.

On the evening of October 15, 1962, Ford was the featured speaker at one of the exclusive Hickory Hill seminars convened at Kennedy's estate and the homes of other administration officials. The seminar that night was hosted by Secretary of Defense Robert S. McNamara and attended by Robert Kennedy and other high-level government figures. However, midway through Ford's talk, his presentation was interrupted by another contingent of CIA officers delivering more urgent information: the first aerial reconnaissance photographs confirming the presence of Soviet nuclear missiles in Cuba.

To that point, Kennedy's response to Ford's presentations had been "very affirmative," and others in attendance also were interested and wanted to learn more. But that night, when the courier arrived with the latest news from Cuba,

Ford's unofficial initiative to generate government support for American cybernetics programs, along with all the other business of the nation, came to a standstill while the missile crisis between the two superpowers played out.

———

During those peak Cold War years, Ford assembled an extensive network of contacts in academia, industry, and government, but he never made official or unofficial contact with Wiener, who remained on the FBI's list of individuals who "should not be contacted in any way by [federal] agents in the course of their investigations." Instead, he met with people at MIT who had worked with Wiener, and with other experts in cybernetics, and he organized an informal discussion group of technical consultants and government officials in Washington who shared his concern that the Soviets were winning the cybernetics race.

However, outside those select circles, other developments in American science and technology were attracting far more attention than cybernetics.

In Cambridge, the Gestalt-trained psychologist Joseph C. R. Licklider, who had been a presenter at the last Macy conference Wiener attended and a "faithful adherent" of Wiener's Tuesday night dinner circle, was conducting his own research in response to Wiener's call for new approaches to systems involving "machines and human beings in joint enterprises." At the Lincoln Lab, he helped write the first human-computer interfaces for the SAGE air defense network. In 1960, he published a landmark paper that combined the latest ideas on computing and artificial intelligence into a new vision of humans and machines working symbiotically in a worldwide "supercommunity" of networked computers. Two years later, Licklider went to Washington to bring his vision to fruition.

In October 1962, the same month as the missile crisis, Licklider was appointed director of the new Information Processing Techniques Office at the Defense Department's Advanced Research Projects Agency. ARPA, as it was called, had been assembled by the military in response to the Soviets' Sputnik shot and modeled after the wartime OSRD. At ARPA, Licklider was given $12 million annually to distribute to scientists across the country to devise a new global command-and-control computer network for the military. In March 1963, in one of its first bequests, his office awarded $2.3 million to MIT's new "Project MAC," directed by MIT information theorist Robert Fano and Licklider's SAGE programming colleague Oliver Selfridge, to develop the interactive, time-sharing network that came to life a few years later as the ARPAnet—the earliest incarnation of the Internet.

From ARPA's initial grant to Project MAC, $1 million went to MIT's new Artificial Intelligence Laboratory to provide "intelligent assistance" to the new computing network. More funds would follow, totaling almost $10 million over the next decade, along with other sums Licklider and his successors awarded to the new AI laboratories at Stanford, Carnegie Tech, the RAND Corporation, and ten other AI research centers. Those generous grants threw the nation's support behind artificial intelligence research and established the new field as a legitimate scientific enterprise.

Marvin Minsky, whose AI Lab at MIT prospered from ARPA's largesse, recalled that heyday of unfettered military funding, and the steady stream of MIT men who flowed through the ARPA pipeline. "Licklider went to Washington. He had been a good friend of mine, so one day Fano and I went down there and talked to him. He said, 'We need time-sharing.' He said, 'Why don't we start a big project that will do AI? I can get $3 million a year.' This was to be administered by a former student of mine and, for ten years, whenever one of them got tired of it we'd get another postdoc to go down there. It was heaven. It was your philanthropic institute run by your students with no constraints and no committees. Of course there was no way to spend that much money, so we built some machines and for the next few years I never had to make any hard decisions whether to fund one project or another because we could just do both."

Through those years of copious funding for computer networking and AI research, no major or minor sum was awarded by ARPA for research in the field of cybernetics. Heinz von Foerster, whose new Biological Computer Laboratory at the University of Illinois was in constant need of funds, watched the climate for cybernetics research grow cold after ARPA and AI came on the scene and "artificial intelligence" became the new buzzword of America's scientific bureaucracy.

According to von Foerster, the embrace of AI by the U.S. military, together with the Soviets' embrace of cybernetics, spelled doom for cybernetics in America during those ideologically charged years. "They wanted to chase out cybernetics as fast as they could. It was not suppressed, but they neglected it and began funneling all their money into intelligence, whether it was artificial or natural," said von Foerster, describing the attitude he encountered among government funding agencies newly attuned to the importance of Cold War intelligence operations. "I talked with these people again and again, and I said, 'Look, you misunderstand the *term*,'" he said of AI. "They said, 'No, no, no. We know exactly what we are funding here. It's *intelligence!*'"

Over time, cybernetics and artificial intelligence became cold and hot buttons, respectively, wherever von Foerster turned within the circles of scientific funding. "At the University of Illinois, we had a big program which was not called artificial intelligence. It was partly called cybernetics, partly cognitive studies, but I was told everywhere, 'Look, Heinz, as long as you are not interested in intelligence we can't fund you.'" Von Foerster charged federal officials and agencies with limited thinking, and a failure to grasp the fundamental points of cybernetics and their value for science and society that had excited scientists and governments in many countries, the Soviets especially.

The story as von Foerster told it presaged the broad decline of cybernetics research in America that began in the early 1960s, owing in part to the internal politics of American science, and in part to the Cold War competition in science and technology. "That more or less gave it the death knell," he said. "People were saying, 'Let's get away from that cybernetics. It undermines our American way of thinking.'"

For years, Wiener had railed against the use of the new science for military purposes and he boycotted such efforts. Now, in an act of retribution by the gods, and apparently by the lords of American science as well, his science was paying a tangible price for its successes on foreign soil, and, maybe, too, for its father's defiant words and actions.

Wiener, who was out of the loop of official decisions but never indifferent to them, continued to steer a separate course. In the sixties, his focus returned to the earliest instincts of his childhood, his horror at the specter of suffering and mutilation, and to his first priority after the war, to "repair the . . . damage done by the weapons of war on which he had worked." His first project with Jerome Wiesner, to build a "hearing glove" for the deaf that would translate speech into tactile sensations, sank in the morass of Wiener's split with McCulloch and with Wiesner himself. But the new prospect cybernetics made possible—to design an electronic "sensory prosthesis" that worked symbiotically with the electrical signals of the human nervous system—continued to tug at his imagination. Then, one day, Wiener took another leap that led to his last great project of cybernetics in the service of humankind.

In September 1961, a year after he returned from the Soviet Union, while wending his way through Building 7, Wiener tumbled down a flight of stairs and landed across the river, at Mass General, with a broken hip. The doctors could not believe their good fortune when Wiener wound up in traction in their midst. The hospital's best orthopedists had recently returned from Moscow, where they had witnessed the unveiling of the first triumph of Soviet biocybernetics: a prosthetic hand powered by electronics and precisely controlled by sensors, servomotors, and other cybernetic mechanisms. The Americans were awestruck by the demonstration, and struck again when the Soviets told them, as one doctor reported, "Look, you people must know all about this, because we got all the ideas from Wiener." The team came back determined to track down Wiener and pick his brain for their own prosthetic project, when they happened upon him in the VIP wing of their hospital. "Lo and behold, he was already their captive," Amar Bose recalled. "They all converged on him. 'What about all this stuff?' And he said it was basically what he had told them ten years before."

Indeed, in the early 1950s, Wiener had given a speech at Harvard Medical School about the future of electronics in medicine. He said there was no longer any need for primitive biomedical devices like mechanical limbs and iron lungs that worked passively on people. He explained that live electrical signals could be picked up from a limb or a lung, even if the nerve endings had been severed, and harnessed to control intelligent electronic machines. "At the time, the speech was ignored as a crazy scheme by some wild-eyed mathematician from MIT," Bose recalled. But a decade later, after viewing the Soviets' prosthetic hand, some of those same doctors had a delayed reaction. Standing over his bedside, they asked Wiener to steer them through their own project to build not just a prosthetic hand, but a prosthetic arm to ease the lives of "high" amputees.

Bose was not present at that meeting, but he was more than a witness to the events that followed. He was yanked into the "Boston Arm" project the same way Wiener had conscripted him into several earlier joint ventures. "Some Mass General surgeons called me and said, 'The first meeting is this week,'" he recalled. "I said, 'Meeting of *what?*' They said, 'What do you mean? You're the head of it!' Then I found out Wiener had volunteered me."

With his trusted postdoc and their new teammates assembled at his bedside, Wiener laid out a detailed design for the first cybernetic arm. In the device he envisioned, the prosthetic arm would be attached to the wearer with straps and electrical sensors, and the controlling signals would be picked up through the skin from nerves firing in the arm at points above the amputation. The mechanics and electronics would be controlled bionically—by the patient's thoughts alone—and refined progressively by feedback training. Wiener's vision powered the project. Bose oversaw the venture and guided their team of doctors, electrical engineers, and biomedical technicians through its formative stages. Research facilities were provided by Mass General, MIT, and Harvard Medical School, with additional funding from the Boston-based Liberty Mutual Insurance Company, which took a major interest in the venture on behalf of the many disabled workers its policies covered.

After two years of research and development, when the team was readying its first test of the device, a young doctor from Harvard who had come late to the project took Bose aside. "He said, 'Wiener's leaking this stuff. You've got to tell him we don't want any leaks. They would be bad for the project.'" Bose did what the doctor asked. "I went to Wiener and he was like a child. He was so apologetic. He said, 'I can't think of anybody I talked to, but maybe I have, and

I'm so sorry.' It turned out, he never had, but this doctor wanted him to suppress everything until he could make the release himself."

Finally the big day came. "We actually produced an arm, and we found an amputee with a high amputation," Bose recalled. "We attached the arm—I can remember the reaction very clearly—the man was sitting down and the arm came up, and the man jumped and said, '*My god, it's chasing me!*' But in ten minutes time he was able to wear it beautifully." The team players were elated, most of all Wiener.

Then, in December 1963, the first press report on the project appeared in the *Saturday Review*. The article made passing references to Wiener as someone whose early theories were "poetically appropriate," but it named the young doctor from Harvard, who also worked in Liberty Mutual's clinic, as the one who "from the scientific point of view . . . deserves the applause for putting the Wiener theory to work." Bose recalled Wiener's muted response to the slight. "I took the magazine to Wiener. He read the whole thing and just put it aside. Didn't say a word. Not a word, no criticism, no nothing."

Bose remembered the media circus that ensued when the arm was first demonstrated publicly. "When it came out in *The New York Times* the whole thing was credited to this doctor and it was a hoax. He hadn't contributed to the project at all. He was there and the last day, when we had a man with the electrodes on and the thing was working, he appeared with the press and got his picture taken."

The final chapter in the Boston Arm project played out like a scene from *The Tempter*, Wiener's fictionalized tale of scientific betrayal and corporate intrigue. Several years later, a patent was granted for the first wearable Boston Arm and assigned to Liberty Mutual, which proceeded to manufacture and market the device. From the beginning of the project, Wiener had stipulated that "if any patents are taken out [it] will merely be to put them into good hands for manufacturing, not to make any profit," but all that arm-twisting by the Harvard doctor and his patrons turned Wiener's humanitarian effort into a wholly commercial enterprise.

Wiener's design of the Boston Arm was a coup for cybernetics, and a vivid demonstration of the power of his science to foster man-machine interactions with tangible benefits for people's daily lives. Wiener did not profit in any way from the device, or from the lucrative industry in electronic prostheses that blossomed in the years that followed, but he held to his own ethical standards and took pride in his serendipitous achievement. "I have seldom seen Wiener so happy as when he told how he turned the mishap of his fall into a victory for the handicapped," Dirk Struik recalled.

———

By 1963, Wiener's hip was better, but he was not in the best of health. He had gained more weight—topping out at just over 200 pounds. His doctors diagnosed him with Type 2 diabetes and prescribed a diuretic that relieved him of much of the extra fluid he was carrying, but his heart was not strong. He never fully recovered from his earlier bout with angina, and he was put on digitalis for that condition. His hearing was failing, too. He bought a hearing aid, but it was a primitive electronic device, noisy, with no filters, and he turned it off frequently, a move that would exasperate Margaret.

Margaret remained steadfast in her devotion to her husband, but she was no longer his tower of strength. Early in the 1960s, she was diagnosed with colon cancer. She underwent a successful colonectomy, but her continuing debility weighed on them both. Her cancer hit Wiener hard. "Norbert used to come over to our house and his tears would be running down his face just at the thought of losing her," Mildred Siegel recalled. Fagi Levinson, too, saw Wiener's suffering. "When he talked about her, his face welled with tears. It was hard for him. He felt very responsible."

Despite their infirmities, 1964 started off with great promise. Wiener learned that he had been chosen to receive the National Medal of Science, America's highest scientific honor. The medal marked the recognition of a grateful nation for his contributions to science in wartime and peacetime, and it bestowed something even greater on Wiener: the recognition of his peers. A panel of the country's most distinguished scientists and mathematicians had nominated him, and President Kennedy himself had made the final selection.

It was a high point of Wiener's career as a singular American scientist who had been steeped and tempered in all the world's scientific traditions, and who had distinguished himself and his country on the stage of international science. And not a bad day's work for someone who still had an open security file at the FBI.

Kennedy was cut down before Wiener could shake his hand, and in January 1964 Wiener, Margaret, and Peggy traveled to Washington for the award ceremony at the White House presided over by the nation's new president, Lyndon B. Johnson. Barbara was pregnant with her fifth child and unable to attend, although her relations with both her parents had soured too much for her to join in the festivities under any circumstances. The family convened in the Old Executive Office Building next to the White House, in the office of the Presidential Science Advisor Jerome Wiesner, who had been appointed by JFK and would serve only a few months more under LBJ. Wiener had had almost no contact with Wiesner for more than a decade, but on this occasion they greeted one another genially. Then Wiesner led the group on a tramp through the network of underground tunnels that connected the executive offices to the White House—"Dad got a big kick out of that," Peggy recalled—and into the White House library for the presentation.

It was one of the most extraordinary collections of scientific minds that had ever gathered at the White House—and the most awkward juxtaposition of American scientists assembled anywhere in a long while. Along with Wiesner, who attended in his official capacity, another old friend and one old foe of Wiener's also had been selected to receive medals that year: his MIT colleague and wartime boss Vannevar Bush, who had suppressed Wiener's early digital computer designs and later shooed him out of the scientific war effort; and John R. Pierce, the head of communications research at Bell Laboratories, who dismissed Wiener's contributions to information theory in preference to those of his Bell Labs colleague Claude Shannon.

Wiener stood stiffly at attention beside his fellow recipients, then he stepped forward as Johnson called his name, presented his medal, and read from the official citation. In his slow Texas drawl, Johnson extolled Wiener for his:

> . . . marvelously versatile contributions, profoundly original, ranging within pure and applied mathematics, and penetrating boldly into the engineering and biological sciences.

Flashbulbs burst in Wiener's lenses as he posed with the president and his peers and, for one brief moment, basked in the spotlight. But he did not look well that day. He was pale and drawn from his recent health ordeals, and from worrying about Margaret after so many years of her worrying only about him. Now, standing shoulder to elbow with the towering LBJ, he looked at once great *and* small, but no longer childlike. At sixty-nine, his fine white hair thinned to a wispy comb, his tufted goatee almost invisible against his blanched chin, for the first time in his long walk, he looked old—and tired.

But a month later he was back in action, walking to his own beat on a course that would take him out of the country and far from the turf wars and ideological battles that were enveloping his science. He and Margaret had made plans to travel to Europe in February 1964, where Wiener was to spend the spring term as a visiting professor and the honorary head of neurocybernetics at the Netherlands' Central Institute for Brain Research.

Peggy said bon voyage to her parents in Washington, but Barbara did not say goodbye to her father or her mother. She had had no contact with either of them for several years and, as it happened, she picked the worst possible moment to break things off for good. On the eve of their departure, Wiener called his favorite daughter one more time. Still deeply hurt by events from the past, she would not come to the phone. "I just couldn't face him," Barbara admitted. "I felt he would have pleaded for me to come back, to be friends again, and I wouldn't have been able to refuse him. I felt it was better just not to talk." A few days later, Wiener and Margaret flew to Amsterdam.

Wiener was excited about his appointment in Holland, and by the chance to pursue the kind of brain research essential to cybernetics that had been foreclosed to him when he broke with McCulloch and his group at the RLE. He was also looking forward to spending time with his old friend Dirk Struik. In the chill that lingered long after the McCarthy era ended, Struik had been refused an emeritus post at MIT and was unable to find another teaching position in the United States, so he had moved back to his homeland, where he was welcomed at the University of Utrecht. The two men were planning a grand reunion with their wives. However, soon after Wiener got settled in Amsterdam, he had to leave the Low Countries for a few weeks to deliver a series of lectures in Norway and Sweden. Before he left, he telephoned Struik. "He said, 'I have several jobs in Scandinavia. When I come back, we'll have that dinner together.'" But it was not to be.

Wiener's critics would say later that he went to Sweden "to lobby to get the Nobel prize," which was awarded in many categories but not mathematics. Like Claude Shannon and other world-class mathematicians who made fundamental contributions to communication theory, Wiener felt the injustice personally and for his field. But Wiener went to Sweden, as he went everywhere in his later years, as an educator and ambassador of cybernetics, not to beg or berate the Nobel committee. After lecturing for four days in Trondheim, Norway, he arrived in Stockholm in mid-March to deliver a lecture at the Royal Academy of Sciences. The next day, he attended a luncheon in his honor with his Swedish colleagues and led a lively discussion of his work. Afterwards, he went to the Royal Institute of Technology to see the institute's new communications laboratory.

His breathing was labored and his heart started to race as Wiener and his Swedish host made their way up the long steps of the Royal Institute. Then, suddenly, Wiener seized, collapsed, and lost consciousness. There, on the stairs, he went into cardiac arrest and stopped breathing. He was pronounced dead on arrival at a nearby hospital at 3:30 P.M. on March 18, 1964. Because of the abrupt circumstances and Wiener's reputation, an autopsy was performed and the cause of death was attributed to a pulmonary embolism. The frequently fatal affliction may have been a delayed effect of his broken hip and the long convalescence that followed, or a consequence of sitting too long in his travels around Scandinavia.

Back at MIT, word of Wiener's death flashed down the infinite corridor and over to the plywood palace of the RLE. Work came to a halt as people gathered to share the news and their memories, and the institute's flags were lowered to half staff in honor of the fallen institute professor who had roamed its halls for forty-five years.

That night, a select group met at Joyce Chen's for one last session of Wiener's supper club. Someone tore a sheet of filler paper out of a binder and scratched out a few words. Twenty-one people—including Wiener's first graduate student Y. W. Lee, the founder of MIT's Servomechanisms Laboratory Gordon Brown, physicist Jerrold Zacharias who had been the Rad Lab's liaison to Bell Labs' fire control team during the war, the first director of MIT's Lincoln Laboratory Albert Hill, the founder of the RLE's Communications Biophysics Lab Walter Rosenblith, the information theorist Robert Fano, Jerome Wiesner who had recently returned to MIT from Washington, MIT's President Julius Stratton, Warren McCulloch, and Joyce Chen—signed their names to the simple statement of fact they would send on to Margaret:

We loved him.

When the formal inquest into Wiener's death was complete, Margaret arranged a small funeral service in Stockholm. Wiener's body was cremated and his ashes were sent back to the United States, while she went on to Germany to visit with her relatives before returning home. In keeping with his wishes, Wiener's remains were laid to rest in a far corner of the Vittum Hill Cemetery in South Tamworth, beside the proud New Englanders whose sober dignity and legendary reticence he loved, in the shade of a stand of sugar maples. Margaret arranged another small service for Wiener's interment, which was presided over by a clergyman from the local Episcopal church.

Several weeks later, back in Belmont, Margaret organized a public memorial service at the Unitarian church she and the girls had frequented. It drew a good crowd from MIT and the greater Boston community, but something about all those traditional Christian ceremonies did not sit right with Margaret. Amid her grieving and growing unease, she picked up the telephone and called Swami Sarvagatananda at the MIT chaplain's office.

"What a man we have lost!" said the Swami to Margaret, expressing his condolences on behalf of all of MIT's chaplains. But something else was weighing on Margaret's mind.

"Mrs. Wiener called me and said, 'Swami, Norbert is not happy,'" the Swami remembered. "I was shocked to hear that. 'What do you mean, Norbert is not happy?' She got a sign, she said. I think she had some nightmares."

Margaret appealed to the Swami. "You know, you did not attend our family's memorial service and he loved you most," she said. "They spoke in English and he loved Sanskrit." She asked him to perform a memorial service in the MIT chapel. "I'll be very happy. Norbert will be very happy," said Margaret.

The date was set for a Friday. The Swami invited all of MIT's chaplains—Jewish, Catholic, and Protestant—to join him. The turnout was unprecedented. "Do you know that day the whole chapel was full, the *hall* was full, outside.

Outside! I looked, there was no place for people to come. The Catholic father could not get through the crowds."

On June 2, 1964, Swami Sarvagatananda presided over the memorial service at MIT in remembrance of Norbert Wiener—scion of Maimonides, father of cybernetics, avowed agnostic—reciting in Sanskrit from the holy books of Hinduism, the *Upanishads* and the *Bhagavad Gita*. Wiener's final sendoff came with the prayers of clergymen from three world religions joined in recognition of his universal humanism, and the crush of the MIT community.

The tributes poured in. Wiener's image and obituary appeared on the front page of *The New York Times*, in wire service news stories, and in publications as varied as *The Bulletin of the American Mathematical Society*, *The Journal of Nervous and Mental Disease*, and *The New York Review of Books*, as his admirers and detractors alike tried one last time to get a rope around the Elephant's Child. One source in the *Times* described Wiener as a man with a "view of life so critical and lacking in humor as to set him apart from his colleagues." *Time* magazine eulogized him as an interdisciplinary thinker who "drifted from university to university like a medieval scholar but remained almost a stranger in the vast world outside the classroom." *Newsweek* said he "wasn't brought up, he was programmed like some human Univac."

His MIT associates did a better job of capturing Wiener. Warren McCulloch, cast out of Wiener's inner circle for all those years, came as close as anyone to preserving the essence of the man who was a myth before he ever met him:

> Norbert had not only the imagination to invent but the burning desire to share his notions of the useful and the good. . . . We are too near the man to see his greatness in perspective. Genius like his rarely takes the time or trouble to grow armor to shield it from the roughness of the world, and retains the charm of childhood throughout life. . . . One thing is certain: Neither Medicine nor Engineering nor Mathematics will ever be the same as if he had not been.

———

When his death notice appeared in the *Boston Globe*, agents in the FBI's Boston Field Office put the clipping in Wiener's file and closed the security investigation they had opened seventeen years before. His fate was in history's hands now, but the fate of Wiener's brainchild, cybernetics, was not so secure.

In the days after his presentation at Robert Kennedy's Hickory Hill seminar, at the request of President Kennedy's Special Assistant Arthur Schlesinger, Jr., John Ford, the CIA's expert on Soviet science and technology, had prepared a summary report on Soviet cybernetics that Schlesinger took personally to the president.

Kennedy and his aides were deeply divided over its significance. When the Cuban Missile Crisis abated, Kennedy instructed his Science Advisor, Jerome Wiesner, to organize a special "cybernetics panel" within the President's Science Advisory Council to make an independent evaluation of Ford's findings and their own assessment of the Soviet threat. In their first meetings, the panel expressed near-unanimous agreement with Ford's concerns, but before any substantive policy decisions could be made, Kennedy's assassination brought an end to his administration, and to the influence of its progressive thinkers on American science and Cold War foreign policy.

Finally, in February 1964, after five years of internal squabbling and a year of wrangling with the President's Science Advisory Council, the CIA circulated the first of Ford's classified reports on Soviet cybernetics—its first official statement on the subject—to a hundred recipients at the Defense Department, the State Department, NASA, the Atomic Energy Commission, the top-secret National Security Agency, and to the CIA's own office of counterintelligence and psychological warfare. However, Ford's reports drew scant response from government officials. While the military offices of research and development were generously funding projects on artificial intelligence, interactive computing, and industrial automation, officials in both military and civilian agencies became openly hostile to cybernetics. Most did not understand the first thing about it. Some dismissed it out of hand. Others gave every indication that, in the eyes of the American government, cybernetics had taken on a disagreeable red tinge.

As Ford complained in vain to his superiors, American military and civilian officials displayed a short-sighted preoccupation with cybernetic machinery and, in their "state of hardware-oriented false euphoria," he believed, they remained largely ignorant of cybernetics' wider biological and social dimensions, which the Soviets were developing hand-in-hand with their cybernetic technology. Privately, Ford told his colleagues and family, "They just don't get it."

Ford was not the only one who was frustrated and alarmed by the spreading reaction against cybernetics at high levels of American science and government. Soon after Wiener's death, Warren McCulloch joined Ford's cybernetics circle in Washington, knowing that, with Wiener no longer on the scene, he and other leading figures in the field were the only ones who could take cybernetics forward in the United States. McCulloch brought in other chieftains in the far-flung tribe of American cyberneticists, including Heinz von Foerster, Margaret Mead, and Wiener's former collaborators Y. W. Lee, Julian Bigelow, and Arturo Rosenblueth.

In July 1964, Ford and his new cohort formally incorporated and founded the American Society for Cybernetics, for the purposes of "fostering development of the discipline, anticipating the impact of cybernetics, and providing current information on cybernetics." The group sought especially to encourage the "youngsters"—as McCulloch called them—to study and undertake research in cybernetics. McCulloch was well aware of Wiener's objections to uses of his science by "would-be social reformers," and of his own position as an outcast from Wiener's circle. But he and other ASC founders saw their role, to the contrary, as an effort "to forestall the opportunists and the do-gooders," and as a chance to win recognition for cybernetics as Wiener had envisioned it, as a unifying force among the sciences and a new field with valuable knowledge to offer the whole of society. "Perhaps we will succeed," McCulloch said hopefully. "The Russians have."

In the end, cybernetics did not give the Soviet Union the winning hand in the Cold War. As Wiener himself had forewarned during his visit to Moscow in 1960, the socialist system's creed of centralized planning and rigid, top-down, authoritarian rule ran counter to the most basic principles of self-governing cybernetic systems, and that structural flaw in the Soviet façade ultimately would bring down the entire edifice.

Three decades later, when the Soviet system failed catastrophically for many reasons, historians confirmed the painful lessons of Soviet cybernetics, that "[freedom] of information and . . . decentralization of control" are essential to social systems as well as technical ones, and that "computerization of a society strengthens local . . . tendencies . . . to preserve and . . . develop in every possible way the relative independence of information processes" and the "degrees of freedom . . . and . . . reaction available" to people throughout the system. The fall of the Soviet Union and the breakup of the whole Eastern bloc took place, in large part, because those fundamental information age imperatives were ignored or defied, as Soviet leaders realized too late that the new communications equipment—personal computers, vast data banks, unified information networks—and the people who use them could not be controlled with an iron fist.

Those fatal flaws explained the collapse of the Soviet system and, with it, the decline of Soviet cybernetics as a force for technological progress and social change in the Eastern bloc. But there were flaws in the American system, too, that contributed to the decline of cybernetics concurrently in the United States. In the late 1960s and 1970s, Margaret Mead and her peers in the American Society for Cybernetics worked strenuously to persuade government officials to support applications of cybernetics for peaceful purposes, to promote interdisciplinary research in the field, cross-cultural cooperation between the superpowers, and more open communication among all the world's peoples. But, as Soviet cybernetics grew and the Soviets' Unified Information Net began to take form, Mead saw the new fear being spread in U.S. government circles, "that the Soviet system may become totally *cyberneticized* . . . with thousands of giant computers linked together in a system of prodigious and unheard-of efficiency." She pleaded with the lords of American science: "If we continue to discuss the computerization of the Soviet [system] in terms of emulation and dread, cybernetics as a way of thought will cease to be ideologically free." She urged them instead to look critically at the needs of American society and to use cybernetics to develop more sophisticated ways of handling American systems that, she believed, were "in dire need of attention."

In Mead's mind, cybernetics became politicized and fell victim to the Cold War between the sciences at home and abroad. For Heinz von Foerster, the prevailing funding strategies of the period—particularly the favored position of artificial intelligence labs at MIT and elsewhere, which had promised to provide their military patrons with systematic new sources of "intelligence"—dealt a crippling blow to researchers in cybernetics, who by and large were pursuing more humane projects in biology, bionics, the social sciences, and cross-cultural communication. Four decades later, von Foerster was still appalled when he looked back on that tipping point for the flagship of the new communication sciences. "A tremendous research project collapsed. Isn't it fantastic what happened? These cybernetics notions are of great importance to understanding oneself and others. They could be very helpful for ethical behavior and cooperative interaction. And at that time it was eclipsed."

Cybernetics lost the funding war and the turf war between the two disciplines, but as von Foerster saw it Wiener's science won the battle for hearts and minds on the wider terrain of the new technological age. "Cybernetics introduced a way of thinking which is implicit in so many fields but it is not explicitly referred to as cybernetics," he pointed out. "Nobody will call it cybernetics, but they understand it's a holistic . . . integrative form of thinking. . . . I would say cybernetics melted, as a field, into many . . . other fields." John Dixon, a charter member of the American Society for Cybernetics, concurred. "The concepts were transmitted and transmogrified into other areas. The word cybernetics dropped out, but Wiener's work continues under other names. Look at the development of brain research, mathematical modeling, computers, networking. All of this you could claim is cybernetics."

For a decade after Wiener's death, cybernetics survived and flourished in some segments of American science and society, not by any organized effort, but by the power of its new conceptual tools and practical problem-solving abilities, and by Wiener's persisting presence in the public mind and on the nation's bookshelves. Cybernetic principles continued to infuse the ferment of interdisciplinary study and dialogue, and Wiener's uncompromising ethical principles spoke loudly to younger generations, and to many scientists and scholars who swore their own vows of noncooperation with government policies in the Vietnam war.

By the late 1970s, cybernetics was on the wane in the West, and, with its decline, Wiener's name and legacy began to fade from the popular consciousness, but his wisdom and warnings would not be denied. Many of Wiener's wildest forecasts began to break just slightly behind the ambitious schedule he had laid out for them decades before. His predictions about the new technology and its social impact proved to be, as Steve J. Heims attested, "on the whole prophetic and ahead of their time."

In the 1960s, the numbers of American industrial workers began their historic decline. In the first wave, more than a million factory workers lost their jobs to automation, including 160,000 members of Walter Reuther's United Automobile Workers union. With the advent of microchip technology in the 1970s, the push by industries to automate and depopulate became "a virtual stampede," as one historian described it. Job losses piled up in manufacturing and, then, spread into service industries and professional and managerial positions, as American society veered down the road to the workerless future Wiener had foretold.

Wiener's predictions for biology and brain science were even more prescient. His hunches about neurohormones and other "to-whom-it-may-concern messages" were confirmed by the discovery of hundreds of new neurotransmitter molecules that travel irregular paths through the brain and bloodstream. And many of his later predictions about future applications of cybernetics to medicine, which were viewed at the time as pure fantasy, also came to pass. In an interview with the British magazine *New Scientist*, two months before he died, Wiener forecast new methods of medical cybernetics that would detect illness "by sensing devices within the body." He predicted that "living materials would, by 1984, be used as part of computers" and that "complex nucleic acids that carry genetic information in living cells would be used in machines." He was only off by a decade: In the 1990s, medical researchers began testing diagnostic cameras the size of a pill that patients could swallow, and marketing hybrid silicon "biochips" that used DNA snippets to detect genetic defects and an array of other "bioinformatics" markers.

By that time, new laser and fiberoptic technologies had made Wiener's early vision of optical computing an everyday reality, personal computers were ubiquitous, and the spread of interactive computing networks into the public domain had fulfilled Wiener's technical prophecies. Wiener's editor Jason Epstein received his boldest pronouncement about the future of computing one afternoon in the early 1960s over milk and potato chips in Walker Memorial. "Wiener predicted that within a decade or less, computers, which were then room-sized machines, would be miniaturized as solid-state devices replaced vacuum tubes. These miniaturized machines—he held out the palm of his hand to indicate their eventual size—would be linked by wireless or telephone lines to libraries and other sources of information so that everyone on earth could, in theory, have access to all but limitless data in an all-encompassing feedback loop, endlessly correcting and updating itself." Back then, Epstein dismissed the vision. "My failure to take Wiener's prophecies seriously reflected the limitations of my own worldview . . . and that of my intellectual friends who . . . felt that the fate of Western civilization depended upon the positions they took in . . . their dinner-party conversations. . . . I should have seen that Wiener was describing an even more profound technological shift than either movable type or the internal combustion engine, but . . . because Wiener was not one of us, his prophecies seemed unreal to me and I ignored them."

A quarter century later, the young science fiction writer William Gibson, in a nod to Wiener's foresight and his science, coined the new word that embodied the cosmic explosion of the Internet—*cyberspace*. In his novel *Neuromancer* Gibson defined cyberspace in terms even more vivid than Wiener's vision, as *"A consensual hallucination experienced daily by billions of legitimate operators, in every nation. . . . A graphic representation of data abstracted from the banks of every computer in the human system. Unthinkable complexity. Lines of light ranged in the nonspace of the mind, clusters and constellations of data. Like city lights, receding. . . ."*

Wiener's brainchild went into decline before his vision was fulfilled. He knew there was still much work to be done on the core formulations of cybernetics, and on his quest to define humankind's nature and purpose in new terms for the new technological age. Yet, as the young British cyberneticist Gordon Pask perceived in the 1950s, "Wiener . . . realised there was another step to take, but did not know how to do so. He was waiting for others to pick up the baton and run with it, to complete the forming of the subject he had begun."

In death, as in life, he continued to rack up impressive tributes. His last nonscientific work, *God & Golem, Inc.*, was published posthumously in 1964 and won the National Book Award for Science, Philosophy and Religion. He also garnered an array of eponymous awards and prizes, including the Norbert Wiener Prize in Applied Mathematics, awarded jointly by the American Mathematical Society and the Society of Industrial and Applied Mathematics, the Norbert Wiener Medal for Cybernetics, established by the American Society for Cybernetics, and the Norbert Wiener Award for Social and Professional Responsibility, given annually by the public interest group Computer Professionals for Social Responsibility. In 1970, when lunar orbiters mapped the surface of the moon, the International Astronomical Union named a crater on the far side Wiener—its girth measured 234 miles.

———

Wiener's colleagues and loved ones met a miscellany of fates, some triumphant, some tragic.

Warren McCulloch stayed on in the Neurophysiology Lab at the RLE, but he never got over his break with Wiener. By 1968, he was an old man even in his own eyes. Mary Catherine Bateson saw him that year as "a curious blend of glee and grief, of belligerence and gentleness." Oliver Selfridge said, "He probably consumed too much ethanol, and his consumption went up, I fear."

Walter Pitts spent the 1960s in seedy bars, so beset with delirium tremens that he could not speak two sentences without shaking uncontrollably. MIT retained him as a lecturer, though he never set foot on the campus, and even agreed to grant Pitts his Ph.D. if he would only sign his name to a piece of paper acknowledging his acceptance. He refused. Pitts died alone in his Cambridge rooming house in May 1969, at age forty-six, from complications of acute alcoholism. McCulloch died quietly that September, at seventy, on his farm in Old Lyme, Connecticut.

Margaret Wiener lived out her days in South Tamworth, but she did not find the serenity she longed for throughout her marriage. She managed the consequences of her colon cancer with dignity, and she seldom complained about her health problems, but she never stopped badmouthing her daughters to her New Hampshire neighbors. She died in 1989, at age ninety-five, and her remains were laid to rest beside her husband's in the Vittum Hill Cemetery.

Several years later, the last few cartons of Wiener's papers were delivered to the Institute Archives at MIT. In a box of family effects was Margaret's pink, clothbound journal in which she kept the family's accounts and an informal diary of her thoughts. On one page, Margaret wrote a little aphorism that seemed to describe the philosophy she had lived by, and the strategy she followed in her role as Wiener's *frau-professor*:

> "One way to arrive at the aristocracy if you aren't born there is to eschew all forms of liberalism."

The adage gave one last clue to the puzzle of the Wiener-McCulloch split and helped to explain why the liberated bohemians of the McCulloch group posed such a threat to Wiener's wife and her dream of attaining a high social position.

Wiener's daughter Peggy received her Ph.D. in toxicology and worked as a forensic toxicologist in the New York State Police Crime Laboratory. In 1988, she appeared on the television quiz show *Jeopardy!* and won the Senior Tournament, along with $22,000 in cash and prizes. She died of cancer in December 2000. Barbara received her Ph.D. in developmental biology and taught as an assistant professor at Northeastern University. In December 2003, she and her husband Tobey Raisbeck celebrated their fifty-fifth wedding anniversary, their five children, eleven grandchildren, and the birth a year earlier of their first great grandchild.

Wiener's eldest grandchild and namesake, Michael Norbert Raisbeck, became a lawyer and software engineer on Boston's high-tech highway, Route 128. Now in midlife, with two young sons of his own, he pondered the lingering questions about Leo Wiener's experiment, and the fate of his grandfather and other famous prodigies whose lives led on to greatness and to disaster. "Is the world better off because you do that to ten kids and nine of them crack up and one of them turns out brilliant?" he asked. "Part of me says you should never do that to a kid. But now step back and look at the development of civilization and where we are going. Does humanity need some real points of brilliance in order to advance? If everybody is raised to be just a happy medium, a bunch of happy farmers, what happens when humanity is confronted by some really serious challenge: a global environmental disaster, a plague, the next Hitler, whatever it is? If you have created the kind of world where everyone is not all that bright but wonderfully well-adjusted, will you have the tools necessary to confront that kind of challenge?"

Michael never got to debate those questions with his grandfather, but Wiener did leave behind the answer to his grandson's question. The Most Remarkable Boy in the World, the Elephant's Child who roamed the banks of the river Charles and ranged far and wide over the world with a 'satiable curtiosity, left behind remarkable ideas that changed the world, fundamentally and irreversibly, and may yet tip the balance toward humankind's continued existence on this earth. He was a dark hero in the highest sense, a restless, rebel soul who waged a scientific and technological revolution predicated on the transforming power of communication—and on his unwavering belief that people are more important than machines.

His brilliant mind, indomitable spirit, and abiding love for that little spark in every human being still hovers above the revolution he ignited and lives on in all the new knowledge and technologies that serve us, in every reach of the global information society, and out across the vast expanse of cyberspace.

EPILOGUE

TIME FUTURE: SURVIVING THE GLOBAL SOCIETY

> I wonder greatly what will happen to our bottle feeding of all possible disruptive inventions. We have so tied ourselves up with . . . demands for goods as a hedge against a business depression that it will not be easy for us to move again at more reasonable levels. . . . I dread to think of the amount of individual misfortune and desolation which may come of all this. . . . It will be a long and arduous task to fit ourselves again into a system . . . adequate for our better understood needs. It must be done. Let us hope that it can be done.
>
> **—Norbert Wiener, "A Scientist's Dilemma in a Materialistic World"**

IN THE GLOBAL SOCIETY OF the twenty-first century, Wiener's distant visions have become everyday realities, while new realities reaffirm his role as the dark hero of the information age.

Half a century after *Cybernetics* was published and Wiener first spoke his mind in *The Human Use of Human Beings*, the accuracy of his warning shots has become clear: in the explosion of new technologies descended from prototypes he pioneered; in the conflicts rising within and between nations over the work performed by human beings and machines; and in the outbreak of human crises and spiritual turmoil he foretold as people struggle to survive and adapt to life in a new technological age.

Many of Wiener's early technical designs are still cutting edge in the twenty-first century, from the latest advances in optical computing, a field he conceived in the 1920s, to progress in bionics, another field he fathered, to the first practical applications of the daring idea he developed with Walter Pitts in the 1940s to model electronic circuits and networks in three dimensions like the brain's own neural networks. Fifty years after Wiener broke with Pitts and their project was abandoned, computer theorists and manufacturers have at last begun to incorporate three-dimensional structures into their designs for faster and more versatile electronic circuits and silicon microchips. "It's quite clear that the [next] paradigm will be the third dimension," famed computer wizard Ray Kurzweil declared early in the new millennium. He predicted that the next wave of three-dimensional circuits and computing devices could achieve technical performance speeds a million times faster than the human brain and sustain progress in the electronics industry for decades to come.

Some of the most important insights that have flowed from Wiener's ideas have been in the field of neuroscience itself. In the years since Wiener and his colleagues first identified the action of the brain's complex analog information processes, researchers have fleshed out the workings of the organ's myriad chemical transmitters and hormonal messengers, and the fleeting brain waves that Wiener first made sense of scientifically in the 1950s. Their findings only underscore Wiener's early warnings that the omnipresent "electronic brains" that have taken charge of so many human tasks are not like human brains at all, and that we cannot always count on them to do our bidding.

More evidence of that reality is coming to light. Wiener would still be at swords' points with the American military over its increasing reliance on computerized weapons systems of unproven reliability, as he was from the first application of cybernetic technology to the weapons of atomic warfare. The military's advanced radar-guided antiaircraft and antimissile systems deployed in Middle East war zones since the early 1990s—the direct descendants of the automated antiaircraft guns Wiener helped to design during the Second World War—have repeatedly misidentified friendly aircraft as enemy targets and fired on them indiscriminately, after giving their human supervisors only seconds to detect the error and override their computerized programs. Other "smart" weapons in America's arsenal also have gone astray in the field and killed friendly troops and innocent civilians, and the military's planned national missile defense shield to protect the home front threatens far more serious failures. One outspoken MIT professor who voiced those concerns incurred the wrath of the military, and was given the Norbert Wiener Award by Computer Professionals for Social Responsibility.

———

By many other measures, the global society is thriving. The burst of knowledge and invention Wiener's work touched off is showering the world with many of the benefits Wiener and his contemporaries hoped their efforts would bring: an increased ease and convenience of daily life, an abundance of mass-produced goods, unlimited quantities of information, and new biomedical technologies that have improved and extended people's lives. The revolution has created millions of new jobs, whole new industries and professions, and collapsed borders and barriers to trade between nations. It has changed societies from within and without, and transformed the way people the world over communicate with one another. But those technical advances are only part of Wiener's legacy. He was far more concerned about the human factors in the equation.

Wiener foresaw the wrenching changes cybernetic technology would bring to the world's populations and to the need for human workers in every enterprise. His most dire prediction in 1950—that the advent of intelligent machines would "produce an unemployment situation, in comparison with which . . . even the depression of the thirties will seem a pleasant joke"—was not realized in the twentieth century. But, when the nineties' technology bubble burst, his warning of a new depression that would ruin many industries, even those built upon the new technologies, did not seem so farfetched any longer. The collapse of the technology-driven stock market early in the new millennium, and the global shift toward offshoring of jobs in manufacturing and the new technology industries themselves, brought stark reminders, as Wiener had warned, that technology alone will not create a utopia for the world's populations or even for most Americans, and that the most promising technologies may exact sizable human costs.

Today employers, economists, and media commentators seldom speak about those harsh realities as consequences of automation. They speak instead about the impressive rise in "productivity"—the unit output of goods and services per hour of human labor—that technology has brought across the board and the benefits of that improved efficiency for companies and nations fighting to survive in the new era of global competition. But for millions of idled workers, productivity has become a socially and politically acceptable euphemism for job elimination—and, for those still employed, for their mounting levels of job stress and uncertainty—as the twin forces of technology and global competition are giving rise to chronic joblessness and structural changes for workers in all the industrialized nations. And that may be only the first ripple of the global wave of unemployment and misery Wiener foretold. Even in the most robust sectors of the information economy—computer programming and technical services, which are themselves becoming subject to automated technology and programming techniques—millions of highly educated, experienced workers are becoming superfluous and prohibitive in the new cost equations of international enterprise.

Wiener's hope in the 1950s, when he advised the Indian government to train a new generation of "non-commissioned officers of science and technology," was that the new technology industries would provide a source of opportunity for India, or any developing nation, to travel an easier path to prosperity than any of its alternatives. He only dimly envisioned the extent to which the cybernetics revolution itself, and the rise of an immense, low-paid, high-skilled, worldwide work force, linked by global telecommunications, would become a source of economic conflict between nations.

To advance and prosper, Wiener knew, each country would have to build a stable, cybernetically sound, technological society on the foundation of its own unique history and culture, and many emerging nations are doing just that. India and China are using the new technologies to lead their people out of centuries of poverty and material deprivation, while the former Soviet bloc countries are only beginning to recoup the losses they suffered in the twentieth century. To date, India's engineers and entrepreneurs have had the most success following the path Wiener charted for their country's advancement, and while their numbers are still small compared to the whole of their population, they are reaping many of the benefits Wiener envisioned without the drawbacks of older models of industrialization. China's entrepreneurs and government bureaus are traveling two roads: developing the new technologies and technicians at a breakneck pace and, at the same time, walking the older path of development in new garb, by mobilizing their population to work primarily in more conventional industries and factories equipped with the latest automated manufacturing methods. Both approaches are proving productive, but they are not without pitfalls.

The Chinese approach, heavily dependent on the mass production of material goods, may lead China and the entire global economy into a greater crisis Wiener foresaw decades ago: a crisis of overproduction provoked by mass-produced products pouring from so many sources that the world's consuming economies will be unable to absorb the output. Like the bewitched brooms in Goethe's "Sorcerer's Apprentice," the glut of production has begun to swamp the global economy with surplus goods, straining supplies of energy and raw materials, pushing up prices of basic commodities and, at the same time, acting as a deflationary force on prices in the marketplace—producing the volatile mix of oscillating forces that Wiener knew could tear any system apart.

That delicate balance of economic forces exemplifies the cybernetic nature of the new global society, its dependence on reciprocal actions among all its diverse components, and the dangers of "very great feedback" that worried Wiener from his first wartime investigations. Yet those material effects on production and prices give only a glimpse of the next wave of problems Wiener foresaw that has already begun to affect workers and enterprises in the information sectors of the global economy, and of the brewing crisis in the value of information itself and all the new technologies and human activities associated with it.

Wiener urged society to rethink and reappraise entirely the economic values it assigned to every form of information, but like the other directives he issued, that work was largely neglected, and the failure has come back to haunt information entrepreneurs everywhere. The reduction of all modes of information, knowledge, and human experience itself to easily transportable, freely reproducible bits has reduced large segments of the information economy—from the recording and motion picture industries to the computer software industry—to vendors of intangible commodities that simply cannot be produced, protected, or marketed like material goods and services. Without a new consensus on the value of information, information technology, the legal rights and protections for information products—and the value of the human beings who produce them—even highly skilled professionals and technically advanced industries may learn too late, as Wiener forewarned, that "information is information, not matter or energy," and that "no materialism which does not admit this can survive at the present day."

His warnings about information apply equally to automation and globalization. All three factors will press on people in industry, services, and even the arts in countries whose governments and enterprises continue to weigh the costs and benefits of the new technologies and employment strategies on the scale Wiener repudiated long ago, only "in the terms of the market, of the money they save." The way forward is no different today than it was when Wiener laid down a new bottom line. He called on societies to reorder their priorities on a base of human values beyond buying and selling, and to arrive at that society today, as in Wiener's day, still requires much foresight and planning—and struggle.

Amid that struggle, another human peril hangs over the global society that has become a tangible factor in deliberations among nations. Wiener was sensitive to the problems of traditional cultures in transition, and to their collisions with modern technology and mores, but that scarcely would have prepared him for the global wave of religious-political terror that has erupted in the twenty-first century. The new terrorism has echoes of the horror he felt when the atomic bomb ushered in a new era in which technology had increased the power of "a limited group of a few thousand people to threaten the absolute destruction of millions," and even darker forebodings for civilization.

Surely he would be appalled by the way modern-day Muslim extremists, descended from the learned culture of mathematicians, astronomers, and master builders who embraced his famous ancestor Moses Maimonides in Cairo, have turned the technologies of the West and the open communication networks of the global society into implements in their holy war against America and other nations. Their usurping of new communication technologies, and their looming use of advanced weapons, go far beyond the profound moral and spiritual challenges Wiener portended in his last enigmatic work, *God & Golem, Inc.: A Comment on Certain Points where Cybernetics Impinges on Religion*.

The new terrorism is a product of complex historical, religious, and political influences, but the phenomenon can be illuminated from the new communication perspective Wiener and his colleagues in the social sciences pioneered. From the outset, in *Cybernetics*, Wiener focused on "control of the means of communication" as the most important factor governing the stability—or instability—of modern societies. That controlling factor has been demonstrated repeatedly in the explosive societies of the Middle East, where satellite broadcast networks, the Internet, and the informal channels of communication in Arab cultures have been the terrorists' principal instruments for propagating their crusade. Wiener's insights into the intimate relation between communication and mental health help to explain how extremist masterminds disseminate their mindset, and how the drone of information and indoctrination, in twisted scriptures, political propaganda, and other distorted messages, can "build itself up into a process totally destructive to the ordinary mental life." The armies of unquestioning holy warriors and religious martyrs such communication practices may produce are the newest and most dangerous examples of those "machines of flesh and blood" Wiener described decades ago that may arise "when human atoms are knit into an organization in which they are used, not in their full right as responsible human beings, but as cogs and levers and rods . . . in a machine," and in which they become, for all intents and purposes, "an element in the machine."

Among its many profound causes and implications, the new terrorism, in its modes of operation and the ease with which it is capturing new recruits, is one extreme reaction to the emergence of the global society and, at once, a product of it. It also provides further proof of the destructive potential of communication technologies deployed without regard for human beings or human values. On that count, the lessons of the new terrorism and its ultimate solutions are clear, and not unlike those Wiener spelled out as the cure for communism in the 1950s. Vulnerable societies in the Middle East and other disenfranchised populations must be provided with viable pathways to modernization, education, unbiased communication channels, and a more rewarding existence that will enable them to share in the benefits of a global technological society and, at the same time, to bring forward in their own ways and at their own pace the human values embodied historically in their religious and cultural traditions.

In this combustible climate of colliding technical and human communication potentials, the newest technologies to emerge from Wiener's scientific ideas are providing further proof that the ethical guidelines he laid down are essential for humankind's survival. Those technologies are not digital devices but, rather, advances in the analog domain Wiener worked in throughout his career and preferred personally over digital technology. And they are some of the most promising and, potentially, the most dangerous technologies ever devised.

The latest breakthroughs in the analog domain—which today spans everything from biotechnology and genetic engineering, to robotics, to the diminutive realm of molecular and atomic-scale nanotechnology—mark the reemergence in the twenty-first century of technologies that embody information physically and perform tangible actions in the world (in contrast to digital devices that code information in abstract strings of ones and zeroes and communicate mainly with one another). The new analog inventions make use of powerful sensor and effector technologies, improved versions of the "elements of the nature of sense organs" and "effectors . . . which act on the outer world" that Wiener foresaw coupling with computers as early as 1950. In their new incarnations, they are revolutionizing medicine, improving diagnostics in the laboratory and inside the body, enhancing prosthetics, and aiding—and in some cases even performing—surgery. They are making wide inroads into industry, where they are ushering in a new era of "*hyper*automated cybermanufacturing" that is speeding up production lines, coordinating operations within factories and across continents, and setting new standards of quality control. And they have begun their advance into the commercial marketplace. Engineered biotech products, inexpensive sensors, effectors, and robots of one species or another are cropping up in a new generation of smart appliances and consumer products, from sentient robot vacuum cleaners that navigate around objects in a room, to quick-thinking collision-avoidance systems for automobiles, to new styles of smart clothing and household items that adapt automatically to their wearers' activities and changes in the weather.

These analog technologies are having sweeping effects on the digital revolution itself. According to one leading forecaster, the new analog inventions have begun "the erosion of the entire digital order that we now take for granted." Their tangible actions give them decisive advantages over digital bits in practical applications, and they portend a coming turn in technology when digital methods will no longer be sufficient to accomplish many desired technical goals. In the near term, hybrid analog-digital technologies that combine the best of both domains will flourish, but in the longer term, the new analog avatars predict, "digital will seem just a bit dull" and may decline altogether in its importance. The discarded analog paradigm will again be the wave of the future, prompting scientists and technicians to dig back through the technical records of the mid-twentieth century for the new keys to the twenty-first.

That bold prediction, which sounded preposterous when it was made only a few years ago, is rapidly becoming another fact of life in the global society, as the resurgent analog domain is providing new reasons to revisit Wiener's words and warnings. No doubt he would view today's innovations with a mix of fascination and dread. He would take paternal pride in the new robots and minimachines that have evolved from the first cybernetic "bedbugs" and "barflies" he and his peers put together, but he would be deeply concerned for the next generation of human labor the new analog devices are displacing. He would marvel at the breakthroughs in biotechnology that have grown out of the cybernetic perspective he injected into the life sciences in the 1950s. But knowing the capriciousness of molecules and matter at the atomic level, as he did, he would be deeply fearful of manipulating life at its most fundamental levels, and of the disruptive and potentially deadly organisms the new biotechnologies could unleash.

As the digital information revolution approaches its own physical limits on the speed and size of its information-processing technology, the new analog technologies may prove to be Wiener's ultimate legacy for better and for worse—and the greater peril to which all his warnings and wisdom-rich parables were alluding. The new technologies make possible, for the first time in practically attainable terms, Wiener's worst-case scenario: the specter of autonomous machines, made of organic or inorganic materials, that might mutate and replicate without limit or any form of human control. They are at once more complex and much simpler than their digital brethren. Many do not even require fully articulated programs but, instead, operate as autonomous agents and life-like cellular automata, using statistical principles of probability and biologically based rules, and future forms will be self-programming entirely. Moreover, by their nature, many analog systems are not just fast but *instantaneous* as they self-organize and respond continuously to their changing inputs and environmental conditions.

And therein lies their danger. One of Wiener's greatest concerns was that the speed and complexity of intelligent technology would outstrip people's capacities to respond to their machines and to keep them fully under human control. Those thresholds have already been crossed with many digital systems and programs deployed in critical arenas of society—from the Internet to the stock markets to the military's command-and-control operations—and the new analog systems will push those potentials to their limits. The fast actions and dispersed operations of engineered organisms and nanoscale devices will leave little time for error-correction by their own mechanisms, and even less time for corrective action at the speed of human responses. With analog systems, very often, it will be impossible to know exactly what the system is doing at any given moment, and what decisions it is making in the process of carrying out its assigned tasks.

Given their ominous possibilities, the new analog developers conjuring in corporate laboratories, military research centers, universities, and private-sector startups would be wise to take Wiener's warnings to heart along with his scientific insights, and to exercise the utmost caution and stringent controls over developing technologies that could endanger people and the earth as a whole.

———

Today's technologies, both digital and analog, pose unprecedented challenges, not only for engineers and entrepreneurs, but for people in their daily lives, and for a world that is at once more connected and more divided than ever before. Those challenges require new modes of thinking and new conceptual tools to grasp and subdue their technical and human complexities, and Wiener's scientific principles and ethical precepts provide a few of the new tools that are needed to do that work.

As he made clear from the outset, the universal processes and principles of cybernetics—information, communication, feedback, "circular causality" or reciprocal influence, and "teleology" or purposeful, goal-directed action—apply equally to technology, biology, and all the complex systems of society. The biological and social dimensions of cybernetics were widely overlooked in Wiener's day and in the decades since his death, as technology has loomed ever larger. Yet those neglected aspects of Wiener's science hold some of the most powerful insights cybernetics has to offer, and they are as important as any technical tool for understanding the complex forces that shape and influence all our lives.

The conceptual tools of cybernetics can help people to think in more effective and productive ways, to create, innovate, and perhaps even begin to envision at the levels to which Wiener raised his most talented disciples that enabled them to "see over the fence," as he did, and down the road to the end result of any effort. Cybernetics, its sister sciences of information theory and system theory, and their descendants in the new sciences of complexity and human communication offer scientists and nonscientists alike new ways to think systematically and strategically, to solve problems, paint scenarios, and identify potential trouble spots before disaster strikes. As the systems of life speed up and grow more complicated, acquiring those skills and applying them in day-to-day affairs, and to the larger problems of the world's economies and cultures, may be the best move any individual can make, and the next step on the learning curve of the global society.

Wiener's grim predictions may still strike some twenty-first century observers as alarmist, or perhaps, as one computer scientist reflected, "it was his timely warnings that saved us from some of the troubles that he foresaw." Yet his most important warnings have gone unheeded. Many of the technical and human perils that concerned him are only now coming to bear. Long-range plans are still needed for the future of work and the livelihood of workers. Difficult choices may have to be made between some developing lines of technology and the wider interests of humankind. And just as scientists and technicians are now confronting the consequences of their creations, similar ethical choices abound in everyday domains of communication and culture.

Every person today shares in the life of the global society and participates in its vital processes, directly and indirectly, through the messages they exchange in every medium and their uses of new information and communication technologies. Each choice and action they make sends a message that may influence the actions of governments and corporations half a world away, and that power must be exercised responsibly. Even the young whiz kids and mischief makers prowling in cyberspace—the virus writers, website hackers, intellectual property thieves, and more malevolent cybercriminals—and their extremist counterparts acting out offline—the techno-terrorists and anti-globalization anarchists—would do well to make Wiener their dark hero and exemplar, and to use their talents and technology constructively in their battles on the barricades of the global society.

Americans especially have new opportunities and responsibilities on the global stage. As the place where Wiener's revolution was born, America pioneered the innovations that made the new society and economy possible. Now that leadership role must be shared and expanded to shape the next rounds of domestic and global development, the new technical and human skills, and the supporting social and economic structures that will be needed to ensure a peaceful, productive era of global communication, commerce, and culture. Some of that necessary work is now getting under way, and there are hopeful signs on the horizon, as Wiener liked to call them. Individuals and organizations in the United States and other countries are jointly taking up far-reaching issues of scientific responsibility, the ethical use of technology, and the safety of new biotechnologies and other analog technologies that are making their way into the environment and the public domain. Some are aggressively defending the rights and privacy of computer users and citizens of cyberspace, while others are working to expand access to the new technologies' benefits beyond the wealthiest individuals and nations.

Surely Wiener would applaud today's heroes who are using the new technologies for benevolent ends: the young visionaries who brought the military's Internet into the public domain and created the World Wide Web, then refused to take a penny for it; the insurgent programmers of the "open source" software movement, who share his ethic that new knowledge and technologies should not be proprietary products but public property for the use and benefit of all. He would be gratified to see some of his twenty-first century colleagues ringing alarms about the perils of genetic engineering and nanotechnology.

Wiener knew that all the new technologies to come would not provide the solutions to the problems and dangers humankind would face in the future. But he laid a foundation that can enable people to foresee the choices that lie ahead and the likely outcome of their efforts. He showed us the limits of our knowledge and technology, and the flaws in the dominant institutions of our societies—our governments, corporations, and militaries. He showed the errors of running an information society on the basis of matter-and-energy principles and economic values. He showed the arrogance and blindness of our cultural and religious biases. And he left a clear directive to learn which of our uniquely human capacities must be preserved and protected in a world of intelligent technology. He knew that people either would rise to meet these challenges and build a global society on a universal base of human values or that they would create a world in which technology will take control by default and human beings as we have known them, and life itself, will not survive.

At this juncture in the ongoing story of Wiener's legacy and humanity's progress, if people generally and Americans especially resolve to improve their own societies and, at the same time, to make the interests of humankind their own, we can survive the turmoil of the moment and the new technologies at the door, and bring about a safer, saner, and more prosperous global society. And, quite possibly, in the process, we may come to know our nature and purpose, as Wiener wished so devoutly, and create a world that embraces as its goal and highest good the human use of human beings.

ACKNOWLEDGMENTS

In his autobiography, Wiener made a plea for privacy in the events of his personal life, except where those events were "directly relevant to the incidents of my career as a scientist." Wiener's personal bond with his wife, Margaret, remains a matter between them, but Margaret Wiener's role in Wiener's life went well beyond that private arena, affecting his scientific work and the lives of those closest to him personally and professionally. Those events have had far-reaching consequences on the history of science, and on the course of the information age that affects all our lives.

This book, eight years in the making and a dozen more in the planning, would not have been possible without the trust and cooperation of Barbara Wiener Raisbeck, her husband Gordon "Tobey" Raisbeck, Peggy Wiener Kennedy, and Michael Raisbeck, who gave us unrestricted access to Wiener's private archives, photographs, and family records. We express our deepest gratitude for their memories and their participation in this project. Wiener's daughters made a plea of their own for the story of Wiener's fateful break with his colleagues Warren McCulloch, Walter Pitts, and their circle of scientists to come out at last. "The truth matters," Barbara wrote in a letter in 1998, with her wish that the other survivors of Wiener's inner circle "will be encouraged to tell all they know, flattering or unflattering to my family." Peggy agreed. "Serious unanswered questions remain concerning Dad's life and relations with his colleagues. It is very important to tell the whole story, whatever it is—not to mention my 'satiable curtiosity."

We are grateful to Jerry Lettvin for his candor and recollections in his office during the last days of MIT's legendary Building 20. Wiener's other friends and colleagues, and their children, gave testimony to his genius, his dark side, his lighter side, and historic events they witnessed: John Barlow, Julian Bigelow, Amar Bose, Rudy Carlson, Morris and Marion Chafetz, Helen Chen, Pauline Cooke, Ivan Getting, Taffy McCulloch Holland, Jean King, Maggie Lettvin, Fagi Levinson, Benoit Mandelbrot, David McCulloch, Marvin Minsky, Karl Pribram, Paul Samuelson, Swami Sarvagatananda, Oliver Selfridge, Claude and Betty

Shannon, Armand and Mildred Siegel, Dorothy Setliff, Dirk Jan Struik, and Heinz von Foerster. Others offered memories of key figures in cybernetics' early years, and during its developmental phase in the 1950s and 1960s: Noam Chomsky, John Dixon, Ioana Dixon, Charles Fair, J. Patrick Ford, Morris Halle, Angela Maddux, Robert Mann, Paul Pangaro, Donald Theall, and Terry Winograd.

Thanks also to Mary Catherine Bateson, Cornelia Bessie, Elizabeth Bigelow, Jerry and Maggie Lettvin, the McCulloch family, and Tom von Foerster for their recollections and family photographs. Rabbi Michael Azose, Rick Ford, and Shavit Ben-Arie provided genealogical research and leads on Maimonides' descendants in the Middle East and Eastern Europe.

We are most grateful to Steve J. Heims for his portraits of Wiener, John von Neumann, and the players in the Macy Conferences on Cybernetics, and for his enlightened writings on that ferment in the modern history of science which he was the first to chronicle. Our longtime mentors, Alfred G. Smith, Carl Carmichael, and Fred Crowell, gave us historical perspective on the development of the communication sciences. Larry Augustin, Richard Stallman, Stuart Umpleby, and David Wolpert offered their distinctive views on today's science and technology; Amy Dean and John Curtain gave insights into the domestic and global labor scene; and Reggie Kriss offered clinical insights into the stresses of life in high-technology venues.

We are beholden to the many archivists who cooperated with this project and gave us their personal attention, foremost among them, Mary Eleanor (Nora) Murphy, Reference Archivist, Institute Archives and Special Collections, MIT Libraries, and Deborah G. Douglas, Curator of Science and Technology, MIT Museum. Thanks also to Elisabeth Kaplan, Sylvia Mejia, and Jeff Mifflin, Institute Archives and Special Collections, MIT Libraries; Jenny O'Neill, MIT Museum; Melanie M. Halloran, Harvard University Archives; Alison M. Lewis, American Philosophical Society; Judith L. Macor, Ed Eckert, Irene Lewicki, Bunny White, Romaine Abbot, Bell Laboratories/AT&T Corporate Archives; Erwin Levold, Rockefeller Foundation Archives; Barbara Field, *IEEE Control Systems Magazine*; Robert D. Colburn, IEEE History Center; Linda Arntzenius, Institute for Advanced Study, Princeton; Mary Wolfskill, Yvonne Brooks, and Bonnie Coles, Library of Congress; Arlene Shaner, New York Academy of Medicine; Patricia Yee, Institute for Intercultural Studies; Paul Lukasiewicz, Yale University Mathematics Library; Karen Fitzgerald, Yale University Mathematics Department; Jennifer S. Blue, U.S. Geological Survey; and Kimberly McAllister, FBI (Boston Field Office).

Our thanks to many other scientists, engineers, and historians of science who provided documentation and guidance in their fields of expertise: Stuart Bennett, University of Sheffield; Edmund F. Robertson, University of St. Andrews; Paul N. Edwards, University of Michigan; Miklos Redei, Lorand Etvos University, Budapest; Dr. Pedro Antonio Reyes López, Instituto Nacional de Cardiología, Mexico City; Susana Quintanilla, Mexico City; Dr. Kazi Kabir Hossain, Indian Statistical Institute, Calcutta; and Jay Hauben, Gerald Holton, John Hutchins, Larry Owens, Louis Slesin, Sandra Tanenbaum, and John Townley. We're grateful to John P. Luneau, S. Eric Rayman, and N. Lindsey Smith for their guidance on legal and media matters; and to Leonard Greenberg and George Pelesky for their professional assistance.

Thanks especially to our primary support team in New York: Noel Adams; Mohamed Amen, Antonia Artuso, Michelle Ballew, Ovidio Biaggi, David Bynoe, John Campbell, John Canoni, Gloria Cruz, Kevin Currenti, Miledys Diaz, Laysa Diaz, Mary Hamm, Ali Hassan, Roger Johnston, Moustapha Kone, Nancy Jacobs, Roseann Lentin, Vincent Mariano, Rich Marsillo, Tim Mercado, Glenn Krinsky, Mitchele Lewis, Barbara Melser, Sandy Olson, Rhonda Pomerantz, Jessica Porier, Lisa Rittel, Jose Rodriguez, Lana Rosenberg, Art Rosner, Jose Sanabria, Iris Sanchez-Hernandez, and Safokles Tsouros.

We thank our families and friends for their care and patience with us through this project, and for innumerable favors: Maureen Arslanian, Margaret Barela, Patricia Barron, Catherine, Gordon, and James Clark, Bob Conway Jr. and Virginia Conway, Christine and Robert Vincent Conway, Holly Conway and Mike Sheerin, Kacey Conway and Patrick Green, Loretta Conway, Nelle McElravy Conway, Susan DeFlora, Brad and Linda Demsey, Bob Emmons and Nena Lovinger, Ellen Fair, Tom Feran, Mike Feury, Tom Foster, Dana Hetherington, Susan Horton, Kathleen Hudson, Joan and Hillard Lazarus, Jeffrey and Adam Lazarus, Dan Polster, Jean-Claude and Emma Rancoud-Guillon, Roger Repohl, Don Ross, Lauren Rubin, Ray Smelich, Richard Weingarten, and Bill Wise.

This book could not have been written without the hospitality, generosity, and human intelligence of Eleanor Clark; and without the friendship and support of Lynne Forrester. Sid Buck, our colleague at Stillpoint Press, aided us greatly through the first years of this undertaking. Brad and Elayne Searles provided essential support to Stillpoint Press.

Special thanks to Marc Jaffe, who respected our priority on this project; Peg Cameron, whose lessons we continue to follow; Lisa Drew, for her interest in the book's early stages; Bruce Harris, for his publishing insights. We thank Kurt Vonnegut for his friendship and support in countless ways for this book and all our ventures.

We are much obliged to Chad Bantner for providing the phrase that became the book's title, and to Bill Stiffler, Nancy Mittleman, and Amber Darragh for helping us make that connection.

We're grateful to the literary agents who helped us through each phase of the process: Michael Carlisle and Neal Bascomb at Carlisle & Company, Melissa Chinchillo at Fletcher Parry, and Peter Robinson at Curtis Brown U.K. To our agent Christy Fletcher, our heartfelt thanks for her masterful efforts on our behalf.

This book has benefited from the work of two outstanding editors. We are grateful to Amanda Cook for her dedication to this book and her astute work on the original manuscript; and to Jo Ann Miller for her swift and keen eye on the final draft. Thanks also to Ellen Garrison, Norman MacAfee, Amiee Munro, Rick Pracher, Christian Purdy, Iris Richmond, Carolyn Savarese, and Jennifer Thompson at Basic Books.

Finally, we thank our parents, Robert and Helen Conway and Leonard and Arline Siegelman, for their unwavering support, and for their love that continues to sustain us and lift our spirits.

FLO CONWAY AND JIM SIEGELMAN

New York, N.Y.
July 2004

IN MEMORIAM

JULIAN BIGELOW	CLAUDE SHANNON
ROBERT P. CONWAY, SR.	ARMAND SIEGEL
HELEN CONWAY	LEONARD P. SIEGELMAN
JOHN DIXON	DIRK JAN STRUIK
IVAN GETTING	HEINZ VON FOERSTER
PEGGY WIENER KENNEDY	

NOTES

Wiener's unpublished papers and correspondence are located in the Norbert Wiener Papers, MC 22, Institute Archives and Special Collections, MIT Libraries, Cambridge, Massachusetts ("MIT") unless otherwise noted. Other correspondence and unpublished materials were obtained from members of the Wiener family and are identified collectively as Wiener Family Records ("WFR"). The authors gratefully acknowledge the assistance of the Institute Archives and the Wiener family in providing documentation for this book. The works of Steve J. Heims (Heims 1980; Heims 1991) and Pesi R. Masani (ed., Wiener 1976, 1979, 1981, 1985; Masani 1990) have informed and greatly enriched this book.

—F.C. & J.S.

ABBREVIATIONS

Archives and Frequently Cited Published Works

Cyb	*Cybernetics: or Control and Communication in the Animal and the Machine* (Wiener 1948a).
ExP	*Ex-Prodigy: My Childhood and Youth* (Wiener 1953c).
G&G	*God & Golem, Inc.: A Comment on Certain Points where Cybernetics Impinges on Religion* (Wiener 1964a).
HUHB	*The Human Use of Human Beings: Cybernetics and Society* (Wiener 1950a, page references are from Avon edition unless otherwise noted).
IAM	*I Am a Mathematician: The Later Life of a Prodigy* (Wiener 1956b).
LoC	Library of Congress, Washington, D.C.

McC CW	*Collected Works of Warren S. McCulloch* (McCulloch 1989, vol. # I, II, III, IV).
NYT	*The New York Times.*
NW CW	*Norbert Wiener 1894–1964 (Collected Works)* (Wiener 1976, 1979, 1981, 1985, vol. # I, II, III, IV).
WFR	Wiener Family Records.
YP	"Yellow Peril" (Wiener 1942/1949b).

Frequently Cited Interviews and Personal Communications

BWR	Barbara Wiener Raisbeck, personal communication.
	Barbara Wiener Raisbeck & Gordon "Tobey" Raisbeck.
B&TR1	1st int., Cape Negro, Nova Scotia, Canada, Aug 27–28, 1997.
B&TR2	2nd int., Portland, ME, Apr 23, 1998.

B&TR3	3rd int. (telephone), May 17, 1998.
B&TR4	4th int., Portland, ME, Dec 10, 1999.
PWK	Peggy Wiener Kennedy, personal communication.
PWK1	1st int., Lake Oswego, OR, Dec 28, 1997.
PWK2	2nd int. (telephone), May 3, 1998.
PWK3	3rd int. (telephone), May 17, 1998.
PWK4	4th int. (telephone), June 8, 1998.
PWK5	5th int. (telephone), Aug 21, 1999.
PWK6	6th int. (telephone), Aug 29, 1999.

	Julian Bigelow
JB1	1st int., Princeton, NJ, Apr 28, 1999.
JB2	2nd int. (telephone), June 27, 1999.
JB3	3rd int., Princeton, NJ, Oct 30, 1999.
JB4	4th int. (telephone), June 14, 2000.

	Jerome "Jerry" Y. Lettvin
JL1	1st int., Cambridge, MA, Nov 24, 1997.
JL2	2nd int., Cambridge, MA, Apr 22, 1998.
JL3	3rd int. (telephone), May 31, 1998.
JL4	4th int., Cambridge, MA, Dec 12, 1999.

EPIGRAPH

viii Epigraph: Eliot.

PROLOGUE

xi Two academic biographies . . . memoirs . . . autobiography: Heims 1980 and Masani 1990; Norman Levinson 1966 and Rosenblith & Wiesner; Wiener 1953c and Wiener 1956b.

xiv "watch your hat and coat": Wiener and Campbell 1954c.

xiv "a two-edged sword, and sooner or later it will cut you deep": G&G, 56.

PART 1 CHAPTER 1

3 Epigraph: "The Elephant's Child," in Kipling.

3 "Hey, mother . . . isn't it time to go to college?": New York World Magazine, 10.7.1906.

6 "I am myself overwhelmingly of Jewish origin": ExP, 8.

6 Ex-Prodigy: Wiener 1953c.

6 "attitude toward life": ExP, 9.

6 Talmudic scholars in the Wiener line: Raven, 42; ExP, 10. Among them were Rabbi Joseph Ettinger of Jaworow, a liberal activist who fought the spread of orthodox Hassidism in southern Poland in the mid-1700s; Rabbi Pollack of Brody, who wrote the famous religious commentary "The Heart of the Lion," published in 1820; and Rabbi Akiba Eger, the Grand Rabbi of Posen from 1815 to 1837, a traditionalist who, to Wiener's dismay, staunchly opposed the new wave of secular learning that infused European Jewry under the influence of the German-Jewish philosopher Moses Mendelssohn.

6 A family legend . . . Maimonides: The link to Maimonides was in the family of Wiener's great-grandmother Rosa Zabludowska. BWR, 8.30.00.

6 "After so much passage of time . . . shaky legend": ExP, 10.

7 A cousin of Wiener's: Raven, 11–12.

7 several hundred modern descendants of Maimonides: Inquiries by the authors on the genealogical website www.jewishgen.org found multiple links to Maimonides in that region on the Polish-Lithuanian border: in a small Polish town southeast of Lublin, in Kielce, Poland, and in Sereje, in southwest Lithuania. A young Israeli was descended from a "very large family tree of hundreds of people that come from the Maimon family," many of whom left Lithuania for Palestine before World War II. Lublin, Kielce, and Sereje (now called Seirijai) were joined by old roads that met at Bialystok.

8 "frenzied fury": Raven, 12.

8 more than forty languages: Harvard Gazette (Mar 1940) claimed thirty; Wiener said "some forty," IAM, 48; Raven said forty-two. Raven, 11.

9 "young Slav engineering student": Unless otherwise cited, details of Leo's career and early travels are from Leo Wiener 1910.

9 "vegetarian humanitarian socialist commune": Raven 15.

10 In a Balcony: ExP, 21, although Wiener mistakenly refers to the work as On a Balcony.

11 "incurables": ExP, 32.

11 "horrible and hair-raising" tracts: ibid., 15.

12 "moderate inconvenience": ibid., 41.

12 still counting on his fingers . . . multiplication tables: ibid., 45, 66.

12 "in order to instill in him something of the scientific spirit": "The Case of the Wiener Children," typescript of unknown authorship and origin, circa 1913 (perhaps a partial transcript of Addington), MIT, box 33b, folder 903.

13 "tactful compulsion": "The Case of the Wiener Children," op. cit.

13 "the blessedness of blundering": "Harvard's Four Child Students," Boston Sunday Herald, 11.14.1909.

13 "the child must be made, in a kindly manner": "The Case of the Wiener Children," op. cit.

13 "systematic belittling": ExP, 70.

13 "My father would be doing his homework for Harvard": Amar Bose int., Framingham, MA, 11.26.97.

13 "Brute!" "Ass!" "Fool!" and "Donkey!": *ExP*, 67.

13 "austere and aloof figure": ibid., 34.

13 "He would begin the discussion in an easy, conversational tone": ibid., 67.

14 "juvenile ineptitudes . . . morally raw all over": ibid., 68.

14 "I relearned the world . . . still totally astonishing": Amar Bose int.

14 "analogous to the . . . fall of a train of blocks": *ExP*, 64.

14 the wild idea that he could turn a doll into a baby: ibid., 83.

15 "quasi-living automata": ibid., 65.

15 "fight or flight": Cannon 1929.

15 "homeostasis": Cannon 1932.

15 the works of Tolstoy, all twenty-four volumes: Tolstoy 1904–1905.

16 "The Theory of Ignorance": Wiener 1905.

17 "Futile as it was . . . the slightest danger": *ExP*, 99–100.

17 "Little as I wished to grow up . . . not nearly out of the woods": ibid., 100.

17 "an outsider at the feast": ibid.

18 "no good": PWK1.

18 "crabbed": *ExP*, 104.

18 "drawing was a bugbear": "The Case of the Wiener Children," op. cit.

18 his mind worked faster than his body: *ExP*, 130.

18 "effectors": *HUHB*, 213.

18 "at probably the greatest cost in apparatus": *ExP*, 105.

18 James . . . co-founded . . . pragmatism: James 1907. James named the doctrine and identified his Harvard colleague Charles S. Peirce as its founder.

18 Leo . . . incorporated James's education theories: James, like Leo, denounced "the philosophy of tenderness in education" and saw genius as "nothing but a power of sustained attention." James 1899, 51–52, 78.

18 Norbert . . . admired James's colorful style far more than his logic: *ExP*, 110.

19 "a field in which one's blunders . . . with a stroke of the pencil": ibid., 21.

19 his math professor turned the class over to him: "I used to sit in the front row while he worked at the board. It was easier that way." Professor William Ransom, *Boston Globe*, 2.2.64.

19 "I could not stop the wheels from going around": *ExP*, 115.

19 *"What should I do in the future"*: ibid.

19 "doubt as to whether the future of an infant prodigy": ibid., 116.

19 "This was the first time . . . a freak of nature": ibid.

19 "the child who makes an early start": ibid., 117.

20 "rather it petered out": ibid., 121.

20 "a sword with which I could storm the gates of success": ibid., 122.

20 Harvard . . . in service to the ministry: An early brochure (ca. 1643) proclaimed Harvard's mission: "To advance Learning and perpetuate it to posterity; dreading to leave an illiterate Ministry to the Churches." www.harvard.edu.

21 "the slicks": *ExP*, 119.

21 "who were eager to sell my birthright at a penny a line": ibid., 118.

21 "Harvard's Four Child Students": *Boston Sunday Herald*, 11.14.1909.

21 "My children are *not* anomalies . . . could have been ready for Harvard at 8": ibid. (emphasis added).

21 "a conscious blunder is a grand thing": ibid.

21 "sober dignity, reserve . . . reticence": *ExP*, 141.

22 "the sharing of a precocious school career": ibid., 139.

22 Berle . . . Roosevelt's Brain Trust: Berle served as Roosevelt's Assistant Secretary of State for Latin American affairs (1938–44) and ambassador to Brazil (1945–46).

22 Sessions . . . three Pulitzer Prizes: Two for his Concerto for Orchestra and a special Pulitzer citation for his life's work.

22 William Sidis soared briefly: After graduating with great fanfare, Sidis began his career as a mathematics professor, but he was mocked by his students for his immaturity. After eight months, he quit teaching and disavowed mathematics entirely. He suffered a nervous breakdown, went into seclusion, and spent his adult life doing menial work and writing ponderous books on obscure subjects. In 1937, a reporter for *The New Yorker* dragged him back from obscurity (Manley, www.sidis.net/newyorker3.htm). Other views suggest Sidis was no failure. He published a theory of black holes fifteen years before the idea was accepted by astronomers (Sidis 1925) and a "100,000-year history of North America" (Sidis 1982). See also Sperling; Wallace; www.sidis.net.

22 "a cruel and quite uncalled-for": *ExP*, 134.

22 "You Can Make Your Child a Genius": *This Week Magazine (Boston Sunday Herald)*, Mar 1952; *ExP*, 135.

22 "So you can make your child a genius, can you?": ibid., 136.

23 "a dear sweet bit of arsenic and old lace": PWK1.

23 "much earlier philosopher": *ExP*, 143.

24 "disinherited": ibid.

24 "where it really hurt . . . internal spiritual security": ibid., 147.

24 "As I reasoned it out to myself": ibid., 148–149.

24 "To be at once a Jew": ibid., 153–156.

24 "One thing became clear very early": ibid., 153–154.

25 "a turbid and depressing pool": ibid., 152.

25 In an article in the popular *American Magazine*: Bruce.

25 "When this was written down": *ExP*, 159.

26 His philosophy papers during those formative years: Wiener 1912a, 1912b, 1913a.

26 oral exams in a "trance" of dread: *ExP*, 172.

26 Prodigies . . . as omens "of impending change in the world": Feldman and Goldsmith, 4.

26 "fabulous monsters": *ExP*, 288.

26 a greater "gambit" on nature's part: Feldman and Goldsmith, 11.

26 "exemplar and beacon": ibid., x.

26 "may have something to tell us . . . existence on this earth": ibid., 213.

26 "omnibus prodigy . . . extreme precocity": ibid., 234.

26 "curious almost beyond belief": ibid., 239.

26 "rarely do the young sprigs blossom . . . fundamental and irreversible": ibid., 15–16.

27 "Nubbins": B&TR1.

27 *a small or imperfect ear of corn: American Heritage Dictionary*. New York: American Heritage, 1975.

CHAPTER 2

28 Epigraph: Goethe.

28 "conflicting and self-contradictory": *ExP*, 182.

29 "a heavenly relief": ibid.

29 "flirtation of an old sea captain's daughter": ibid.

29 "Russell's attitude seems to be . . . utter indifference mingled with contempt": NW to Leo, 9.30.13.

30 "My course-work under Mr. Russell is all right": NW to Leo, 10.18.13.

30 His blustering attacks: Russell was a formidable scholar, but he was described by friends and detractors alike as a callous, egotistical, sadistic, and at times suicidal figure. Ayer; Macrae; Monk.

30 "I have a great dislike for Russell": NW to Leo, 10.25.13.

30 "infant prodigy named Wiener": Russell to Lucy Donnelly, 10.19.13, in Grattan-Guiness 1974.

30 "a slovenly, mean little woman": *ExP*, 185.

30 "hopelessly and utterly lonely": ibid., 186.

31 "in such a way to comb a Yankee's hair the wrong way": ibid., 184.

31 "happier and more of a man": ibid., 197.

31 "a very close and permanent bond between myself and England": ibid., 184.

31 "the perfect March Hare": ibid., 195.

31 "with his pudgy hands": ibid.

31 Harvard . . . "has always hated the eccentric": ibid., 197.

31 Hardy's course was "a revelation": ibid., 190.

32 Wiener thanked Hardy, not Russell: *IAM*, 190–191.

32 Lebesgue integral: ibid., 22–23.

32 his first published paper: Wiener 1913b.

32 "Looking back . . . I do not think it was particularly good": *ExP*, 190.

32 His next paper for Hardy . . . significant contribution to the field: Wiener 1914a. H. E. Kyburg called it "fundamental and justly famous." NW CW I, 33.

32 "humbug": *ExP*, 189.

32 his special theory of relativity: Einstein 1905c.

32 his quantum theory of photoelectricity: Einstein 1905a.

32 In that third paper: Einstein 1905b (see also Einstein 1906).

33 "It appears to me unlikely that . . . Mr. Russell's set of postulates": Wiener 1915a.

33 Kurt Gödel's Incompleteness Theorem: Gödel.

33 "It is not very easy for me . . . to write of . . . Russell": *ExP*, 193.

33 "Nevertheless he turned out well": Grattan-Guinness 1975.

34 "the one really universal genius of mathematics": ibid., 209.

34 he did "not have the type of philosophical mind . . . at home in abstractions": ibid., 214.

35 "I soon became aware that I had something good": ibid., 211–212.

35 The paper he produced: Wiener 1915b.

35 "To see a difficult, uncompromising material take living shape and meaning": *ExP*, 212.

35 "more or less by itself": ibid.

35 "long, moist, and harmonious": ibid., 206.

36 "a not too hilarious Christmas dinner": ibid., 220; Masani, NW CW IV, 73.

36 "The Highest Good" . . . ."Relativism": Wiener 1914a; 1914b.

36 "You seem to be doing philosophy rather than math": Eliot to NW, 1.6.15.

36–37 "no knowledge is self-sufficient": Wiener 1914b.

37 "*All* philosophies are . . . relativisms": ibid. (emphasis added).

37 "The hypothesis that a highest good exists may well be doubted": Wiener 1914a.

37 Eliot was a devout realist: although he would later become a devout convert to the Church of England.

37 "In a sense . . . philosophising is a *perversion of reality*": Eliot to NW, 1.6.15 (emphasis added).

37 "I am quite ready to admit . . . avoid philosophy": ibid.

37 Eliot . . . gave Wiener a rave review: ibid.

37 sowed seeds that would blossom in his work decades later: see Masani in NW CW IV, 68–75.

38 "took it green up to the bridge": *ExP*, 221.

38 "emancipation": ibid., 216.

38 "much more a citizen of the world": ibid., 215.

38 "the cream of the intellectual crop": ibid., 216.

38 "Slough of Despond": ibid., 215. The phrase is from John Bunyan, *Pilgrim's Progress*, Part I.

38 "completely lacking in tact": *ExP*, 222.

39 "low point": ibid., 226.

39 "men of the American scene": ibid., 208.

39 "facile . . . a certain thinness of texture": ibid., 224.

39 "pontifical": ibid., 228.

39 "youthful loquacity . . . gift for gab": ibid.

39 "I was protected by my very inexperience": ibid.

40 "inconsiderate": ibid., 127.

40 "an aggressive youngster," "not a very amiable young man": *IAM*, 28.

40 "certainly no model of the social graces": *ExP*, 132.

40 Anti-Jewish attitudes at Harvard: ibid., 131. In 1922, Lowell proposed the use of quotas for Jewish college students. He later retracted the statement, but the precedent was set and quotas were widely instituted by American universities.

40 "a star of the first magnitude": ibid., 230.

40 Birkhoff's "special antipathy": *IAM*, 28. Masani defended Birkhoff against Wiener's charge of anti-Semitism but acknowledged that "There is no evidence, however, of Birkhoff's opposition to the Jewish [quotas]." Masani, 363. Einstein, too, believed that "Birkhoff is one of the world's great academic anti-Semites." Nasar, 55.

40 resentments aimed at his father: *ExP*, 231–232.

40 "without the benefit of any foreign training whatever": ibid., 230.

40 "my eyesight . . . to hit a barn out of a flock of barns": ibid., 238.

41 "a strapping lot of young farmers and lumberjacks": ibid., 240.

41 "my eyes were against me everywhere": ibid., 244.

41 "shamefaced resignation": ibid., 248.

41 "to dare to contravene his orders": ibid.

42 "utter balderdash": ibid., 251.

42 "useless fumblers with symbols": ibid., 255.

43 "dust, flies, heat—everything but water": NW to Leo/Bertha, 7.29.18, WFR.

43 "I see more shell and shrapnel bursts in a day": ibid.

43 "not to be a slacker": NW to Leo/Bertha, undated, WFR.

43 "I have a special talent for making blunders": ibid.

43 "I was appalled by the irretrievability": *ExP*, 259.

43 Two days later, the Armistice was signed: ibid., 260.

44 "strained by my corpulence": ibid., 259.

44 "Dear Dad: I am well and happy": NW to Leo/Bertha, 10.6.18, WFR.

44 "Dear Conta: I'm O.K.": NW to Constance, 10.10.18, WFR.

44 "Dear Ma: I am well and happy here": NW to Bertha, 11.21.18, WFR.

44 "Dear Ma: . . . I have delayed in writing to you . . . rather depressed": NW to Bertha, 1.13.19, WFR.

45 "there were many good candidates": *ExP*, 271.

45 "many blunders": *IAM*, 29.

45 "Americanization . . . completely out of touch with the educated elements": *ExP*, 267.

45 "not unready to leave the paper": ibid., 268.

46 "did not have a particularly high opinion of me or the job": ibid., 271.

CHAPTER 3

47 Epigraph: Browning.

48 A single, labyrinthine corridor . . . gerrymandered textile mills: Hapgood, 54.

48 "infinite corridor": ibid.

48 Tech brooked no nonsense . . . had no chapel: *ExP*, 281–282; Hapgood, 39.

48 "running interference for the engineering backs": *ExP*, 277.

49 "Tech boys" wanted to work: ibid., 282.

49 "How could one bring to a mathematical regularity . . . the water surface?": *IAM*, 33.

49 "problem of the waves": ibid.

49 The problem had been around for two centuries: Swiss mathematician Daniel Bernoulli is credited with founding the science of fluid dynamics in his *Hydrodynamica*, published in 1738.

49 the writings of G. I. Taylor: Wiener read Taylor's writings at Trinity. NW CW I; Masani, 4; Taylor.

50 "*I came to see that the mathematical tool for which I was seeking*": *IAM*, 33.

50 *Elementary Principles of Statistical Mechanics*: Gibbs.

51 "an intellectual landmark in my life": *IAM*, 34.

51 Wiener's favorite example—the path of a drunken man: ibid., 35, 37.

51 "essential irregularity of the universe": ibid., 323.

52 Gibbs's statistical principles, of which Einstein was unaware: www.in-search-of.com/frames/biographies/einstein_albert/einstein.shtml.

52 his major work on Brownian motion: Wiener 1921.

52 "Wiener measure": K. Itô, NW CW I, 513–515.

52 "Wiener space": a probability measure over an infinite dimensional space. NW CW IV, 372.

53 "new and powerful communication techniques of Oliver Heaviside": *ExP*, 281.

53 "shot effect": The shot effect was first analyzed by W. Schottky in 1918. Ott.

53 "discreteness of the universe": *IAM*, 40.

54 "everything from a groan to a squeak": ibid., 75.

54 "harmonic analysis . . . generality and even rigour": www-groups.dcs.st-and.ac.uk/00history/Mathematicians/Fourier.html.

54 Wiener reformulated Fourier's work: Wiener acknowledged that he was not the only clarifier of communication engineering during that period. He wrote later: "From . . . 1900 until 1930, Heaviside's methods dominated . . . communications engineering technique, and their rigorous mathematical justification was a moot question. . . . Toward the end of this period . . . several avenues [for formalizing] Heaviside calculus were found. . . . The construction of a comprehensive table of Fourier transforms by Drs. Campbell and Foster of the Bell . . . Laboratories almost at once replaced the Heaviside calculus . . . as the method of choice in communications engineering." *YP*, 7–8. See also Bush 1929 (Appendix by Wiener); Campbell and Foster.

55 electronic signals could be captured and frozen in time: Wiener's "great contribution to harmonic analysis . . . was translation invariance or stationarity." Masani, NW CW II, 809.

55 "stillborn": *IAM*, 40.

55 "to clear the tracks for the two doctoral candidates": ibid., 84.

55 "thundered": ibid., 85.

55 "I was more or less repelled . . . too dangerous to be ignored": ibid., 87.

56 "a complete right of veto": ibid., 32.

56 Hedwig . . . as many careers as Leo Wiener: Hedwig worked as a cook in lumber camps in Utah and Montana, ran a boardinghouse in Salt Lake City, and later a dry-cleaning establishment. B. Raisbeck 1986; BT1.

57 "was beautiful in a pre-Raphaelite way": *ExP*, 282–283.

57 "were not silent in their approval": ibid., 284.

57 Lebesgue himself . . . lauding Wiener's work: *IAM*, 92; Masani, NW CW I, 5.

57 "the whole system of . . . physics must be restructured from the ground up": *NYT*, 12.12.2000.

58 "My talk in Göttingen, like quantum theory . . . no music at all": *IAM*, 105–107.

58 Heisenberg . . . was sitting in Wiener's seminar that day: Masani, 117.

59 "presented to the Göttingers . . .": *IAM*, 107.

59 Born sought Wiener's help: ibid., 108.

59 "notable contribution" in the advance of quantum theory: Armand Siegel, "Comment," NW CW III, 531; Wiener and Born 1926a.

59 "hardly assimilated": Born did not see the quantum wave dynamics implicit in his work with Wiener. "We worked together hard on this project . . . but we just missed the most important point in a way which makes me ashamed even to this day," he wrote later. "Thus we were quite close to wave mechanics but did not reach it, and Schrödinger is rightly considered its discoverer." Born, 226–230.

59 "his statistical interpretation of the wavefunction": www.almaz.com/nobel/physics/1954a.html.

59 "an excellent collaborator": www.nobel.se/physics/laureates/1954/born-lecture.pdf. In his memoirs, Born recalled the start of his collaboration with "a remarkable young man, Norbert Wiener . . . a mathematician of repute" (Born, 226). Wiener recalled Born's initial skepticism: "Born had a good many qualms about the soundness of my method and kept wondering if Hilbert would approve of my mathematics. Hilbert did, in fact, approve of it" (*IAM*, 108–109). Soon after he returned to Göttingen, Born unveiled a new statistical interpretation of Schrödinger's wave mechanics. He asserted that Schrödinger's "electron wave" was not a physical wave but a "probability wave"—a remapping onto the quantum plane of Wiener's probability measure of Brownian motion. But Born claimed the concept as his own. He later insisted that "the idea of electron waves was familiar to me when Schrödinger's papers . . . appeared," and that, "At that time it was clear that the proper interpretation of quantum mechanics must be of a statistical type" (Born, 231). However, he acknowledged that, in his work with Heisenberg and others before he teamed with Wiener, "we never contemplated waves, and all they imply" Born, 220.

59 Cecilia Helena Payne: Zipporah "Fagi" Levinson int., Cambridge MA, 8.24.98.

59 "a person more like himself . . . with hope of matrimony": Heims 1980, 24.

59 "a bit too intermittent to suit us": *IAM*, 110.

60 "I was in a very exulted mood . . . boasting and gloating": ibid., 112–113.

60 "Dear Parents: Marguerite and I are having a wonderful honeymoon": NW to Leo/Bertha, 3.30.26, WFR.

60 "Gretel": NW to Leo/Bertha, Constance, Fritz, beginning with NW to Leo & Bertha, 11.27.26, WFR.

60 "Your (no longer) dutiful son Norbert." NW to Leo and Bertha, 4.11.26, WFR.

60 Nationalist extremists dogged his . . . interactions: NW to Constance & Phil Franklin, from Italy, summer 1926, WFR.

61 Courant turned against Wiener: "Courant was trying to seesaw between me and the government. . . . I felt the cards were stacked against me": ibid.

61 to the verge of a nervous breakdown: "the edge of a nervous breakdown." *IAM*, 115.

61 "blues": ibid.

61 "black depression": ibid., 119.

61 "partly to share in my supposed success": ibid., 117.

62 "to work out our problems of adjustment": NW to Bertha/Leo, 8.3.26, WFR.

62 "really afraid . . . a nervous breakdown. . . . Norbert needs absolute quiet": Margaret to Leo/Bertha, Aug 1926, WFR.

CHAPTER 4

63 Epigraph: Sophocles.

63 "to a life of handyman domesticity": *IAM*, 124.

63 "was never my métier": ibid.

63 every young woman in the United States: www.pbs.org/wgbh/amex/telephone/filmmore/transcript/index.html.

65 "*the size of the message*": Hartley 1926 (emphasis added).

65 "*transmission of intelligence*": Nyquist (emphasis added).

65 "*the commodity to be transported by a telephone system*": Hartley 1926 (emphasis added).

65 There simply was no rigorous science of communication: "Heaviside's methods worked by black magic—they gave the right answers, but nobody knew why." Gordon Raisbeck, personal communication, 7.23.99. Raisbeck contrasted the days of circuit design before Wiener and his doctoral student Y. W. Lee as the era of "cut-and-try methods," in contrast to "a systematic tailoring [of circuits] to a set of criteria chosen in advance."

65 The distinction between strong and weak currents: "What the Germans call . . . *Starkstromtechnik* and *Schwachstromtechnik*." YP, 3.

66 the fundamental assumption of Gibbs' statistical mechanics . . . not merely inadequate, but impossible": The "ergodic hypothesis" was first stated by Gibbs and Boltzmann and later refined by Wiener's nemesis G. D. Birkhoff and by Wiener himself. Wiener 1938. "Wiener was . . . shocked and surprised that . . . Gibbs had obtained valid results by methods that were mathematically hopeless." Dennis Gabor, "Comments," NW CW III, 491.

66 "years of growth and progress": *IAM*, 124.

66 more than a dozen papers: The most important of these were Wiener 1926b, 1926c, 1926d.

66 "Generalized Harmonic Analysis": Wiener 1930.

67 "noise": Wiener's papers on harmonic analysis contributed to a rigorous mathematical theory of "haphazard disturbances" such as radio-telephone "tube noises." Wiener 1926d. See also Wiener 1926c; Wiener 1930/NW CW II, 4–5, 133, 159, 298, 372–374.

67 "a valuable means of measuring the electronic charge": Wiener 1926d.

67 "series of data": Wiener 1930/NW CW II, 259.

67 "an infinite sequence of choices . . . binary": Wiener 1930/NW CW II, 276.

68 a new "movement" in scientific thinking: Wiener 1938/NW CW II, 800. See Masani's further comments, NW CW II, 807–810.

68 ahead of its time by a decade or two . . . three: Masani, NW CW II, 377.

68 "Norbert does the math and I do the arithmetic": Dorothy Setliff telephone int., 9.9.98.

68 "very demonstratively affectionate": PWK1.

68 "my new personality as a married man": *IAM*, 128.

68 "had taken a great deal too much for granted . . . family captivity": *ExP*, 285.

69 "made him look as if he had been painted by Rembrandt": *IAM*, 125.

69 "when my life is a simple alternation": ibid., 127.

70 "give our children the experience of the country": ibid., 126.

70 Barbara . . . from the first mood of the first figure of the syllogisms: BWR, 5.14.01.

70 "pure universal": "A pure universal syllogism contains three universal propositions. . . . The AAA–1 syllogism is the highest form of syllogistic argument." www.gibson-design.com/philosophy/Concepts/organon-dictionary-$.html#barbara_1.

70 "*Barbara, Celarent, Darii, Ferio*": From the medieval poem, "The Rhyme of Reason," www.phil.gu.se/johan/ollb/Syllog.machine.html.

70 "as a very clumsy pupil in the art of baby sitting": *IAM*, 127.

70 "Everything is going first rate here": NW to Leo, 7.30.28.

71 "*One fine morning . . . this 'satiable Elephant's Child"*: Kipling.

71 "featly tread": Horace, *Carmina Odes*, 1.37. The phrase is from Conington's 1870 translation.

71 "I can still hear his Latin verse": BWR, 7.8.97.

71 Pascal . . . Leibniz . . . Babbage: For background see the excellent biographies of the School of Mathematics and Statistics, University of St. Andrews, Scotland: www-groups.dcs.st-and.ac.uk/00history/BiogIndex.html.

71–72 "Difference Engine . . . Analytical Engine": www-groups.dcs.st-and.ac.uk/00history/Mathematicians/Babbage.html; www.fourmilab.ch/babbage/contents.html.

72 all the basic components of a modern computer: including a "mill" or central processing unit, a "store" for holding the partial results of arithmetic operations, a logical "control" for deciding among different options as results were obtained, and a press to print the end product of the computation. fourmilab.ch/babbage.baas.html.

72 A hundred years later, Vannevar Bush: Zachary.

72 "closely associated . . . on my own account": *IAM*, 111.

72 "The number of operations . . . is simply enormous": ibid., 138.

73 "I was visiting the show at the old Copley Theatre": ibid., 112.

73 Wiener's earlier work on . . . light waves: Wiener 1925. See also Wiener 1926c, 1926d; Wiener 1930.

73 an all-electronic *optical* computer: *IAM*, 137–138.

73 "The idea was valid": ibid., 112.

73 "got a nice letter from Bush about my machine & his": NW to Leo/Bertha, Breslau, 11.27.26, WFR.

73 "Product Integraph . . . Cinema Integraph": See Bennett 1994a; Masani, 162–163.

74 Bush did not opt for optical computing: Later Bush saw the light. In the 1930s, he became an advocate of optical technology and, with an MIT colleague, developed an optical data-sorting device for microfilm systems. In the mid-1940s, he proposed an expanded optical system he called a "memex," which has been praised as the prototype for hypertext information storage and retrieval. Zachary, 75–76; Bush 1945a.

74 The machine could compute in fifteen minutes: ftp.arl.mil/00mike/comphist/61ordnance.chap1.html.

74 In 1929 Bush . . . book on circuit theory for electrical engineers: Bush 1929.

74 "of considerable practical value": *IAM*, 139.

74 Bush sought Wiener's advice on many of the chapters: ibid.

74 "the fun we had in working together": ibid.

74 "did not know an engineering and a mathematician could have such good times together": Bush 1929; v-vi.

74 "the high-speed computing machines of the present day follow very closely": *IAM*, 138.

74 "The future development of computing machines . . . borne me out in this opinion": ibid., 137–138.

75 The "Wiener-Hopf" equations: Wiener and Hopf 1932b. See also T. Kailath, NW CW III, 63–64.

76 "His steadiness and judgment": *IAM*, 133.

76 "adjustable electronic circuit": ibid.

76 "at the cost of a great wastefulness of parts": ibid.

76 "reduced a great, sprawling, piece of apparatus": ibid.

76 the best circuitry of its kind . . . physically attainable: The Wiener-Lee network consisted of a latticework of lumped linear circuit elements (resistors, capacitors, and inductors) and functioned as a sensitive bandpass filter that separated and further refined in each successive layer the specified band of frequencies to be passed along by the network. Its layers of progressive filtering components, in effect, translated Wiener's method of harmonic analysis into a practical form that, in principle, could be expanded to an infinite number of layers. Gordon Raisbeck, personal communication, 7.23.99.

76 a patent application was filed: "Electrical Network System," Patent No. 2,024,900, *Official Gazette*, U.S. Patent Office, Dec. 17, 1935, 688.

76 "Now Lee wasn't a mathematician . . . no one believed it": Amar Bose int., Framingham, MA, 11.26.97.

76 "When Lee made his doctoral presentation . . . They pounced all over him": Bose was referring specifically to the real and imaginary components of Wiener's harmonic analysis theory and Lee's application of it.

77 Lee's dissertation . . . a landmark: Lee 1930/1932; www.eecs.mit.edu/great-educators/lee.html.

77 over the $5,000 mark in annual income: MIT raised NW's salary to $6,000 in 1931. MIT to NW, 11.11.31.

77 J. D. Tamarkin: *IAM*, 129–130.

77 "I began to be heard of in this country": ibid., 130.

77 Even politically isolated mathematicians in the strife-torn Soviet Union: ibid., 145.

77 "made it possible to allay some part of the hostility": ibid., 128.

78 "Another university is apparently quite desirous of your services": Compton to NW, 12.12.30.

78 Princeton's . . . new Institute for Advanced Study: NW-Veblen letters, Feb 1931.

78 "wear the arms of Trinity College . . . only on the left hand": *IAM*, 150.

78 his first book-length work: Wiener 1933.

78 "tall, powerfully built, beetle-browed man": *IAM*, 160.

79 Haldane "had used a Danish name": ibid.

79 "I have never met a man with better conversation": ibid., 161.

79 Haldane . . . controversial stands . . . most ardent and articulate communists: Clark, 75–77, 209–210, 290–293.

79 "wild, long-haired Russian": Leo Wiener, 1910.

79 the probabilities of genetics . . . statistical methods: In ten papers. Haldane 1924–1934; Clark, 71–73, 306–308.

79 "discoursing loudly at table": Sheehan.

79 "Following his example, I smoked a cigar": *IAM*, 162.

80 "a reasonably routine life": ibid., 164.

80 "extreme sensitivity . . . a pupil, a friend, a colleague": Struik 1966.

81 "It took me some time to come back to a mental equilibrium": *IAM*, 170.

81 "Paley-Wiener criteria for a realizable filter": Wiener and Paley 1934; Masani, NW CW III, 4.

CHAPTER 5

82 Epigraph: Goethe.

82 "as he tossed peanuts into the air and caught them in his mouth": Heims drew the classic image of Wiener as he was frequently described by students and colleagues, "walking the long, industrial halls, leaning backward in his ducklike fashion, looking up as he tossed peanuts into the air and caught them in his mouth—a skill he had perfected." Heims 1980, 381.

82 "safaris": Jackson.

82 "At MIT he used to walk all the corridors": Dirk Struik int, Belmont MA, 8.28.98. See also Struik 1966; Struik 1994.

82 he seemed oblivious to the world around him: Recollections of Morris Halle and others, Heims 1980, 380–381; David Cobb, a former MIT student, reported seeing him walking across campus in his tie and jacket "unaware of the snowstorm raging around him." MIT 1994 (NW centennial program).

83 "He stopped me halfway . . . already had my lunch": Ivan A. Getting telephone int., Coronado, CA, 5.2.99.

83 Sometimes he seemed wholly unaware of the person to whom he was talking: Heims 1980, 380–318.

83 "Wiener needed to write, so he walked . . . into the nearest office": Donald Spencer in MIT 1994.

83 "famously bad": Recollections of Hans Freudenthal, Masani 1990, 353.

83 "In class, while presumably deriving a theorem on the blackboard": Heims 1980, 333.

83 "He could range among the worst and among the best": Struik 1966.

83 five-foot-long equations . . . without notes: "Master Mind," *MD*, June 1975.

84 Wiener's students had to hunt him down: Recollections of Asim Yildiz, Heims 1980, 384.

84 picking his nose "energetically": Recollections of Armand Siegel, ibid., 382.

84 On at least one occasion, he strode into the wrong classroom: Recollections of Albert Hill, Heims 1980, 381.

84 wrote a large "4" on the blackboard: Recollections of David Cobb in MIT 1994.

84 "Every student . . . got an A": Zipporah "Fagi" Levinson int., 8.24.98.

84 "a most stimulating teacher . . . secretary for a young student": Norman Levinson 1966.

84 "Norman came from a very poor family": Zipporah Levinson.

85 "He would lean back and start to snore": Gordon "Tobey" Raisbeck, B&TR1.

85 "Actually, he could fall asleep *and* talk!": Doob 1994.

85 "right-hand man": *IAM*, 170.

85 Both men viewed science as a collaborative venture: ibid., 171.

85 "a field into which few real mathematicians had strayed": Rosenblith and Wiesner.

85 "Ultimately the theme of the many discussions": *IAM*, 171–172.

85 "no man's land between . . . established fields": *Cyb*, 2.

86 "sales resistance . . . more than we could overcome": *IAM*, 135.

86 "was cut out for us . . . some of whom went through my hands": ibid., 174–175.

86 Karl Menger . . . Notre Dame: Menger later found a home at the Illinois Institute of Technology.

87 "desperate straits": *IAM*, 180.

87 "I have never felt the advantage of European culture over . . . the Orient": ibid., 182.

87 "affected by the rigidity that so often taints a university": ibid., 184.

87 "to report our every word and attitude to the management and . . . the police": ibid.

88 Lee's wife, Betty: *IAM*, 186; BWR, 5.20.01.

88 "There was common to almost all that love of the whole world": *IAM*, 197.

88 for the modest sum of $5,000: AT&T to NW, 9.26.35.

88 "the great advantage . . . for purposes of mass production": U.S. Patent # 2,124,599, July 26, 1938.

88 "All this effort we had made went into a paper patent": *IAM*, 135.

89 What was lacking in our work . . . feedback . . . failure was the consequence": ibid., 189–190.

89 "mixture of glamor [*sic*] and squalor": ibid., 194.

89 "It was intriguing to walk down ill-paved alleys": ibid.

90 "Margaret and I now had a large stock of common experiences": ibid., 207.

91 "He stopped me in a hall and asked me if I played tennis": Ivan Getting telephone int., 5.2.99; Masani 1990, 349.

91 seven miles on snowshoes: Bose heard Wiener went in on skis. Bose int., 11.26.97. Barbara heard snowshoes. BWR, 12.5.98. Wiener used both. *ExP*, 242–243; 284.

91 "He liked to recite Greek and Latin . . . I couldn't do anything to make him stop": B&TR1.

92 "He'd come around . . . when nobody was dressed": Recollections of Armand Siegel, Heims 1980, 389–390.

92 "unburdening himself . . . they sent him home": ibid., 385.

92 "He was a friend who could get on your nerves": Dirk Struik int., 8.28.98.

92 painted into a corner: Dorothy Setliff telephone int., 9.9.98.

92 "But you *drove* down!": B&TR1.

92 "When we moved to the Cedar Road house": B&TR1.

92–93 "It is entirely possible . . . totally *im*possible": PWK1.

93 "Before I should even think of subjecting any child": *ExP*, 292–293.

93 "I find my father's often-repeated claim": BWR, 10.21.97, 7.5.00, 8.18.00.

94 "desperate not to have happen to us what happened to him": PWK1.

94 "Princess Sabbath" . . . tore Wiener apart: *IAM*, 215.

94 "He would stand in the front hall shouting and crying": BWR, 7.8.97.

95 "I do not remember my father possessing any sort of firearm": BWR, 10.20.97, 10.27.97.

95 "temper tantrums . . . I was terrified": PWK1.

95 "My father always required vast amounts of praise and reassurance": BWR, 7.6.01.

95–96 "After the storm crested . . . his 'unnatural' feelings for me": BWR, 7.8.97.

96 "emotional deafness . . . the first magic that came to her mind": B&TR1.

96 "My father's responses were *perfectly* natural": PWK1.

96 "internal storm . . . psychoanalytic help": *IAM*, 212–213.

96 "resistance . . . my spiritual make-up . . . what made me tick": ibid., 214–215.

97 the psychiatric diagnosis of manic depression: Kraeplin.

97 "a very exulted mood": *IAM*, 112.

97 Like depression . . . common among creative minds: Among mathematicians especially depression and/or mania have been attributed to Newton, Russell, Cantor, Klein, and Courant (respectively, Manuel, 70; Macrae, 104; Cantor: www-groups. dcs.st-and.ac.uk/00history/Mathematicians/Cantor.html; www-groups.dcs.st-and.ac.uk/00history/Mathematicians/Klein.html; www.nap.edu/readingroom/books/biomems/rcourant.html). Heaviside suffered depression and extreme agoraphobia (Nahin). In the worst tragedy, Boltzmann, who was chronically depressed over his peers' early attacks on his work in statistical mechanics, hanged himself in a hotel room in Trieste in 1906, shortly before his theories were verified experimentally: www-groups.dcs.st-and.ac.uk/00history/Mathematicians/Boltzmann.html.

97 steered by some underlying neurochemical component: Nemeroff.

97 "a disorder of the brain": www.ndmda.org (National Depressive and Manic Depressive Assn.).

97 Severe apnea . . . a trigger of manic behavior: "Disruptions in Sleep May Lead to Mania in Bipolar Disorder," Report on 2nd Intl Conference on Bipolar Disorder, June 1997, www.mentalhelp.net/articles/bipolar1.htm.

98 "We are really very much alike": NW to Fritz, 12.26.26.

98 "My father had a real terror of turning out like Fritz": B&TR1.

98 "I think he did believe that, and I rather think Mother nurtured it": PWK, 6.8.99.

98 "He would come to her, obviously looking for approval or support": B&TR4.

98 "I don't think she knew what she was doing": ibid.

98 "Dad was a very warm and generous person": PWK5.

98 cyclothymia: www.ndmda.org/biover.htm.

99 "there were loose pages with mathematical notes": BWR 7.6.01.

99 "at a level of consciousness so low . . . in my sleep": ExP, 46.

99 "Very often these moments seem to arise on waking up": ibid., 213.

99 "when I think, my ideas are my masters rather than my servants": ibid., 46.

99 "he lived in fear that ideas would lose interest in him": BWR, 10.27.97.

100 *"Haw, haw, haw"*: Mildred Siegel int. Brookline, MA, 8.24.98.

100 "She was an intelligent woman . . . a bit lost": PWK1.

100 "Her sense of humor tended more towards *Schadenfreude"*: PWK, 10.27.00.

100 "One day she told us . . . *Judenrein'"*: B&TR1.

100 "She told me once that Jesus was the son of a German mercenary": BWR, 10.15.01.

100 "Two books decorated her dresser . . . copies of *Mein Kampf"*: BWR, 7.8.97.

100 "very painful": B&TR1.

101 "Margaret spoke very openly about . . . her relatives in Germany": Zipporah "Fagi" Levinson int., 8.24.98.

101 "to whom national and racial prejudice have always been as foreign as they have been to me": IAM, 182.

101 "the fact that Nazism threatened to dominate the": ExP, 211–213.

101 as Hitler's troops were massing on the Polish border. B. Raisbeck 1986.

101 humanitarian values . . . he took from Tolstoy: ExP, 69–70.

101 "*élan* . . . and the emotions thereof": ibid., 74.

101 "that scholarship is a calling and a consecration": ibid., 292.

101 uncompromising intellectual honesty: ibid., 70.

101 "fierce hatred of all bluff and intellectual pretense": ibid., 292.

101 "It was because of this, because my taskmaster was . . . my hero": ibid., 74.

102 "It was an experience curiously reminiscent of that time twenty-four years before": IAM, 225.

102 "to take stock of our emotions and expectations": ibid.

CHAPTER 6

103 Epigraph: Diderot, www.quotationspage.com.

104 "the organization of small mobile teams of scientists": IAM, 231.

105 The other was Wiener: Lee, Winkler, and Smith.

105 "apparatus of the computing machine should be numerical . . . electronic tubes . . . scale of two": *Cyb*, 4. See also Wiener 1940.

106 a method Wiener himself had first applied ten years before: Wiener 1930.

106 "That the entire sequence of operations . . . on the machine itself": Wiener called for the program to be "built into the machine itself," as opposed to von Neumann's later "stored-program" concept which maintained the program actively in the computer's memory. Ferry and Saeks write that Wiener's proposed machine "exhibits every characteristic of the modern digital computer except for the stored program." They note that Wiener's machine was "a textbook example of a Turing machine. . . . [His] realization of Turing's 'tape' . . . in which the 'bits' of a binary number are recorded simultaneously on parallel tracks came . . . a quarter century ahead of the industry." Ferry and Saeks. "Wiener's proposed machine employs . . . a discrete . . . algorithm . . . binary arithmetic and data storage [and] an electronic . . . logic unit. [His] conceptions were to be borne out brilliantly, but only a decade and a half later." Masani, 173–175.

106 Alan Turing . . . "universal" machine: Turing 1936/1937. In 1936, Turing, a young British mathematician, developed his idea for a symbol-processing "universal machine . . . which can be made to . . . carry out any piece of computing, if a tape bearing suitable 'instructions' is inserted into it." However, Turing made no attempt to build such a machine. Lee, Winkler and Smith; www-groups.dcs.st-and.ac.uk/00history/Mathematicians/Turing.html.

106 John V. Atanasoff: The Atanasoff-Berry Computer (ABC) was conceived in 1937 by Atanasoff and developed with his grad student Clifford Berry. It was the first digital machine with vacuum-tube computing elements, a rudimentary memory, and special "logic circuits." However, the ABC used mechanical inputs (punched paper cards) and did not have the internal programming capacities Wiener envisioned. Work on the project ended after Pearl Harbor and the prototype was dismantled. Mollenhopf; www.scl.ameslab.gov/ABC/Biographies.html (Iowa State Ames Laboratory); Lee, Winkler, and Smith.

106 Bush . . . not yet convinced of his design's practicability: Bush to NW, October 1940.

106 Bush turned down Wiener's proposal: "turning Wiener down flat." Zachary, 266.

106 "it is undoubtedly of the long-range type": Bush to NW, 12.31.40 (Zachary, 266). As Wiener recalled it, Bush deemed his conception "too far in the future." IAM, 239.

107 "I hope you may find some corner of the activity": NW to Bush, 9.21.40.

107 Bush . . . believed defense against air attack was . . . America's foremost problem: Zachary, 96.

108 "ducks on the wing": *IAM*, 240.

108 "usurp a specifically human function": *Cyb*, 6.

109 the new British microwave radar: See Buderi, 59–64.

109 another MIT engineer: Richard Taylor of the Electrical Engineering department. Bennett 1994b.

109 "He said, 'You *can't* go into the service'": Bigelow quotes are from JB1-JB4 unless specifically noted.

110 A crew of up to fourteen men: Bennett 1996.

111 "To some extent this is a purely geometrical problem": *IAM*, 241 (text adapted with permission).

111 "will probably zigzag, stunt, or in some other way take evasive action": *Cyb*, 5.

111 "only a prophet with the knowledge of the mind of the aviator": *IAM*, 241.

111 "The pilot does *not* have complete freedom": *Cyb*, 5 (original emphasis).

111 "There are . . . in fact, means . . . to accomplish the minor task": *IAM*, 241.

113 "interesting and exciting, and in fact not unexpected": ibid., 244.

113 "violent oscillation": ibid.

113 "Perhaps this difficulty is in the order of things": ibid.

113 "If then we could not . . . develop a perfect universal predictor": ibid.

113 "mean square error" method: "For the actual distribution of curves which we wanted to predict, or . . . the actual distribution of airplanes that we wanted to shoot down . . . we took the square of the error of prediction at each time. . . . We then took the average of this over the whole time of the running of the apparatus. This average of the square error was what we were trying to minimize." ibid., 244–245.

114 Bell's M-6 mechanical antiaircraft predictor: Bell Labs began developing a radar-guided fire control device with manual tracking controls in late 1937 at its Whippany, NJ laboratory. Buderi, 131.

114 Wiener offered the Bell team his new statistical prediction method: As the official record described it: "to predict non-uniform curvilinear performance of the target [using] a knowledge of the probable performance of the target during the time of shell flight and . . . a statistical analysis of the correlation between the past performance . . . and its present and future performance." Bigelow 1941.

114 "which did not involve statistical concepts in any form": ibid.

114 "We didn't think it was so hot. . . . They told us very little": Bigelow's report stated that, "The MIT group was favorably impressed," and that Wiener was "greatly pleased," particularly by the two teams' "identical concepts of realization by electrical means." However, neither Wiener nor Bigelow accepted Bell Labs' explanation that "no statistical approach was necessary." Bigelow explained the discrepancy between his memory of the meeting and Wiener's words in their report to the NDRC command. "Wiener was careful of what he said in public." JB3.

115 That month, the Bell Labs and Rad Lab teams linked up: In May 1941, Jerrold Zacharias and another Rad Lab staff member began an eight-month assignment at Bell Labs in New Jersey. Buderi, 132.

115 "in the metal": *IAM*, 245.

115 "little laboratory": ibid., 248.

115 The initial funding . . . totaled $2,325: NDRC to S. H. Caldwell for work on Wiener's Project #6 from 12.1.40 to 5.31.41. Later, another $10,000 went to Caldwell for unspecified work through 6.30.42. It "seems to be a second contract for the Wiener work but I cannot be certain of this." S. Bennett, personal communication, 6.1.00.

115 Rad Lab . . . first-year budget of $815,000: With another $500,000 underwritten by MIT. Buderi, 50, 113.

116 "as much more of a shame and a humiliation than a surprise": *IAM*, 272.

116 "something was about to blow off in Japan": ibid.

116 "The Extrapolation, Interpolation and Smoothing of Stationary Time Series": Wiener 1942 (*YP*).

116 "Yellow Peril": Wiener's *YP* was preceded by another "Yellow Peril" well known to engineering students during the war years, technical publications by Springer that also were bound with yellow covers. JB3. No deliberate effort was made to link that phrase to the anti-Asian sentiment it acquired during the initial wave of Chinese immigration to the United States early in the 20th century, although obviously a similar level of fear was intended for a humorous effect.

116 he formally distinguished . . . power engineering . . . from . . . communication engineering: *YP*, 2–3.

117 "the study of messages and their transmission": ibid., 2.

117 "array of measurable quantities distributed in time": ibid.

117 "a message . . . corrupted by a noise": ibid., 117. Hartley (1928) was first to address abstract considerations of the transmission of messages, to describe messages in terms of possible choices, and

to develop "the concept of noise as an impediment in the transmission of information." Aspray. But his perspective was not statistical in the mathematical sense which Wiener pioneered.

117 "a conscious human effort for transmission of ideas": YP, 2.

117 "servomechanism . . . is also a message": ibid.

117 all communication operations "carried out by electrical or mechanical . . . means": ibid.

117 "fundamental unity of all fields of communication": ibid.

117 *"measure or probability of possible messages"*: ibid., 4 (emphasis added).

117 *"such information will generally be of a statistical nature"*: ibid., 9 (emphasis added).

118 A week after Wiener delivered his Yellow Peril, Stibitz: G. R. Stibitz, "Note on Predicting Networks," Feb 1942. Stibitz note is accompanied by a longer paper, dated Feb 8, 1942, "Prediction Circuits a la Wiener." In an earlier report to Section D–2, computer pioneer John V. Atanasoff praised Wiener's prediction theory as a "brilliant and important piece of work." Atanasoff to Weaver, 11.1.41. Bennett 1994b.

118 "did some work on the statistical approach to fire control . . . made no sense": Ivan Getting telephone int., 5.2.99.

118 "big shots": NW to Edward L. Bowles, professor of electrical engineering, MIT, 3.22.42.

118 "knew little to nothing about the microwave electronics": Statement by Lee A. DuBridge. Zachary, 134.

118 "the highly chaotic and anarchic regime . . . in the radiation laboratory": NW to Bowles, 3.22.42.

118 The next day, the Army ordered 1,256 units: Buderi, 134.

118 "herky-jerky": ibid., 131.

119 "theory group": ibid. MIT engineers Albert C. Hall and Ralph S. Phillips headed the Rad Lab's fire-control theory group under Getting's direction. JB3; Getting telephone int.

119 "[Rad Lab] people are coming to us all the time": NW to W. Weaver, 2.22.42.

120 a half second *before* it arrived at the targeted coordinates: IAM, 250.

120 a "miracle" . . . "was it a useful miracle?": Bennett 1996.

120 But to be of practical value in the field . . . time would have to be doubled or tripled: ibid.

121 The Greeks invented automatic wine dispensers and water clocks: See Kelly (www.kk.org/outofcontrol/ch7-a.html); F. L. Lewis.

121 "slave-motors" or servomechanisms: Bennett 1994a.

121 The problem persisted until 1927: Bell Labs engineer Harold S. Black conceived the idea of negative feedback on Aug 6, 1927 while traveling to his lab on West Street in New York City and sketched the first negative feedback circuit on a page of that morning's New York Times ("Retired AT&T Bell Laboratories Researcher, Inventor of Negative Feedback, Dies," Bell Labs News, 1983). His discovery was quickly applied to Bell's telephone, radio, and sound-recording devices, but it was not reported in the engineering literature until Jan 1934. "Historic Firsts: The Negative Feedback Amplifier." Bell System Technical Journal, Dec 1943. See also Black.

121 Thousands of servomechanisms were in commercial use: Bennett 1994a.

121 there was scant theory beneath those diverse feedback inventions: "There was little theoretical underpinning." Bennett 1994a. Nicolas Minorsky, of the General Electric Company, made the first attempt to model and analyze a modern servosystem. Minorsky 1922. Harold L. Hazen, an MIT electrical engineer and protégé of Vannevar Bush, published two seminal papers on servomechanisms in 1934. Hazen 1934a, 1934b.

122 "purpose tremor": Cyb, 8.

122 "to regulate their conduct by observing the errors": IAM, 252.

123 Rosenblueth's insights helped them to . . . make improvements: Cyb, 9.

123 "pathological conditions of very great feedback": IAM, 253.

123 only a ten percent improvement . . . a marginal gain: Wiener and Bigelow, Final Report to NDRC, Sec 2, Div D, Dec 1942. Bennett 1994b; Masani, 190.

124 By February 1944, the first sets . . . saved Allied lives at Anzio: Buderi, 222.

124 V-1 "buzz bomb" attacks: ibid., 223, 226–228.

124 In December . . . the Battle of the Bulge: ibid., 227.

124 one of the American war machine's "greatest success stories": ibid., 223. The first sets reached the Pacific theater in 1944 in advance of MacArthur's return to the Philippines, where they helped to down nearly 300 Japanese planes. ibid., 234.

124 Rad Lab theorists drew directly on Wiener's theories: "Wiener's work . . . was picked up by R. S. Phillips, who was working on the auto-follow radar systems and was seeking criteria . . . that would . . . minimize the errors due to noise." Bennett 1994b; Phillips 1943. Bennett, a British science historian and expert on control engineering, spent a year combing the NDRC's records and concluded that, "The reduction of Wiener's results to practical usable form was considered vital to the

war effort." Bennett 1994b; Bergman; Norman Levinson 1942; 1944; Blackman, Bode and Shannon. In a textbook he co-authored after the war, Phillips reported that he had applied Wiener's work on generalized harmonic analysis to the Rad Lab's signal-smoothing problem and acknowledged that the statistical approach used in his textbook "was inspired by . . . Wiener's [*YP*]." James et al.

124 The Bell Labs team, too, used Wiener's work: Bennett did not know of any technical improvements that were made to the M-9 as a direct result of Wiener's work, yet he conceded that "hearing about Wiener's work . . . may have helped and clarified what [the Bell people] were trying to do [and] it showed that the M-9 was within ten percent of the best achievable." S. Bennett, personal communication, 6.23.00. See also Blackman, Bode, and Shannon.

124 another mathematician at Princeton: Peter G. Bergman. Bergman; Bennett 1994b.

124 Wiener's war work . . . spelled out . . . the physical limits: Bennett 1994b.

124 as the Rad Lab theorists . . . acknowledged: In their textbook, Phillips and his Rad Lab co-authors stated, without mentioning Wiener directly, that work during the war had taken "the theory of servomechanisms in a new direction . . . intended to deal with [devices] of known statistical character in the presence of interference with known statistical character." James et al., Preface. "The Wiener-Bigelow project . . . significantly influenced most future work on radar, noise filtration, and servo designing in communication engineering, military and nonmilitary." Masani and Phillips, 170–171.

124 At the Philadelphia Navy Yard one day late in the war: "Master Mind," *MD*, June 1975. "The story could have originated from Warren Weaver." S. Bennett, personal communication, 6.1.00.

125 "effectiveness . . . estimating which messages are frequent and which are rare": *YP*, 4.

125 Claude E. Shannon . . . in the Bell Labs fire-control group: Bennett 1994b.

125 He would later publish a seminal paper: Shannon 1948.

126 "In the time I was associated with Wiener, Shannon would come and talk": JB1, JB3.

126 "Wiener was very kindly": Zipporah "Fagi" Levinson int., 8.24.98.

126 Through the later years . . . Wiener continued to submit ideas to the NDRC and OSRD: Owens notes that Weaver tried to find more work for Wiener in the war effort, but Wiener's interests remained more theoretical than practical. "Bigelow . . . reported that Wiener . . . 'simply will not hammer out [a] solution with the cheap-and-nasty tools of the everyday applied mathematician.'" Bigelow to Weaver, 4.22.44 (Owens, 291). Wiener was termi-

nated from the war effort three weeks later. Bush to NW, 5.12.44.

126 government's security concerns . . . about Wiener's emotional stability: "There is a letter from . . . Weaver to . . . J. C. Boyce of MIT . . . referring to Wiener's nervous condition and his failure to submit promised reports. . . . Boyce replied that he had advised Wiener to take a few days rest, and he was hopeful that 'this part of the zoo will be quiet again.'" Masani, 192.

127 At one point . . . brink of breakdown: *IAM*, 249.

127 Nazis "almost to a man": B&TR2.

127 "not over our project, but over a feeling of powerlessness": JB3.

127 "war fatigue": NW to John von Neumann, 4.27.45, von Neumann papers, LoC.

127 "detested the arrogance displayed in the use of the bomb against Asians": Struik 1966. Wiener wrote later, "I was acquainted with more than one of these popes and cardinals of applied science, and I knew very well how they underrated aliens of all sorts in particular those not of the European race." *IAM*, 301.

127 "assistance in the early phases of the war . . . splendid cooperation": Bush to NW, 5.12.44.

128 "general breakdown of the decencies in science": *IAM*, 272.

128 "I found that among those I was trusting": ibid.

128 "Tell me, am I slipping?": Norman Levinson quotes Wiener saying this even before the war. N. Levinson 1966.

PART 2, CHAPTER 7

131 Epigraph: Malory.

131 Josiah Macy Jr. Foundation: The foundation, established in 1930, made its mission "the search for new methods and ideas" in biomedical research, but by the 1940s its reach had broadened considerably (Heims 1991, 164). R. H. Macy, the department store magnate, was a third cousin of Josiah Macy, Jr., but the foundation had no connections, financial or otherwise, to that line of the family or to Macy's Department Store.

132 Rosenblueth's wholly unexpected presentation: Heims confirms that Wiener, Rosenblueth and Bigelow 1943 was "the written version of Rosenblueth's talk." Heims 1991, 289.

132 "burly, vigorous man of middle height": *IAM*, 171.

134 Twelve years before, Lawrence Kubie . . . had published one of the first papers: Kubie.

134 Rafael Lorente de Nó had confirmed . . . circular neural networks: Heims 1991, 122; Lorente de Nó.

134 For McCulloch, the new communication perspective. . . raised exciting prospects: Heims 1991, 16.

134 Bateson quickly grasped the human implications of a science of communication: ibid., 17.

134 "I did not notice that I had broken one of my teeth": Mead 1968; Heims 1991, 15.

135 "the cybernetics group": Heims 1991.

135 "Behavior, Purpose and Teleology": Wiener, Rosenblueth and Bigelow 1943.

135 "final cause . . . the purpose means being the best of things": Aristotle, *Physics*, II.3.

136 Born in Orange, New Jersey: McCulloch's background and early work: Heims 1991, 31–36; Lettvin 1989a, 1989b.

136 "*What is a number that a man may know it*": McC CW I, 21–22.

136 His suspicion was . . . logical relations described in the *Principia*: Northrop/McC CW I, 367.

137 McCulloch's paradox of "nets with circles" exceeded his grasp: "It required a familiarity with modular mathematics which I did not have." Arbib 1989/McC CW I, 341.

138 Walter Pitts was an awkward, painfully shy, boy-wonder: Pitts background: JL1–JL4; Lettvin 1989b/McC CW II, 514–529; Heims 1991, 40–45.

138 Russell directed Pitts to study with Rudolf Carnap: JL2; McC as told by Manuel Blum/McC CW I, 339.

138 "Walter would attend classes occasionally": JL2.

138 "Carnap had just written a book on logic": JL2; Carnap.

139 "Walter once came into a science survey class . . . a true-false final exam": JL4.

139 "to understand how [the] brain could so operate": Lettvin 1989b/McC CW I, 516.

139 "A Logical Calculus of Ideas Immanent in the Nervous System": McC and Pitts 1943/McC CW I, 343–361.

140 "could compute any logical consequence of its input": ibid.; McC 1949/McC CW II, 585.

140 "The whole field of neurology and neurobiology ignored the . . . message": Lettvin 1989a/McC CW I, 17.

140 the founding work . . . of artificial intelligence: ibid., 8, 17.

141 By 1943, McCulloch already knew . . . about Wiener and his work: McC 1974/McC CW I, 39.

141 "To me he was a myth before I met the man": McC 1965b/McC CW III, 1350.

141 "I was amazed at Norbert's exact knowledge": McC 1974/McC CW I, 39.

141 His co-intern, a young cousin of Wiener's: Alden Raisbeck, older brother of Gordon "Tobey"

Raisbeck, was Wiener's first cousin, once removed, on his mother's side. B&TR1.

141 "I said, 'I know a mathematician you would like'": JL4.

142 "Pitts has told me of your offer. I'm delighted": McC to NW, 8.27.43.

142 "He is without question the strongest young scientist": NW to Henry A. Moe (Guggenheim Foundation), Sept 1945 (Heims 1991, 40–41).

143 "We used to sneak beef stock into the soup": Oliver Selfridge int., Cambridge, MA, 4.20.98.

143 Pitts relocated to New York . . . a petrochemicals company: the Kellex Corporation, a subsidiary of M. W. Kellogg Company, designed the gaseous diffusion plant for the Manhattan Project.

143 "He was wonderful company": JL4.

144 Born in 1903, the son of a wealthy Budapest banker: background on von Neumann: See Heims 1980 especially; also Ulam 1958; Ulam 1976; Halmos; Macrae; Poundstone; Wigner.

144 "theory of games": von Neumann 1928.

144 "a kindly milquetoast uncle": Poundstone, 21.

144 "a perfect instrument": Wigner, 260.

144 "conservative, bankerlike attire . . . exuded Hungarian charm": Heims 1980, 353; Nasar, 80.

144 Wiener and Margaret came to Princeton: for four days in April 1937. JvN to NW, 3.28.37.

144 "Gentleman Johnny": BWR, 1.11.98.

144 "ABC" Computer: See note in chap 6.

145 Aiken . . . IBM Mark I: Aiken's machine also used paper punched cards to store and input data. Its design was more primitive than the ABC: It ran numbers in decimal, not binary, form and used electromechanical relays, not vacuum tubes. See I. B. Cohen.

145 Wiener . . . principal consultant on computation: Ferry and Saeks/NW CW IV, 137–139.

145 von Neumann . . . proved the viability of the implosion method: Heims 1980, 192; www.lanl.gov/worldview/welcome/history/21_implosion.html.

145 "obscene interest" in computing: Macrae, 276.

145 Bush's new 100-ton analog computer . . . was already being overwhelmed: Zachary, 73; Macrae, 276–277.

145 ENIAC . . . Mauchly . . . Eckert: For details on ENIAC, see McCartney; Goldstine; Augarten. The two men claimed they had worked out their design over ice cream and coffee in a Philadelphia restaurant, but in truth Mauchly had taken much of ENIAC's design from Atanasoff's computer. See Mollenhopf; www.scl.ameslab.gov/ABC/Trial.html.

145 In August 1944, he joined the ENIAC group . . . mathematics of atomic fission: Macrae 284–286.

146 It was initiated, not by von Neumann, but by Wiener: *IAM*, 269.

146 "a group of people interested in communication engineering": Aiken, von Neumann, and Wiener to Princeton conference invitees, 12.4.44.

146 a mere seven invitees: The other invitees were S. S. Wilks from the Princeton University mathematics department, W. E. Deming from the U.S. Census Bureau, Leland E. Cunningham from Ballistics Research Laboratory, Aberdeen Proving Ground, and Ernest H. Vestine from the Carnegie Institute.

146 In a second letter . . . "be known as the Teleological Society": At von Neumann's request, an eleventh participant was added, Captain Herman Goldstine, an Army mathematician at Aberdeen, with a Ph.D. in mathematics from the University of Chicago, who was the military's liaison to the ENIAC project. Aiken, von Neumann, and Wiener to invitees, 12.28.44.

147 their group would not solicit, or be beholden to . . . corporations: ibid.

147 Wiener's position was seconded by Aiken . . . even greater frustrations . . . with IBM: See I. B. Cohen; Harry R. Lewis.

147 Wiener believed—or wished to . . . von Neumann's views . . . in harmony with his: Heims suggests that von Neumann had his own agenda as a founder of the Teleological Society. "He realized that for the impending hydrogen bomb project . . . better and faster [computing] machines were needed. . . . There appeared to be . . . parallels between the brain and computers [but] he had little preparation in physiology. . . . It would be useful . . . to be in close touch with an experimental physiologist such as . . . McCulloch so . . . he could [have access to] experimental information concerning the nervous system." Heims 1980, 187.

147 unanimous in their feeling that . . . IBM, RCA, Bell Labs: "Memorandum to Mr. Killian," (signed) R.M.K., 10.26.44, MIT Institute Archives, Collection AC 4, Office of the President, Records 1930–1959, box 238, folder 3.

147 The tiny Teleological Society met . . . on January 6 and 7, 1945: *Cyb*, 15; Heims 1980, 185.

148 "Lorente de Nó and I . . . were asked to consider . . . two hypothetical black boxes": McC 1974/McC CW I, 40.

148 "Very shortly we found that people . . . were beginning to talk the same language": *IAM*, 269.

148 "a great success": NW to AR, 1.24.45.

148 Neumann . . . appeared to be fully on board Wiener's train: JvN memo to Princeton conf. participants, 1.12.45.

149 he put some distance between himself and Wiener: JvN to NW, 2.1.45.

149 "our little [Teleological Society] fits perfectly into the picture": NW to JvN, 3.24.45.

149 "the Princetitute": ibid.

149 "would go through on wheels": ibid.

149 "a very slick organizer": NW to AR, 1.24.45.

149 "the best way to get 'something' done . . . propagandize everybody": JvN to NW 4.21.45.

149 ENIAC . . . the largest agglomeration of electronic circuitry ever assembled: ftp.arl.mil/00mike/comphist/61ordnance/chap2.html.

149 "First Draft of a Report on the EDVAC": von Neumann 1945. Von Neumann received sole credit for the 101-page report and the writing was almost entirely his work, but EDVAC was designed largely by Mauchly and Eckert, who conceived the stored-program concept before von Neumann joined the project. Goldstine, 191. See also Augarten (www.stanford.edu/group/mmdd/SiliconValley/Augarten/Chapter5.html); Macrae, 288.

150 Von Neumann cited only . . . McCulloch-Pitts paper: "Johnny was 'enormously impressed' with it. . . . [It] helped explain why, when shown an electronic computer, Johnny mentally hit the ground running." Macrae, 283. "Certainly he developed a great interest in neurophysiology around the time the paper appeared." Goldstine, 275.

150 "neuron analogy": von Neumann 1945, 4.

150 "the *associative* neurons . . . *sensory* or *afferent* and the *motor* or *efferent* neurons": Von Neumann 1945, 3 (original emphasis).

150 Language like this had never been used before: Mauchly and Eckert dismissed von Neumann's EDVAC report as a summary of their concepts translated into the neural language of McCulloch-Pitts which, as Macrae observed, "they plainly regarded as pretty weird." Macrae, 286–287.

150 striking similarities to . . . neural analogies in Wiener's wartime paper: Wiener, Rosenblueth, and Bigelow (1943) linked "input" and "output" mechanisms in machines to "sensory" or afferent nerve impulses and "efferent . . . pathways" in the brain and nervous system, and spoke interchangeably of "the sensory receptors of an organism [and] the corresponding elements of a machine."

151 Von Neumann's . . . computer . . . incorporated key features of Wiener's . . . error-correction . . . feedback principles: JB2; von Neumann 1945, 1, 14, 24, 33.

151 Wiener probably conveyed all, or nearly all, his ideas . . . to von Neumann: "Wiener and von Neumann . . . talked to each other a good deal

about their work, with the exception of von Neumann's work on the . . . bomb." Heims 1980, 188.

151 D. K. Ferry: At the time (1985) Ferry was director of the Center for Microelectronics Research at Arizona State University, Tempe. NW CW IV, v.

151 "Most of the elements of the von Neumann machine," save the stored program: Ferry and Saeks/NW CW IV, 137–139.

151 "It really looks . . . the appointment and his acceptance were in the bag": NW to AR, 7.2.45.

151 "Johnny was down here. . . . He is almost hooked": NW to AR, 8.11.45.

151 "superbomb": The thermonuclear "superbomb" was the brainchild of von Neumann's Hungarian colleague at Los Alamos, physicist Edward Teller, but the two men worked collaboratively on the math and physics that brought the superbomb to fruition, and von Neumann was a leading advocate for the bomb within the government's bureaucratic circles. Heims 1980, 247; Ulam 1976, 209.

151 Wiener may not . . . have known the details of von Neumann's . . . EDVAC report: There is no record of the report reaching Wiener. Von Neumann probably did not give Wiener a copy of what he "thought . . . was a working document . . . clarifying the way he and [the Moore School team] should move ahead in what . . . he still regarded as a crucial wartime project." Macrae, 288.

151 von Neumann was quietly negotiating a better deal: "Von Neumann, with the MIT offer in hand . . . sought to persuade the [IAS] to permit him to build [his] computing machine at the institute itself." Heims 1980, 189. See also Macrae, 299–300.

151 Late in November 1945, von Neumann . . . formally declined: JvN to NW, 11.20.45 (JvN papers, LoC). "The [IAS] and [RCA] . . . have decided to undertake a joint high-speed, automatic, electronic computer. . . . I have been offered the over-all direction of this project." The language suggests this was von Neumann's first mention of his EDVAC project to Wiener.

152 Von Neumann had other new partners: Goldstine, 243; Heims 1980, 189; Macrae, 300–304.

152 von Neumann got his man . . . Bigelow: IAM, 243.

152 The IAS computer's vital organs . . . complex feedback mechanisms: JB2.

152 Bigelow was the perfect person to implement . . . mechanisms: "Without his leadership it is doubtful that the computer would have been a reality." Goldstine, 306; also 308–309.

153 "great heresy": McC 1954/McC CW III, 881.

CHAPTER 8

154 Epigraph: Carroll.

154 Fremont-Smith . . . "supper club": Heims 1991, 166, 164–165.

154 "This meeting is going to be a big thing for us and our cause": NW to McC, 2.15.46.

155 Bateson prevailed: Heims 1991, 17.

155 Nattily attired, as always, in his vested suit: Heims 1980, 205.

155 "could compute any computable number or solve any logical problem": McC 1947.

155 "duet": ibid.

156 "the computing machine of the nervous system": ibid.

156 "this statement is false . . . and vice versa": ibid. See also Heims 1991, 21.

156 After lunch, Wiener and Rosenblueth took the floor: McC 1947. See also Heims 1991, 21–22.

156 "took cognizance of the world about it": McC 1947.

157 That night, Gregory Bateson and Margaret Mead: Heims 1980, 203.

157 "more limbs and height than he knew what to do with": Mary Catherine Bateson 1984, 20.

157 Born into Britain's scientific gentry: Bateson background: Heims 1991, 55–59; Lipset; www.oikos.org/baten.htm.

157 Mead . . . a Quaker family: Mead background: Heims 1991, 67–72; Mary Catherine Bateson 1984, 18; Howard. By the time of the Macy meetings, Mead was assistant curator, and later curator, of ethnology at the American Museum of Natural History in New York. She was one of only two women in the core group of conferees. The other was Molly Harrower, a South African-born Gestalt psychologist and brain researcher who was a prominent figure in early psychological and neurological research along the lines of the conferences' scientific themes. For more on Harrower, see Heims 1991, 138–139.

157 *Coming of Age in Samoa*: Mead 1928.

157 one of the most controversial figures in American science: Mead found that teenage girls in Samoa, who engaged in guilt-free sexual relations with many young partners as a matter of custom, entered adulthood happier and healthier than their counterparts in cultures that sought to repress those urges. Her conclusions set off a firestorm of debate that was still burning at the end of the twentieth century. See Mary Catherine Bateson 1984, 224; www.dispatch.co.za/2000/01/24/features/FOCUS4.HTM.

157 "achieved stability by inverse feedback": McC 1947.

157 a comical ritual transvestite ceremony: Heims 1991, 24.

158 Heinrich Klüver, a German-born psychologist: background on Klüver, Heims 1991, 224–234.

158 Klüver appealed to the conferees . . . how the mind perceives *forms*: ibid., 224.

158 Klüver's challenge: Klüver's question "sparked a period of active research in . . . perception and . . . cognition . . . that continues today." ibid., 225.

158 Wiener would ponder the problem from a communication perspective: Heims 1991, 231; *Cyb*, 133–143.

158–159 Kurt Lewin . . . "electric effects": Bradford et al., 82–83. Some of those effects were discovered after Lewin arrived in the U.S. For background on Lewin, see Heims 1991, 209–220; Ullman; Mark K. Smith.

159 Paul Lazarsfeld: For background, see Heims 1991, 187–193.

159 "looking to mathematics and engineers working on communications": McC 1947.

159 "Of our first meeting Norbert wrote . . . *'Don't stop me when I am interrupting.'*": McC 1974/McC CW I, 40–41 (emphasis added). Wiener's quote is from *Cyb*, 19.

160 "the study of the nervous system as a communication apparatus": Wiener, Rosenblueth, and García-Ramos/NW CW IV, 466–510.

160 In their multi-year plan funded in part by the Rockefeller Foundation: The foundation funded Wiener's time in Mexico but not Rosenblueth's time in Boston. According to Heims, the foundation wanted to keep Rosenblueth in Mexico "as part of its policy to promote science in Latin America." Heims 1991, 50.

160 "indications of a new way of living with more gusto": *IAM*, 276.

160 in keeping with the mandate of the Instituto . . . electrical signals . . . of the heart: Wiener and Rosenblueth 1946a.

160 "pathological conditions of very great feedback": *IAM*, 253.

161 "I want nothing more to do with the nervous system. To hell with it": JL1.

161 "is possible only in a logarithmic system": *Cyb*, 20.

161 Their numbers were almost identical to . . . as far back as the 1920s: ibid., 20–21. Wiener cited MacColl (1945) as the source of his insights. MacColl, in turn, cited Nyquist (1924). Wiener found that the same logarithmic relation appeared in physiological studies dating back to the 1860s. Ibid., 20.

161 Robert Merton: Heims 1991, 187.

161 Clyde Kluckhohn: ibid., 184–186.

161 Talcott Parsons . . . especially "became an enthusiast": ibid., 184, 312.

162 Pitts . . . was preparing . . . his doctoral dissertation: "Walter agreed with Norbert as to the im-

portance of activity in random nets. Walter had found a way of computing this by means of the probability of a neuron being fired in terms of the probability of activity in the neurons affecting it." McC 1974/McC CW I, 43–44. For more on Pitts's doctoral project, see Cowan 1989a; 1989b.

162–163 "the interplay between a leader . . . and a group": Heims 1991, 218.

163 He had laid out his findings in a long article: Lewin 1947. Lewin died suddenly of a heart attack in February 1947, a month before the third conference. His approach proved elusive for some conferees, who found his subjective Gestalt framework and interpersonal "field" theories to be short on mathematical rigor and physical evidence. Yet, in his brief appearances at the Macy conferences, Lewin forged a permanent connection between the new communication theories and the advancing sciences of social, industrial, and organizational psychology; and he mapped out a program for future research that his disciples, and a generation of social scientists he inspired, would follow. For more, see ibid., 209–220.

163 "Teleological Mechanisms": New York Academy of Sciences conference, Oct 21–22, 1946. Frank et al.

163 "is not a regressive movement . . . a new frame of reference": ibid.

164 "the notion involves *entropy* as well": Wiener 1946c/Frank et al., 203 (original emphasis). The connection between entropy and information in the technological sense had no history in the literature before Wiener's New York Academy address, but it had a prehistory. Weaver states that, according to von Neumann, Bolzmann made the first link between entropy and "missing information" in 1894; and that both Szilard and von Neumann himself had treated information in the context of particle physics and quantum mechanics (Shannon and Weaver, 3). However, Finnemann, who read Boltzmann in German, found no mention of the term "missing information" in Boltzmann (Finnemann 1994/1999). Similarly, Kay reports that Szilard's paper on Maxwell's demon showed "the relation between entropy and . . . 'some kind of memory,' or 'intelligence,'" but that Szilard "never used the term *information*." Kay, 65, original emphasis.

164 In the first statement on record . . . information . . . conceived today: Hartley (1928) first stated the logarithmic rule for measuring information, but he expressed no preference for one base over another. Wiener made the choice for electronic technology and information theory generally that the logarithmic measure should be made "to the base 2." Wiener1946c/Frank et al., 203.

164 "Entropy here appears . . . *information is the negative of entropy*": ibid. (emphasis added).

164 "In fact, it is not surprising that entropy and information are negatives": ibid.

165 "The number of digits . . . will be the logarithm . . . to the base 2": ibid. (emphasis added).

165 "the information furnished by such a machine": ibid., 212.

165 "the coupling of human beings into a larger communication system": ibid.

165 "emotional non-verbal communication": ibid., 218–219.

165 "forecast of several extensions": McC 1946/Frank et al., 259–288.

165 Bateson laid out an ambitious program: G. Bateson 1946; Heims 1991, 249–250. Bateson's paper was not published in the Academy's *Annals*. Heims found a preprint in the Margaret Mead papers (box 104) at LOC.

165 "business cycles, armaments races . . . checks and balances in government": Heims 1991, 250.

165 G. E. Hutchinson, the ecologist, Hutchinson/Frank et al., 221–258. Hutchinson was then a zoologist at Yale and a colleague of Mead's at the American Museum of Natural History. His address that day, and his ongoing applications of the new communication framework at Yale, would redefine ecology and environmental science. Bryant.

166 "regard it as a personal communication . . . you may find fruitful in your own field": form letter from McC, Aug 1947.

166 Erwin Schrödinger: Schrödinger's 1944 book, *What Is Life?*, started Wiener thinking about the physical nature of information and its role in the cycles of living systems. Schrödinger, in turn, found Wiener's paper to be "very interesting, at least I think so. . . . The fact that I am not acquainted with the concepts of 'Communication' makes it very difficult to understand for me." Schrödinger to McC, 10.11.47. Schrödinger had made no mention of information in his theory of negative entropy. He was speaking only about purely physical sources of negative entropy obtained by organisms through the metabolism of "foodstuffs." ("The device by which an organism maintains itself stationary . . . consists of continually sucking orderliness from its environment . . . in more or less complicated organic compounds.") Kay notes that Schrödinger later rejected the information-negentropy connection on the grounds that it did not correspond quantitatively to the physics of entropy (Kay, 65), but he may not have fully understood Wiener's and Shannon's information concepts.

166 "The Bell people are fully accepting my thesis": NW to McC, 5.2.47.

166 "You see to what extent your ideas on . . . information are taking hold": McC to NW, 4.24.47.

166 "had appreciated the connection between entropy and information even before Wiener": Heims 1980, 208.

166–167 "Johnny had a genuine admiration for Wiener's mind": Macrae, 107.

167 Wiener was more comfortable than von Neumann with the uncertainties: Later, von Neumann would confess to physicist George Gamow, "I shudder at the thought that highly efficient purposive organizational elements should originate in a random process." JvN to Gamow, 7.25.55; Heims 1980, 154.

167 "we selected . . . the most complicated object under the sun": JvN to NW, 11.29.46.

167 his own thinking was . . . gravitating to . . . biology: Wiener and Rosenblueth 1946a; Wiener 1946c/Frank et al., 210.

167 "At subsequent meetings . . . there was a noticeable coolness": Heims 1980, 208.

167 "When Wiener was lecturing, von Neumann sat in the front row": ibid.

167 "pouring all the stuff out and punching with both fists": Bateson to Heims, ibid., 205.

167 "the sharpness of his intellectual arguments and show of contempt": Heims 1991, 45.

168 "She kept the sessions going"...Wiener was the "papa": Heinz von Foerster, 1st telephone int. (Pescadero, CA) 2.8.98

168 "in his role as brilliant originator of ideas . . . *enfant terrible* . . . evidently enjoyed the meetings": Heims 1980, 206. "irrepressible in his enthusiasm": Heims 1991, 28.

168 "a perpetual conference and in the wash of speculation": Mary Catherine Bateson 1972, 8.

168 "floating dreamily in the water . . . binary discriminations were key": Mary Catherine Bateson 1984, 43–47; Heims 1991, 68.

168 Wiener . . . "used to stop by, smoking smelly cigars": Mary Catherine Bateson 1984, 38, 48.

169 "enormous sense of delight, fun, joy": Heims 1991, 37.

169 He never threw a tantrum . . . while he was there: JL3; Taffy McCulloch Holland and David McCulloch int., Old Lyme, CT, 11.28.97.

169 "I am absolutely certain he never told Mother . . . skinny dipping": PWK1. Casual nudity notwithstanding, there were no reports of lewd behavior at Old Lyme. Lettvin's wife, Maggie, a frequent summer guest, confirmed that the fun was exceptionally clean. "People went skinny dipping in the lake, but they didn't behave sexually toward one another. It was nice, healthy skinny dipping." JL4.

169 Other scientists shared bohemian lifestyles: Most notably, the tribe of molecular biologists whose work and play catalyzed the discovery of DNA. Watson 2002.

169 "the elaboration, critique, extension, refinement": Heims 1991, 18.

169 "played a significant historical role . . . the human and the natural sciences": ibid., vii. However, as Heims notes, not everyone was enamored of the Macy conferences and some disciplines were not even represented at them. Behavioral scientists were largely absent and some Freudians were disturbed by the proceedings, which at times were openly hostile to their orthodoxy. No economists or political scientists were invited to the explicitly nonpolitical conferences. Ibid., 18–19.

169 "The first five meetings were intolerable": McC 1974/McC CW I, 40–41.

170 "Nothing that I have ever lived through . . . like the first five of those meetings": McC CW III, 856.

170 "unit of cultural micro-evolution": Heims 1991, 80.

170 "evolutionary cluster": Mead 1964, 248, 265, 272; Heims 1991, 80.

170 Wiener was that "one irreplaceable individual": Heims 1991, 80.

CHAPTER 9

171 Epigraph: *Cyb*, 132.

171 Such a cosmic convergence . . . once before in recorded history: as it happened, soon after Leo Wiener set sail from Europe for America in the spring of 1882. umbra.nascom.nasa.gov/eclipse/010621/text/saros-history.html (NASA). www.mreclipse.com/SENL/SENL9905/SENL905ca.htm (Solar Eclipse Newsletter).

171 "entirely ripe for the assimilation of the new": *IAM*, 315.

172 "the engineering work excellent": *Cyb*, 23.

172 "one of the most interesting men I have ever met": *IAM*, 315.

172 "as nearly free from the motive of profit": ibid., 131. 316.

172 "Why don't you write a book on the theories": Latil, 11.

172 "we sealed the contract over a cup of cocoa": *IAM*, 317.

172 "When Wiener left, Freymann smiled and said, 'Of course he'll never'": Latil, 11.

173 "It became clear to me almost at the very beginning . . . of the universe, and of society": *IAM*, 325.

173 He wrote longhand in pencil . . . "somewhat manic mood": BWR, 1.5.02, 4.30.02.

174 "I went to work very hard on this, but . . . what title": *IAM*, 321–322.

174 The classical texts . . . *kubernêtai*: Homer, *Odyssey*, book 11, line 10; Sophocles, *Oedipus Tyran-*

nus, 922; Euripides, *Suppliants*, 879; Aeschylus, *Suppliant Women*, 769.

174 "The art of the steersman saves the souls of men and their bodies": Plato, Gorgias, 11, www.ilt.columbia.edu/publications/Projects/digitexts/plato/gorgias/gorgias.html. See also Aristotle, *Politics*, 1279a.

174 "art of government": Ampère.

174 "but at that time I did not know it": *IAM*, 322.

174 *gubernātor*: e.g., Caesar, *De Bello Civili*, I:58:1; Cicero, *Epistulae ad Familiares*, 2:6; Suetonius, *De Vita Caesarum*, 10:4; Plautus, *Miles Gloriosus*, 4:4:40.

174 "On Governors": Maxwell.

174 the first automatic, feedback-controlled steering engines: Bennett 1994a; 1996.

174 "it was the best word I could find": *IAM*, 322.

175 "an air-mail package arrived at the Rue de la Sorbonne": Latil, 11. Latil may have compressed the timing a little.

175 "There's something to get on with. . . beat American efficiency twice over." Latil, 11.

175 "the MIT authorities . . . to publish the book in America": *IAM*, 331.

175 "could not let the work of one of its own professors": Latil, 12.

175 Selfridge claimed Wiener had confused the two copies: Selfridge int. with Heims, Heims 1991, 293.

175 the book "came out in a rather unsatisfactory form": *IAM*, 332.

175–176 "I had only an hour's chance . . . 12-year-old seeking approbation." McC 1974/McC CW I, 41–42.

176 *Cybernetics: or Control and Communication in the Animal and the Machine*: Wiener 1948a (*Cyb*).

176 "memorable and influential" works of twentieth century science: Morrison and Morrison.

176 "a patron saint for cybernetics": *Cyb*, 12.

176–177 "no-man's land . . . boundary regions . . . blank spaces": ibid., 2–3.

177 "the essential unity of . . . communication [and] control": ibid., 11.

177 the jargon of the engineers "became contaminated . . . a common vocabulary": ibid., 15.

177 "it had become clear to all that there was a substantial common basis of ideas": ibid.

177 "fundamental notion of the message": ibid., 8.

177 "*statistical theory of the amount of information*": ibid., 61 (emphasis added).

177 "discrete or continuous": ibid., 8.

178 "control by informative feedback": ibid., 113–114.

178 "something catastrophic": ibid., 102.

178 purpose tremor . . . Parkinson's . . . hapless drivers on icy roads: ibid., 95–96, 107–108, 113.

178 Under his influence the term would leap into . . . the repertoire: "No single word for that general idea seems to have existed in the English language before *feedback* was introduced in the context of cybernetics." Heims 1991, 271.

178–179 Building on the insights of Gestalt . . . and . . . McC and Pitts: *Cyb*, 140–141.

179 electronic scanning mechanism . . . brain waves in the visual cortex: ibid., 137–139. Computer visual displays first appeared on MIT's Whirlwind in 1951, but computer vision systems using television technology were not developed until the 1960s. www.rl.af.mil/History/1960s/1960s_DPandDisplay.html.

179 "functional . . . diseases of memory . . . secondary disturbances of traffic": *Cyb*, 147; 121, 129–130.

179 "malignant worry . . . no way to stop": ibid., 147.

179 "overload": ibid., 151.

179 "a point will come . . . very possibly amounting to insanity": ibid.

179 "homeostatic processes": ibid., 158.

179 "small, closely knit communities have a very considerable measure of homeostasis": ibid., 160.

180 the modern glut of information . . . "constriction of the means of communication": ibid., 158, 161. Wiener's social concerns were informed by his cybernetic insights, his awakening social conscience, and his lifelong respect for the uncertainty inherent in all knowledge. He acknowledged the limits of cybernetic applications in the social sciences, owing to the complexity of the subject matter and the problems of obtaining "verifiable . . . information which begins to compare with that . . . in the natural sciences" (ibid., 164); yet he believed such applications were essential to cybernetics and to society. "We cannot afford to neglect them; neither should we build exaggerated expectations of their possibilities." Ibid., 162–164.

180 "hucksters": ibid., 159–160.

180 "Any organism is held together . . . by the possession of . . . information": ibid., 161.

180 "Of all of these anti-homeostatic factors in society . . . control of the means of communication": ibid., 160.

180 "Information is information, not matter or energy": ibid., 132.

180 "the human arm by . . . the human brain . . . nothing . . . worth anyone's money to buy": ibid., 27–28.

181 "Those of us who have contributed to the new science of cybernetics thus stand": ibid., 28–29.

181 *Scientific American*: Wiener 1948b.

181 "a true boy genius . . . fast-talking professor of mathematics": *Newsweek*, 11.15.48.

181 "Once in a great while a scientific book is published that sets bells jangling wildly": *Time*, 12.27.48.

181 The book flew off the shelves: Latil reported, "The book sold twenty-one thousand copies" in its initial sale, a modest number by today's mass-marketing standards but a major coup for a scientific book in those days, and that figure may have represented sales of the French edition only. Latil, 11–12.

182 "Freymann had not rated the commercial prospects of *Cybernetics* very highly": *IAM*, 331.

182 "a book which . . . would appeal to only a small technical audience": *Business Week*, 2.19.49.

182 "amazed dealers by [its] sales . . . rhetorically impeccable": Davis.

182 "seminal books . . . Rousseau or Mill": John R. Platt, *New York Times* (n.d.), quoted on back cover of *Cyb*, 2nd ed.

182 "automation": coined in 1935 by Delmar S. Harder, a General Motors factory manager.

183 Mechanical engineers were only dimly aware: MIT engineer Harold Hazen published the first formal theory of servomechanisms (Hazen 1934a), but "at the time . . . Hazen was unaware of . . . Black's work on the negative feedback amplifier. . . . He 'did not recognize . . . the . . . fundamental interconnection between this and the [servomechanism] approach,' and he associated it only with communication network theory." H. L. Hazen int., Bennett 1994a.

183 "designed, built, and manufactured . . . without any clear understanding of the dynamics:" Bennett 1994a. "There was a lack of theoretical understanding with no common language . . . and . . . no . . . easily applied analysis and design methods."

183 "There was no connection between the Bell Labs work . . . and . . . servomechanisms": Ivan Getting telephone. int., 5.2.99.

183 published with a long appendix: written by Wiener's protégé Norman Levinson. N. Levinson 1942.

183 *Extrapolation, Interpolation and Smoothing*: Wiener 1949b (*Time Series*).

183 Together, the two books . . . took the postwar engineering world by storm. "The ideas contained in [*Time Series*] have had a tremendous impact on communication engineering." Masani, NW CW III, 5.

184 "Within a year or two of *Cybernetics*'s publication": Kelly, 120.

184 *Life . . . Fortune, The New Yorker . . . Time*: *Life*, 1.9.50, 5.29.50, 12.18.50; *New Yorker*, 10.14.50; *Fortune*, Dec 1953; *Time*, 1.23.50.

184 The French newspaper *Le Monde* . . . Sweden: Freymann to NW, 12.29.48; Wallman to Wiener, 1.4.49; Dubarle.

184 "too popular. . . stolen by my friends": McC to NW, 12.9.48.

185 "I arrived in New York from Vienna and, three weeks later": Heinz von Foerster, 1st telephone int, 2.8.98.

185 Shannon, like Wiener, had toyed with electronics since his youth: Background on Shannon: Claude and Betty Shannon int. Winchester MA, 10.19.87. See also Liversidge; Waldrop; Gleick.

185 The ten-page thesis, written two years after Turing's landmark theoretical paper: Shannon 1938. The thesis, which laid the logical and technical foundation for digital computing, has been called "the most important master's thesis of the twentieth century" (Waldrop). However, the early digital computing pioneers, including Bell Labs' Stibitz, Iowa State's Atanasoff, Harvard's Aiken, Penn's Eckert and Mauchly, and von Neumann at Princeton, make no reference to Shannon's thesis in their designs and theoretical contributions. Stibitz's computer was conceived a year before the thesis appeared. There is no mention of it in accounts by Atanasoff, Aiken, Eckert, and Mauchly, or in von Neumann's EDVAC report. "Atanasoff knew of Boolean algebra . . . but he did not know of . . . the development by . . . Shannon . . . of a general method for . . . the synthesis of electromechanical relay switching circuits." Burks and Burks, 31.

185 "hit on an idea which even then showed a profound originality": *IAM*, 178.

186 "coming to pluck my brains": Zipporah "Fagi" Levinson int., 8.24.98.

186 Shannon's two-part paper, "A Mathematical Theory of Communication": Shannon, 1948/Shannon and Weaver 1949. Page references refer to Shannon and Weaver (*MTC*). *MTC* drew heavily on Nyquist (1924) and Hartley (1926, 1928), and on Wiener's *YP* (see below).

186 ideas Wiener had introduced two years earlier: McCulloch sent copies of Wiener's NY Academy address to Shannon and others at Bell Labs in the summer of 1947, although the paper did not appear in print until October 1948 (Frank et al.). Wiener claimed, "The Bell people are fully accepting my thesis," in May 1947. NW to McC 5.2.47.

186 logarithms in base 2 . . . information and entropy: Shannon correctly traced the first use of logarithmic measures to Hartley's 1928 paper, but Hartley's perspective was not statistical in the strict sense and made no reference to probability measures taken over a "set" or "ensemble" of possible messages, as Wiener (1938; *YP*, 4) and, then, Shannon did (*MTC*, 31, 81–96).

186 Shannon declared information . . . equivalent of entropy, not negative entropy: In a letter to Wiener, Shannon remarked on that divergence in their respective theories which had left him "somewhat puzzled. . . . I consider . . . the larger the set the more information. You consider the larger uncertainty . . . to mean less . . . information." CS to NW, 10.13.48. Wiener's response was gracious. "I think that the disagreements between us on the sign of entropy are purely formal, and have no importance whatever." NW to Shannon, n.d., ca. Oct 1948. In fact negative entropy as Schrödinger (1944) defined it was essential to Wiener's concept of information and would factor into his philosophy of communication and of life itself. *HUHB*, 130; *IAM*, 324–328.

186 "binary digit . . . bit": Tukey also received a copy of Wiener's New York Academy paper in August 1947. Around that time, he proposed the term "bit" to Shannon, who was the first to use it in print. *MTC*, 32; Waldrop.

187 "These semantic aspects of communication are irrelevant": *MTC*, 31. The statement echoed a statement by Hartley: "It is desirable . . . to eliminate the psychological factors. . . . Hence . . . we should ignore the question of interpretation . . . and . . . set up a . . . quantitative measure of information based on physical considerations alone." Hartley 1928, 536–538.

187 That dictum flew in the face . . . since the fourteenth century: The word "information" first appeared in print in 1390 and signified "communication of the knowledge or 'news' of some fact or occurrence." Aspray, 117.

187 "Communication theory is heavily indebted to Wiener": *MTC*, 84–85.

187 "elegant solution . . . considerably influenced the writer's thinking in this field: ibid., 115. He was referring specifically to Wiener's solution to "problems of filtering and prediction of stationary ensembles."

187 a slim volume Shannon co-authored with . . . Weaver: Shannon and Weaver (*MTC*). Weaver's descriptions in his introductory essay were strikingly similar to Wiener's. He said information was "to be measured by the logarithm of the number of available choices. It being convenient to use logarithms to the base 2" (ibid., 9), where Wiener had written, "the number of significant decisions between two alternatives which we make in our measurement is the logarithm to the base 2" (Wiener 1946/Frank et al., 203). He offered a similar apology for complicating the theory of information with abstruse logarithms. "It doubtless seems queer, when one first meets it, that information is defined as the logarithm of the number of choices." (*MTC*, 10) As Wiener had written, "It may not be

obvious, at first sight, why the notion of logarithm occurs in the measurement of the amount of information" (Wiener 1946/Frank et al., 203). Weaver even used the same vivid example of information choices that Wiener had introduced in the Yellow Peril, citing "a man picking out one of a set of standard birthday greeting telegrams" (*MTC*, 10), as Wiener three years earlier had described the choice, "if I send one of those elaborate Christmas or birthday messages favored by our telegraph companies." Wiener 1946/Frank et al., 202.

187 "exploded with the force of a bomb . . . a bolt out of the blue . . . a revelation": Waldrop.

188 Wiener voiced the "highest regard for Dr. Shannon . . . his personal integrity": NW to Francis Bello, technology editor of *Fortune* magazine, 10.13.53. Others were not so generous. In a review of *MTC*, probability theorist Joseph L. Doob wrote: "It is not always clear that the author's mathematical intentions are honorable. The point of view is that stressed by Wiener in his [Yellow Peril]." Doob 1949. Gordon Raisbeck suggests one motive for Wiener's generosity with regard to Shannon: "Wiener must have known that he had a reputation for being less than generous about the professional work of his contemporaries." However, Raisbeck notes that for the most part Wiener "did little to ameliorate" that reputation (G. Raisbeck, 1973). Shannon, for his part, gave little credit to Wiener in a later interview. "I don't think Wiener had much to do with information theory. He wasn't a big influence on my ideas [at MIT], though I once took a course from him.'" Shannon 1993, xix; Mindell et al.

188 "one of the major spirits behind the present age": *IAM*, 179.

188 "a work whose origins were independent of my own work": Wiener 1950b.

188 "the Shannon-Wiener definition . . . (. . . belongs to the two of us equally)": *IAM*, 263.

188 "all of the procedures by which one mind may affect another. . . all human behavior": *MTC*, 3.

189 "I complained about the use of the word 'information' . . . there was no information at all": Heinz von Foerster, 1st telephone int., 2.8.98.

189 "In the first place, you called it a theory of *communication*": Claude and Betty Shannon int. Winchester, MA, 10.19.87.

190 Wiener's conception . . . was greater than Shannon's or Weaver's: As Wiener stated in a blunt letter to the technology editor at *Fortune* magazine, who wrote a major piece on information theory and its originators (Bello 1953): "Dr. Shannon is an employee of Bell Telephone Company, and is committed to . . . developing communication notions within a . . . limited range confronting the interests

of the company. . . . He must work much more . . . towards immediately usable results than I do, and he has been both industrious and prolific. . . . On the other hand, I am a college professor, and . . . I have found the new realm of communication ideas a fertile source of new concepts not only in communication theory, but in the study of the living organism and in many related problems." NW to Francis Bello, 10.13.53.

190 *a measure of the degree of organization in a system*: *Cyb*, 11 (emphasis added).

190 *organon*, organ of the body. www.geocities.com/etymonline (Online Etymology Dictionary).

190 "the harmonious interaction of . . . parts that gave meaning to the whole": Kay, 46; Weiss, 102.

190 "essential unity": *Cyb*, 11.

191 von Neumann "spent hours after the meetings and other long sessions with me": McC 1974/McC CW I, 44.

191 "the possibilities of building reliable computers of unreliable components": ibid.

191 He met with other Macy conferees, and with physicists . . . the new cell biology: McCulloch describes von Neumann's long conversations with von Foerster. ibid. See also Kay, 106–108.

191 self-organization, "of things essentially chaotic becoming organized": McC 1974/McC CW I, 44.

191 "cerebral mechanisms": Hixon Symposium on "Cerebral Mechanisms in Behavior," Sept 20–25, 1948. The full text of von Neumann's remarks can be found in Jeffress and in von Neumann's *Collected Works*, V.

191 McCulloch was there, and eight other Macy group conferees: The list is in McC CW III, 820–827.

191–192 "automaton whose output is other automata": von Neumann 1948/Jeffress, 315.

191 "probabilistic logic" inspired by the model . . . Pitts was developing under Wiener: "Johnny . . . saw its importance." McC 1974/McC CW I, 43–45.

192 Haldane . . . the first . . . to apply Wiener's theories . . . to . . . genetics: In 1950, Haldane wrote a paper on "population cybernetics" and later reported to Wiener that he had "worked out the total amount of . . . information . . . in a fertilized egg, and . . . similar points." Haldane to NW, 7.13.50; Haldane to NW, 5.6.52.

192 "A Cybernetical Aspect of Genetics": Kalmus; Kay, 87. The text quoted is from Kay, not Kalmus.

192 "general system theory": Bertalanffy 1950a, 1950b, 1968. For the fullest treatment of living systems dynamics, see James Grier Miller.

192 Bertalanffy . . . fiercely possessive: Davidson, 19.

192 "cybernetics upstaged . . . systems thought": Davidson said Bertalanffy "damned cybernetics with faint praise, and occasionally he just damned it." ibid., 204–205. In fact, soon after *Cybernetics* appeared, Bertalanffy wrote to Wiener praising his "pioneering work" and seeking "a relationship between our ideas which is perhaps worthwhile [*sic*] to be further developed." Wiener replied cordially, but, amid the many requests he was receiving, he did not follow up on Bertalanffy's invitation. Years later, Wiener arranged to address the Boston chapter of the Society for General Systems, but he came down with a fever and had to cancel the engagement. Bertalanffy to NW, 5.17.49; NW to Bertalanffy, 6.6.49; William Gray, M.D. (secretary-editor, Boston Systems Group) to Bertalanffy, 1.19.64 (in NW papers, MIT, box 24, folder 336).

193 *Communication: The Social Matrix of Psychiatry*: Bateson and Ruesch 1951.

00 *The Human Use of Human Beings*: Wiener 1950a (*HUHB*).

193 "Cybernetics . . . was born in 1943 [and] christened in 1948": McC 1974/McC CW I, 49.

193 "I think that cybernetics is the biggest bite out of the . . . Tree of Knowledge": G. Bateson 1972, 471, 476.

CHAPTER 10

195 Epigraph: Goethe.

195 "he was a familiar sight, standing splayfoot": McC 1965/McC CW III, 1350.

195 "He had a habit of writing on the blackboard and then leaning against it": BWR, 8.6.01.

195 a pair of inverted bifocals: JL1.

196 "He wrote and talked with the clarity of the late nineteenth-century Englishman": JL2.

196 "pretty damn incomprehensible": Oliver Selfridge int., 4.20.98.

196 "He came in and started writing on the blackboard": Mildred Siegel int., Brookline, MA, 8.24.98.

196 Armand, a physicist with whom Wiener had become acquainted: And with whom he would later collaborate on research in the mathematics of quantum theory. Wiener and Siegel 1953e; 1956a; Wiener et al. 1966.

196 "Whatever was on his mind . . . Wiener's visit was one of the high points": Wiesner.

196 "Wiener Early Warning System": Recollections of Edmond Dewan, Heims 1980, 383.

196 Fagi Levinson knew of one colleague . . . under his desk: Zipporah Levinson.

196–197 "to affirm in the strongest terms the great excellence . . . There was an instructor at MIT": N. Levinson 1966, 26.

197 "what others at MIT thought of him": Heims 1980, 207.

197 "his first question was 'What do they think of my work?'": Gordon "Tobey" Raisbeck, B&TR1.

197 "After their visit, Norbert was driving Pólya to the airport": ibid.

197 "One time he said Norbert's name": ibid.

197 "He used to come to the physics department colloquia . . . drop on the newspaper and catch him alight?" Stephen Burns int., Cambridge, MA, 11.24.97.

198 "He was in Urbana at the dedication . . . fallen out of his chair": Doob 1994.

198 "he'd get up from the table as soon as he had had enough to eat": Gordon Raisbeck, B&TR1.

198 "immature": N. Levinson 1966, 25; "petulant": Masani, 16; "infantile": BWR, 7.8.97.

198 "It was like caring for triplets": Mildred Siegel int., 8.24.98.

198 "No, that she said every day": Zipporah "Fagi" Levinson int., 8.24.98.

198 "Margaret Wiener had a big stake . . . he got along just fine": Gordon Raisbeck, personal communication, 7.23.99.

198 "he was quite competent": JL1.

199 "I would pretend to be too stupid to understand them": B&TR1.

199 "even to a comment from our father that we were growing up": PWK4.

200 "Look what happened to Uncle Fritz. We'll put you away!": B&TR1.

200 "telling lies that ruined men's reputations": BWR, 7.8.97; B&TR1.

200 "The next day was Sunday. I went to church": B&TR1.

200 Margaret's "paranoia" over sexual matters . . . "very, very uncomfortable": PWK1, PWK2.

201 "She accused me of trying to 'seduce' my father": B&TR1.

201 "She misinterpreted grossly": PWK1, PWK2.

201 "prurient view of everything": Taffy McCulloch Holland int., Old Lyme, CT, 11.28.97.

201 "very valuable to the girls in enlarging their outlook": NW to AR, 1.6.47.

201 Pitts, "like the Beats, spent a good bit of time 'on the road'": Heims 1991, 46.

201 "Walter is such a migratory person": NW to McC, 8.27.46.

201 "He must get [his] paper on the statistical mechanics of the nervous network . . . ready for publication": NW to AR 1.6.47.

202 "I clung to the boys . . . didn't involve any hidden demand for sex": BWR, 1.5.02.

202 "The laird of the castle appeared and disappeared . . . incredibly sophisticated and knowledgeable": BWR, 7.8.97.

202 "Rook was a saint . . . a little *too* saintly": JL2.

202 McCulloch's cavortings, which were common knowledge: "McCulloch believed in an unrestricted sex life. His was what a later generation labeled an 'open' marriage": Heims 1991, 132.

203 "I had just eaten supper with Rook and the children": BWR, 7.8.97.

203 "The next morning I was out—of a job, out of the house": ibid.

203 "lost priority on some important work": NW to McC 4.5.47. As he wrote to Rosenblueth in the midst of the mess: "This meant that the paper of one of my competitors, Shannon of the Bell Telephone Company, is coming out . . . before mine." NW to AR, 4.16.47.

204 "conspiracy of silence": NW to McC, 4.5.47.

204 A few more letters set matters straight . . . He apologized to McCulloch: NW to McC, 4.10.47; McC to NW, 4.24.47.

204 "the three boys together, or any two of them": NW to McC, 5.2.47.

204 "I saw that the damage to Walter would be considerable. . . . I was under a cloud": JL4.

204 the only paper Wiener co-authored with Pitts: Wiener et al. 1948d.

204 Selfridge solved a mathematical problem . . . his first published paper: Selfridge 1948.

205 "he practiced being blind by burying his face in a book": MIT 1994 (NW centennial program).

205 "the automatic barfly": BWR, 10.3.98. Wiener's "barfly" emulated the work of his British colleague W. Grey Walter. See also *IAM*, 320; Masani, 90; and note on Grey Walter in Chapter 14.

205 "After the first meeting, one of us would take the lead . . . role of communications in the universe": rleweb.mit.edu/rlestaff/p-wiesj-dp.HTM; See also Wiesner.

205 "sitting at the table . . . singing De Contemptu Mundi": PWK1; PWK, 4.1.98.

206 *Urbs Sion aurea, patria lactea*: Bernard de Morlaix, Monk of Cluny, ca. 1140.

206 Wiener would later paraphrase the poem: *HUHB*, 254.

206 "went well beyond the ambit of a theoretical man like me": *IAM*, 275.

206 Lee also wrote technical papers and books . . . "interpreter to the engineering public": ibid., 274; Lee, 1950; Lee 1960; www-eecs.mit.edu/greateducators/lee.html.

206 "I think it was the first time . . . the recognition he deserved": PWK1.

207 "She was his *frau-professor*": B&TR4.

207 "Peggy said, on more occasions than one, 'I'm tired of being Norbert Wiener's daughter'": *IAM*, 223.

207 "Dad was not malicious. . . . That wasn't his style at all": PWK1.

207 "My father's vocabulary was enormous . . . worth listening to as a literary product": B&TR1.

207 "Sometimes he was disturbed: someone had challenged one of his ideas": Wiesner.

208 one witness said he resigned from the department fifty times: Amar Bose int., Framingham, MA, 11.26.97.

208 "They knew what was happening . . . make sure he didn't go off the rails": Gordon Raisbeck, B&TR1.

208 Warren McCulloch scrambled to save the Macy conferences: After the first Macy conference, when the foundation failed to fully cover his expenses, he demanded to "take my name off the list." NW to McC, 5.10.46. The grievance was redressed and Wiener's participation was secured, until he threatened to quit again a year later. NW to McC, 4.5.47.

208 "a self-perpetuating . . . irresponsible body of men": NW to Dr. Frank B. Jewett, President, National Academy of Sciences, 9.22.41. Masani writes: "It should be noted that Wiener's criticism of internal politics within the Academy was not entirely groundless . . . the Academy 'employs a wondrously arcane electoral process that has all the attributes of a papal election except smoke.'" Masani 1990, 360–361. See also Greenberg.

208 "bad catering . . . tedious and expensive dinners": NW to Jewett, 9.22.41.

208 "It seemed to us whenever we saw Wiener elsewhere . . . he was jovial": JL3.

209 "She unbalanced him. She would induce his depressions": B&TR1.

209 "Wiener had this problem": JL1.

209 "I'm no good. I'm no good": PWK1.

209 *"Ich bin muede"*: BWR, 7.6.02.

209 "I'm just doing the best I can": Morris and Marion Chafetz int., Washington, DC, 5.14.98.

209 "a little sheepish, a little embarrassed about it": B&TR1.

209 "Then you stayed out of his way . . . during a depression": JL1, JL2.

209 Dr. Janet Rioch: Along with Harry Stack Sullivan and Erich Fromm, Rioch and her brother David were co-founders of the New York-based William Alanson White Institute, "the premier representative of the 'loyal opposition' to the psychoanalytic establishment in the United States." wawhite.org/abriefh.html.

209 "by the dream book . . . in rapport with me as a human being": *IAM*, 215–216.

210 "A few days later, Mrs. Wiener showed up in the math department": JL1.

211 "On the flight over I had bad pressure in my ears": PWK1, PWK2.

211 "How she had children I'll never know": PWK2.

211–212 "to know a little bit about the good restaurants and cafés of Paris": *IAM*, 335.

212 My uncle spoke of Wiener in extraordinarily glowing terms": Benoit Mandelbrot int., Scarsdale, NY, 7.8.00.

CHAPTER 11

213 Epigraph: Shakespeare, *Macbeth*, Act II, Scene II.

213 "might not like my views when they found out what they were": *IAM*, 335–336.

214 "my overexertion in lecturing": ibid., 336.

214 "a reasonably well man but dead tired": ibid., 337.

215 the RLE's . . . director: Albert G. Hill, the veteran Rad Lab physicist.

215 an automatic language-translation machine: See Hutchins.

216 "When Jerry Wiesner invited us to RLE": Lettvin 1989b/McC CW II, 519–20.

216 "We'd all been brought in . . . with the idea . . . the brain was a supercommunicator": Wall 1993.

216 "The McCulloch laboratory . . . cared about . . . cellular events and . . . mind": Gesteland 1989/McC CW III, 1007.

216 the RLE's work "was inspired by . . . Wiener and his exciting ideas": rleweb.mit.edu/rlestaff/p-wiesj-dp.HTM. rleweb.mit.edu/rlestaff/p-wiesj-dp.HTM; see also Wiesner.

216 "daily rounds of the laboratory to investigate everyone else's research": Edwards, Chapter 8.

217 "arose from exacerbation of some silly dispute": Masani, 218.

217–218 "The blow-up arose in part from the two men's differing temperaments": Heims 1991, 138.

218 "a personal misunderstanding . . . a violent rift." Lettvin 1989b/McC CW II, 520. Observers farther from the action could only guess at what happened. Neil Smalheiser, in the sole biographical work written about Pitts, attributed Wiener's actions to "an alleged slander" involving one of his family members (Smalheiser). Michael Arbib, a brain researcher who worked with both Wiener and McCulloch at MIT, claimed a "tone deaf" and "emotionally challenged" Wiener had "spent three years of his life" working on a new theory of brain function that he took literally from one of McCul-

loch's colorful metaphors, had it "shot down" at an important "physiology congress," and concluded "that McCulloch had set him up—and thus the fury." Arbib attributed this version to Pat Wall Arbib 2000.

218 "We arranged to have lunch at a local restaurant": Morris and Marion Chafetz int., 5.14.98.

218 "a book of almost wholly American interest": Stanley Unwin, Allen & Unwin Ltd, to NW, 12.10.51.

219 "Dear Arturo and Norbert: Know, o most noble, magnanimous and puissant lords": Pitts/Lettvin to NW/AR, 11.14.51.

219 "IMPERTINENT LETTER RECEIVED FROM PITTS AND LETTVIN": NW to Killian, 12.2.51.

220 "You are now in the midflight of a brilliant life": Frederick G. Fassett/Technology Press to NW, 11.28.51.

220 "for a long time been distracted and worried about . . . cybernetics at MIT": NW to Killian, 12.2.51.

221 "feeling of discipleship . . . knowingly followed any course unfriendly to you": Killian to NW, 12.11.51.

222 "Walter and I meant nothing but a wonderful enthusiasm": JL3.

223 "That's the point. He *never* was concerned": ibid.

224 "I'm sure she disapproved of him": Taffy McCulloch Holland and David McCulloch int., Old Lyme, CT, 11.28.97.

224 "It would be impossible for . . . Margaret Wiener to like Warren": Pauline Cooke int., Cambridge, MA, 11.25.97.

224 "If this was the kind of household they had, Mother would not have understood it": PWK1, PWK, 1.18.98.

224–225 "loth to . . . damn a colleague": NW to Killian, 12.2.51.

225 his wife "collaborated" in his writing activities: Wiener acknowledged in *ExP*, written during that period, that "the greater part of the . . . manuscript was dictated to my wife [who] has collaborated with me through the entire book." *ExP*, Foreword. Margaret also "served as his secretary writing letters for him . . . when they were traveling abroad together. She did know all about his correspondence because she had typed it." BWR, 4.6.98.

225 "For years I've thought . . . the real motivating force . . . her social position": PWK, 5.1.99.

225 "I didn't know what happened until Arturo told me ten years later": JL2.

226 Margaret alleged . . . "more than one" of the boys had seduced JL2. Lettvin's account was confirmed by other witnesses. Lettvin's wife, Mag-

gie, who was with Lettvin at the dinner with Rosenblueth, corroborated his testimony (JL2). Morris Chafetz and his wife, Marion, provided further corroboration. Morris Chafetz said, "Arturo was a remarkable human being. He would not have lied about that. If he told you that happened, that happened." Marion Chafetz said, "Margaret told me . . . what a bad influence [McCulloch] was on Norbert and how he should separate from him." She also "heard this ugly story about this awful mass seduction or whatever." Morris and Marion Chafetz int., 5.14.98.

226 "She made it all up!": JL2. Barbara confirmed that she never made any sexual allegations about McCulloch's group or said anything Margaret could have misconstrued. "I never had any sexual contact with any of the McCulloch group. I never claimed any such contact." BWR, 2.10.02. And Wiener harbored no such suspicions. In fact, in a letter to McCulloch's wife written soon after Barbara returned to Boston, Wiener thanked Rook McCulloch and her family for their "kindness to Barbara" during her time in Chicago. NW to Rook McC, 5.6.47. Margaret, too, continued to interact amicably with McCulloch and his wife for more than a year after Barbara moved back from Chicago. "If you or Mrs. McCulloch should be in town anytime please let us know." Margaret Wiener to McC, 3.3.48.

226 "Margaret wanted Wiener to think Warren was stealing cybernetics": JL3.

227 "Professor McCulloch . . . has taken such measures to aggrandize his role in cybernetics": NW to J. Z. Young, University College, London, 12.14.52. See also NW to Fremont-Smith, 3.25.53.

227 "There she was, feeding poison into the king's ears": Pauline Cooke int., 11.25.97.

227 "It is the sort of story that would spring to Mother's mind": PWK2.

227 "The question is, did Mother deliberately make up this story . . . or . . . really believe what she said": PWK, 5.30.98.

227 Barbara . . . further embittered toward her mother . . . not entirely forgiving of her father: "My father should have known me better and I think he did, but I would guess that . . . he saw the situation primarily as a threat to . . . his career. . . . He was, in fact, exceptionally kind and considerate to me when he came back from Mexico." BWR, 5.1.98, 6.7.98.

227 "The affair with 'the boys' is still rather nasty": NW to Morris Chafetz, 3.1.52.

227 "The Wiesner-McCulloch-Pitts-Lettvin imbroglio stinks . . . anginal attack": NW to AR, 3.10.52.

227 "The McCulloch thing is settling itself essentially in my favor": NW to Chafetz, 4.7.52.

228 "in a rather nervous state": "It is clear also that he thinks . . . people are stealing his ideas." D. A. Sholl to McC, 7.22.52.

228 enable the deaf to "hear" by touch: See Heims 1980, 214; Wiener 1949c; Wiener and Levine 1949d.

228 "You will realize that neither now nor in the future is any collaboration possible": NW to Wiesner, 11.17.52.

228 "Walter and I didn't believe it for a long time": JL3.

228 "He wouldn't talk and we didn't know why": JL2.

228 "Warren kept on going, but . . . he also was devastated": JL1.

228 McCulloch unleashed a rant . . . so splenetic: McC 1952/McC CW II, 761–787.

228 "was going through a disturbed episode": Kubie to Franklin McLean, 3.19.54/Heims 1991, 137. See also Kubie to McLean 2.20.52/Heims 1991, 136.

228 "something about his talk or behavior was erratic": Heims 1991, 137.

228 "One force which may be relevant to [McCulloch's] upset": Kubie to John Fulton, Yale University, 3.21.53/Heims 1991, 138.

228 "Walter suffered monstrously": JL3.

229 "He burned everything he'd ever written": JL1, JL2.

229 "Remember, mutatis mutandi, he was a Roundhead; I, a cavalier!": McC 1974/McC, CW I, 40–41.

229 "destined to have an effect on the history of cybernetics": Heims 1991, 138.

229 "neuroscience": In the early 1960s, MIT launched its Neuroscience Research Project directed by Francis O. Schmitt, who was said to have coined the term "neuroscience." Kay, 304.

229–230 The analog mode . . . first electronic media: Even the telegraph, with its two-figure code of dots and dashes, was not strictly digital and depended on various wave forms, transmission frequencies and, after 1900, radio waves. See Nyquist.

230 "cleavage": JL1.

231 "elements of both procedures, digital and analogy [sic], are discernible": von Neumann 1948/Jeffress.

231 Von Neumann restated a point Wiener had made: Wiener 1946c,/Frank et al., 210; Cyb, 129–130, 156.

231 Von Neumann's bombshell began a debate . . . would continue for decades: See Rumelhart and McClelland, 1986; Hillis, 1985, 1987. These theorists began in the 1980s to design "massive distributed parallel processing" neural network computers that sought to embody von Neumann's idea of "reliable computers/computing from unreliable compo-

nents" by using many processors connected in parallel, like the connected neurons in the brain.

231 "electronic brains": The first UNIVAC was delivered to the U.S. Census Bureau on Mar 31, 1951. www.computinghistorymuseum.org.

232 "There's nothing that you deal with . . . where digital and analog are in combat": JL2, JL4.

232 "multi-valued logic . . . probabilistic logic": Lettvin 1989b/McC CW II, 518, 529.

232 "messages which go out generally into the nervous system": Cyb, 129.

232 "Walter was ahead of his time": JL2, JL3, JL4 (edited with permission).

233 "They were halcyon days": JL4.

234 They made primary contributions to . . . the evolution of Wiener's thinking: "Both McCulloch and Pitts played an absolutely positive role in the evolution of Wiener's ideas in neurophysiology . . . on the problems of logical manipulation, Gestalt or pattern-recognition, gating, brain rhythms and sensory prosthesis." Masani, 218.

PART 3
CHAPTER 12

237 Epigraph: Kipling.

237 "signified the beginning of a new and terrifying period in human history": Struik 1966.

238 "a new world . . . with which we should have to live ever after": IAM, 299–300.

238 "the practical certainty that other people will follow": IAM, 303.

238 "the lords of the present science": HUHB, 173; IAM, 304.

238 "Behind all this I sensed the desires . . . to see the wheels go round": IAM, 305.

238 In October 1945 . . . Wiener vowed to remove himself . . . quitting . . . science altogether: "I have no intention of letting my services be used in such a conflict . . . I have seriously considered the possibility of giving up my scientific productive effort because I know no way to publish without letting my inventions go to the wrong hands." NW to Giorgio de Santillana, 10.16.45 (Heims 1980, 188). Two days later, Wiener drafted a letter to MIT's president declaring his intention "to leave scientific work completely and finally. I shall try to find some way of living on my farm in the country. I am not too sanguine of success, but I see no other course which accords with my conscience." NW to Karl Compton, 10.18.45. However, Heims believed the signed letter in Wiener's files "was probably never sent." Heims 1980, 189.

238 "We had expected that after this war": IAM, 306–307.

239 Under the master plan Bush set forth: Bush 1945b; Mazuzan.

239 government paying more money to large corporations: "Most Federal R&D funding was channeled through private corporations, even at the peak of university support." Sent, Esther-Mirjam, "The Economic Value(s) in and of Science." www.uab.edu/ethicscenter/sent.htm (adapted from Mirowski and Sent, Introduction).

239 "push-button warfare . . . The whole idea . . . has an enormous temptation": IAM, 305.

239 "I thus decided that I would have to turn from . . . the greatest secrecy": ibid., 308.

240 "Since the termination of the war I have highly regretted": NW to George E. Forsythe, Boeing Aircraft Company, 10.31.46; IAM, 296.

240 "I do not expect to publish any future work": Wiener 1947a.

240 "every thoughtful, well-meaning and conscientious human being": Einstein to War Resisters' International, 1928. www.san.beck.org/GPJ23-Einstein.html.

240 "If I had thought out fully how I was thus subjecting myself to a deep moral commitment": IAM, 296–297.

241 "semimilitary project on mechanical computation": ibid., 297.

241 "It was in fact to be under the Bureau of Standards. . . . My hand was forced": ibid., 297–298.

241 "I went to Aiken and tried to explain the situation": ibid., 298.

241 Symposium on Large-Scale Digital Calculating Machinery: Held in Cambridge, MA, Jan 7–12, 1947. For further details, see Goldstine, 251.

242 "M.I.T. Scientist 'Rebels' At War Research Talk" . . . New York Times: Boston Traveler, 1.8.47; NYT, 1.9.47.

242 "I greatly admire and approve . . . of Professor Wiener": Einstein statement of 1.20.47, Overseas News Agency, in Nathan.

242 "I am . . . giving up all work on the computing machine": NW to McC, 1.8.47.

242 "not work on any project . . . the . . . death of innocent people": Wiley Bulletin, John Wiley & Sons, Nov 1948.

243 "the nation's largest non-industrial defense contractor": with 75 separate contracts worth $117 million. libraries.mit.edu/archives/exhibits/midcentury; Buderi, 255.

243 its ties to industry were second to none: The Rad Lab alone had contracts with 70 companies, including defense contractors General Electric, RCA, Raytheon, Westinghouse, Western Electric, Philco and Sperry. rleweb.mit.edu/Publications/currents/4–2cov.HTM.

243 Tech's faculty and administrators . . . were embarrassed: "MIT had all these contracts with the [military], and the president of MIT worked with the generals in Washington, so it was very embar-

rassing for the administration. They tried to make him moderate his objections but it had no effect." Zipporah "Fagi" Levinson int., 8.24.98.

243 "In the first place, it is clear that the degradation of the . . . scientist": Wiener 1948c.

243 "megabuck science": Wiener 1958b.

243 "I wondered whether I had not got into a moral situation": IAM, 295.

243 "The automatic factory could not fail to raise new social problems": IAM, 295–296.

244 He made contact with two union leaders . . . Congress of Industrial Organizations. ibid., 308; Cyb, 28.

244 "we were here in the presence of . . . good and for evil": Cyb, 27 (emphasis added).

244 "It may very well be a good thing for humanity to have the machine remove from it the need": ibid.

245 *The answer, of course, is to have a society based on human values*": ibid., 28 (emphasis added).

245 "This thing will come like an earthquake": "THINKING MACHINE SEEN REPLACING MAN: MIT SCIENTIST SEES DAY WHEN UNSKILLED LABOR WILL BE OBSOLETE," *Boston Traveler*, 4.17.49.

245 "a very dangerous thing socially": ibid.

245 National Association of . . . "Malefactors": Cyb, 27; BWR, 7.6.01.

245 General Electric . . . he refused both requests: Noble, 75.

245 "show a sufficient interest in the very pressing menace": NW to Reuther, 8.13.49.

246 "DEEPLY INTERESTED IN YOUR LETTER": Reuther to NW, 8.17.49.

246 Finally, in March 1950, Reuther came to Boston: NW-Reuther meeting March 14, 1950. Heims 1980, 343.

246 "labor-science council": ibid., 343.

246 "found in Mr. Reuther and the men about him": IAM, 309.

246 The early postwar years . . . a record number of strikes: Zachary, 351.

246–247 "machines without men": Leaver and Brown.

247 Air Force . . . leading the way . . . to bring automation: Noble, especially Chapter 6.

247 The ideal of writing a popular work: Knopf to NW, 1.11.49; NW to Knopf, 1.19.49; Knopf to NW, 1.24.49; NW to Knopf, 1.26.49.

247 "mercurial . . . unpredictable . . . touchy": Brooks, 55–56.

248 *Pandora* and *Cassandra*: He had two earlier working titles, *The Communication State* and *Man, Progress and Communication*. Davis; NW to McC. 8.10.49.

248 "would, in the opinion of everyone here, kill the book dead": Brooks to NW, 11.10.49.

248 "protest against this inhuman use of human beings": ibid.

248 "What a book!": ibid.

248 "real change of point of view": HUHB, 13. Quotes in text are from 2nd edition, published by Houghton Mifflin in 1954. Page references are to the Avon Books paperback, 1967.

248 "in working science, and . . . in our attitude to life in general": ibid., 19–20.

248 "society can only be understood through . . . messages and . . . communication facilities": ibid., 25.

248 "cement . . . which binds its fabric together": ibid., 39.

248 The process of receiving and of using information": ibid., 27.

249 "does not represent a universal basis of human values": ibid., 154.

249 "leads to the misunderstanding and the mistreatment of information": ibid., 155.

249 "content": ibid., 26.

249 "unhampered exchange": ibid., 166.

249 *It is perfectly clear that this will produce an unemployment situation*": ibid., 220 (emphasis added).

250 "The [automatic machine] is not frightening . . . what man's nature . . . and . . . built-in purposes are": ibid., 247–250.

250 "calmly transfer to the machine . . . the responsibility": ibid., 252–253.

250 "Any machine constructed for the purpose of making decisions": ibid., 253–254.

251 *The hour is very late, and the choice of good and evil knocks at our door*: ibid., 254 (emphasis added).

251 "a book of enormous importance . . . since civilisation began": Russell 1951.

252 When the Cold War commenced, von Neumann enlisted . . . two dozen . . . organizations: Including the Los Alamos Scientific Laboratory, Oak Ridge National Laboratory, the Armed Forces Special Weapons Project, the Army Bureau of Ordnance, the Navy Bureau of Ordnance, the Air Force Scientific Advisory Board, Nuclear Weapons Panel and Strategic Missiles Evaluation Committee, the General Advisory Committee of the Atomic Energy Commission, the Central Intelligence Agency, the National Security Agency, the RAND Corporation, and, in the corporate sector, Standard Oil and IBM. Macrae, 333–334; Kay, 102.

252 "preventive" atomic attack: "If you say why not bomb them tomorrow, I say why not today? If you say today at five o'clock, I say why not one o'-clock?" Heims 1980, 247.

252 Under his influence . . . game theory . . . became the centerpiece in America's arsenal: ibid., 313.

252 At the RAND Corporation . . . "think tank" . . . a major player in . . . the nation's weapons policies: ibid., 315.

252 MAD—mutual assured destruction: "Game-theoretic thinking played a major role in evolving and justifying the policy of deterring attack by the threat of massive retaliation." ibid., 319.

252 "Von Neumann's picture . . . a perversion of the facts": *Cyb*, 159–160.

252 "hucksters . . . only too true a picture of the higher business life": ibid.

253 "In the long run, even the most brilliant and unprincipled huckster": ibid., 159.

253 "people are selfish and treacherous as . . . laws of nature": JvN quoted in Wigner, 261.

253 "No man is either all fool or all knave": *Cyb*, 160.

253 "cooperative games . . . a logical circle": Nasar quotes Dixit and Nalebuff's description of realistic games that progress by principles of circular causality. "A game with simultaneous moves involves a logical circle. . . . Poker is an example of, 'I think he thinks that I think that he thinks that I think.'" Nasar, 97.

253 Finally, in January 1952, Reuther invited Wiener: Reuther to NW, 2.5.52; NW to Reuther, 2.7.52.

CHAPTER 13

255 Epigraph: Swift, Jonathan, "Thoughts on Various Subjects," 1706/1726.

255 "NORBERT WIENER a.k.a. NORBERT WEINER": FBI #B2, 1.10.47. Documents from Wiener's FBI headquarters (Washington, DC) file #100–348294, released to the authors under the Freedom of Information Act on Apr 9, 1999. Additional files released on appeal on Mar 2, 2001. Boston Field Office file #100–18619 released on Mar 27, 2000. Some other files released by Army and Navy intelligence agencies as designated. FBI headquarters files designated with FBI # only; Boston files designated with #B prefix.

255 "Security Matter – C . . . persons suspected of subversive activities": The phrase was used by Roosevelt in 1940 to authorize wiretapping of suspected subversives.

255 "key facility": FBI #21, SAC Boston to Hoover, 1.14.54.

256 SAC Boston sent his first report: FBI #1/#B3, SAC Boston to Hoover, 1.16.47.

256 "subject . . . was of Russian-Jewish extraction . . . essential in winning the war": FBI #1.

256 "There were two very strong 'Communists'": withheld in FBI #1, released in #B38.

257 Those groups . . . deemed to be "subversive": American Friends of the Chinese People reported during 1942 screening for OSRD, copied in #B70.

257 "had expressed a desire to talk to Wiener": #B5, 1.17.47; #B9, 2.25.47.

257 "a member of the Communist Political Association . . . a protege of . . . WIENER": FBI #3/#B38, 10.29.48.

257 Haldane . . . "a guest of Professor . . . WEINER": #B8, 1.18.47.

257 Members of Congress called for federal prosecution of Albert Einstein . . . concern to the FBI since 1932: www.amnh.org/exhibitions/einstein/global/mccarthy.php (American Museum of Natural History).

257 director of the National Bureau of Standards: Physicist Edward U. Condon, NBS director, was the target of repeated congressional security investigations. Wang 1999; Wang 2001.

257 the head of the Harvard Observatory: Harlow Shapley, director of the observatory since 1921, lost his security clearance based on anonymous charges. eee.uci.edu/programs/humcore/essayeighttips.html (Univ Calif Irvine); www.aas.org/publications/baas/v25n4/aas183/abs/S104.html (American Astronomical Society).

257 the editor of *The Bulletin of the Atomic Scientists*: Russian-born chemist Eugene Rabinowitch, who worked in the Manhattan Project's Chicago Metallurgical Laboratory and later co-founded the *Bulletin* and served as its editor for twenty years, was one of many scientists who "faced bizarre, kafkaesque circumstances" for their advocacy of international control of atomic energy. Wang 1999; www.thebulletin.org/issues/1999/ja99/ja99reviews.html [sic].

258 "seventy of the most distinguished citizens of Massachusetts": #B17, 4.24.47.

258 a shot that made the FBI see red: #B20, copy of SAC Chicago to Hoover, 8.20.47; #B34, 9.27.48.

258 "The Armed Services are not fit almoners for education and science": Wiener 1947b.

258 "a distinct New York Jewish accent": #B13, 3.29.47.

258 Madame Irène Joliot-Curie: Madame Joliot-Curie was the daughter of physicists Pierre and Marie Curie, pioneers in the study of radioactivity, and her own research was important in the discovery of uranium fission. After the war, she became an advocate for the international control of atomic energy and the abolition of nuclear weapons. www.nobel.se/chemistry/laureates/1935/joliot-curie-bio.html.

258 "Prof. NORMAN WEINER" [sic]: #B25, 3.27.48. See also #B23, 3.10.48; #B24, 3.27.48; #B26, 3.30.48; #B27, 3.30.48.

258 "Red Dean . . . Prof. ROBERT WEINER": #B40, 11.18.48.

259 The institute's faculty, students, and even its president were under suspicion: Nasar, 152–153. "Karl Compton . . . president of MIT [whose] family had been missionaries in China and . . . were sympathetic with the communists there . . . felt that he would be next." Zipporah "Fagi" Levinson int., 8.24.98.

259 Levinson . . . William Ted Martin . . . in local communist groups: Nasar, 153.

259 at least five MIT professors, administrators and secretaries, fed information to the FBI: FBI #3/#B38, 10.29.48, released on appeal, lists Confidential Informants T6, T11, T14, T15, and T16 as individuals associated with MIT.

259 "extremely erratic . . . naïve politically": Confidential informant "T-8, a colleague of Prof. WIENER." #B62, 5.23.51.

259 "a complete egotist": Confidential informant "T-14," another MIT source. FBI #3/#B32, 5.28.48;

259 "a screwball": Confidential informant "T-16 . . . was attached to the staff of MIT": FBI #3/#B33, 6.1.48;

259 "worried about Wiener, as 'he is known to be nuts'": #B3; #B28, 3.31.48.

259 "indicating a disloyal attitude toward the United States": Confidential Informant "0." Report on NW remarks at dinner with members of the staff of the Worcester Foundation of Biological Research, Worcester, MA. FBI #3/#B70, full text released on appeal.

259 Project Paperclip: In Sept 1946, President Truman authorized "Project Paperclip" to bring selected German scientists to work for America during the Cold War, although Truman excluded anyone found to have been a Nazi party member or active supporter of Nazism. CIA Director Allen Dulles had scientists' dossiers cleansed of incriminating evidence. By 1955, more than 760 German scientists, including former members of the Gestapo who had conducted medical research in concentration camps, had become U.S. citizens and risen to prominent positions in American science. Hunt.

260 The same source reported that WIENER . . . would commit suicide: FBI #23/#B70, 11.30.51, released by the Navy.

260 "at the first appropriate occasion . . . a derogatory manner upon MIT": FBI #23/#B70, 11.30.51, released by the Navy.

260 Satisfied that the threat to the nation had been . . . neutralized . . . FBI closed its . . . investigation: The investigation was closed on Dec 29, 1953. FBI Communications Section to SAC Boston, 12.29.53, FBI #18.5 (decimals added to unnumbered documents between FBI #17 and #19)

260 Internal Security Act . . . "world Communist movement": history.acusd.edu/gen/20th/ coldwarspies.html; www.multied.com/documents/ McCarran.html.

260 Julius and Ethel Rosenberg were arrested in New York City: foia.fbi.gov/roberg.htm.

261 The government also stepped up its campaign against . . . Einstein: foia.fbi.gov/einstein.htm. See Jerome; D. Overbye, "New Details Emerge From the Einstein Files," NYT, 5.7.02.

261 "to resist the inroads of 'tyranny'": "Scientist Scorns 'Ivory Tower' Life: Mathematician Urges Entry into Political Quarrels—Warns of 'Tyranny,'" report on NW address to Philosophy of Science Assn. meeting at Columbia University. NYT, 3.13.49; #B53, 7.21.49.

261 "WIENER or a member of his family . . . British Communist, J. B. HALDANE": FBI #4, John Clements Associates, NYC to Clyde Tolson, FBI, 12.20.50.

261 "TO REOPEN THIS [MATTER] AND CONDUCT THE NECESSARY INVESTIGATION": FBI #5/#B58, Hoover to SAC Boston, 12.29.50.

261 "British Intelligence Authorities": FBI #6, Hoover to Legat London, 12.29.50, released on appeal.

262 Neither MI-5 nor the FBI had the foggiest notion: FBI #7, Legat London to Hoover, 2.20.51.

262 "None of the WIENER family have come to our notice": FBI #9, Legat London to Hoover, 4.3.51, released on appeal.

262 "reported to be one of the world's foremost mathematicians": FBI #12, Legat London to Hoover, 6.7.51, released on appeal.

262 "racking headache": IAM, 336.

262 "they displayed anti-American tendencies": #B68, SAC Houston to Hoover, 10.31.51.

263 He sent new orders to the State Department, and to the CIA's . . . Special Operations: FBI #13, Hoover to Donald L. Michelson, Department of State; cc: Lyman B. Kirkpatrick, Assistant Director for Special Operations, Central Intelligence Agency, 11.30.51.

263 "Marks: Scar on left forefinger": #B70, SAC Boston to SAC WFO (Washington Field Office), 11.30.51. ONI copy released by Navy.

263 Once again, Hoover's panic and paranoia were misplaced . . . Wiener had returned . . . without incident: #B71, SAC Boston to Hoover, 12.20.51; FBI #16/#B76; report from SAC Boston, 3.21.52; #B77; SAC Houston to Hoover, 4.23.52.

263 "to advocate, advise, counsel, and incite the overthrow by force and violence": Jackson.

263 "I know Struik to be a person of the highest character and honesty": NW (from Mexico) to Killian, 9.13.51.

263–264 "Wiener was far more upset than I was": Dirk Struik int., Belmont, MA, 8.20.98.

264 "the current political situation . . . the informer seems to be running wild": NW to J. Rioch, 6.22.50.

264 The House Un-American Activities Committee . . . "name names": huac.tripod.com.

264 "Security Index": The FBI created this detention list in the 1940s, even before legislation was passed providing any statutory authority (the Emergency Detention Act of 1950). www.eff.org/CAF/civil-liberty/freedom.essay (Electronic Frontier Foundation). The list at FBI headquarters included 11,982 names; FBI field office lists contained over 200,000 names. Halperin et al.

264 "RETURNING NYC . . . GREATLY FATIGUED . . . WELCOME YOU HOME": Text of telegrams, August 1951, MIT Institute Archives, Collection AC 4, Office of the President, Records 1930–1959, box 238, folder 3.

264 "These are extremely difficult times, and one wonders what the outcome will be": ibid.; NW to Killian 9.27.51.

265 "abnormally large percentage of communists at MIT": Statement by Robert L. Kunzig, HUAC counsel, 4.22.53. Nasar, 152.

265 "MIT was turned topsy-turvy . . . pressure to name names": Zippporah "Fagi" Levinson int., in Nasar, 153.

265 Martin "gave a pathetic, frightened performance": ibid.

265 Levinson held his ground: "He stipulated that he would not disclose any names of individual who had not already been revealed to the committee." Zipporah "Fagi" Levinson int., 8.24.98.

265 Martin "shattered and deeply depressed" . . . Levinson's teenage daughter: Nasar, 153–154.

265 "Wiener was incredibly loyal": Zipporah "Fagi" Levinson int., 8.24.98.

265 "the revolutionary way of non-cooperation": NYT, 6.12.53. See also Jerome; foia.fbi.gov/einstein.htm.

265 "This individual has been described as a genius . . . does not appear practicable": #B79, SAC Boston to Hoover, 6.5.52.

265 SAGE air defense system: for Semi-Automatic Ground Environment. www.nap.edu/readingroom/books/far/ch4_b1.html (National Research Council).

266 "Security Squad . . . Security Index . . . or an Interview . . . other interested intelligence agencies": FBI #18.5/#B82, 12.29.53; #B83, 12.31.53.

266 Files on dozens of individuals and groups were searched . . . Wiener's daughter: FBI File No. 100–17997, Subject: Margaret [a.k.a. Peggy] Wiener. #B83, 12.31.53.

266 "No Association . . . Project Lincoln . . . any U.S. Government . . . Research": #B85, SAC Boston to Hoover, 12.29.53.

266 "In view of the possible interest of . . . McCarthy": FBI #19/#B87, Hoover to SAC Boston, 1.11.54.

266 SAC Boston filed his final report: FBI #20/#B93, FBI #21/#B88, #B92, SAC Boston to Hoover, 1.14.54.

266 "on numerous occasions . . . attempted to . . . recruit him . . . completely unsuccessful": FBI #21/#B88, #B92, details released on appeal.

266 "No unusual subversive activity noted": FBI #20/#B93, 1.14.54.

266 "Because of WIENER'S temperament . . . unwise at this time": FBI #21;#B92.

266 After further consideration . . . recommend him for inclusion on the security index": FBI #21/#B88; #B92.

267 "the cream of Soviet science": IAM, 347.

267 "needling": FBI #22. Confidential Foreign Service Despatch [sic], American Consul General, Bombay, to U. S. State Dept., Washington (cc: CIA), 1.25.54.

267 "ribbing": "I ribbed the Russians." IAM, 350.

267 J. Robert Oppenheimer . . . "a hardened Communist": Blackwell, Jon, "Security risk" (The Trentonian) www.capitalcentury.com/1953.html. See also www.aps.org/apsnews/0601/060106.html (American Physical Society).

268 "The professor said . . . representatives from the Soviet Embassy . . . the proper course of action?": FBI #22.

268–269 "WIENER mentions how he hopes to perfect artificial limbs . . . he also felt that Wiener was harmless . . . no further action will be taken in relation to him": #B98, SAC Boston to Hoover, 4.13.54.

269 "half reminiscent of Nazi Germany, half of Alice in Wonderland": Struik 1993.

269 The National Academy of Sciences and other associations: See Wang 1999; Wang 2001; Walker 1999; Walker 2002/2003; Wittner.

270 "I do not think that the average American has much idea . . . we shall deserve to perish": NW to Reuther, 7.26.50.

CHAPTER 14

272 Epigraph: Goethe.

272 "Ulam was here at the time": JL1.

273 "He was not focused at all and he was extraordinarily unhappy": Benoit Mandelbrot int., Scarsdale NY, 7.8.00.

273 He had received an offer to publish . . . "The Bent Twig": Henry W. Simon to NW, 1.17.52; Simon to NW 1.31.52.

273 "animadversions . . . dispensable": Simon to NW, 4.14.52.

274 *Time* and *Newsweek*: *Time*, Mar 30, 1953; *Newsweek*, Mar 30, 1953.

274 the *Today* show: NW appeared on *Today* Mar 27, 1953.

274 *"Magnificent!"*: Simon to NW, 3.30.53 (emphasis added).

274 The memoir was especially hurtful to his mother: B&TR1.

274 "The story is shot through with pain": Mead 1953.

274 "There were four desks pushed together in the center of the room": Arbib 2000.

275 "Pieces of him appeared, a hat, a jacket": JL quoted in Taffy McCulloch Holland and David McCulloch int., Old Lyme, CT, 11.28.97.

275 "After '52 it was totally downhill": JL2.

275 "Walter was so totally devastated by what Wiener did": Maggie Lettvin in JL3. When packaged goods and patent medicines no longer sufficed, Pitts held up in the neurophysiology laboratory, where he synthesized potent long-chain alcohols and "novel analogues of barbiturates and opiates." The potions gave Pitts blackouts and, later, seizures. Smalheiser.

275 Pitts . . . hailed by *Fortune* magazine: Bello 1954.

275 five of the young scientists . . . would win Nobel prizes: Feynmann, Schwinger, Watson, Lederberg, and biochemist Robert Woodward.

275 "that the brain functions more 'analogically' than 'digitally'": paper presented at 7th Macy conf. by Ralph Gerard, who would become McCulloch's successor at the University of Illinois Neuropsychiatric Institute. Heims 1991.

275 Shannon from Bell Laboratories: attended as a guest at the 7th, 8th, and 10th Macy confs.

275 Roman Jakobson: attended as a guest at 5th Macy conf.

275 Max Delbrück: attended as a guest at 5th Macy conf.

275 his sudden, unexplained departure blew a hole in the proceedings: "For me it was a . . . hole that was left." Heinz von Foerster, 1st telephone int., Pescadero CA, 2.8.98.

275 "psychological deterioration": Heims 1991, 155.

276 discussions of pressing social and cultural issues were studiously avoided: Heims 1991, 76.

276 "to improve the human condition or alleviate and prevent misery": ibid., 28.

276 "babel of laboratory slangs . . . I wish Wiener were still with us": McC 1955; Heims 1991, 277.

276 "developmental" stage: JL1.

276 "rewritten in terms of information, communication, feedback and systems": Heims 1991, 127.

276 "double bind": Bateson et al. 1956/G. Bateson, 1972, 207.

276 His theory traced back to Wiener's concerns . . . was the reigning model: *Cyb*, 144–154; Heims 1991, 151. Bateson's work at the Langley Porter Psychiatric Clinic and his later studies at the Palo Alto VA Hospital and with colleagues at Stanford put a firm foundation under his "theory of communication, adapted to the human situation." His new psychology, "based on the premise that all actions and events have communicative aspects" and the new awareness that, "as human beings and members of a society, we are biologically compelled to communicate" (Bateson & Reusch, 6–7) became a rallying cry for psychology and all the human sciences in the 1950s. Fired by Bateson's enthusiasm on the West Coast, and Mead's commanding presence back east, the human communication perspective transformed anthropology and clinical psychology. It also gave rise to new domains of communication science: nonverbal communication and its subdisciplines "kinesics" (or as it became known "body language"); "proxemics," the study of spatial relations as a factor in communication; "paralinguistics," the study of the vocal effects that accompany human speech; and the broad realm of intercultural communication. Bateson and Mead worked with fellow anthropologist Raymond Birdwhistell, a pioneer of nonverbal research (Mary Catherine Bateson 1984, 109; Birdwhistell; Fast). Mead and Birdwhistell influenced their colleague Edward T. Hall, a pioneer of proxemics and intercultural communication (Hall 1959; 1966). Hall's colleague George Trager led the field of paralinguistics (Hall and Trager 1953; Trager 1958). Since the 1980s, Bateson's double bind theory has been widely considered to be discredited, but several recent efforts have been made to revive and reframe it in the new organic context of clinical psychology and neuroscience. See Koopmans; Roberts.

277 Other social scientists in the 1950s . . . small groups and large organizations: Kurt Lewin's disciples branched out from MIT and established pioneering group research centers. The National Training Laboratory in Bethel, Maine, founded just after Lewin's death in 1947, became the incubator of influential "t-groups," "encounter groups," and "sensitivity training" methods predicated on providing group participants with immediate, unfiltered feedback on their personal responses and group in-

teractions. www.ntl.org/about-history.html. See also Leavitt and Mueller; Rosenberg and Hall; Cadwallader.

277 the new "humanistic" psychologists: "What we sought was a new paradigm for humanistic psychology, a new set of metaphors, a new underlying structure. . . . Bateson's contribution . . . was and will be great." May in Brockman, 77. Other support came from the next wave of Gestalt and existential psychologists who arrived from Europe after the war, and from Macy psychiatrists Lawrence Frank and Lawrence Kubie. "This victory was the door by which therapy by psychologists became legal throughout the whole nation . . . with the help of . . . Frank . . . Kubie, and other far-sighted psychiatrists." May, xi.

277 Kenneth E. Boulding . . . "missionary" effort . . . the dismal science: Boulding to NW, 1.12.54; Boulding 1950 (the last chapter addresses the application of cybernetics to economic ideas and methods); Boulding 1953; 1956.

277 Donald Theall . . . handed McLuhan copies of *Cybernetics*: D. Theall, personal communication, 4.19.03.

277 "The relevance of Wiener in McLuhan's mind": ibid.

277 "culture and communication": The Center for Culture and Technology opened in 1963 with McLuhan as its director. www.mcluhan.utoronto.ca/marshal.htm.

277 "the medium is the message . . . global village": McLuhan 1964; 1968.

277 "the transportation of messages . . . an extension of man's senses" . . . "society can only be understood . . . messages and . . . communication facilities": *HUHB*, 133, 25. McLuhan did not cite Wiener or his work in his books. Theall took issue with his "habits of borrowing" and his loose approach to "the way we credit those we have used. . . . Cybernetic issues remained central in McLuhan's thought, even if concealed." D. Theall, personal communication, 4.19.03. See also Theall.

278 Cercle d'Etudes Cybernétiques . . . Divisione di Cibernetica . . . Intl Assn for Cybernetics: Mindell et al.; pespmc1.vub.ac.be/IAC.html, www.iiass.it/caianiel.html.

278 W. Grey Walter . . . built the first mobile robot: www.epub.org.br/cm/n09/historia/greywalter_i.htm; www.epub.org.br/cm/n09/historia/turtles_i.htm; www.ias.uwe.ac.uk/goto.html?walterbot.

278 W. Ross Ashby . . . cybernetic theory after Wiener: See Ashby 1952, 1956.

278 Gordon Pask: www.pangaro.com/Pask-Archive/Pask-Archive-listing.html. See also Pask 1957, 1960a, 1960b.

278 Stafford Beer . . . "management cybernetics": Beer 1959.

278 Canada, Mexico, Uruguay . . . the entire economy of Chile: Beer 1973; Beckett.

278 reinvented biology in the light of cybernetics . . . Watson and . . . Crick: "reinvented biology," *NYT*, 2.25.03.

278 "the possible future importance of cybernetics at the bacterial level": Watson 1953.

278 Crick "formalized information . . . remarkably similar to . . . Wiener's": Kay 173–174. Crick 1957/1958.

279 "cybernetic groundswell": Kay, 64. Kay places equal, if not greater, emphasis on what she calls the "information discourse." (Kay, xvi) However, she often uses cybernetics and information theory interchangeably, and she tends to disparage both disciplines in her critique of the influence of information concepts in biology and related sciences from the 1950s on.

279 "cellular automata": Von Neumann introduced the concept at Hixon. Burks coined the term after von Neumann's death, when he completed and co-edited von Neumann's papers on the subject. Burks 1970; von Neumann 1987.

279 *Cybernétique Enzymatique*": Kay, 221; Monod 1970, 45. "Both [Monod and Jacob] credited . . . Wiener . . . and [other] 'founding fathers' of information theory . . . for this profound reorientation of biology." Kay, 17.

279 "teleonomy": Monod, 9. See also Monod and Jacob.

279 "cognitive science": Gardner saw Wiener, Rosenblueth, and Bigelow (1943) and Wiener's larger "cybernetic synthesis" as "an integrated vision" that contributed to the foundations of cognitive science. (Gardner 19–21) He noted that, "The basic ideas for cognitive science were . . . heatedly debated at the Macy conferences" (Gardner, 26) and concluded, "Though Wiener's synthesis was not ultimately the one embraced by cognitive science . . . it stands as a pioneering example of the viability of such an interdisciplinary undertaking." Gardner, 21.

279 The new cognitive science was born in 1956 . . . at MIT: Gardner, 28–29.

279 *Plans and the Structure of Behavior*: Miller et al.

279 "a cybernetic approach to behavior": Gardner, 32–33.

280 The new subdiscipline . . . artificial intelligence: In 1952, McCarthy convened a small private conference in New Jersey, in conjunction with Shannon at Bell Labs, on the nascent theory of machine intelligence (Shannon and McCarthy). A year later, Minsky and Selfridge organized the first public conference on the subject in Los Angeles. That meeting and a later trip by Selfridge planted the seeds of AI among incipient computer programmers on the West Coast, including the young Stanford

mathematician, Allen Newell, and Herbert Simon, an expert on organizational decision-making from Carnegie Tech, who were then consulting with the RAND Corporation in Los Angeles. In 1956, Mc-Carthy, Minsky, Selfridge, Shannon, Simon, Newell, and others formally launched the new field at a summer conference at Dartmouth. AI had other forerunners: McCulloch and Pitts (1943) provided a blueprint for many early computer programmers. Turing (1950) proposed the first systematic test of machine intelligence. In 1952, Grace Hopper, a former assistant to Aiken at Harvard, then working on the UNIVAC at Remington Rand, conceived the first high-level programming language that raised the process above binary logic and enabled programmers to assemble "pre-written code segments" into programs (Lee et al; Hopper). The Air Force's SAGE nationwide air defense system in progress at MIT's Lincoln Laboratory gave a "massive momentum" to the push for improved programming methods. Edwards, Chap. 8.

280 Selfridge . . . a pivotal figure in the evolution of AI . . . programming whiz: Selfridge wrote some of the first programs for the Whirlwind in 1953. The same year, he and Minsky organized the Western Joint Computer Conference. O. Selfridge int., 4.20.98. Newell described his meeting with Selfridge in Sept 1954 as a "conversion experience." Simon 1997 (in Edwards).

280 "Pandemonium": Selfridge 1955; 1958/1959. "Pandemonium and its predecessors resembled the neural nets of McCulloch and Pitts." Edwards.

280 Pitts was indifferent to AI and computers: Smalheiser speculated: "Though he and McCulloch worked closely with . . . pioneers of AI, they never felt attracted personally towards the quest for machine intelligence. Ultimately, [they] were humanists, interested in the fundamental nature of man rather than of machines." Smalheiser.

280 Marvin Minsky: Minsky co-founded MIT's AI Lab in 1958 with John McCarthy, who came to MIT that year.

280 "Wiener didn't think about artificial intelligence": Marvin Minsky int., MIT Media Lab, Cambridge, MA, 7.11.00.

281 Shannon and von Neumann helped . . . to launch the field: Shannon and McCarthy (eds.) 1956; von Neumann 1956.

281 "record-playback" system: Noble, 82–83. The system was developed by engineers in GE's Industrial Control Division, based at the company's main facility in Schenectady, NY. Noble, 154.

281 However, another system . . . MIT's Servomechanism Laboratory . . . "by the numbers": Noble, Chaps. 5–6. The numerical control ("NC") venture reached widely into the MIT community. Servo Lab engineers joined forces with MIT's new

Digital Computer Laboratory to develop the first programming language and control tapes for numerical control machines (Noble, 140–143). MIT's School of Management even mounted a campaign to promote the Servo Lab's NC technology to executives in industry (Noble, 133–135). Some Servo Lab staffers started commercial NC ventures of their own, but MIT officials grew uncomfortable with their growing ties to the private sector. In 1955, President Killian requested the lab and its staff to scale back on their "industrial projects," and he formed a committee to examine charges that Tech faculty were, in Noble's words, "using the Institute's position and name to aid and promote . . . business ventures," and that they were rife with "alleged conflict of interest [among] Institute personnel who sat on government advisory boards . . . and . . . were themselves recipients of [government] contracts." Noble, 138–139, 199–200.

281 "though catalyst is a lukewarm description of his role": Wiesner.

282 "with the aid of up-to-date electronics and Wienerian . . . techniques": Rosenblith.

282 Spurred on by Wiener's enthusiasm and curiosity: "Powerfully spurred on by . . . Wiener's interest, curiosity, and hopes." Rosenblith.

282 "correlator": rleweb.mit.edu/aboutrle/comphist/others.htm; rleweb.mit.edu/groups/g-audhst.htm.

283 The correlator was Wiener's baby: John S. Barlow int., Mass General Hospital, Boston, 7.12.00. See also Barlow 1997.

282 "chattered incessantly among themselves": Barlow, "Analog Correlator System for Brain Potentials," rleweb.mit.edu/aboutrle/comphist/correlat.htm.

282 "find the Rosetta stone for the script of brain waves": *IAM*, 289.

283 Wiener . . . published their findings . . . second edition of *Cybernetics*: Wiener 1956c, 1957; Wiener 1984a/1961, Chapter X.

283 "most significant . . . in physiology": *IAM*, 288–290.

283 "I was drafted by Jerry Wiesner": Amar Bose int., Framingham, MA, 11.26.97.

286 *Nonlinear Problems in Random Theory*: Wiener 1958c.

286 "Thus, I have in my own family exemplified that peculiar genetics": *IAM*, 333. Other examples of prominent father-in-law mathematicians include Richelot (Kirchhoff), Hermite (Emile), and Landau (Schoenberg). Klein was married to Hegel's granddaughter.

286 "He could say 'What is your cheapest cigar?'": B&TR1.

286 "He would just get completely worn out": Mildred Siegel int., 8.24.98.

286 "I remember seeing Wiener when . . . tears would roll down his face": Zipporah "Fagi" Levinson int., 8.24.98.

287 "When the international scene would become unsettled . . . 'Do you think there will be a war?'": John S. Barlow int., Boston MA, 7.12.00.

287 "Margaret is well and happy": NW to AR, 3.10.52.

287 she seemed "so cheerful": JL2.

287 "last ditch fight for total control": BWR, 6.8.98.

287 "One of the problems cybernetics had in the early days": John S. Barlow int., 7.12.00.

287 "He was even skeptical of those who were making *cybernetics* a cause": Struik 1966 (emphasis added).

287 "He was very worried . . . 'I'm not a *Wienerian*'": Struik 1994.

287 *Player Piano*, lauded Wiener as a prophet of . . . automation: Vonnegut, 21–22.

288 "he cannot with impunity . . . play fast and loose": NW to Hope English, Charles Scribners Sons, 7.17.52.

288 "indictment of science as it is being run today": Vonnegut to NW, 7.26.52.

288 *Tech Engineering News . . . Fantasy and Science Fiction*: Wiener 1952a, 1952b.

288 sprinkled with whimsical asides: "Miracle of the Broom Closet" (Wiener 1952b) had a conceit about a fish who brought back photographs of fishermen to his undersea friends "and the pride of the fish is in the weight and size of the American who appears beside him."

288 "where I have run into a . . . suspense and horror movie . . . you are expert": NW to Alfred Hitchcock, 2.4.52.

288 "philosophy of invention": Wiener 1993, xii.

289 "the American lust for standardization": *ExP*, 257.

289 "to decerebrate the scientist": *IAM*, 363.

289 Heaviside . . . "loading coil": The Heaviside-Pupin story had been burning in Wiener for three decades. In the early 1930s he made his first attempt to write a book on the incident (*IAM*, 207). In June 1941, he pitched the story to Orson Welles in Hollywood but Welles did not respond (NW to Welles, 6.28.41). Wiener charged that Pupin took the idea from Heaviside via George Campbell, the lead scientist in the Bell Company's Boston research center. Nahin documents Heaviside's origination of the loading coil concept in 1887 and its descent through Campbell to Pupin. Nahin, 148, 263, 275–276.

289 "a purely expository book on invention": NW to Epstein, 8.2.57; Wiener 1993, xiii.

289 *The Tempter* . . . received surprisingly good reviews: *Saturday Review* (Nov 1959) praised the book as "straightforwardly told in direct and effective prose." *Science* (Mar 4, 1960) said "Wiener has certainly done a service by pointing . . . out" the moral problems arising from interactions among "science, engineering, and business."

290 "Then tell me what happens in the section called '1908'": MIT 1994 (NW centennial program).

290 "He came to me with a review that was not too favorable": Dirk J. Struik int., 8.20.98.

290 "to-whom-it-may-concern" messages: *HUHB*, 97–98.

290 he first wrote about in *Cybernetics*: *Cyb*, 146–151.

290 The new understanding of mental illness . . . play on Wiener's mind: Dr. John Lyman, a neuroscientist and professor of bioengineering at UCLA, where Wiener spent three summers in the late 1950s and early 1960s, remembered speculating at length with Wiener about the neurochemical causes and effects of his manic-depressive states, and possible treatments for them. John Lyman int., Westwood CA, Feb 1977. However, Wiener's daughters confirmed that he never seriously considered taking any of the new medications.

290 sweeping the floor in a pickle factory: Zipporah "Fagi" Levinson int., 8.24.98.

291 "embraced him enthusiastically and encouraged Nash's . . . most important work": Nasar, 136.

291 "I feel that writing to you there I am writing to . . . a ray of light": Nash to NW, 12.9.59. Nasar, 277.

291 "My problems seem to be essentially problems of communications": Nash to Wiener, May 1963. Nasar, 307.

291 *A Beautiful Mind*: Nasar.

291 "the circulating information kept by the brain . . . secondary disturbances of traffic": *Cyb*, 146–151.

292 With expert support from the RLE's technicians, Lettvin and . . . Pat Wall: JL1-JL4; Wall 1993.

292 The pair . . . and . . . their small crew . . . forefront . . . studies: Howland et al., Gesteland et al. 1955; Gesteland 1989. See also Wall 1989.

292 "What the Frog's Eye Tells the Frog's Brain" . . . rethinking . . . cognitive operations: Lettvin et al. 1959/McC CW IV. "The [paper] . . . may well be the most important seminal work which has led to the present explosion of understanding of the visual system and. . . sensory systems in general." Wall 1989/McC CW III, 1015–1016.

292 "was not always predictable . . . insult him in scathing language": Heims 1980, 382.

292–293 At the faculty club . . . a visiting Greek dignitary: Recollections of Henry Zimmerman, Heims 1980, 382.

293 During a lunch at . . . Joyce Chen's: Recollections of Edmond Dewan, Heims 1980, 389–90.

293 In the mid-1950s, he moved . . . bean curd with mushrooms in brown sauce: Helen Chen telephone int., Cambridge, MA, 6.10.02.

293 "made it a habit to regale him . . . goodhumoredly": Recollections of Karl Wildes, Heims 1980, 333.

293 "His tongue was never far from his cheek": "Master Mind," *MD*, June 1975.

293 "Was he absentminded? . . . *like a fox*": Mildred Siegel int., Brookline, MA, 8.24.98.

293 "if I want to contribute anything more to science": *IAM*, 332.

293 "It's difficult to describe": Morris and Marion Chafetz int., 5.14.98.

293 "séances . . . I think I was functioning at a different level": John S. Barlow int., 7.12.00.

294 "He'd come into my office . . . 'Insatiable curiosity'": Amar Bose int., 11.26.97.

CHAPTER 15

295 Epigraph: Saxe. The parable appears in the *Udana*, a canonical Hindu scripture.

295 "rapid industrial growth under the . . . automatic factory": *IAM*, 356.

296 "we need the Orient more and more to supplement a West": ibid., 339.

296 Tata Institute for Fundamental Research: In Bombay, Wiener joined forces with Pesi Masani on a vexing problem in matrix mathematics and "was luckily able to close the books on it." *IAM*, 352. Masani, who had studied at Harvard and Princeton, moved to the U.S., where he and Wiener collaborated on papers in prediction theory and stochastic processes. Wiener and Masani, 1957c, 1958a. Later he served as Wiener's academic biographer and the editor of his collected papers. Masani 1990; Wiener (Masani, ed.) 1976, 1979, 1981, 1985.

296 In Delhi, Wiener lectured extensively: mainly at the National Physical Laboratory and the University of Delhi.

296 "the significance of the automatic factory for the future of India": *IAM*, 354.

296 "Indian scientists are the intellectual equals of those in any country": ibid.

296 "the special problems of countries . . . a truly international scientific life": ibid., 339.

296 "dark satanic mills": *Cyb*, 27. The reference is from William Blake's poem "Jerusalem."

296 "a chance to capitalize . . . Indian famine and Manchester drabness": *IAM*, 355.

297 "might well be an easier avenue towards a prosperous . . . country": ibid., 356.

297 "the class of skilled technicians . . . noncommissioned officers of science and technology": ibid., 354.

297 "its demands on human efforts not at the bottom . . . within a matter of decades": ibid., 355–356.

297 the academic year 1955–1956 . . . Indian Statistical Institute . . . sixty lectures: mospi.nic.in/arep0002_chapter7.pdf.

297 some of the world's brightest minds . . . nearly half of . . . Silicon Valley companies: Michael Lewis, describing the views of Jim Clark, founder of Silicon Graphics and Netscape, called Indian engineers "some of the sharpest technical minds" and "most sought-after corporate employees on the planet." Lewis cites statistics from a study by Berkeley sociologist AnnaLee Saxenian. Michael Lewis, 68–76.

298 "when one realizes that I am . . . an outcast": *IAM*, 351.

298 "we speculated much on the lives . . . our grandchildren might live": ibid.

299 Wiener knew that the purely contemplative life: ibid., 345.

299 "no country can make adequate use of motives and modes . . . passed on to it": ibid.

299 "modern people": ibid.

299 "how aristocratic a simple wool shawl can look": ibid.

299 "In India . . . the same groups and even the same villages": Wiener 1957a (emphasis added).

300 For a while in the mid-1950s, Wiener was . . . optimistic: A. L. Samuel, "Comments," NW CW IV, 689.

300 "If we accept the primacy of man over his means of production": Wiener 1953b.

300 "an evil effect on scientific research all down the line": Wiener 1958b.

300 "into the childish error of worshiping the new gadgets": Wiener 1953/ NW CW IV, 678.

300 "cult of gadget-worshippers . . . eager beavers": Wiener and Campbell, 1954c NW/CW IV, 680.

301 "example of devotion and an inner call": Wiener 1957a.

301 "Some Moral and Technical Consequences of Automation": Wiener 1959/1960.

301 "failsafe" mechanisms: *G&G*, 63–64.

301 Von Neumann died . . . exposure to radiation: Poundstone, 189; Casti, 158.

302 *self*-programming automated machines: Wiener described machines empowered by "a programming technique of automatization" and "the programming of programming." Wiener 1959/1960.

302 "If we use, to achieve our purposes, a mechanical agency": ibid.

303 "If we want to live with the machine . . . we must not worship the machine": Wiener 1954a.

303 "if you are given three wishes, you must be very careful what you wish for": HUHB, 251.

303 The Monkey's Paw: Jacobs.

303 "wished for a hundred pounds . . . his son's death at the factory": HUHB, 253. In HUHB Wiener says £100. In his later writings, he says £200, as Jacobs did in his story. G&G, 58–59.

303 "Any machine constructed for the purpose of making decisions": HUHB, 253–254.

303 "the worship of the machine as a new brazen calf": ibid., 221.

304 "Religion in the Light of Science and Philosophy": Previous lecturers included John Dewey, Carl Jung, Reinhold Niebuhr, and Paul Tillich. www.yale.edu/terrylecture/Lecturer_list.htm.

304 "not religion and science as a whole . . . the communication and control sciences": G&G, 1.

304 "some of the most important moral traps": ibid., 13–14.

304 "knowledge . . . power . . . worship . . . subject to human investigation": ibid., 2.

304 "the scientist . . . the intelligent and honest man of letters . . . clergyman as well": ibid., 5.

304 "unpleasant realities and dangerous comparisons": ibid., 3.

304 "Squeamishness is out of place here": ibid., 4.

304 "made in His own image": ibid., 12.

304 "are very well able to make other machines in *their* own image": ibid., 13 (emphasis added).

304 "an uncanny canniness . . . unexpected intelligence": ibid., 21–22.

305 "The machine . . . is the modern counterpart of the Golem": ibid., 95 (emphasis added).

305 Golems in industry . . . a devastating bounty . . . uncontrolled mass production: ibid., 33, 64, 72, 86.

305 "is a two-edged sword . . . it will cut you deep": ibid., 56.

305 "No, the future offers very little hope ... robot slaves": ibid., 69.

305 "He used to say, 'They blow neither hot nor cold'": B&TR1. The reference is from Rev. 3:14–21 (NASB/New American Standard Bible).

306 "to think over my scientific work": IAM, 353–354.

306 "a new type of chapel . . . which should not look like a church": Swami Sarvagatananda int., Boston, 11.28.97.

306 "a self-contained, inward-feeling . . . serene island of contemplation": web.mit.edu/evolving/projects/chapel/index.html.

306 "When they built it, he said, . . . *'Oh, my god!'*": Mildred Siegel int., Brookline, MA, 8.24.98.

307 Swami Sarvagatananda . . . Ramakrishna Vedanta Society: a Hindu order founded by Sri Ramakrishna, an ascetic nineteenth-century sage who preached the harmony of all religions.

308 "a sense of being connected to generations and times past": Feldman and Goldsmith, 186–187.

307 "mystical, metaphysical or otherwise . . . odd": "The reports cannot be dismissed out of hand," Feldman maintained, although he added that "no self-respecting psychologist would embrace [them] without comment or skepticism." Feldman and Goldsmith, 186–193.

308 "religious counsellor" [*sic*]: Masani 1990, 370.

309 "something to tell us that might help tip the balance. . . ": Feldman and Goldsmith, 213.

310 "even deeper than our simple Jewishness . . . the Orient": ExP, 155.

310 "I can't believe that": B&TR1.

310 "I never saw that": Dirk J. Struik int., 8.20.98.

310 "so outstanding . . . I told him that Wiener was a little closer to God": Struik 1966.

310 "There were maybe small glimpses, but I did not see them": Amar Bose int., 11.26.97.

310 "what man's nature is and what his built-in purposes are": HUHB, 250.

310–311 "is the touchstone of our . . . identity . . . *patterns that perpetuate themselves*": ibid., 130 (emphasis added).

311 "We are swimming upstream against a great torrent of disorganization": IAM, 324–328.

311 "At the age of sixty, I do not find myself at the end": ibid., 365.

CHAPTER 16

312 Epigraph: Heaviside in Berg and Nahin.

312 "He was absolutely delighted when our son was born": B&TR1.

312 "He and I would go walking": Michael Norbert Raisbeck int., Chelmsford, MA, 11.22.97.

314 the whole rogues' gallery . . . Wiener reviled: including the U.S. Army Signal Corps, the U.S. Air Force Office of Scientific research, Air Research and Development Command, the U.S. Navy Office of Naval Research, and Bell Labs. Lettvin et al., 1959/McC CW IV, 1161.

314 Wiener's reconciliation plan . . . "worrying about nothing": B&TR1; BWR, 6.7.04.

314 "legitimate": Heims 1980, 380.

314 "When he came in the office, I asked him, 'What does this mean?'": Amar Bose int., 11.26.97.

314 "a broad curiosity and the integrity": Heinz von Foerster, "Comments," NW CW IV, 253.

315 "bourgeois perversion": Pav, 777.

315 "Cybernetics clearly reflects one of the basic features of the bourgeois worldview": Materialist (in Pav, 778–779). Graham claimed "the initial Soviet hostility toward cybernetics . . . has been exaggerated outside the Soviet Union" (Graham, 268) and that "In the early 1950s Soviet ideologists were definitely hostile to cybernetics, although the total number of articles opposing the field unequivocally seems to have been no more than three or four (Graham, 272). Gerovitch maintained that, "in the early 1950s, nearly a dozen sharply critical articles appeared in Soviet academic journals and popular periodicals, attacking cybernetics and information theory as products of American imperialist ideology and totally ignoring Russian traditions in these fields" (Mindell et al). For more on the internal dynamics and politics of Soviet cybernetics, see Gerovitch 2001a, 2001b, 2002a, 2002b; web.mit.edu/slava/homepage.

315 "a capitalist warmonger . . . cigar-smoking slave of the industrialists": FBI #22; Masani, 251.

315 "half right": BWR, 6.24.03.

315 the farce of Lysenkoism: See Sheehan (www.comms.dcu.ie/sheehanh/lysenko.htm); Graham.

315 Within a year, Soviet scientists began vigorously propounding cybernetic principles: Graham, 274.

315 Engineers initiated crash programs to produce computers and . . . automated machinery: Gerovitch made clear that the Soviet military embraced cybernetics throughout the official "anti-cybernetics" period and made the new communication and control sciences available to computer developers for the same purposes their counterparts were pursuing in the West—to aid in the design of atomic weapons, guided missile design and control systems, and antiaircraft weaponry—while formally separating the "philosophical" and "man-machine" dimensions of cybernetics. Gerovitch 2001a, 2001b, 2002b.

315 "a new science providing the key to literally every form of the existence of matter": Graham, 276.

315 "the range of cybernetics . . . a possible rival to Marxism": ibid.

315 "One can find no other moment in Soviet history": ibid., 270–271.

316 "virtually stated that it was criminal . . . to denounce the founder of cybernetics": Mikulak, in Dechert, 138.

316 he received the star treatment from *Pravda* and . . . the Party press: "He was lionized." Pav, 780.

316 "cheap Russian caviar and champagne": B&TR4.

316 "science must be free from the narrow restraints of political ideology": Pav, 780.

316 "that through-the-looking-glass world": G&G, 53. See also G&G, 83–84; Wiener 1961.

316 "one of the major tools of the creation of a communist society": Graham, 271.

316 "It is *imperative* to organize wider application of cybernetics": Ford 163–164; CIA #1, 3 (see ref. note below) (original emphasis).

316 "Wiener is the only man I know who conquered Russia, and single-handed at that": Struik 1966.

317 "flabbergasted": "Soviet Exhibit of Automation 'Shatters' Western Experts," Associated Press, 6.29.60.

317 The FBI's informants . . . *Cybernetics* translated into Russian: FBI #B108, 4.8.59.

317 "hour of maximum danger": Kennedy, in his inaugural address, Jan 20, 1961. www.jfklibrary.org/j012061.htm

317 "The Meaning of Cybernetics in the USSR": "Intelligence Memorandum: The Meaning of Cybernetics in the USSR" (Confidential) ("CIA #1"), CIA Office of Scientific Intelligence, Feb 26, 1964, 2. The CIA reports cited here were released to the authors under the Freedom of Information Act in Nov 2000. Some material was withheld under claimed FOIA exemption (b)(1), allowed for "material which is properly classified pursuant to an Executive order in the interest of national defense or foreign policy." After an appeal under the FOIA, no additional records pertaining to Soviet cybernetics were identified or released by the Agency. CIA to authors, 4.17.03. Other records obtained by the authors indicate that the CIA's "First Draft Report on Cybernetics in the USSR" (PD 0–9) was initiated in Sept 1958 and distributed internally for comment on Aug 21, 1959. According to those records, the Moscow conference on control and automation "caused a furor in DCI's [Director of Central Intelligence Agency] office when Soviet progress in the field was reported publicly for the first time." Two months later, a full-time task force, designated "Project Rudder," was authorized, with John J. Ford as its head. Its first report on "The Role of Behavioral Science in Soviet Strategy" (PD 5–43) was circulated on Sept 21, 1960. In Jan 1961 the project was given its own office within the Agency and named the Control Systems Branch (later renamed the Cybernetics and Behavioral Sciences Branch) with Ford as its chief and fifteen staff positions. John J. Ford, "Intelligence on Cybernetics in the USSR: Chronology of Events (1957–64)" CIA, Mar 9, 1965. See also Ford 1964/1966.

317 "in industry, government, labor, finance, and the academic world": "Scientific Intelligence Re-

search Aid: The Features of the Soviet Cybernetics Program through 1963, A Research Source Book" (Official Use Only) ("CIA #2"), CIA Office of Scientific Intelligence, Jan 5, 1965, Preface.

317 Ford learned that the Soviets' conception of cybernetics . . . much broader than the . . . American sense: It included "all those notions borrowed from Wiener" and "a few extensions" reminiscent of British and European developments that gave the Soviets "a comprehensive view of cybernetics . . . as a tool for . . . managing complex systems, not just in the domain of technology, but also in government . . . politics and philosophy . . . up to national economies and societies." CIA #1, 2.

317 "new Communist man": CIA #1, 8.

317 "thousands of subsections . . . one hundred research and development facilities": ibid., 5–6.

317 biggest strides in industrial automation: ibid., 8.

317 "technology for the optimal control of the economy": ibid., 1.

317 "when 'thinking cybernatons' will revolutionize . . . service technology": ibid., 8.

317 Ford recognized the West's commanding lead . . . but he saw signs . . . narrow the gap: Ford reported that production of computing and industrial control equipment was estimated at six times higher in the US than in the USSR, and the number of general-purpose computers in the US was estimated at ten times the number in the USSR. But he reported that the output was "growing 25–30% annually in both countries." ibid., 9.

318 "Unified Information Net": ibid.

318 A year later, in another classified report . . . defense, space vehicle guidance, and urban planning: CIA #2, 2, 4. See also CIA #1, Fig. 1.

318 "autointelligence": CIA #2, 2, Fig. 1. The Soviet program sought to provide solutions to the society's problems "with the use of high-speed computer equipment," "information machines" for "automatic . . . coding, retrieval and abstracting," and "self-adaptive systems" for "pattern recognition" and even "human creativity." As Ford noted, "It is obvious from this . . . that . . . cybernetics . . . and 'artificial intelligence' have far more in common than is usually recognized." CIA #2, 5, 17, 26, 45–46.

318 chaos and complexity theory: building on biologically based theories in which "order arises spontaneously from the chaos of a huge quantity of microelements." CIA #2, 11.

318 von Neumann–style cellular automata: "the construction of reliable systems which consist of comparatively unreliable elements" ibid., 11.

318 And it uncovered a surge of activity in . . . medicine . . . "cyborgs": "Thousands of projects are devoted to . . . medical cybernetics [including de-velopment of] instrumentation and computer aids within . . . newly created laboratories for biocybernetics" ibid., 21–24, 47.

318 man-machine interactions . . . psychological and social effects of automation: ibid., 6.

318 "a higher state . . . of social evolution": ibid., 4.

318 "cybernetic methodology . . . socialist over the capitalist system": Declaration of Admiral A. I. Berg, chairman, Soviet Cybernetics Council. Ibid., 43.

318 "'military cybernetics'": "a particularly important field" that included air- and missile-defense systems, "electronic countermeasures," and cybernetic methods for military training and logistics. According to one CIA source, "military cybernetics is already capable of constructing algorithms for the control of weapons and troop movements in a manner closely approximating the optimum." ibid., 18, 40.

318 "a distinct possibility that the Soviet Union will gradually assume supremacy": ibid., 27–28.

319 "the New Soviet Man of the day after tomorrow": ibid., 20.

319 Ford's team confirmed . . . a new line of digital computers: "Scientific and Technical Intelligence Report: Major Developments in the SovBloc Cybernetics Programs in 1965" (Secret) ("CIA #3"), CIA Office of Scientific Intelligence, Oct 3, 1966, 1–2.

319 "the path to solution . . . which US investigators have left unanswered." CIA #3, 9–10.

319 Ford acknowledged that . . . the Soviets were still well behind the United States: "Soviet descriptions of automation . . . indicate a level of knowledge on a par with that found internationally. In practical applications, however, the Soviets demonstrate a definite backwardness." CIA #2, 6–7.

319 the dangers of Western complacency . . . an escalating cybernetics race: "If a race were on to demonstrate superiority of Western knowhow, there would certainly be room for Western complacency . . . however, it is poor consolation. . . . The Soviets . . . probably will go far beyond the state of the art . . . by combining their native talent with whatever they get from the West." Ibid., 25.

319 "influence their paths of future development along lines inimical . . . to U.S. policy": CIA #3, 3.

319 On the evening of October 15, 1962: John Dixon telephone int., Washington D.C., 3.5.00 (1st) and 5.6.01 (2nd); J. Patrick Ford telephone int., 11.20.03; Detzter, 93–94.

319 his presentation was interrupted by . . . the first aerial reconnaissance photographs: Dixon 1st & 2nd telephone ints. See also Detzter, 96 (the location is in error, and perhaps the time as well).

319 "very affirmative": John Dixon, a U.S. State Department officer who met Ford in the early 1960s and was privy to his meetings with Kennedy and other government officials, described RFK's response as "very affirmative . . . Bob Kennedy was always very attentive and supportive." Dixon 2nd telephone int., 5.6.01.

320 individuals who "should not be contacted in any way . . . unknown inquirer": FBI #B108, 6.7.62. Ford may have been the "unknown inquirer" who received that reminder from the FBI in June 1962.

320 an informal discussion group . . . in Washington: The group, which included employees of the State Department, the Office of Naval Research and the U.S. Patent Office, named itself the Washington Cybernetics Society but remained "an unchartered social institution." www.asc-cybernetics.org/ organization/history.htm.

320 "faithful adherent": Aspray and Norberg. See also, J. R. Hauben, 1996; Michael Hauben and Ronda Hauben, 1994/1997.

320 "machines and human beings . . . joint enterprises": Wiener 1959/1960; G&G, 13–14, 71–73.

320 At the Lincoln Lab . . . interfaces for the SAGE air defense network: See Edwards; www.columbia.edu/00jrh29/years.html; www.orangepeel.com/history/licklider.htm.

320 worldwide "supercommunity" of networked computers: Licklider 1960. See also Licklider 1968.

320 ARPA: For background, see Edwards.

320 $12 million annually: Norberg and O'Neill.

320 "Project MAC": MAC's multi-purpose acronym meant "Man and Computer," "Multi-Access Computing," and "Machine-Aided Cognition." Project MAC became the centerpiece of MIT's new Laboratory for Computer Science and received $25 million in total from ARPA in the 1960s. National Research Council; Norberg and O'Neill; Reed et al.; Fano; Edwards; www.lcs.mit.edu/about/about.html.

320 $1 million went to MIT's new Artificial Intelligence Laboratory: Marvin Minsky int., 7.11.00. See also Norberg and O'Neill; Reed et al.; National Research Council (www.nap.edu/readingroom/ books/far/ch4_b2.html).

320 "intelligent assistance": National Research Council.

320–321 $10 million over the next decade: Other recipients of ARPA funding included Licklider's former employer, Bolt, Beranek, and Newman, the Systems Development Corporation (a spinoff of RAND), and the independent Stanford Research Institute. Edwards.

321 Those generous grants . . . established [AI] as a legitimate scientific enterprise: "provided the bulk of the nation's support for AI research and . . . helped to legitimize AI as an important field of inquiry." National Research Council (www.nap.edu/readingroom/books/far/ch9.html). ARPA remained the "primary patron for the first twenty years of AI research." Even in the 1980s, ARPA—which was renamed DARPA (for Defense Advanced Research Projects Agency) in the 1970s—"typically provided between 50 and 80 percent . . . by far the largest share" of government funding for AI research. Edwards.

321 "Licklider went to Washington . . . we could just do both": Marvin Minsky int., 7.11.00. Allen Newell, co-founder of Carnegie Tech's AI lab, concurred. "The DARPA effort . . . had not been in our wildest imaginings." National Research Council (www.nap.edu/readingroom/books/far/ch9.html).

321 "They wanted to chase out cybernetics . . . It's *intelligence!*": Heinz von Foerster 2nd telephone int., 3.12.00.

322 "That more or less gave it the death knell": In the years that followed, von Foerster encountered "tremendous difficulty" finding funding for his laboratory's studies of cybernetic processes in living systems. He secured some modest grants from the National Institutes of Health and the National Science Foundation, and the Air Force Office of Scientific Research distributed some small sums for cybernetics research. Heinz von Foerster 2nd telephone int., 3.12.00.

322 "repair the . . . damage done by the weapons": Heims 1980, 214.

322 "sensory prosthesis": Wiener 1949c, Wiener and Levine 1949d, Wiener 1951.

322 "Look, you people must know all about this": Amar Bose int., 11.26.97 and all Bose quotations below.

323 Indeed, in the early 1950s, Wiener had given a speech at Harvard Medical School: probably a rendering of ideas developed in citations above. See also *IAM*, 287–288.

323 Boston-based Liberty Mutual Insurance Company: the nation's largest underwriter of workers' compensation policies. www.libertymutual.com.

324 "poetically appropriate . . . putting the Wiener theory to work": "Under Poetic License," *Saturday Review*, Dec 7, 1963. Initial research and development work on the engineering aspects of the Boston Arm was conducted at MIT, primarily by Bose and his graduate student Ralph Alter, who wrote his dissertation on the possibilities of cybernetic limbs. Early in 1964, Robert W. Mann and his graduate student Ronald Rothchild took the lead in R&D on the biomedical engineering aspects of the arm. R. W. Mann int., Lexington, MA, 12.11.99. See also Mann.

324 *The New York Times*: "New Process Will Help Amputee to Control Limb with Thought," *NYT*, 8.16.65.

324 Several years later, a patent was granted: U.S. Patent No. 3,557,387.

324 Liberty Mutual . . . proceeded to manufacture and market the device: Mann, 408. Dr. Mann, a professor emeritus of biomedical engineering at MIT, who led the Boston Arm engineering team after Wiener and Bose departed from the project, and whose former students were hired by Liberty Mutual to complete design and development work on the project, corroborated Bose's account. "Liberty had patented versions of the Arm based on theses performed at MIT under my supervision, with no citations of the theses or publications based on them or reference to me or MIT" (R. W. Mann, personal communication, 10.25.00). "I can say categorically . . . that all the work that created the Boston Arm was done at MIT in the mechanical engineering department. [My student Ronald] Rothchild's master's thesis *was* the Boston Arm." Mann cited a statement by Scott Allen, Asst. Vice-President for Public Relations for Liberty Mutual: "I have never heard of anybody stating or being accused of denying Wiener his prognostications about the potential for a thing like the Boston Arm." But Mann emphasized that "Wiener described something farther along than just a possibility." (Robert W. Mann int., 12.11.99. See also Mann.) In an independent assessment for the U.S. Congress's Office of Technology Assessment, Sandra Tanenbaum, who received her Ph.D. in political science from MIT, found, in Mann's words, that "Liberty wasn't interested in disseminating this, to keep it proprietary for their disability insurance and other reasons." Mann int. See also Tanenbaum; www.libertymutual.com/libertytechnology/products.html.

324 "if any patents are taken out . . . not to make any profit": Wiener 1963.

324 "I have seldom seen Wiener so happy": Struik 1966.

325 "Norbert used to come over . . . and his tears would be running down his face": Mildred Siegel int., 8.24.98.

325 "When he talked about her, his face welled with tears": Zipporah "Fagi" Levinson int., 8.24.98.

325 the award ceremony in the White House: The ceremony took place on Jan 13, 1964.

326 John R. Pierce . . . dismissed Wiener's contributions to information theory: Pierce alternately praised Wiener's work in *Cybernetics* (Pierce 1972; Hauben and Hauben, 1994/1997) and dismissed his contributions to information theory. "Wiener's head was full of his own work. . . . Competent people have told me that Wiener, under the misapprehension that he already knew what Shannon had done,

never actually found out. . . . *Cybernetics* . . . is . . . irrelevant to information theory in the sense in which Shannon proposed it" (Pierce 1973). Gordon Raisbeck replied: "Although I do not doubt that competent people have made such reports to Dr. Pierce, I believe that what they reported is untrue. I worked as an editorial assistant [on] . . . Wiener's . . . *Time Series* . . . and prepared . . . the second edition of . . . *Cybernetics*. These . . . contacts gave me ample opportunity to talk about . . . information theory with him. Any gap in his knowledge as big as ignorance of what Shannon had done . . . would . . . have been quite conspicuous to me." Raisbeck 1973.

326 fellow recipients: Another recipient that year was physicist Luis Alvarez, who had played a leading role in wartime radar research at the Rad Lab and, later, developed the detonators for von Neumann's implosion-method plutonium bomb.

327 "He said, 'I have several jobs in Scandinavia'": Dirk Struik int., 8.20.98.

327 "to lobby to get the Nobel prize": Macrae, 107.

327 Like Claude Shannon . . . Wiener felt the injustice personally and for his field: CS: "You know, there's no Nobel in mathematics, although I think there should be." Claude and Betty Shannon int., Winchester, MA, 10.19.87. "Norbert always felt it wasn't reasonable that there's no Nobel prize in mathematics." Mildred Siegel int., 8.24.98.

327 He was pronounced dead . . . on March 18, 1964: Press Release, Office of Public Relations, MIT, 3.18.64.

328 "What a man we have lost!": Swami Sarvagatananda int., 11.28.97.

329 the front page of *The New York Times*. *NYT*, 3.19.64.

329 *Bulletin of the American Mathematical Society*: *Bulletin of the AMS*, 72: 700, Jan, 1966.

329 *The Journal of Nervous and Mental Disease*: *JNMD*, 140:1, 1965.

329 *The New York Review of Books*: *NYRB*, 9.24.64. See Toulmin 1964.

329 "view of life so critical and lacking in humor": *NYT*, 3.19.64.

329 "drifted from university to university like a medieval scholar": *Time*, 3.27.64.

329 "wasn't brought up, he was programmed like some human Univac": *Newsweek*, 3.30.64.

329 "Norbert had not only the imagination to invent": McC 1965b.

330 In their first meetings . . . near-unanimous agreement with Ford's concerns: "Report on Meeting of the Panel on Cybernetics of the President's Scientific Advisory Board: Criticism of [CIA] Draft Report, 'Long-Range Soviet Scientific Capabilities, 1963–70': Proposal for the Formation of a Task

Force to Make an Independent Evaluation of the Soviet Cybernetics Effort" (Secret-Internal Use Only), CIA: Washington, DC, Apr 8, 1963.

330 Finally, in February 1964 . . . a hundred recipients: including 76 at the Defense Intelligence Agency, 10 at the State Department, 5 at NASA and 2 at the Atomic Energy Commission. CIA #1.

330 However, Ford's reports drew scant response from government officials: In its response to the authors' Freedom of Information Act request and appeal, the CIA did not release or identify correspondence from any party commenting on Ford's reports.

330 "state of hardware-oriented false euphoria": Ford to Asst. Director, Office of Scientific Intelligence, CIA, 7.14.60.

330 "They just don't get it": J. Patrick Ford telephone int., 11.20.03.

330 In July 1964 . . . American Society for Cybernetics: www.asc-cybernetics.org/organization/history.htm. McCulloch served as the society's first president and Ford served as its executive director.

330 "youngsters": Minutes of 6.24.66 meeting. McC Papers (Ford/ASC folder), American Philosophical Society, Philadelphia.

331 "Perhaps we will succeed. . . . The Russians have": McC CW I, 46.

331 As Wiener himself had forewarned Moscow: G&G, 83–84. See also, Wiener 1961; Pav, 780–783. Ford and his CIA team reached the same conclusion: "Further centralization holds less prospect for improvement than would a trend in the opposite direction." CIA #1, 9. Through the mid-1970s, as Graham, chronicled, cybernetics was "a positive rage in the USSR," where it "enjoyed far more prestige . . . than in any other country in the world." But in the late 1970s and early 1980s, "its status diminished considerably," as the new cybernetic ideals of Soviet scientists were quashed by a sudden upsurge of Cold War rhetoric and heightened tensions between the superpowers. In the backlash that followed, cybernetics was forcibly "subordinated to Marxism," reduced by party propagandists merely to "one science among many, operating on a lower plane than the general laws of the dialectic." Graham, 266; Pav, 782.

331 "[freedom] of information . . . and . . . reaction available": Graham, 289–292.

331 "that the Soviet system may become totally *cyberneticized*": Mead 1968 (emphasis added).

332 "A tremendous research project collapsed . . . it was eclipsed": Heinz von Foerster 2nd telephone int., 3.12.00.

332 "Cybernetics introduced a way of thinking . . . into many . . . other fields": Franchi et al. (int with Heinz von Foerster).

332 "The concepts were transmitted and transmogrified": John Dixon 1st int., 3.5.00.

332 scientists and scholars . . . in the Vietnam war: Heims chronicled the protest that began in March 1969 with a work stoppage and teach-in by dissident students and faculty members at MIT and spread to thirty other universities and technical schools. A decade later, Wiener's banner was raised again by scientists and engineers in government and the private sector to encourage whistle blowing and other protests against the lack of adequate safety precautions in the nuclear power industry. Heims 1980, 345–46.

333 "on the whole prophetic and ahead of their time": Heims 1989.

333 In the first wave, more than a million factory workers lost their jobs to automation: Noble, 249–250.

333 "a virtual stampede": ibid., 324.

333 "to-whom-it-may-concern messages": *HUHB*, 97.

333 "by sensing devices within the body . . . living cells would be used in machines": *New Scientist*, 1.23.64.

333 pill that patients could swallow . . . "biochips . . . bioinformatics": Meron; "The Healthy Promise of Biochips," *Business Week*, 1.21.04 (www.ebi.ac.uk European Bioinformatics Institute).

333 "Wiener predicted . . . computers . . . linked by wireless or telephone lines": Epstein (quoted in Lombreglia).

334 "A consensual hallucination experienced daily": Gibson 1984, 51.

334 "Wiener . . . realised there was another step to take": Glanville.

334 Norbert Wiener Prize in Applied Mathematics: www.ams.org/prizes/wiener-prize.html.

334 Norbert Wiener Medal for Cybernetics: In recognition of "outstanding achievements or contributions in the field of cybernetics." www.asc-cybernetics.org/organization/awards.htm.

334 Norbert Wiener Award for Social and Professional Responsibility: www.cpsr.org/cpsr/wiener.html.

334 International Astronomical Union . . . crater . . . Wiener: "Report on Lunar Nomenclature," Intl Astron. Union, 1971.

334 McCulloch . . . an old man even in his own eyes: Heims 1991, 153.

334–335 "a curious blend of glee and grief, of belligerence and gentleness": Mary Catherine Bateson 1972, 24.

335 "He probably consumed too much ethanol": Oliver Selfridge int., 4.20.98. Heims confirmed: "He was indeed frail, his hair and beard were white, his teeth were bad, and he drank too much." Heims 1991, 153.

335 Pitts . . . delirium tremens . . . shaking uncontrollably: Arbib 2000.

335 Pitts his Ph.D. . . . He refused: JL1; Smalheiser; Heims 1991, 46.

335 "One way to arrive at the aristocracy": MIT, NW Collection MC 22, box 25A, folder 345c.

335 *Jeopardy!*: PWK1. See also Trebek.

336 "Is the world better off": Michael Norbert Raisbeck int., Chelmsford, MA, 11.22.97.

EPILOGUE

337 Epigraph: Wiener 1957a.

338 "It's quite clear . . . the third dimension": Kurzweil echoed Wiener's vision and Pitts's Ph.D. thesis at a symposium at Stanford on future technology (Kurzweil). Soon after, IBM announced a new initiative to build three-dimensional integrated circuits that would "interconnect separate layers directly at thousands or even hundreds of thousands of points." *NYT*, 11.11.02.

338 the brain's myriad chemical transmitters . . . and . . . fleeting brain waves: Neuroscientist Karl H. Pribram, a pioneer in the field of cognitive science, has found evidence that the brain's hundred billion neurons operate, not like digital on-off switches, but like little "harmonic analyzers" that analyze their inputs much as Wiener did in his pioneering work on the harmonic analysis of electronic signals (Pribram, 1986; 1991, 1997). McCulloch, the consummate logician, found evidence predictive of those findings: "We have now been obtaining . . . differences of response according to frequency from deeper structures. Moreover, they are . . . attributable to . . . filters that pass impulses at one frequency to one structure and at another frequency to another structure somewhere downstream. . . . It means that a new day has dawned in physiology." McC to NW, 4.24.47.

338 The military's advanced radar-guided antiaircraft and antimissile systems: In the 1991 Gulf War, the widely hailed Patriot antiaircraft/antimissile systems misfired routinely and hit only nine percent of their targets, at most (Cirincione). In the 2003 Iraq war, erring Patriots killed American and British pilots in a scenario from "a bad science fiction movie in which the computer starts creating false targets." *60 Minutes*, CBS News, 2.22.04.

338 One outspoken MIT professor: MIT physicist Theodore Postol incurred the wrath of the military—and won the 2001 Norbert Wiener Award from CPSR—for his critique of the Patriot system before the U.S. Congress, and for his outspoken opposition to the military's planned land-based and space-based antimissile systems that failed repeated performance tests. www.pbs.org/wgbh/pages/frontline/shows/missile/etc/postol.html;

www.cpsr.org/conferences/annmtg01/wiener.html. See also Kaplan.

339 "produce an unemployment situation, in comparison . . . a pleasant joke": *HUHB*, 220.

339 "non-commissioned officers of science and technology": *IAM*, 354.

339 He only dimly envisioned . . . worldwide work force . . . economic conflict between nations: Many studies confirm that offshoring is not the primary threat to American jobs. The greater factor by far has been the replacement of human workers by technology. Even in high-tech industries and the knowledge-intensive work of computer programming, American workers have lost more jobs to advances by American companies in the automation of computer operations and programming than they have to technical workers abroad. "Of 2.8 million jobs lost from March 2001 . . . only a third at most (15–35%) were attributable to offshoring," while "productivity improvements at home . . . account[ed] for the great bulk of the job loss." *NYT*, 10.5.03. "'The impact of outsourcing is overblown,' Professor [Erik Brynjolfsson, Sloan School of Management, MIT] said. 'The far larger factor is substituting technology for labor.'" *NYT*, 3.9.04.

340 the former Soviet bloc countries are only beginning to recoup the losses they suffered: See Dyker; Hart.

340 "very great feedback": *IAM*, 253.

341 "information is information, not matter or energy": *Cyb*, 132.

341 "in the terms of the market . . . the money they save": ibid., 27.

341 human values beyond buying or selling . . . struggle: "The answer, of course, is to have a society based on human values other than buying or selling. . . . We need a good deal of planning and a good deal of struggle." Ibid., 28.

341 "a limited group of a few thousand people": *IAM*, 299–300.

341–342 "control of the means of communication": *Cyb*, 160.

342 "build itself up into a process totally destructive to the ordinary mental life": ibid., 147.

342 "when human atoms are knit into an organization . . . an element in the machine": *HUHB*, 254.

343 powerful sensor and effector technologies: Saffo, 1997/2002.

343 "elements of the nature of sense organs . . . effectors": *HUHB*, 212–213. Actually Wiener described "receptors and effectors" four years earlier at the first Macy conference. McCulloch 1947.

343 "*hyper*automated cybermanufacturing": Saffo 1997/2002 (emphasis added).

343 "the erosion of the entire digital order . . . impossible to accomplish . . . with digital technology": ibid.

343 "digital will seem just a bit dull": ibid. Saffo, research director of Silicon Valley's Institute for the Future, goes further and predicts that, in the long run, "digital is dead. . . . We may suddenly discover that a host of insights from the analog 1950s are going to be very relevant in the analog years after 2000." See also Hedger.

344 Wiener's ultimate legacy for better and for worse: Already, the new analog technologies are finding eager customers in the American military where they are being developed for use in unmanned "warbots" and submicroscopic "smart dust" surveillance systems, to name only a few applications (Roos). See also "Little Worries: Invasion of the Nanobots?" *Time Europe*, 5.12.03 (www.time.com/time/europe/magazine/print-out/0,13155,901030512–449458,00.html); robotics.eecs.berkeley.edu/00pister/SmartDust.

345 "over the fence" . . . the end result of any effort: Amar Bose int., 11.26.97.

345 "it was his timely warnings that saved us": Arthur L. Samuel, a Bell Labs veteran, and pioneer in artificial intelligence. A. L. Samuel, "Comments," NW CW IV, 690.

346 "there are hopeful signs on the horizon": *HUHB*, 220. See e.g., Electronic Frontier Foundation (www.eff.org); Cyber-Rights & Cyber-Liberties (U.K.) (www.cyber-rights.org); Global Internet Liberty Campaign (www.gilc.org); ETC (Action Group on Erosion, Technology and Concentration, www.etcgroup.org); Acorn (Association of Community Organizations for Reform Now, www.livingwagecampaign.org); the Foresight Institute (www.foresight.org).

346 his twenty-first century colleagues ringing alarms: See especially Joy 2000a, 2000b.

ACKNOWLEDGMENTS

349 "directly relevant to the incidents of my career as a scientist": *IAM*, 86–87.

349 "The truth matters": BWR, 3.15.98.

349 "Serious unanswered questions remain": PWK, 1.18.98, 3.25.98.

BIBLIOGRAPHY

Ampère, A. M. *Essai sur la Philosophie des Sciences.* Paris: Bachelier, 1845.

Anon. "The Case of the Wiener Children." (typescript, ca. 1913) MIT Institute Archives, MC22, box 33, folder 903.

_____. "Master Mind." *MD*, June 1975.

Arbib, Michael A. "Comments on 'A Logical Calculus of the Ideas Immanent in Nervous Activity.'" In McCulloch 1989, I.

_____. "Warren McCulloch's Search for the Logic of the Nervous System." *Perspectives in Biology and Medicine*, 43.2, 2000 (muse.jhu.edu/demo/pbm/43.2arbib.html).

Ashby, W. Ross. *Design for a Brain: The Origin of Adaptive Behavior.* London: Chapman & Hall, 1952.

_____. *An Introduction to Cybernetics.* London: Chapman & Hall, 1956.

Aspray, William. "The Scientific Conceptualization of Information: A Survey." *Annals of the History of Computing*, 7:2, Apr 1985.

_____ and A. L. Norberg. "Interview of J. C. R. Licklider." (Cambridge, MA, Oct 28, 1988), OH 150, Charles Babbage Institute, Univ. of Minnesota, Minneapolis, MN (www.cbi.umn.edu/oh/display.phtml?id=87).

Augarten, Stan. *BIT by BIT: An Illustrated History of Computers.* New York: Ticknor & Fields, 1984 (www.stanford.edu/group/mmdd/SiliconValley/Augarten/Chapter5.html.)

Ayer, A. J. *Russell.* London: Wm. Collins, 1972.

Barlow, J. S. "The Early History of EEG Data-Processing at the Massachusetts Institute of Technology and the Massachusetts General Hospital." *International Journal of Psychophysiology*, 26: 443–454, 1997.

Bateson, Gregory. "Circular Causal Systems in Society." (New York Academy of Sciences Conf., Oct 21–22, 1946), Library of Congress, Washington DC, Margaret Mead Papers, Box 104.

_____. *Steps to an Ecology of Mind: A Revolutionary Approach to Man's Understanding of Himself.* New York: Chandler/Ballantine, 1972; Chicago: Univ. of Chicago Press, 2000 (with a new foreword by Mary Catherine Bateson).

_____. *Mind and Nature: A Necessary Unity.* New York: Dutton, 1979/Bantam, 1979.

_____ and Jurgen Reusch. *Communication: The Social Matrix of Psychiatry.* New York: Norton, 1951, 1968.

_____, G. D. Jackson, J. Haley and J. Weakland. "Toward a Theory of Schizophrenia." *Behavioral Science*, 1: 1956 (in Bateson 1972, 201–227).

Bateson, Mary Catherine. *Our Own Metaphor.* New York: Knopf, 1972.

_____. *With a Daughter's Eye: A Memoir of Margaret Mead and Gregory Bateson.* New York: Morrow, 1984.

Beckett, Andy. "Santiago dreaming." *The Guardian* (U.K.), Sept 8, 2003. (www.guardian.co.uk/chile/story/0,13755,1037547,00.html).

Beer, Stafford. *Cybernetics and Management.* New York: Wiley, 1959.

_____. *Brain of the Firm: The Managerial Cybernetics of Organization.* New York: Herder and Herder, 1972; 2nd ed. New York: Wiley, 1981.

_____. "Fanfare for Effective Freedom: Cybernetic Praxis in Government." The Third Richard Goodman Memorial Lecture. Brighton Polytechnic (U.K.) Feb 14, 1973. (www.staffordbeer.com/papers/Fanfare%20for%20Effective%20Freedom.pdf).

Bello, Francis. "The Information Theory." *Fortune*, Dec 1953.

_____. "The Young Scientists." *Fortune*, June 1954.

Bennett, Stuart. "A Brief History of Servomechanisms." *IEEE Control Systems*, 14: 2, 75–79, Apr 1994a.

———. "Norbert Wiener and Control of Anti-Aircraft Guns." *IEEE Control Systems*, 14:6, 58–62, Dec 1994b.

———. "A Brief History of Automatic Control." *IEEE Control Systems*, 16:3, 17–25, June 1996.

Berg, E. J. "Oliver Heaviside: A Sketch of His Work and Some Reminiscences of His Later Years." *Journal of the Maryland Academy of Sciences*, 1: 105–114, 1930.

Bergman, Peter G. "Notes on the Extrapolation." Dec 14, 1942, Records of NDRC, Sec 2, Div D, National Archives & Records Service, Washington DC.

Bertalanffy, Ludwig von. "The Theory of Open Systems in Physics and Biology." *Science*, 111, 23–29, 1950a.

———. "An Outline of General System Theory." *Brit. J. Philos. Sci.* 1, 139–164, 1950b.

———. *General System Theory: Foundations, Development, Applications.* New York: George Braziller, 1968.

Bigelow, J. "Conference at Bell Laboratories, Julian H. Bigelow and Professor Norbert Wiener." Confidential memo, June 4, 1941, declassified Aug 2, 1960. Records of NDRC, Sec 2, Div D, Record Group 227, Project #6, National Archives & Records Service, Washington, DC.

Birdwhistell, Ray. *Kinesics & Context: Essays on Body Motion Communication.* Philadelphia: Univ. of Pennsylvania Press, 1970.

Black, Harold S. "Stabilized Feedback Amplifiers." (paper presented at winter convention of AIEE, Jan 23–26, 1934), *Bell System Technical Journal* and *Electrical Engineering*, Jan 1934.

Blackman, Hendrick Bode, and Claude Shannon. *Monograph on Data Smoothing and Prediction in Fire Control Systems.* Feb 1946. Records of NDRC, Sec 2, Div D, National Archives & Records Service, Washington, DC.

Born, Max. *My Life: Recollections of a Nobel Laureate.* New York: Scribner's, 1975.

Bose, Amar G. "Ten Years with Norbert Wiener" (centennial ceremony speech, transcript). Cambridge, MA: MIT, Oct 12, 1994.

Boulding, Kenneth E. *A Reconstruction of Economics.* New York: Wiley, 1950.

———. *The Organizational Revolution: A Study in the Ethics of Economic Organization.* New York: Harper, 1953.

———. *The Image: Knowledge in Life and Society.* Ann Arbor: Univ. of Michigan Press, 1956.

Brand, Stewart. "For God's Sake, Margaret" (interview with Gregory Bateson and Margaret Mead). *CoEvolutionary Quarterly,* June 1976 (www.oikos.org/forgod.htm).

Brockman, John, ed. *About Bateson: Essays on Gregory Bateson.* New York: Dutton, 1977.

Brooks, Paul. *Two Park Street: A Publishing Memoir.* Boston: Houghton Mifflin, 1986.

Browning, Robert. "In a Balcony." 1884. In *Poems and Plays, Vol. II: 1844–1864.* New York: Dutton, 1963.

Bruce, H. Addington. "New Ideas in Child Training." *The American Magazine,* July 1911.

Bryant, Bill. "Nature and Culture in the Age of Cybernetic Systems." (epsilon3.georgetown.edu/~coventrm/asa2000/panel3/bryant.html).

Buderi, Robert. *The Invention That Changed the World: How a Small Group of Radar Pioneers Won the Second World War and Launched a Technological Revolution.* New York: Simon & Schuster, 1996.

Burks, Alice R. and Arthur W. Burks. *The First Electronic Computer: The Atanasoff Story.* Ann Arbor: Univ. of Michigan Press, 1988.

Burks, Arthur W. *Essays on Cellular Automata.* Urbana, IL: Univ. of Illinois Press, 1970.

Bush, Vannevar. *Operational Circuit Analysis* (appendix by N. Wiener). New York: Wiley, 1929.

———. "As We May Think." *Atlantic Monthly,* July 1945a (www.press.umich.edu/jep/works/vbush/vbush-all.html).

———. "Science—The Endless Frontier." Washington, DC: U.S. Govt Printing Office: July 1945b (www.nsf.gov/od/lpa/nsf50/vbush1945.htm).

———. *Pieces of the Action.* New York: Morrow, 1970.

Cadwallader, Mervyn, L. "The Cybernetic Analysis of Change in Complex Social Organizations." *The American Journal of Sociology,* 65:154–157, 1959 (in Smith, A.G.).

Campbell, George A., and Ronald M. Foster. *Fourier Integrals for Practical Applications.* New York: Bell Telephone Laboratories, 1931.

Cannon, Walter B. "Organization for Physiological Homeostasis." *Physiological Review,* 9, 1929.

———. *The Wisdom of the Body.* New York: W. W. Norton Co., 1932.

Carnap, R. *The Logical Syntax of Language.* New York: Harcourt, Brace and Company, 1938.

Carroll, Lewis. *Alice's Adventures in Wonderland & Through the Looking-Glass* (1865, 1871). New York: New American Library, 1960.

Casti, John L. *The One True Platonic Heaven: A Scientific Fiction of the Limits of Knowledge.* Washington DC: Joseph Henry Press/National

Academies Press, 2003 (books.nap.edu/books/0309085470/html/158.html).

Cirincione, Joseph. "The Performance of the Patriot Missile in the Gulf War: An Edited Draft of a Report Prepared for the Government Operations Committee, U.S. House of Representatives." Carnegie Endowment for International Peace, Oct 1992/Nov 2003 (www.ceip.org/files/projects/npp/resources/georgetown/PatriotPaper.pdf).

Cohen, I. B. *Howard Aiken: Portrait of a Computer Pioneer*. Cambridge, MA: MIT Press, 1999.

Cohen, Louis, *Heaviside's Electrical Circuit Theory* (introduction by M. I. Pupin). New York: McGraw-Hill, 1928.

Cowan, Jack. "Epilogue." 1989a (in McCulloch 1989, I).

———. "Neuronal Nets." 1989b (in McCulloch 1989, III).

Clark, Ronald W. *JBS: The Life and Work of J. B. S. Haldane*. New York: Coward-McCann, 1968, 1969.

Crick, F. H. C. "The Biological Replication of Macromolecules" (paper presented to Society for Experimental Biology, 1957). *Symp. Soc. Exp. Biol.* 12: 138–163, 1958.

Davidson, Mark. *Uncommon Sense: The Life and Thought of Ludwig von Bertalanffy (1901–1972), Father of General Systems Theory*. Los Angeles: J. P. Tarcher, 1983.

Davis, Harry M. "An Interview with Norbert Wiener." *New York Times Book Review*, 4.10.49.

Dechert, Charles, R., ed. *The Social Impact of Cybernetics* (Papers presented at a Symposium on Cybernetics and Society, Washington, DC, Nov 1964, under the sponsorship of Georgetown University, American University, and George Washington University, with the co-operation of the American Society for Cybernetics). Notre Dame, IN: Univ. of Notre Dame Press, 1966/New York: Clarion/Simon & Schuster, 1967.

Detzer, David. *The Brink: Cuban Missile Crisis, 1962*. New York: Crowell, 1979.

Deutsch, Karl W. "Mechanism, Teleology and Mind." *Philosophy and Phenomenological Research*, 12: 1951a.

———. "Mechanism, Organism and Society." *Philosophy of Science*, 18: 1951b.

———. *The Nerves of Government: Models of Political Communication and Control*. London: Free Press of Glencoe, 1963.

Dixit, Avinash K., and Barry J. Nalebuff. *Thinking Strategically: The Competitive Edge in Business, Politics and Everyday Life*. New York: Norton, 1991.

Doob, Joseph L. "Review of C. E. Shannon. 'The Mathematical Theory of Communication.'" *Mathematical Reviews*, 10: 133, Feb 1949.

———. "Norbert Wiener Centennial Speeches" (transcript). Cambridge, MA: Royal East Restaurant, Oct 9, 1994.

Dubarle, Dominique. "A New Science: Cybernetics." *Le Monde*, 12.28.48.

Dyker, David. "The Computer and Software Industries in the East European Economies: A Bridgehead to the Global Economy?" www.sussex.ac.uk/spru/publications/imprint/steepdps/27/steep27.doc.

Edman, Irwin. "Mind in Matter." *The New Yorker*, 10.14.50.

Edwards, Paul N. *The Closed World: Computers and the Politics of Discourse in Cold War America*. Cambridge, MA: MIT Press, 1996 (www.stanford.edu/group/mmdd/SiliconValley/Edwards/ClosedWorld1995.book).

Einstein, A. *"Über einen die Erzeugung und Verwandlung des Lichtes betreffenden heuristischen Gesichtspunkt"* ("On a Heuristic Point of View about the Creation and Conversion of Light"), *Annalen der Physik*, 17: 132, March 1905a.

———. *"Über die von der molekularkinetischen Theorie der Wärme geforderte Bewegung von in ruhenden Flüssigkeiten suspendierten Teilchen"* ("On the Motion of Small Particles Suspended in a Stationary Liquid According to the Molecular Kinetic Theory of Heat"). *Annalen der Physik*, 17: 549, May 1905b.

———. *"Zur Electrodynamik bewegter Korper"* ("On the Electrodynamics of Moving Bodies"), *Annalen der Physik*, 17: 891, June 1905c.

———. *"Zur Theorie der Brownschen Bewegung."* *Annalen der Physik*, 19: 371, 1906 (translated as "Investigations on the Theory of the Brownian Movement," 1926).

Eliot, T. S. "Four Quartets." In *Collected Poems 1909–1935*. New York: Harcourt, Brace, 1936.

Epstein, Jason. *Book Business: Publishing Past, Present, and Future*. New York: Norton, 2001.

Fano, Robert M. *Transmission of Information: A Statistical Theory of Communication*. Cambridge, MA: MIT Press, 1961.

Fast, Julius, *Body Language*. New York: M. Evans/Lippincott, 1970.

Feldman, David Henry with Lynn T. Goldsmith. *Nature's Gambit: Child Prodigies and the Development of Human Potential*. New York: Basic Books, 1986; Teachers College Press, 1991.

Ferry, D. K., and R. E. Saeks. "Comments." Wiener 1979/NW CW II, 137–139. Reprinted in *Annals of the History of Computing*, 9: 183–197, 1987.

Festinger, Leon. *A Theory of Cognitive Dissonance.* Evanston, IL: Row Peterson, 1957.

Finnemann, Niels Ole. *Thought, Sign and Machine: The Computer Reconsidered (Tanke, Sprog og Maskine).* Copenhagen: Akademisk Forlag, 1994; translated by Gary Puckering for e-text ed., rev. and abridged by the author, Mar 15, 1999. www.hum.au.dk/ckulturf/pages/publications/nof/tsm/contents.html.

Ford, John J. "Soviet Cybernetics and International Development." In Dechert.

Franchi, Stefano, Güven Güzeldere, and Eric Minch. "Constructions of the Mind" (interview with Heinz von Foerster). *Stanford Humanities Review*, 4: 2, June 26, 1995 (shr.stanford.edu/shreview/4–2/text/interviewonf.html).

Frank, Lawrence K., G. E. Hutchinson, W. K. Livingston, W. S. McCulloch, and N. Wiener. "Teleological Mechanisms" (New York Academy of Sciences Conf., Oct 21–22, 1946). *Annals of the New York Academy of Sciences*, 50: 4, Oct 1948.

Gardner, Howard. *The Mind's New Science: A History of the Cognitive Revolution.* New York: Basic Books, 1985, 1987.

Gerovitch, Slava. "'Mathematical Machines' of the Cold War: Soviet Computing, American Cybernetics and Ideological Disputes in the Early 1950s." *Social Studies of Science*, 31: 2, 253–287, Apr 2001a.

———. "'Russian Scandals': Soviet Readings of American Cybernetics in the Early Years of the Cold War." *Russian Review*, 60: 4, 545–568, Oct 2001b.

———. "Love-Hate for Man-Machine Metaphors in Soviet Physiology: From Pavlov to 'Physiological Cybernetics.'" *Science in Context*, 15: 2, 339–374, 2002a.

———. *Newspeak to Cyberspeak: A History of Soviet Cybernetics.* Cambridge, MA: MIT Press, 2002b.

Gesteland, R. C., J. Y. Lettvin, and W. H. Pitts. "Chemical transmission in the nose of the frog." *Journal of Physiology* (U.K.): 181: 525–559, 1965.

———. "The Olfactory Adventure." In McCulloch 1989, III.

Gibbs, J. Willard. *Elementary Principles of Statistical Mechanics.* New York: C. Scribner's Sons, 1902.

Gibson, William. *Neuromancer.* New York: Ace/Berkeley, 1984.

Gilbert, Edgar N. "History of Mathematics at Bell Labs." 00cm.bell-labs.com/cm/ms/center/history.html.

Glanville, Ranulph. "A Cybernetic Musing: In the Animal and the Machine." *Cybernetics & Human Knowing*, 4: 4, 1997.

Gleick, James. "Bit Player." *NYT*, 12.30.2001.

Gödel, Kurt. "*Über formal unentscheidbare Sätze der Principia Mathematica und verwandter Systeme, I*" ("On Formally Undecidable Propositions"). *Monatshefte für Mathematik und Physik*, 38: 173–198, 1931.

Goethe, J. W. von. *The Sorrows of Young Werther* (1774). New York: Modern Library/Random House, 1971, 1993.

Goldstine, Herman H. *The Computer from Pascal to von Neumann.* Princeton: Princeton Univ. Press, 1972.

Graham, Loren R. *Science, Philosophy, and Human Behavior in the Soviet Union.* New York: Columbia Univ. Press, 1987.

Grattan-Guiness, I. "The Russell Archives: Some New Light on Russell's Logicism." *Annals of Science*, 31, 1974.

———. "Wiener on the Logics of Russell and Schröder. An Account of his Doctoral Thesis, and of his Discussion of it with Russell." *Annals of Science*, 32, 1975.

Greenberg, D. S. "The National Academy of Sciences: Profile of an Institution." *Science*, Apr 14, 21, 28, 1967.

Haldane, J. B. S. "A Mathematical Theory of Natural and Artificial Selection." *Transactions of the Cambridge Philosophical Society*, 1924–1933 (parts 1–9); *Genetics*, 1934 (part 10).

Hall, Edward. T. *The Silent Language.* New York: Doubleday, 1959.

———. *The Hidden Dimension.* New York: Doubleday, 1966.

———, and George Trager. *The Analysis of Culture* (Foreign Service Institute training manual). Washington, DC: U.S. State Department, 1953.

Halmos, P. "The Legend of John von Neumann." *American Mathematical Monthly*, 80, 1973.

Halperin, Morton, Jerry Berman, Robert Borosage, and Christine Marwick. "The Bureau (FBI) in War and Peace." Excerpt from *The Lawless State: The Crimes of the U.S. Intelligence Agencies.* New York: Penguin Books, 1976, www.thirdworldtraveler.com/NSA/Bureau_War_Peace_LS_html.

Hapgood, Fred. *Up the Infinite Corridor: MIT and the Technical Imagination.* Reading, MA: Addison Wesley, 1993.

Hart, Shane. "Computing in the Former Soviet Union and Eastern Europe" www.acm.org/crossroads/xrds5–3/soviet.html (Association for Computing Machinery).

Hartley, R. V. L. "Transmission Limits of Telephone Lines." *Bell Laboratories Record*, 1: 6, 225–228, Feb 1926.

_____. "Transmission of Information." *Bell System Technical Journal*, 7: 535–563, 1928.

Hauben, J. R. "Norbert Wiener, J. C. R. Licklider and the Global Communications Network." 1996, www.columbia.edu/~jrh29/licklider/lick-wiener.html.

Hauben, Michael, and Ronda Hauben. "Cybernetics, Time-sharing, Human-Computer Symbiosis and Online Communities: Creating a Supercommunity of Online Communities." Chapter 6 in *The Netizens and the Wonderful World of the Net: On the History and the Impact of the Internet and Usenet News* (online manuscript, Jan 10, 1994, www.columbia.edu/~hauben/netbook; latest version: www.columbia.edu/~rh120. Published as *Netizens: On the History and Impact of Usenet and the Internet*. Los Alamitos, CA/Hoboken, NJ: IEEE Computer Society Press/John Wiley & Sons, 1997.

Hazen, Harold L. "Theory of Servomechanisms." *Journal of the Franklin Institute*, Sept 1934a.

_____. "Design and Test of a High Performance Servomechanism." *Journal of the Franklin Institute*, Nov 1934b.

Hedger, Leigh. "Analog Computation: Everything Old Is New Again." www.indiana.edu/~rcapub/v21n2/p24.html.

Heims, Steve J. *John von Neumann and Norbert Wiener: From Mathematics to the Technologies of Life and Death*. Cambridge, MA: MIT Press, 1980.

_____. Introduction to *The Human Use of Human Beings*. London: Free Association Books, 1989.

_____. *Constructing a Social Science for Postwar America: The Cybernetics Group, 1946–53*. Cambridge, MA: MIT Press, 1991, 1993.

Hillis, W. Daniel. *The Connection Machine*. Cambridge, MA: MIT Press, 1985.

_____. "The Connection Machine." *Scientific American*, Vol. 256, 108–115, June 1987.

Holton, Gerald. "From the Vienna Circle to Harvard Square: The Americanization of a European World Conception." In Stadler, F., ed. *Scientific Philosophy: Origins and Developments*. New York/Dordrecht (The Netherlands): Kluwer Academic Publishers, 1993.

Hopper, Grace Murray. "The Education of a Computer." *Proc. ACM Conference*, reprinted in *Ann. Hist. Comp.*, 9:3–4, 271–281, 1952.

Howard, Jane. *Margaret Mead*. New York: Ballantine, 1984.

Howland, B., J. Y. Lettvin, W. S. McCulloch, W. H. Pitts, and P. D. Wall. "Reflex inhibition by Dorsal Root Interaction." *Journal of Neurophysiology*, 19: 1–17, 1955.

Hunt, Linda. *Secret Agenda: The U.S. Government, Nazi Scientists and Project Paperclip*. New York: St. Martin's Press, 1991.

Hutchins, John. "'From First Conception to First Demonstration: The Nascent Years of Machine Translation, 1947–1954. A Chronology." *Machine Translation*, 12:3, 1997a, 195–252 (ourworld.compuserve.com/homepages/wjhutchins/PPF–2.pdf).

Hutchinson, G. E. "Circular Causal Systems in Ecology." (New York Academy of Sciences Conf., Oct 21–22, 1946), in Frank et al.

Itô, Kiyosi. "On Stochastic Processes (Infinitely Divisible Laws of Probability)." *Japanese Journal of Mathematics*, 18: 1942.

Jackson, Allyn. "Dirk Struik Celebrates his 100th." *Notices of the AMS*, 42:1, Jan 12, 1995.

Jacobs, William Wymark. "The Monkey's Paw." In *The Lady of the Barge*. New York/London, Harper & Brothers, 1902.

James, H. M., N. B. Nichols, and R. S. Phillips. *Theory of Servomechanisms*. New York: McGraw-Hill, 1947.

James, William. *William James Talks to Teachers on Psychology and to Students on Some of Life's Ideals*. New York: H. Holt, 1899; Cambridge, MA: Harvard Univ. Press, 1983.

_____. *Pragmatism, A New Name for Some Old Ways of Thinking*. New York: Longmans, Green, 1907.

Jeffress, L. A., ed. *Cerebral Mechanisms in Behavior: The Hixon Symposium*. New York: John Wiley & Sons, 1951.

Jerison, David and Daniel Stroock. "Norbert Wiener." *Notices of the American Mathematical Society*, 42:4, Apr 1995.

Jerome, Fred. *The Einstein File: J. Edgar Hoover's Secret War Against the World's Most Famous Scientist*. New York: St. Martin's, 2002.

Joy, Bill. "Why the Future Doesn't Need Us." *Wired*, Apr 2000a.

_____. "Will Spiritual Robots Replace Humanity by 2100?" Symposium organized by Douglas Hofstadter, Symbolic Systems Program, Stanford Univ., Apr 1, 2000b, technetcast.ddj.com/tnc_play_stream.html?stream_id=258.

Kalmus, H. "A Cybernetical Aspect of Genetics." *Journal of Heredity*, 41: 19–22, 1950.

Kaplan, Fred. "How Smart Are Our Smart Bombs? They're better than ever, but they still won't topple Saddam." *Slate*, 10.17.02 (slate.msn.com/id/2072709).

Kay, Lily. *Who Wrote the Book of Life? A History of the Genetic Code*. Stanford, CA: Stanford Univ. Press, 2000.

Kelly, Kevin. *Out of Control: The New Biology of Machines, Social Systems, and the Economic World.* Reading, MA: Perseus Books, 1994.

Kipling, Rudyard. *Just So Stories.* New York: Penguin Books, 1974.

Koopmans, Matthijs. "From Double Bind to N-Bind: Toward a New Theory of Schizophrenia and Family Interaction." *Nonlinear Dynamics, Psychology, and Life Sciences,* 5:4: 289–323, Oct 2001.

Kraeplin, Emil. *Clinical Psychiatry: A Textbook for Physicians.* New York: Macmillan, 1913.

Kubie, Lawrence. "A Theoretical Application to Some Neurological Problems of the Properties of Excitation Waves Which Move in Closed Circuits." *Brain,* 53: 166–178, July 1930.

Kurzweil, Ray. "Will Spiritual Robots Replace Humanity by 2100?" Symposium organized by Douglas Hofstadter, Symbolic Systems Program, Stanford Univ., Apr 1, 2000, technetcast.ddj.com/tnc_program.html?program_id=82.

Lane, Edward William, trans. *Stories from the Thousand and One Nights.* New York: Collier & Son, 1909.

Latil, Pierre de. *Thinking by Machine: A Study of Cybernetics* (Y. M. Golla, trans.). London: Sidgwick and Jackson, 1956 (orig. *Le Pensée Artificielle.* Paris: Gallimard, 1953).

Leaver, Eric W., and Brown, John J. "Machines Without Men." *Fortune,* Nov 1946.

Leavitt, Harold J., and Ronald A. H. Mueller. "Some Effects of Feedback on Communication." *Human Relations,* 4:401–401, 1951 (in Smith, A. G.).

Lee, J. A. N., Stanley Winkler, and Merlin Smith. "Key Events in the History of Computing" (summary prepared for IEEE Computer Society 50th Anniversary), 1996, ei.cs.vt.edu/~history/50th/30.minute.show.html. Virginia Tech.

Lee, Yuk Wing. *Synthesis of Networks by Means of Fourier Transforms of LaGuerre's Functions.* Dissertation for Sc.D. Elec. Eng., MIT (completed in 1930), *J. Math, and Physics,* 11: 261–278, 1932.

———. *Applications of Statistical Methods to Communications Problems.* Cambridge, MA: MIT Research Laboratory of Electronics, 1950.

———. *Statistical Theory of Communication.* New York: Wiley, 1960.

Lettvin, J. Y. "Introduction." McCulloch 1989, I, 7–20 (1989a).

———. "Warren and Walter." McCulloch 1989, II, 514–529 (1989b).

———, H. R. Maturana, W. S. McCulloch, W. H. Pitts. "What the Frog's Eye Tells the Frog's Brain." *Proceedings of the IRE,* 47:11, 1940–51, Nov 1959 (in McCulloch 1989, IV, 1161–1172).

Levinson, Norman. "Prediction of Stationary Time Series by a Least Squares Procedure." Report produced under U.S. Army Air Force Air Corps Meteorological contract about Mar 1942. Records of NDRC, Sec 2, Div D, Record Group 227, Project #6, National Archives & Records Service, Washington, DC.

———. "Report of Conference on the Methods of N. Wiener, Oct 3, 1944." Statistical Research Group/Division of War Research/Columbia University. Records of NDRC, Sec 2, Div D, National Archives & Records Service, Washington, DC.

———. "Wiener's Life." *Bulletin of the American Mathematical Society,* 72: 700, Jan 1966.

Levinson, Zipporah (Fagi). "Norbert Wiener Centennial Speeches" (transcript). Cambridge, MA: Royal East Restaurant, Oct 9, 1994.

Lewin, Kurt. "Frontiers in Group Dynamics." *Human Relations,* 1: 5–153, 1947.

Lewis, F. L. *Applied Optimal Control and Estimation.* New York: Prentice-Hall, 1992 (www.theorem.net/theorem/lewis1.html).

Lewis, Harry R. "Computing's Cranky Pioneer" (review of I. B. Cohen). *Harvard Magazine,* May–June 1999.

Lewis, Michael. *The New New Thing: A Silicon Valley Story.* New York: Norton, 1999.

Licklider, J. C. R. "Man-Computer Symbiosis." *IRE Transactions on Human Factors in Electronics,* HFE–1, 1960 (memex.org/licklider.pdf).

———. and Robert Taylor. "The Computer as a Communication Device." In *Science and Technology: For the Technical Men in Management,* 76: Apr 1968(memex.org/licklider.pdf).

Lipset, David. *Gregory Bateson: The Legacy of a Scientist.* Englewood Cliffs, NJ: Prentice Hall, 1980.

Liversidge, Anthony. "Father of the Electronic Information Age" (Claude Shannon interview). *OMNI,* Aug 1987.

Lombreglia, Ralph. "The Believer" (rev. of Epstein). *Atlantic Unbound,* 1.18.01 (www.theatlantic.com/unbound/digital-reader/dr2001–01–18.htm).

Lorente de Nó, Rafael. *A Study of Nerve Physiology.* New York: Rockefeller Institute, 1947.

MacColl, LeRoy A. *A Fundamental Theory of Servomechanisms.* New York: Van Nostrand, 1945.

Macrae, Norman. *John von Neumann: The Scientific Genius Who Pioneered the Modern Computer,*

Game Theory, Nuclear Deterrence and Much More. New York: Pantheon, 1992.

Malory, Sir Thomas. *Le Morte Darthur*. etext.lib.virginia.edu.

Mandrekar, V. R. "Mathematical Work of Norbert Wiener." *Notices of the AMS*, 42:6, 664–669, June 1995.

_____. with Pesi R. Masani, eds. *Proceedings of the Norbert Wiener Centenary Congress* (Michigan State University, Nov 27-Dec 3, 1994). American Mathematical Society, 1997.

Manley, Jared (rewrite by James Thurber). "Where Are They Now? April Fool!" *The New Yorker*, Aug 14, 1937 (www.sidis.net/newyorker3.htm).

Mann, R.W. "Sensory and Motor Prostheses in the Aftermath of Wiener." *Norbert Wiener Centenary Congress, Proceedings of Symposia in Applied Mathematics* (Ann Arbor, MI centennial). Providence, RI: American Mathematical Society, 52: 401–439, 1997.

Masani, Pesi R. *Norbert Wiener 1894–1964* (Vita Mathematica Series). Boston: Birkhauser, 1990.

_____. and R. S. Phillips. "Antiaircraft Fire-Control and the Emergence of Cybernetics." In Wiener 1985/NW CW IV, 141–179.

Mason, Stephen F. *History of the Sciences*. New York: Collier-Macmillan, 1962 (originally published as *Main Currents of Scientific Thought*. Abelard-Shuman Ltd, 1956).

Massachusetts Institute of Technology. "The Legacy of Norbert Wiener: A Centennial Symposium" (program notes prepared with the assistance of Tony Rothman). Cambridge, MA: Oct 8–14, 1994.

Materialist (pseudonym). *Komu Sluzhit Kibernetika?* ("Whom Does Cybernetics Serve?"). *Voprosy Filosofii* (*Problems of Philosophy*) 7: 210–219, 1953 (Pav, in Wiener 1981/NW CW III, 778–779).

Maxwell, James Clerk. "On Governors." *Proc. Roy. Soc*. London, 16: 270–283, 1868.

May, Rollo. *Psychology and the Human Dilemma*. New York: Norton, 1967, 1979.

Mazuzan, George T. "NSF- The National Science Foundation: A Brief History." (NSF 88–16) 1994 (www.nsf.gov/pubs/stis1994/nsf8816/nsf8816.txt).

McCartney, Scott. *ENIAC: The Triumphs and Tragedies of the World's First Computer*. New York: Walker, 1999.

McCulloch, Warren Sturgis. "A Recapitulation of the Theory, with a Forecast of Several Extensions" (New York Academy of Sciences Conf., Oct 21–22, 1946), in Frank et al.; McCulloch 1965a; McCulloch 1989, II.

_____. "An Account of the First Three Conferences on Teleological Mechanisms." New York: Josiah Macy, Jr. Foundation, Oct 1947.

_____. "Through the Den of the Metaphysician" (lecture at the Univ. of Virginia, 1948). *Brit. J. Phil. Sci*. 5: 1954 (also in McCulloch 1965a; McCulloch 1989, III).

_____. "The Brain as a Computing Machine" (address to AIEE winter general meeting, NY, NY, Jan 31-Feb 4, 1949), in McCulloch 1989, II.

_____. "The Past of a Delusion" (speech to Chicago Literary Club, Jan 28, 1952), in McCulloch 1965a/McCulloch 1989, II.

_____. "Summary of the Points of Agreement Reached in the Previous Nine Conferences on Cybernetics." *Transactions of the Tenth Conference* (Apr 22–24, 1953). New York: Josiah Macy, Jr. Foundation, 1955 (also in McCulloch 1965a; McCulloch 1989, III).

_____. *Embodiments of Mind*. Cambridge, MA: MIT Press, 1965a.

_____. "Norbert Wiener and the Art of Theory." *Journal of Nervous and Mental Disease*, 140:1, 1965b (in McCulloch 1989, IV).

_____. "Recollections of the Many Sources of Cybernetics." *ASC Forum*, VI: 2, Summer 1974 (in McCulloch 1989, I).

_____ (Rook McCulloch, ed.). *Collected Works of Warren S. McCulloch*, Vols. I-IV. Salinas, CA: Intersystems Publications, 1989.

McCulloch, W. S., and Walter H. Pitts. "A Logical Calculus of Ideas Immanent in the Nervous System." *Bulletin of Mathematical Biophysics* 5: 115–133, 1943 (in McCulloch 1965a/McCulloch 1989, I).

_____. "How We Know Universals: The Perception of Auditory and Visual Forms." *Bulletin of Mathematical Biophysics*, 9: 127–147, 1947 (in McCulloch 1965a/McCulloch 1989, II).

McLuhan, Marshall. *Understanding Media: The Extensions of Man*. New York: McGraw-Hill, 1964; Signet/New American Library, 1966.

_____. with Quentin Fiore. *War and Peace in the Global Village: An Inventory of Some of the Current Spastic Situations That Could Be Eliminated by More Feedforward*. New York: McGraw-Hill, 1968.

Mead, Margaret. *Coming of Age in Samoa: A Psychological Study of Primitive Youth for Western Civilisation*. New York: W. Morrow & Company, 1928.

_____. (review of *Ex-Prodigy*), *Virginia Quarterly Review*, Summer 1953.

_____. *Continuities in Cultural Evolution*. New Haven: Yale Univ. Press, 1964.

_____. *Soviet Attitudes Toward Authority: An Interdisciplinary Approach to Problems of Soviet Character.* New York: McGraw-Hill, 1951; Westport, CT: Greenwood Press, 1979; New York/Oxford, U.K.: Berghahn Books, 2001.

_____. "Cybernetics of Cybernetics." In *Purposive Systems: Proceedings of the First Annual Symposium of the American Society for Cybernetics* (Washington, DC, Oct 25–27, 1967). New York: Spartan Books, 1968.

Meron, Gabi. "The Development of the Swallowable Video Capsule (M2A)." *Gastrointestinal Endoscopy*, 52: 6, 2000.

Mikulak, Maxim W. "Cybernetics and Marxism-Leninism" (in Dechert).

Miller, George A., E. Galanter, and K. Pribram. *Plans and the Structure of Behavior.* New York: Holt, Rinehart & Winston, 1960.

Miller, James Grier, *Living Systems.* New York: McGraw-Hill, 1978.

Mindell, David A. "Opening Black's Box: Rethinking Feedback's Myth of Origin." *Technology and Culture*, July 2000.

_____, Jérôme Segal, and Slava Gerovitch. "From Communications Engineering to Communications Science: Cybernetics and Information Theory in the United States, France, and the Soviet Union." In Walker 2002/2003 (jerome-segal.de/Publis/science_and_ideology.rtf).

Minorsky, Nicolas, "Directional Stability of Automatically Steered Bodies." *J. Am. Soc. Naval Eng*, 34: 284, 1922.

Mirowski, Philip and Esther-Mirjam Sent. *Science Bought and Sold: Essays in the Economics of Science.* Chicago: Univ. Chicago Press, 2001.

Mollenhoff, Clark R. *Atanasoff: Forgotten Father of the Computer.* Ames: Iowa State Univ. Press, 1988.

Monk, Ray, *Bertrand Russell: The Spirit of Solitude, 1872–1921.* New York: Free Press, 1996.

Monod, Jacques. *Chance and Necessity: An Essay on the Natural Philosophy of Modern Biology* (Austryn Wainhouse, trans.). New York: Knopf, 1971/Vintage, 1972.

Monod, Jacques and François Jacob. "Teleonomic Mechanism in Cellular Metabolism, Growth, and Differentiation." *Cold Spring Harbor Symposia on Quantitative Biology*, 26: 389–401, 1961.

Morrison, Philip and Phylis Morrison. "100 or So Books That Shaped a Century of Science." *American Scientist*, 87:6, Nov–Dec 1999.

Nahin, P. J. *Oliver Heaviside: Sage in Solitude: The Life, Work, and Times of an Electrical Genius of the Victorian Age.* New York: IEEE Press, 1988.

Nasar, Sylvia. *A Beautiful Mind: A Biography of John Forbes Nash, Jr., Winner of the Nobel Prize in Economics, 1994.* New York: Simon & Schuster, 1998/Touchstone, 1999.

Nathan, Otto. *Einstein on Peace.* New York: Avenel/Crown 1960, 1981.

National Research Council. *Funding a Revolution: Government Support for Computing Research.* Washington, DC: National Academy Press, 1999.

Nemeroff, Charles B. "The Neurobiology of Depression." *Scientific American*, June 1998.

Noble, David. *Forces of Production: A Social History of Industrial Automation.* New York: Oxford Univ. Press, 1984.

Norberg, Arthur, and Judy E. O'Neill. *Transforming Computer Technology: Information Processing for the Pentagon, 1962–1986.* Baltimore: Johns Hopkins Univ. Press, 1996.

Northrop, F. S. C. "On W. S. McCulloch." McC CW I.

Nyquist, H. "Certain Factors Affecting Telegraph Speed." *Bell System Technical Journal*, 3: 324–346, 1924.

Ott, H. *Noise Reduction Techniques in Electronic Systems.* New York: John Wiley & Sons, 1976.

Owens, Larry. "Mathematicians at War: Warren Weaver and the Applied Mathematics Panel, 1942–1945." In Rowe, D., and McCleary, J., eds. *The History of Modern Mathematics, Vol. II* (287–305). Boston: Academic Press, 1988.

Pangaro, Paul. "Cybernetics: The Center of Science's Future." Address to Philosophical Society of Washington, Feb 2, 1991, www.pangaro.com/abstracts/philos-wash-cybersci.html.

Pask, Gordon. "Automatic Teaching Techniques." *British Communications and Electronics*, Apr 1957.

_____. "Teaching Machines." *Proc. 2nd Cong. Intl. Assn. Cybernetics* (Namur 1958). Paris: Gauthier Villars, 1960a.

_____. "The growth process in the cybernetic machine." *Proc. 2nd Cong. Intl. Assn. Cybernetics* (Namur 1958), Gauthier-Villars, 1960b.

Pav, P. "Soviet Cybernetics. A Commentary." In Wiener 1981/NW CW III, 777–783.

Phillips, Ralph S. "Servomechanisms." Cambridge, MA: MIT, Radiation Laboratory Report No. 372, May 11, 1943.

Pierce, John R. "Communication." *Scientific American*, 227, Sept 1972 (reprinted in *Communications: A Scientific American Book.* San Francisco: W. H. Freeman, 1972).

_____. "The Early Days of Information Theory." *IEEE Transactions on Information Theory*, 19:1, Jan 1973, 3–8.

Poundstone, William. *Prisoner's Dilemma: John von Neumann, Game Theory and the Puzzle of the Bomb*. New York: Anchor Books, 1993.

Prescott, Samuel C. *When MIT Was "Boston Tech."* Cambridge, MA: The Technology Press, 1954.

Pribram, Karl H. "The Cognitive Revolution and Mind/Brain Issues." *American Psychologist*, 41: 507–520, 1986.

_____. *Brain and Perception: Holonomy and Structure in Figural Processing*. Hillsdale, NJ: Lawrence Erlbaum Associates, 1991.

_____. "What Is Mind That the Brain May Order It?" In Mandrekar and Masani, 1997.

Pupin, Michael I. *From Immigrant to Inventor*. New York/London: Scribner's Sons, 1923.

Raisbeck, Barbara. "Viewing." *Columbia*, 11: 1986.

Raisbeck, Gordon. "Comments on 'The Early Days of Information Theory,'" *IEEE Transactions on Information Theory*, 19: 6, Nov 1973.

Raven, Paul. *Deep Roots and Lofty Branches: The History of a Great Family*. London: (privately published), 1980.

Reed, Sidney G., et al. *DARPA Technical Accomplishments, Volume 1: An Historical Review of Selected DARPA Projects*. Alexandria, VA: Institute for Defense Analysis, 1990.

Rheingold, Howard. *Tools for Thought: The History and Future of Mind-Expanding Technology*. Cambridge, MA: MIT Press, 2000.

Ridenour, Louis. "Military Support of American Science, a Danger?" *Bull. of Atomic Scientists*, 3: 8, Aug 1947.

Roberts, John. "Family or Fate? A Critical Evaluation of a Psychogenic and a Genetic Theory of Schizophrenia." www.ahisee.com/content/schiz1essay.html, 2001.

Rogers, Everett M., William B. Hart, Yoshitaka Miike. "Edward T. Hall and the History of Intercultural Communication: The United States and Japan." *Keio Communication Review*, 24: 2002 (www.mediacom.keio.ac.jp/pdf2002/Rogers.pdf).

Roos, John. "WarBots: Eyes and Ears for MOUT [Military Operations in Urban Terrain]." *Armed Forces Journal International*, 139: 4, 2001.

Rosenberg, Seymour, and Robert L. Hall. "The Effects of Different Social Feedback Conditions upon Performance in Dyadic Teams." *Journal of Abnormal and Social Psychology*, 57: 271–277, 1958 (in Smith, A. G.).

Rosenblith, Walter A. "From a Biophysicist Who Came to Supper." In *Research Laboratory of Electronics, R.L.E.: 1946+20*. Cambridge, MA: RLE/MIT Press, 1966 (rleweb.mit.edu/Publications/currents/6–2back.htm).

_____ and Jerome Wiesner. "From Philosophy to Mathematics to Biology." *Bulletin of the American Mathematical Society*, 72: 700, Jan 1966.

Rumelhart, David E., James L. McClelland, and the PDP Research Group. *Parallel Distributed Processing: Explorations in the Microstructure of Cognition*. Vols. 1 and 2. Cambridge, MA: MIT Press, 1986.

Russell, Bertrand, *The Autobiography of Bertrand Russell*. London: Allen & Unwin, Vol. 1, 1967.

_____. "Are Human Beings Necessary?" *Everybody's* (U.K.), Sept 15, 1951.

Saffo, Paul. "Sensors: The Next Wave of Infotech Innovation." www.saffo.com/sensors.html, 1997/2002.

_____. "Smart Sensors Focus on the Future." *CIO Insight*, Apr 15, 2002.

Samuelson, Paul A. "Some Memories of Norbert Wiener." *Proceedings of Symposia in Pure Mathematics*, 60: 37–42, 1997.

Saxe, John Godfrey. "The Blind Men and the Elephant" (in Sillar, FC and RM Meyler, Elephants Ancient and Modern. London: Studio Vista, 1968), www.milk.com/randomhumor/elephant_fable.html.

Schramm, Wilbur. "Information Theory and Mass Communication." *Journalism Quarterly*, 32:131–146, 1955 (in Smith, A. G.).

_____, ed. *The Science of Human Communication: New Directions and New Findings in Communication Research*. New York: Basic Books, 1963.

Schrödinger, Erwin. *What Is Life?* Cambridge: Cambridge Univ. Press, 1944 (home.att.net/~p.caimi/Life.doc).

Selfridge, Oliver G. "Some Notes on the Theory of Flutter." *Arch. Inst. Cardio. Mexico* 18: 177–187, 1948.

_____. "Pattern Recognition and Modern Computers." *IRE Proceedings of the 1955 Western Joint Computer Conference*, 1955.

_____. "Pandemonium, a Paradigm for Learning." *Proc. Symp. Mechanisation of Thought Processes*, Teddington (U.K.), Natl. Physical Lab., Nov 1958 (in D. V. Blake and A. M. Utley, eds. *Proceedings of the Symposium on Mechanisation of Thought Processes*. London: H. M. Stationery Office, 1959).

Shannon, Claude E. "A Symbolic Analysis of Relay and Switching Circuits." *Trans. Amer. Inst. Elec. Eng.* 57: 713–723, 1938.

_____. "A Mathematical Theory of Communication." *Bell System Technical Journal*, 27: 379–423, 623–656, July and Oct 1948.

_____. "The Bandwagon." *IEEE Transactions on Information Theory*, Mar 1956.

_____ (Sloane, Neil, J. A., and Aaron D. Wyner, eds.). *Collected Papers.* Piscataway, NJ: IEEE Press, 1993.

_____ and John McCarthy, eds. *Automata Studies.* Princeton: Princeton Univ. Press, 1956.

_____ and Warren Weaver, *The Mathematical Theory of Communication.* Urbana, IL: Univ. of Illinois Press, 1949.

Sheehan, Helena, *Marxism and the Philosophy of Science: A Critical History: The First Hundred Years.* Atlantic Highlands, NJ: Humanities Press Intl, 1985/1993.

Sidis, William J. *The Animate and the Inanimate.* Boston: Gorham Press, 1925.

_____ (John W. Shattuck, pseud.). *The Tribes and the States.* ca. 1935/Scituate, MA: Penacook Press, 1982.

Simon, Herbert A. "Allen Newell." *Biographical Memoirs,* vol. 71. Washington, DC: National Academies Press, 1997 (stills.nap.edu/readingroom/books/biomems/anewell.html).

Smalheiser, Neil R. "Walter Pitts." *Perspectives in Biology and Medicine,* 43: 2, 2000.

Smith, Alfred G., ed. *Communication and Culture: Readings in the Codes of Human Interaction.* New York: Holt, Rinehart and Winston, 1967.

_____. "The primary resource." *Journal of Communication,* 25: 2, 15–20, 1976.

Smith, Mark K. "Kurt Lewin: Groups, Experiential Learning and Action Research." National Grid for Learning (U.K.), 2001, www.infed.org/thinkers/et-lewin.htm.

Sophocles. *Electra* (Jebb, R. C., trans.). classics.mit.edu/Sophocles/electra.html.

Sperling, Abraham. *Psychology for the Millions.* New York: F. Fell, 1946.

Stibitz, George R. "Note on Predicting Networks." Feb 1942. Records of NDRC, Sec 2, Div D, Record Group 227, Project #6, National Archives & Records Service, Washington, DC.

Struik, Dirk Jan. "Norbert Wiener—Colleague and Friend." *American Dialog,* Mar-Apr 1966.

_____. "The Struik Case of 1951." *Monthly Review,* Jan 1993.

_____. "Norbert Wiener Centennial Speeches." (transcript), Cambridge, MA: Royal East Restaurant, Oct 9, 1994.

Tanenbaum, Sandra. *Engineering Disability: Public Policy and Compensatory Technology.* Philadelphia: Temple Univ. Press, 1986.

Taylor, Geoffrey I. "Turbulent Motion in Fluids." Cambridge (U.K.), 1915.

Theall, Donald. *The Virtual Marshall McLuhan.* Montreal/Ithaca, NY: McGill-Queen's Univ. Press, 2001.

Tolstoy, L. N. *Complete Works* (24 vols.) (L. Wiener, trans.). Boston: D. Estes & Co., 1904–1905.

Toulmin, Stephen. "The Importance of Norbert Wiener." *New York Review of Books,* Sept 24, 1964.

Trager, George. "Paralanguage: A First Approximation." *Studies in Linguistics,* 13, 1–12, 1958.

Trebek, Alex with Peter Barsocchini. *The Jeopardy! Book: The Answers, the Questions, the Facts, and the Stories of the Greatest Game Show in History.* New York: Harper Perennial, 1990.

Turing, Alan M. "On Computable Numbers with an Application to the *Entscheidungs* Problem." *Proceedings of the London Mathematical Society,* 42:2, 230–265, 1936; rev. 1937.

_____. "Computing Machinery and Intelligence." *Mind,* LIX: 236, Oct 1950.

Ulam, S. "John von Neumann, 1903–1957." *Bulletin of the American Mathematical Society,* 64: 3: 2, May 1958.

_____. *Adventures of a Mathematician.* New York: Scribners, 1976.

Ullman, Deborah. "Kurt Lewin: His Impact on American Psychology, or Bridging the Gorge between Theory and Reality." 2000, www.sonoma.edu/psychology/os2db/history3.html.

von Foerster, Heinz, ed. *Cybernetics: Circular, Causal and Feedback Mechanisms in Biological and Social Systems. Transactions of the Ninth Conference.* New York: Josiah Macy, Jr. Foundation, 1953.

_____. *Cybernetics: Circular, Causal and Feedback Mechanisms in Biological and Social Systems. Transactions of the Tenth Conference.* New York: Josiah Macy, Jr. Foundation, 1955.

Vonnegut, Kurt, Jr. *Player Piano.* New York: Charles Scribner's Sons, 1952; Avon, 1967.

von Neumann, John. "*Zur Theorie der Gesellschaftspiele*" (Theory of Parlor Games). *Mathematische Annalen* 100, 295–320, 1928 (in von Neumann 1963, 6: 1–28).

_____. "First Draft Report on the EDVAC." (Moore School of Electrical Engineering, Univ. of Pennsylvania, June 30, 1945) *IEEE Annals of the History of Computing,* 15:4, 27–75, 1993 (qss.stanford.edu/~godfrey/vonNeumann/vnedvac.pdf).

_____. "The General and Logical Theory of Automata" (paper presented at Hixon Symposium, Oct 1948). In Jeffress (crl.ucsd.edu/~elman/Courses/cog202/Papers/vonneumann.pdf).

_____. "Probabilistic Logics and the Synthesis of Reliable Organisms from Unreliable Components." In Shannon and MacCarthy; von Neumann 1961–1963 (*Collected Works*), 1956.

_____. *Theory of Self-Reproducing Automata* (Arthur Burks, ed). Urbana: Univ. Illinois Press, 1966.

_____. *The Computer and the Brain*. New Haven: Yale University Press, 1958.

_____ (A. H. Taub, ed.). *Collected Works*, Vols. I–VI. New York: Macmillan/Pergamon Press, 1961–1963.

_____ (William Aspray and Arthur Burks, eds.). *Papers of John von Neumann on Computing and Computer Theory*, Charles Babbage Institute Reprint Series, Cambridge, MA/Los Angeles: MIT Press/Tomash Publishers, 1987.

Waldrop, M. Mitchell. "Claude Shannon: Reluctant Father of the Digital Age." *Technology Review*, 104: 6, July/August 2001.

Walker, Mark. "Atomic Secrets and the Red Scare" (review of Wang, 1999). *Physics World*, May 1999 (physicsweb.org/article/review/12/5/1).

_____, ed. *Science and Ideology: A Comparative History*. London: Routledge, 2002/2003.

Wall, Patrick. "An Assessment of the Significance of the Physiological Contributions after 1950" (in McCulloch 1989, III).

_____. "Oral History Interview with Patrick Wall" (Aug 10 1993). John C. Liebeskind History of Pain Collection (ms. no. 127.2), Darling Biomedical Library, Univ. of California, Los Angeles, www.library.ucla.edu/libraries/biomed/his/wall-oralhistory.htm, 1993.

Wallace, Amy. *The Prodigy: A Biography of William James Sidis, America's Greatest Child Prodigy*. New York: Dutton, 1986.

Wang, Jessica. *American Science in an Age of Anxiety: Scientists, Anticommunism and the Cold War*. Chapel Hill, NC: Univ. North Carolina Press, 1999.

_____. "Edward Condon and the Cold War Politics of Loyalty." *Physics Today*, 54:12, Dec 2001. (www.physicstoday.org/vol–54/iss–12/p35.html).

Watson, James D. "Terminology in Bacterial Genetics." *Nature*, 171: 701, 1953.

_____. *Genes, Girls & Gamow*. New York: Knopf, 2002.

Weaver, Warren. *Scene of Change: A Lifetime in American Science*. New York: Scribner, 1970.

Weiss, Paul, *Principles of Development*. New York: Henry Holt, 1939.

Wiener, Leo. "Stray Leaves from My Life." *Boston Evening Transcript*, Mar 19, 26, Apr 2, 9, 16, 26 and 30, 1910.

_____. *Africa and the Discovery of America*. Philadelphia: Innes & Sons, 1920–22; Brooklyn: A & B Books, 1992.

Wiener, Norbert. "The Theory of Ignorance." MIT Institute Archives, MC22, box 10, folder 421, 1905.

_____. "The Rationalism of Descartes, Spinoza and Leibniz." MIT Institute Archives, MC22, box 26D, folder 434, circa 1912a.

_____. "The Place of Relations in Knowledge and Reality" (Bowdoin Essay), MIT Institute Archives, MC22, box 27A, folder 448, 1912b.

_____. "Bertrand Russell's Theory of the Nature of Reality." MIT Institute Archives, MC22, box 27A, folder 452, 1913a.

_____. "On the Rearrangement of the Positive Integers in a Series of Ordinal Numbers Greater than that of any Given Fundamental Sequence of Omegas." *Messenger of Mathematics*, 3: 511, Nov 1913b.

_____. "A Simplification of the Logic of Relations." *Proceedings of the Cambridge Philosophical Society*, 27: 5, 1914a.

_____. "The Highest Good." *Journal of Philosophy, Psychology and Scientific Method*, 9: 19, 1914b.

_____. "Relativism." *Journal of Philosophy, Psychology and Scientific Method*, 9: 21, 1914c.

_____. "The Shortest Line Dividing an Area in a Given Ratio." *Journal of Philosophy, Psychology and Scientific Method*, V: 12, 1915a.

_____. "Studies in Synthetic Logic." *Proceedings of the Cambridge Philosophical Society*, 28: 1, 1915b.

_____. "The Average of an Analytical Functional and the Brownian Movement." *Proceedings of the National Academy of Science*, 7: Oct 1921.

_____. "*Verallgemeinerts Trigonometrische Entwicklungen.*" *Gött. Nachrichten*, 1925.

_____ and Max Born. "A New Formulation of the Law of Quantization for Periodic and Aperiodic Phenomena." *Journal of Mathematics and Physics*, 5: 2, Feb 1926a.

_____. "The Operational Calculus." *Mathematical Annals*, 95: 4, Feb 1926b.

_____. "The Harmonic Analysis of Irregular Motion" (1st paper), *Journal of Mathematics and Physics*, 5: 2, Feb 1926c.

_____. "The Harmonic Analysis of Irregular Motion" (2nd paper), *Journal of Mathematics and Physics*, 5: 3, Mar 1926d.

_____. "Generalized Harmonic Analysis." *Acta Mathematica*, 55: 117–258, Sept 1930.

_____. "Back to Leibniz! Physics Reoccupies an Abandoned Position." *Technology Review*, 34: 1932a.

_____ and E. Hopf. "*Uber eine Klasse Singularer Integralgleichungen.*" *Sitzungsber. d. Preussischen Akad. d. Wissensch.*, 696, 1932b.

_____. *The Fourier Integral and Certain of its Applications.* Cambridge: Cambridge Univ. Press, 1933.

_____ and R. E. A. C. Paley, *Fourier Transforms in the Complex Domain.* New York: American Mathematical Society Colloq. Publications, 19: 1934.

_____. "The Historical Background of Generalized Harmonic Analysis." *Amer. Math. Soc. Semicent. Publs.* Vol. II, Semicentennial Addresses, 1938.

_____. "Memorandum on the Scope etc. of a Suggested Computing Machine" (memo to Vannevar Bush, National Defense Research Committee) Sept 1940. MIT Institute Archives, MC22, box 28A, folder 558.

_____. "The Extrapolation, Interpolation, and Smoothing of Stationary Time Series" (monograph), Report 370, Feb 1, 1942. Records of NDRC, Sec 2, Div D, National Archives & Records Service, Washington, DC.

_____, A. Rosenblueth, and J. Bigelow. "Behavior, Purpose and Teleology." *Philosophy of Science,* 10: 18–24, 1943.

_____ and A. Rosenblueth. "The Mathematical Formulation of the Problem of Conduction of Impulses in a Network of Connected Excitable Elements, Specifically in Cardiac Muscle." *Arch. Inst. Cardiol. Méx.,* 16, 205–265, 1946a.

_____, A. Rosenblueth, and J. García Ramos. "Muscular Clonus: Cybernetics and Physiology." (written ca. 1946, not published until Wiener 1989) 1946b.

_____. "Time, Communication and the Nervous System." (New York Academy of Sciences Conf., Oct 21–22, 1946), in Frank et al. 1946c.

_____. "A Scientist Rebels." *Atlantic Monthly,* 79: January 1947a.

_____. "The Armed Services Are Not Fit Almoners for Research." *Bull. of Atomic Scientists,* 3: 8, Aug 1947b.

_____. *Cybernetics: Or Control and Communication in the Animal and the Machine.* Cambridge, MA: The Technology Press and New York: John Wiley & Sons; Paris: Hermann et Cie., 1948a; 2nd ed., 1961.

_____. "Cybernetics." *Scientific American,* 179: 5, Nov 1948b.

_____. "A Rebellious Scientist After Two Years." *Bull. of Atomic Scientists,* 4: 11, Nov 1948c.

_____, A. Rosenblueth, W. Pitts, and J. Garcia Ramos. "An Account of the Spike Potential of Axons." *J. Comp. Physiol.,* Dec 1948d.

_____. "A New Concept of Communication Engineering." *Electronics,* 22: 1, Jan 1949a.

_____. *Extrapolation, Interpolation and Smoothing of Stationary Time Series, with Engineering Applications.* Cambridge, MA: The Technology Press and New York: John Wiley & Sons, 1949b.

_____. "Sound Communication with the Deaf." *Philosophy of Science,* 16: 3, July 1949c.

_____ and L. Levine. "Some Problems in Sensory Prosthesis." *Science,* 110: 2863, Nov 1949d.

_____. *The Human Use of Human Beings: Cybernetics and Society.* Boston: Houghton Mifflin, 1950a; 2nd ed., 1954; New York: Avon Books, 1967; New York: Da Capo Press, 1988; London: Free Association Books, 1989 (with a new introduction by Steve J. Heims).

_____. "The Mathematical Theory of Communication" (review of Shannon and Weaver), *Physics Today,* 3: 31–32, Sept 1950b.

_____. "Problems of Sensory Prosthesis." *Bulletin of the American Mathematical Society,* 56, 1951.

_____ (W. Norbert, pseud.). "The Brain." *Tech Engineering News,* Apr 1952a (reprinted in Conklin, Groff, ed. *Crossroads in Time.* New York: Doubleday, 1953).

_____ (W. Norbert, pseud.). "Miracle of the Broom Closet." *Tech Engineering News,* Apr 1952b (reprinted in *Fantasy and Science Fiction,* Feb 1954).

_____. "The Concept of Homeostasis in Medicine." *Transactions and Studies of the College of Physicians of Philadelphia,* 4: 20: 3, Feb 1953a.

_____. "The Future of Automatic Machinery." *Mechanical Engineering,* Feb 1953b.

_____. *Ex-Prodigy: My Childhood and Youth.* New York: Simon & Schuster, 1953c; Cambridge: MIT Press, 1964.

_____. "Problems of Organization." *Bull. Menninger Clinic,* 17: 34, 1953d.

_____ and A. Siegel. "A New Form for the Statistical Postulate of Quantum Mechanics." *Physical Review,* 91:6, Sept. 1953e.

_____. "The Machine as Threat and Promise." *St. Louis Post Dispatch,* Dec 13, 1953f (Wiener 1985/NW CW IV, 673–678).

_____. "Men, Machines, and the World About." *Medicine and Science,* 16 (New York Academy of Medicine Lectures to the Laity). New York: International Universities Press, 1954a (Wiener 1985/NW CW IV, 793–798).

_____. "Conspiracy of Conformists." (n. r.), May 1, 1954b (Wiener 1985/CW IV, 752).

_____ and Donald Campbell (Assoc. Professor of Elec. Eng., MIT). "Automatization: Norbert Wiener's Concept of Fully Mechanized Industry." *St. Louis Post Dispatch,* Dec 5, 1954c (Wiener 1985/NW CW IV, 679–683).

_____. "On the Factorization of Matrices." *Commentarii Mathematici Helvetici*, 29: 2, 1955.

_____ and A. Siegel. "The 'Theory of Measurement' in Differential Space Quantum Theory." *Physical Review*, 101: 429–432, Jan. 1956a.

_____. *I Am a Mathematician: The Later Life of a Prodigy*. New York: Doubleday, 1956b; Cambridge, MA: MIT Press, 1964.

_____. "Brain Waves and the Interferometer." *J. Physiol. Soc. Japan*, 18: 8, 1956c.

_____. "The Role of the Mathematician in a Materialistic Culture (A Scientist's Dilemma in a Materialistic World)." (Proceedings of the Second Combined Plan Conference, Oct 6–9, 1957), *Columbia Engineering Quarterly*, 22–24, 1957a (in Wiener 1985/NW CW IV, 707–709).

_____. "Rhythms in Physiology with Particular Reference to Encephalography." *Proceedings of the Rudolf Virchow Medical Society in New York*, 16: 109–124, 1957b.

_____ and P. Masani. "The Prediction Theory of Multivariate Stochastic Processes." Parts I and II, *Acta Mathematica*, 98: 111–150, 1957c; 99: 93–137, 1958a.

_____. "Science: The Megabuck Era." *New Republic*, Jan 27, 1958b.

_____. *Nonlinear Problems in Random Theory*. Cambridge, MA: The Technology Press and New York: John Wiley & Sons, 1958c.

_____. *The Tempter*. New York: Random House, 1959.

_____. "Some Moral and Technical Consequences of Automation" (adapted from a lecture to the AAAS Committee on Science in the Promotion of Human Welfare. Chicago, Dec 27, 1959). *Science*, 131: 1355–1358, Dec 6, 1960.

_____. "Science and Society." *Voprosy Filosofii (Problems of Philosophy)*, 7: 1961.

_____. Contribution to: *Proc. of Int'l Symposium on the Application of Automatic Control in Prosthetics Design*. Opatija, Yugoslavia, Aug 27–31, 1962.

_____. "The Mathematics of Self-Organizing Systems." *Recent Developments in Information and Decision Processes*. New York: Macmillan, 1962.

_____. "Introduction to Neurocybernetics" (with J. P. Schade) and "Epilogue." *Nerve, Brain and Memory Models* (in *Progress in Brain Research*, vol 2.). Amsterdam: Elsevier, 1963.

_____. *God & Golem, Inc.: A Comment on Certain Points Where Cybernetics Impinges on Religion*. Cambridge, MA: MIT Press, 1964a.

_____. *Selected Papers of Norbert Wiener, including Generalized Harmonic Analysis and Tauberian Theorems* (with contributions by Y. W. Lee, N. Levinson and W. T. Martin). Cambridge, MA: MIT Press, 1964b, 1965.

_____, A. Siegel, B. Rankin, W. T. Martin, eds. *Differential Space, Quantum Systems, and Prediction*. Cambridge, MA: MIT Press, 1966.

_____. 1894–1964 (Pesi Masani, ed.). *Mathematical Philosophy and Foundations; Potential Theory; Brownian Movement, Wiener Integrals, Ergodic and Chaos Theories, Turbulence and Statistical Mechanics (Collected Works, vol. I)*. Cambridge, MA: MIT Press, 1976.

_____. 1894–1964 (Pesi Masani, ed.). *Generalized Harmonic Analysis and Tauberian Theory, Classical Harmonic and Complex Analysis (Collected Works, vol. II)*. Cambridge, MA: MIT Press, 1979.

_____. 1894–1964 (Pesi Masani, ed.). *The Hopf-Wiener Integral Equation; Prediction and Filtering; Quantum Mechanics and Relativity; Miscellaneous Mathematical Papers (Collected Works, vol. III)*. Cambridge, MA: MIT Press, 1981.

_____. 1894–1964 (Pesi Masani, ed.). *Cybernetics, Science, and Society; Ethics, Aesthetics, and Literary Criticism; Book Reviews and Obituaries (Collected Works, vol IV)*. Cambridge, MA: MIT Press, 1985.

_____. *Invention: The Care and Feeding of Ideas* (with an introduction by Steve Joshua Heims). Cambridge, MA: MIT Press, 1993.

Wiesner, Jerome. "The Communication Sciences." In *Research Laboratory of Electronics, R.L.E.: 1946+20*. Cambridge, MA: RLE/MIT Press, 1966.

Wigner, E. *Symmetries and Reflections*. Cambridge, MA: MIT Press, 1970.

Wittner, Lawrence S. (Review of Wang 1999). *Bull. of the Atomic Scientists*, 55: 4, July/Aug 1999 (www.thebulletin.org/issues/1999/ja99/ja99reviews.html).

Zachary, G. Pascal. *Endless Frontier, Vannevar Bush, Engineer of the American Century*. New York: Free Press, 1997.

INDEX